# OPTOELECTRONICS - MATERIALS AND DEVICES

Edited by **Sergei L. Pyshkin** and **John Ballato**

**Optoelectronics - Materials and Devices**
http://dx.doi.org/10.5772/59334
Edited by Sergei L. Pyshkin and John Ballato

**Contributors**
Fernando de Souza Campos, José Alfredo Covolan Ulson, Rudolf Riehl, Bruno Albuquerque De Castro, José Eduardo Cogo Castanho, Oleksandr Kozhukhar, Hryhoriy Barylo, Mariya Ivakh, Zenon Gotra, Ivanna Makara, Volodymyr Virt, Nobuo Goto, Misaki Takahashi, Hiroki Kishikawa, Shin-Ichiro Yanagiya, Meichun Huang, Jie Zhang, Dongyan Zhang, Daqian Ye, Chenke Xu, Jin-Gang Liu, Hong-Jiang Ni, Wei-Feng Zhou, Shiyong Yang, Zhen-He Wang, Mikhail Nikitin, Viacheslav Kholodnov, Oscal T.C. Chen, Wei-Jean Liu, Mikhail Belkin, Alexander Sigov, Leonid Belkin, Alexey Loparev, Vladimir Iakovlev, Yi Gu, Yong-Gang Zhang, Lung-Chien Chen, Anca Stanculescu, Florin Stanculescu, Sergei L. Pyshkin, Fuxing Gu, Yan Liang, Heping Zeng, Rúben Neto, Henrique Salgado, Luís Pessoa, Pedro Batista, Nicola D'Ascenzo, Valeri Saveliev, Qingguo Xie, Lin Wang, Victor Stuchinsky, Sergey Dvoretsky, Vladimir Vasilyev, Aleksandr Predein, Alexei Vishnyakov, Dmitry Brunev, Alexei Zverev, Jian-Chiun Liou

**CBS,** Edition **2017**

**Published by InTech**
Janeza Trdine 9, 51000 Rijeka, Croatia


**Publishing Process Manager** Iva Lipovic

**Technical Editor** InTech DTP team

**Cover** InTech Design team

Additional hard copies can be obtained from orders@intechopen.com

Optoelectronics - Materials and Devices, Edited by Sergei L. Pyshkin and John Ballato

p. cm.

ISBN 978-953-51-2174-9

INTECH

open science | open minds

# Contents

# Preface

Optoelectronics, the marriage of optics and electronics, has proliferated around the world through a myriad of useful modern conveniences. Their continued growth and utility, however, require the constant global development of new materials and devices that meet next-generation demands.

As with the first book in this series, *Optoelectronics – Materials and Techniques*, edited by Professor P. Predeep in 2011 and the second book, *Optoelectronics – Advanced Materials and Devices*, edited by us in 2013, this newest offering, *Optoelectronics – Materials and Devices*, covers recent global achievements in optoelectronic materials, devices, and applications. With pleasure we note the growing number of countries participating in this endeavor, now including Brazil, Canada, China, Egypt, France, Germany, India, Italy, Japan, Malaysia, Mexico, Moldova, Morocco, Netherlands, Portugal, Romania, Saudi Arabia, South Korea, Switzerland, Ukraine, the United States, and Vietnam.

Comparing the 2013 and 2015 editions with the first one (2011), one quickly notes the growing attention to new device structures as well as new prospects for optoelectronics based on new materials.

An example is our semi-centennial investigation of long-term ordering of impurities in GaP. Novel and useful properties of perfect long-term ordered GaP include its efficient stimulated emission, very bright and broadband luminescence at room temperature, and the creation of excitonic crystal, which provide a unique opportunity to propose a new approach to selection and preparation of perfect materials for optoelectronics and new applications for this novel solid-state host – the excitonic crystal as high intensity light source with expected low threshold for the generation of nonlinear optical phenomena. Our results are proposed for further collaboration with the representatives of electronic science and industry in R&D as an inexpensive, resource-saving, and impactful way to develop optoelectronics through a special transformation of an ordinary semiconductor into the base material for various device structures.

We are grateful to all the authors and hope that the contribution of authors and number of participating countries will continue to grow, while optoelectronics itself will enhance human quality of life.

**Sergei L. Pyshkin**
Professor, Principal Investigator
Institute of Applied Physics
Academy of Sciences of Moldova
Kishinev, Moldova

Adjunct-Professor, Senior Fellow
Clemson University, South Carolina, USA

**John Ballato, FOSA, FSPIE, FACerS**
Professor
Center for Optical Materials Science and Engineering Technologies
Department of Materials Science and Engineering
Clemson University, South Carolina, USA

Chapter 1

# Excitonic Crystal and Perfect Semiconductors for Optoelectronics

Sergei L. Pyshkin

Additional information is available at the end of the chapter

http://dx.doi.org/10.5772/60431

**Abstract**

This chapter demonstrates the growth of perfect and contamination-free gallium phosphide (GaP) crystals and discusses the influence of crystallization conditions on their quality and properties. The long-term ordered and therefore close to ideal crystals replicates the behavior of the best nanoparticles exhibiting pronounced quantum confinement effect. These perfect crystals are useful for application in top-quality optoelectronic devices as well as they are a new object for the development of fundamentals of solid state physics.

Since samples of gallium phosphide doped by nitrogen (GaP:N) were originally prepared by the author in the 1960s, followed by the introduction of the excitonic crystal concept in the 1970s, the best methods of bulk, film and nanoparticle crystal growth have been elaborated. The results of semi centennial evolution of GaP:N properties are compiled here. Novel and useful properties of perfect GaP including its stimulated emission, very bright and broadband luminescence at room temperature were observed. These results provide a new approach to selection and preparation of perfect materials for optoelectronics and a unique opportunity to realize a new form of solid-state host — the excitonic crystal as high intensity light source with expected low threshold for the generation of non-linear optical phenomena.

Using the example of GaP here is proposed as a cheap, resource-saving and impactful way to develop optoelectronics through a special transformation of an ordinary semiconductor into the base material for various device structures.

**Keywords:** GaP, long-term ordering, excitonic crystal, perfect semiconductors for optoelectronics

## 1. Introduction

Single crystals of semiconductors grown under laboratory conditions naturally contain a varied assortment of defects such as displaced host and impurity atoms, vacancies, dislocations, and impurity clusters. These defects result from the relatively rapid growth conditions and inevitably lead to the deterioration of mechanical, electric, and optical properties of the material, and therefore to degradation in the performance of the associated devices. Note, the deterioration of optical properties of any luminescent material for application in optoelectronics may appear in the complete absence of light emission in the spectral region, where the perfect material gives an excellent luminescence, in too narrow emissive band, in a very weak light emission, in impossibility to control its shape and brightness, as well as in fast degradation of a device prepared on the base of this material. In order to partly overcome the noted preceding difficulties, industry uses expensive but only palliative decisions such as limitless extension of the list of materials for the device making or small improvement of technologies for growth and preparation of electronic materials. Huge material, time and mental resources already have spent and will be spent further in our efforts to support or improve achieved parameters and reliability of electronic devices. Therefore, finding of alternative drastic methods for device making is one of the main priorities of electronic industry development. This paper describes the experience of the author in this field.

The pure and doped GaP crystals discussed herein were prepared about 50 years ago [1]. Throughout the intervening decades they have been periodically re-evaluated in order to investigate the marked changes over time in their electro- and photoluminescence, photoconductivity, behavior of bound excitons characteristic for doped GaP, nonlinear optics, and other phenomena. Accordingly, it was of interest also to monitor the change in crystal quality over the course of several decades while the investigated crystals are held under ambient conditions.

Over time, as it is confirmed by the author during 50 years of the relevant experiments, that driving forces such as diffusion along concentration gradients, strain relaxation associated with clustering, and minimization of the free energy associated with properly directed chemical bonds between host atoms result in an ordered redistribution of impurities and host atoms in a crystal. In the particular case of GaP and some other chemical compounds, having in their compositions highly volatile components, any attempt to accelerate these processes through annealing at increased temperatures cannot be successful because high-temperature processing results in thermal decomposition (in GaP — due to P desorption) instead of improved crystal quality. Therefore, successful thermal processing of these compounds can only take place at temperatures below the sublimation temperatures of their volatile constituents, requiring a longer annealing time. For instance, evaluated in the framework of the Ising model, the characteristic time of the substitution reaction during N diffusion along P sites in GaP:N crystals at room temperature constitutes 15–20 years [2]. Hence, the observations of luminescence and some other phenomena in the crystals made in the 1960s–1970s and in the 1980s–1990s were then compared with the results obtained in 2005–2014 under similar experimental conditions.

http://dx.doi.org/10.5772/60431

The long-term ordering of doped GaP and other semiconductors has been observed as an important accompanying process, which can only be studied using the same unique set of samples and the interest to observe them over decade time scales. More specifically, the optical and mechanical properties of single crystalline GaP, and some other semiconductors also grown in the 1960s, have been analyzed. Comparison of the properties of the same crystals has been performed in the 1960s, 1970s, 1980s, and 1990s [1, 3-17] along with those of newly made GaP nanocrystals [18-20] and freshly prepared bulk single crystals [21-24]. Jointly with the references [25-27], this review provides a generalization of the results on long-term observation of luminescence, absorption, Raman light scattering, and microhardness of the bulk single crystals in comparison with the same properties of the top quality GaP nanocrystals. It is shown that the combination of these characterization techniques elucidates the evolution of these crystals over the course of many years, the ordered state brought about by prolonged room-temperature thermal annealing, and the interesting optical properties that accompany such ordering. It is demonstrated that long-term natural stimuli that improve the perfection of crystals prevail over other processes and can lead to novel heterogeneous device systems and new semiconductor devices with high temporal stability.

Additionally, it is worth noting, that semiconductor nanoparticles for optoelectronic applications also were synthesized mainly to avoid limitations inherent to freshly grown bulk semiconductors with a wide range of different defects. For instance, different defects of high concentration in freshly prepared GaP single crystals completely suppress any luminescence at room temperature due to the negligible free path for non-equilibrium electron-hole pairs between the defects and their non-radiative recombination, while the quantum theory predicts their free movement in the field of an ideal crystal lattice. However, the long-term ordered and therefore close to ideal crystals even at 300K demonstrate bright luminescence and stimulated emission equivalent to the best nanoparticles. These perfect crystals, due to their unique mechanical and optical properties, are useful for application in high-quality optoelectronic devices as well as they are a new object for the development of fundamentals of solid state physics, nanotechnology, and crystal growth.

Also noted is the application of GaP/polymers nanocomposites in device structures for accumulation, conversion and transport of light energy that has only recently received attention while bulk and thin GaP films have been successfully commercialized for many years. Therefore, for completeness, during the recent years, since 2005, the author and colleagues continued their efforts on the preparation of GaP nanoparticles in order to improve their quality and to apply their composites with appropriate polymers for advanced light emissive structures [18-20, 28-32].

In preparing this review, the author did not intent to evaluate the works on GaP of the other authors, but their works are cited and used here when it is necessary for explanation and interpretation of new phenomena observed during long-term ordering of impurities and host atoms in the crystal lattice. Elaborating optimal methods of preparation of GaP bulk crystals, nanoparticles and their light emissive composites with compatible polymers, we use our own experience and literature data [33-42]. The main goal of these 50 year efforts and this review are the observation and description of very interesting results of the long-term evolution of GaP properties and the relevant idea to propose for many years ahead an alternative and resource-saving way for the development of electronics, as well as to propose and justify the excitonic crystal [26, 27] as a new optical media for the future optoelectronic devices used in optical data processing, storage, and transmission as well as for the generation of non-linear optical effects at rather modest thresholds for nonlinearities. Interesting and very useful for application results of long-term evolution of GaP properties as well as the unique collection of tested and stored for years pure and doped perfect GaP crystals are demonstrated and proposed to academic researchers, engineers, and managers of electronic industry for intensification of collaboration in patent activity, reorganization of the material and device making processes, reduction in price, improvement of parameters, and reliability of devices.

Perfect GaP single crystals, excitonic crystal on the base of the GaP crystal doped by nitrogen (GaP:N), as well as understanding of properties and available application of these new objects coming these days into science and industry are the result of intense many years work of headed by the author groups of top specialists on crystal growth, investigation of their properties, and application in optoelectronics in Russia, Moldova, the USA, and Italy. This activity was stimulated and followed by the natural processes and phenomena elapsed with time in the crystals.

According to the chosen plan of presentation, this review is divided into the next sections:

1. Introduction
2. Properties of GaP
3. Growth Technology for Perfect GaP Bulk and Nano-Crystals
4. Optical Properties of Perfect, Long-term Ordered GaP:N Crystals
5. Comparison of Properties of GaP Nanocrystals and GaP Perfect Bulk Single Crystals
6. Excitonic Crystal and Its Importance in Optoelectronics
7. Already Discovered and Possible Nonlinear Optical Phenomena in GaP
8. Conclusions
9. Acknowledgments
10. References

## 2. Properties of GaP

GaP crystallizes in zinc blende structure, where Ga and P atoms create two interpenetrating face-centered lattices spaced ¼ of the (111) cube diagonal apart.

Brillouin zone of GaP and other III-V compounds represent the truncated octahedron (Figure 1) having the next high symmetry points:

$\Gamma - \overline{K} = (000)$ — center of Brillouin zone

$\Delta$ — along the (100) axis inside the zone

X – (100) Brillouin zone edge

Σ – along the (110) axes inside the zone

K – (110) Brillouin zone edge

Λ – along the (111) axes inside the zone

L – (111) Brillouin zone edge

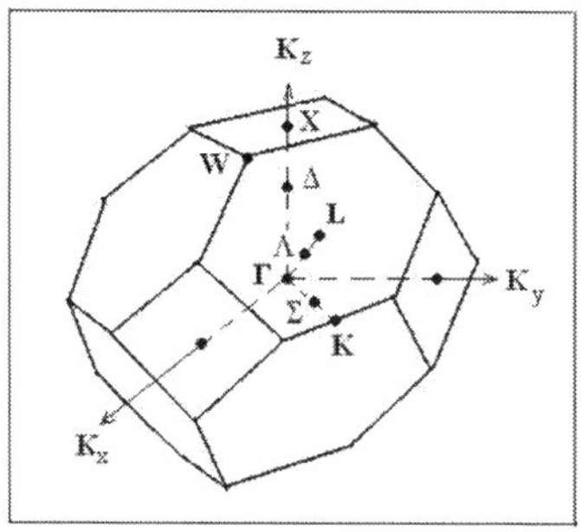


**Figure 1.** Brillouin zone of gallium phosphide.

A concrete band structure for each III-V representative in the limits of common for them Brillouin zone depends on the type of symmetry of the wave functions of valence electrons of the atoms, creating the compound. The most reliable data on GaP band structure were obtained from the experiments on light absorption and reflection as well as using the spectral distribution of photoconductivity in the region of intrinsic absorption.

According to W. Paul's empirical rule [36], the energy gaps equally depend on the pressure for the relevant electron states. Using this rule and experiments on dependence of electron transitions on pressure, the authors of Ref. [37] have proposed the band structure of GaP presented in Figure 2.

Experimental data confirm the details of the GaP band structure. So, the absolute minimum of the conductance band (the X-point) presented in Figure 1 lays at the edge of the Brillouin zone in the (100) direction, while the valence band maximum (the Γ-point) position is the center of the zone. The absolute minimum value of the forbidden gap (Figure 2) corresponds to the indirect optical transition $\Gamma_{15}^{v} \rightarrow X_1^{c}$; this value depends on the temperature changing between 2.354 eV at 4.2K (liquid helium) and 2.328 at 77K (liquid nitrogen) until 2.248 eV at 300K (room temperature) [37]. Minimum gap for direct optical transition $\Gamma_{15}^{v} \rightarrow \Gamma_1^{c}$ at 300K is equal to 2.78eV [37]. The valence band, taking into account its spin-orbit splitting, consists of two confluent bands and another one shifted downward. The other details of GaP band structure are shown in Figure 2.

According to the quantum selection rules for optical transitions, the lattice phonons do not participate in the direct transitions, while in the indirect transition lattice phonons participate, the type and energy of which are determined in Ref. [38] together with the low- and high-frequency dielectric constants, 10.182 and 8.457, respectively. The data on GaP phonon spectrum are widely used at interpretation of its light emissive and absorption spectra. In indirect optical transitions [38, 40] participate transversal and longitudinal acoustic and optic phonons with energies 12.8 (TA), 31.3 (LA), 46.5 (LO), and 50.0 meV (TO). Note, at low temperatures, when the thermal energy, kT, is less than the respective energies of the free and bound exciton creation (10 and 21 meV, respectively, for free and N bound excitons [4, 7]), the indirect optical transitions occur mainly through the excitonic states.

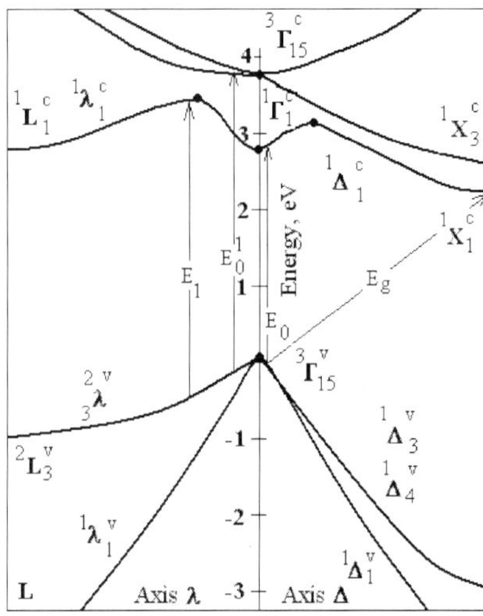


**Figure 2.** Band structure of gallium phosphide.

## 3. Growth Technology for Perfect GaP Bulk and Nano-Crystals

Single crystals of gallium phosphide, in principle, can be obtained in several ways [1, 3, 7]. The method for obtaining gallium phosphide from solution-melt, chosen by us, has several significant advantages:

1. A significant temperature reduction of the process and the presence of large amounts of solvent dramatically reduce crystal pollution by material of the container.
2. The light sources creating on their basis have high efficiency.
3. Due to specifics of the method, gallium phosphide at the appropriate level of the experiment can be obtained in the form of the relatively large lamellar crystals of a definite crystallographic orientation. Note, lamellar crystals are the most convenient and economical material in the manufacture of many semiconductor devices.

Consider the peculiarities of growth of lamellar crystals of gallium phosphide with the set-up properties, formed under uniform cooling of the P solution in Ga. The influence of the following factors has been investigated: 1) quantity and chemical nature of impurities; 2) geometric shape of the container and conditions of heat rejection; 3) accuracy of the temperature control; 4) cooling rate and the law of temperature change in time.

The aim of this work was to elucidate the mechanism of crystallization of lamellar GaP crystals and experimental confirmation stated earlier assumptions about the nature of this process.

I. The solution-melt method of GaP growth is described in detail in Refs.[1, 3, 7]. So we focus here only on the characteristics for this study associated of the crystal growth.

Synthesis, alloying, and crystallization were combined in a single cycle. Experiments on the establishment of the influence of impurities on the growth of GaP crystals was performed using Ga000 and phosphorus B5, industrially prepared in former USSR, the highest purity produced at that time (1960). When the growth process was thoroughly investigated, the growth of perfect crystals has been conducted in a sealed and drained simple container from optical quartz that was flushed with spectral pure argon (Figure 3a). Container 1 was introduced into the programmable furnace with rod 2 fixed on the vibrator to agitate the mixture of Ga, P, and any chosen dopant, to facilitate chemical reaction between components, and to avoid a possible explosion of the container in dangerous temperature points of the process of mass crystallization (Figure 3a). However, for investigation of the growth process some other containers were used. For instance, the container having a good thermal isolation of the walls and a Cu metal rod on its bottom was used to create the center of crystallization for needle-like crystals (Figure 3b). The container, comprising the set of necessary growth components 4, thermoisolating layer of $Al_2O_3$ 6, Ga etalon 7, and thermocouples in the critical places 5, 8, has been used for the investigation of thermal processes during the crystal growth (Figure 3c). At last, the upside-down rotatory container with the thin quartz net 9 was used to see and investigate the crystal forms at different growth regimes (Figure 3d).

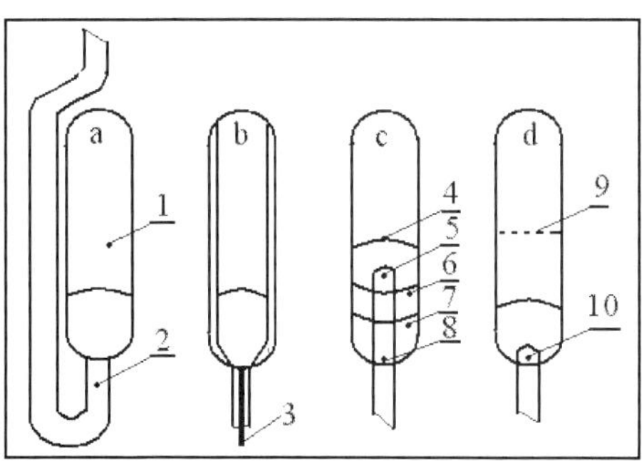


**Figure 3.** The containers for growth of perfect GaP crystals and investigation of the growth process. **a**, **b**, **c**, **d**. Types of containers. 1. Silica ampule. 2. Silica inlet rod, fixing the ampule to the vibrator. 3. Cu metal rod. 4. Container for investigation of thermal processes. Points 5, 8, 10 — locations of thermocouples. 6. Thermoisolating $Al_2O_3$ layer. 7. Ga etalon. 9. Silica net.

The impact of the temperature control accuracy was investigated at the facility, which allows to reproduce with controlled accuracy the necessary law of the temperature change in the growth container [1, 3]. Permissible fluctuation of temperature can be set within ± 0.5 ± 10° C interval. In order to study the form and quality of growing crystals, the crystallization process could be interrupted at any temperature below the point of liquidus. This interruption was achieved with the help of a special designed growth container (Figure 3d) and its programmable heating furnace, which allow to stop the change of temperature of the solution-melt, to identify and examine the crystals grown to this time (see details in [1, 3]).

Let us now describe and explain the peculiarities of temperature regime during the growth of lamellar perfect GaP single crystals from approximately 10 cm$^3$ of 5 at.% P – Ga melt solution (Figure 4). The smoothly heated slightly over 400°C solution must be kept for around 2 hours at this temperature for safe, non-dangerous explosion transformation of red phosphorus into its white modification. Then the heating process may be smoothly prolonged until a temperature of approximately 1200°C that increases the liquidus point for 5% P solution in Ga (1123°C) and creates good conditions for Ga-P reactions. After 1–3 hrs. soaking interval start to gradually cool the solution with a velocity of around 30°C/hrs. until 600–700°C and shutoff of the furnace. For perfect quality of the growing crystals during this 12–15 hrs. cooling of the solution, it is extremely important to support very fine, ±0.5°C temperature control possible with the specially elaborated heating and temperature control installation [3].

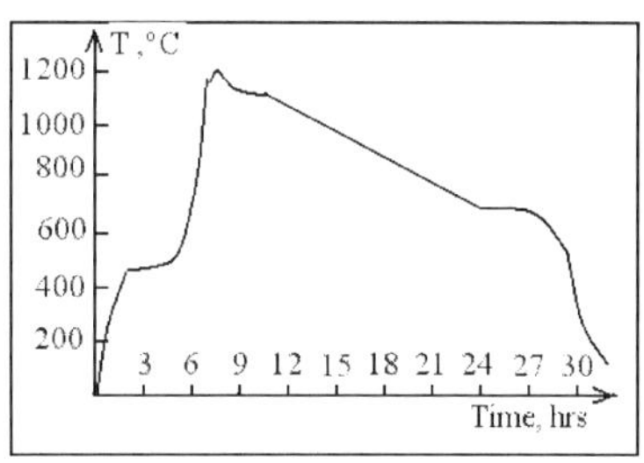


**Figure 4.** Programmable temperature changes in growth container.

With the introduction into the Ga solvent of various impurities, the choice of which was determined by practically important properties of the obtained crystals, at the same time the effects of contaminants on the nature of solidification have been investigated. The following impurities were introduced one by one in the solvent: Cu, Zn, Cd, In, Si, Ge, Sn, S, Se, Te, Ni, Cr, Fe, Co, Sm, La, Gd, and $Sm_2O_3$.

Typical features of the doped GaP crystals (size, color, polarity, lamellar structure, twinning, the dislocation structure, and others) were compared with similarly grown but undoped crystals.

It was found that the properties of the crystals depend on the chemical nature as well as on the quantity of the selected impurities. Their small admixtures did not significantly change the form of the crystals, while significant change of the crystallization environment composition created changes in forms of growth. For instance, the introduction into the solution of large quantities of Sn, the element of another valence compared with Ga, led to the emergence of the crystals having octahedral isometric forms. Isometric forms with simultaneous reduction of the sizes of crystals occurred with the introduction of solvent refractory impurities — samarium oxide ($Sm_2O_3$), solid units of which, apparently, were the centers of crystallization. Introduction into the solution of significant quantities of In, the chemical analogue of Ga, only slightly influenced on the morphology of the crystals, changing the lattice parameter, indicating the formation of 0.01 at. % In-GaP solid solution.

It was established that an increase in concentration of Zn and Gd in the Ga-P solution leads to the increase in the concentration of carriers (holes), reducing the size of crystals and their chemical resistance with a simultaneous increase in microhardness on the plane (111). At the introduction of S (1 at. % and more) the forms of the crystals varied from lamellar to volumetric. By increasing the concentration of tellurium in solution from 0,007 to 0.02 at. % the lattice constant varied from 5.4511 to 5.4524Å. At small quantities of impurities Zn and Te in GaP, until about $10^{19}$ $cm^{-3}$, a direct proportional relationship between the amount of impurities added to the solution and the concentration of charge carriers in the crystals is observed. Doping of the Ga-P solution by small quantities of rare earths and elements of the Fe group did not impact significantly on the forms of crystals, however, led to a significant change in the electrical and luminescent properties of GaP crystals. It is interesting to note that the density of etching pits on the plane (111) of doped crystals sharply increased in comparison with undoped ones.

III. The following factors influenced on the morphology of GaP crystals grown from the Ga-P solution of a certain concentration in conditions of the mass crystallization: the degree of supersaturation, which depends on the temperature of the solution and its first time derivative, the conditions of the heat crystallization removal and the accuracy of the temperature control in the solution.

Changing conditions of crystallization, mainly the rate of cooling of the melt, significantly affects the morphology of crystals. So, increasing the cooling rate reduced the size of the crystals, but the dislocation density was not changed and was primarily the determined concentration and chemical nature of impurities. The number of crystals having isometric forms increased with the decreasing cooling rate at simultaneous improvement of quality of the crystal surface. Uneven removal of heat from walls of the container has led to the formation of up to 30 mm crystal needles, elongated in the direction of the heat removal.

Significant impact on the crystal quality provides more accurate temperature control. The crystals obtained with the accuracy of temperature control ±0.5°C had a perfect plane (111), low dislocation density, and dimensions in 2–3 larger than the crystals obtained at the same conditions, but with the accuracy ±5°C or according to the regimes described in the literature. Crystals reached 25 mm in length and had the dislocation density ~ $10^3 cm^{-2}$, which is significantly less than that of the crystals obtained with low accuracy of the temperature control.

IV. In order to clarify the mechanism of crystallization of lamellar GaP crystals, differential record of temperature change between the standard of pure Ga and 5 at.% Ga-P solution near the point of liquidus as well as interruption of the crystallization process at different temperatures below the point of liquidus were conducted in special containers presented in Figure 3c and 3d (please see details described in Refs.[1, 3]). The record of temperature was conducted in the locations of the Ga standard and the Ga-P solution, interruption of the process was carried out by turning the container with quartz net from the top down. The pattern of temperature change in the solution-melt is presented in Figure 5.

Let us now present an overview of the growth of lamellar GaP crystals basing on the findings of these experiments. Total quantity of the solution-melt in the experiments was 10–12 $cm^3$, 1 $cm^3$ of this solution (5 at.% of P) at a temperature above the line of the liquidus contained $3x10^{21}$ and $6x10^{22}$ of P and Ga atoms, respectively. The first crystals, having the form of dendritic needles with the length of the order of 5 mm and thickness of 0.15 mm, were registered at a temperature of 1107°C at approximately 16°C supercooling. The crystals, found on the net inside the growth container [1] at lower temperatures of the process interrupting, represent thin plates in the form of rhombs, triangles, or hexagons. The twinning on the transverse cross-sections was observed at the study of microsections and cleaved facets. It turned out that the lamellar GaP crystals contain the plane of twinning, parallel to the planes (111).

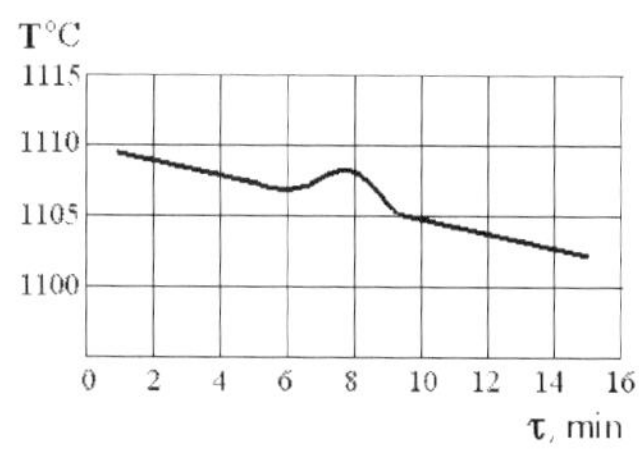


**Figure 5.** Change of temperature in the container when cooling with 30 deg/hour the 5 at.% solution of P in Ga.

The growth process of lamellar crystals can be formally divided into two components: the tangential growth in the plane (111) and layer-by-layer growth in the direction normal to the plane (111). Because the ratio of Ga and P atoms differed significantly from the stoichiometric, it is natural to assume that the process determining the rate of growth is diffusion. According to our estimates, the coefficient of diffusion of phosphorus atoms in Ga at T~1400K is equal $D_p \sim 3x10^{-3}$ $cm^2sec^{-1}$.

Upon cooling the system to 1107°C degrees, the supercooling of the solution was approximately $1.6x10^{20}$ $cm^{-3}$. Accepting the growth of needle-shaped crystals ends during recalescence time about 2 min (Figure 3), we get the diffusion length $L = (D_p \cdot \tau)^{1/2} = 0.6$ cm, and, further, the dimensions of the crystal in diffusion approximation: needle length 6 mm and thickness 0.56 mm; this value is in a good agreement with the experimental data. With further lowering of the temperature, the crystals acquire lamellar form with mainly developed planes (111) having the greatest reticular density and composed of the same atoms. Note that the removal of heat of crystallization is facilitated by the lamellar form of the crystals.

Thus, accepting the above presented two-stage model of lamellar crystal growth, let us consider the ratio of the contributions of the tangential growth in the plane (111) and layer-by-layer growth in the direction normal to the plane (111). The growth in the tangential direction is limited by diffusion and by the size of the previous layer, that is, it is determined by the initial conditions and the shape and size of the plates, formed during the cooling of the solution-melt 10–20 degrees below the point of liquidus. Since the formation of the initial crystals occurs in a very short time, the growth in the tangential direction is difficult to control, as it is in the case of the dendritic crystal growth. The growth in the direction normal to the plane (111) is sensitive to the fluctuations of the degree of supersaturation, resulting from the thermal motion of atoms and temperature fluctuations. In general, uniformity and perfection of the surface of lamellar GaP crystals the better, the less the time ratio needed for formation a new layer on a flat nucleation center to the time for which will be created the next nucleation center. The latter, obviously, in an extreme extent depends on fluctuations of the degree of supersaturation in different points of the growing plane and at different points in time. With the deterioration of precision temperature control, the number of fluctuations increases and bad crystals grow in the conditions of "entanglement" of the above mentioned characteristic times. Small fluctuations of supersaturation also are probably the cause of twinning planes. Indeed, assuming that the probability of fixation of the twins on the plane (111) is $W = \exp(-n\varepsilon/kT)$, where n is the number of atoms in a flat nucleus and $\varepsilon$ is the energy of packing defect per atom, and knowing the ratio of the number twin planes to normal equals $10^{-5.5}$, we obtain $W = \exp(-n\varepsilon/kt) = 10^{-5.5}$, $k = 1.4\times10^{-16}$ erg/degree, T = 1400K, and $n\varepsilon = 1.6$ eV or n = 10, $\varepsilon = 0.16$ eV/atom that under the order of the values is reasonable.

Increase of the accuracy of temperature regulation promotes some reduction in the number of twins and this fact also confirms the notable influence of fluctuations on the twinning.

Thus, deteriorations of quality or shape defects of GaP plates during their growth from Ga – P melt solution are observed in the next cases: 1) at a considerable increase of concentration of impurities in the GaP solution and dependently on the chemical nature of the chosen impurity; 2) at the change of the crystallization conditions, namely, at the sufficient deterioration of the cooling velocity control or in the case of nonuniform heat removal from the container for the crystal growth.

Processes of creation of the dendritic needles and their transformation to platelet crystals take place in a small temperature interval and for a short time, therefore they are hardly controlled, while crystal growth is easily controlled in the direction normal to the plate (111), because it is very sensitive to the external factors such as velocity of the solution cooling and accuracy of the temperature control.

All the above-stated means that only very perfect growth equipment and top quality growth experience will give an opportunity to grow good and high service ability GaP crystals. However, in spite of all precautions and high experience of the specialists, growing the crystals, freshly prepared GaP crystals cannot be successfully applied in electronics due to their low operational data, such as absence of high photosensitivity, bright and broadband luminescence, satisfactory transparency, and mechanical characteristics. Further, it will be shown how to considerably increase the quality and service ability of the freshly prepared GaP crystals.

## 4. Optical Properties of Perfect, Long-term Ordered GaP:N Crystals

It is necessary to note that the very important for optoelectronics long-term ordering and considerable improvement of the semiconductor crystal lattice and accompanying phenomena have been discovered and observed over decade time scales only with the help of the same unique collection of samples. This collection of long-term ordered perfect GaP single crystals gives opportunities to find deep fundamental analogies in properties of the perfect single crystals and nanoparticles as well as to predict and to realize in nanoparticles and perfect bulk crystals new interesting properties and applications.

Investigating the samples grown by the author in the 1960s, it was noted [1, 7], gallium phosphide (GaP) crystals for the first time clear improved their optical and mechanical properties only after 10–12 years since their preparation. It was unusual and interesting in the situation when all non-living matter around us usually deteriorates in time; therefore the author decided to investigate, to understand, and to use in the future this phenomenon. Shortly, the investigation process can be described as following.

Grown about 50 years ago pure and doped GaP crystals throughout the intervening decades have been periodically re-evaluated in order to see and investigate the change over time in their electro- and photoluminescence, photoconductivity, behavior of bound excitons, giving interesting and bright luminescence, nonlinear optical, and other phenomena. Accordingly, it is of interest also to monitor the change in crystal quality over the course of several decades while the crystal is held under ambient conditions.

Figure 6 demonstrates Raman spectra in GaP and GaP:N in 1989–1993 (a) and in 2006 (b, c). Concerning the general differences in the Raman spectra produced by the ordered and disordered forms of the crystals, note in accordance with Refs. [10, 17, 21-24] that the spectrum of the most ordered heavily N doped crystal (spectra 2 in Figure 6b) shows a considerably more intense narrow LO line than the less ordered pure or doped crystals (spectra 3, 4 in Figure 6b). Whereas the distribution and the environment of the P or N anions at a particular site in the unit cell in the ordered crystal are uniform, the great variability in this environment from site to site exists in a disordered or less ordered crystal.

Note that the theory of Raman light scattering in GaP predicts the LO phonon decay into two longitudinal acoustic phonons LA. LA phonons with a frequency LO/2 [41], and 2-phonon processes of 2TO and TO+LO emission also can be observed in perfect crystals [42]. This observation of a multi-phonon process and a decay of LO phonon, having a low intensity, confirm the high quality of the host lattice, uniform impurity distribution, and as a consequence, low noise background in the Raman scattering.

Figure 7 provides the Raman spectra obtained in 2005–2006 from pure and heavily doped GaP at 300K (a) and 80K (b). It can be seen only in the GaP:Bi crystal at 300K that a very narrow LO phonon is slightly shifted (approximately 2 $cm^{-1}$), while the TO and LO peaks maintain their spectral positions at 80K for the pure and doped GaP crystals independently of the type of impurity and its concentration. The LO phonon line is narrower in the doped crystal than in the undoped (pure) one and also is more intense than the TO phonon line. Note that these results are obtained despite the fact that the maximum concentrations of N, Sm, and Bi in these

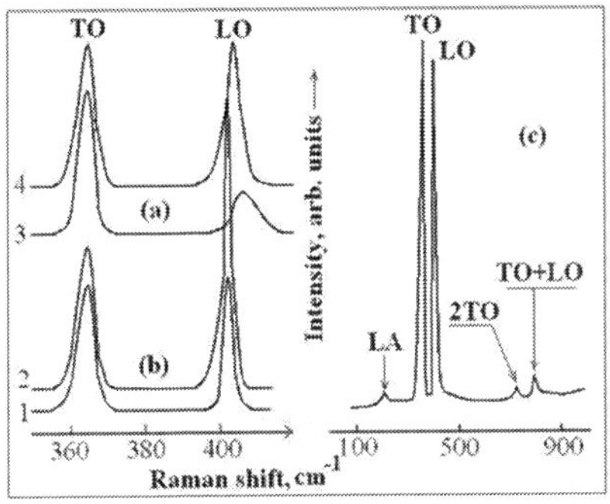


**Figure 6.** Raman spectra in GaP and GaP:N in 1989–1993 (a) and in 2006 (b, c). 1, 4—pure GaP, 2, 3—GaP heavy doped by N. Spectra 3, 4 are taken from Ref. [11] and obtained by one of the authors (SLP) in 1991. A new phenomenon observed in the Raman spectra that has developed in the crystals over 40 years are the peaks denoted by us here as LA, 2TO, and TO+LO (Figure 6c).

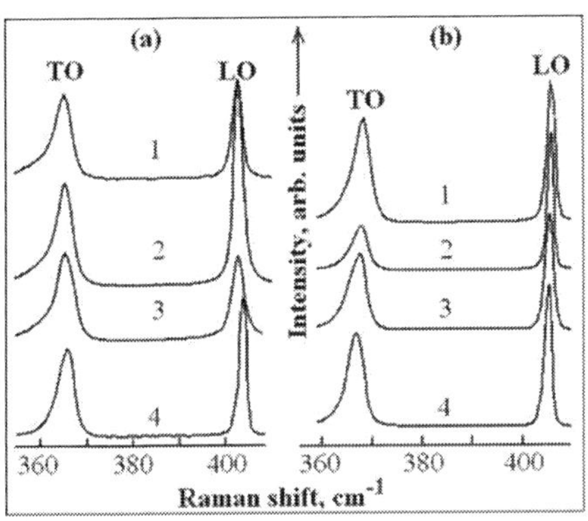


**Figure 7.** Spectra of Raman light scattering of pure and doped GaP at 300K (a) and 80K (b). 1. Pure. 2-4: GaP doped by N, Sm, and Bi, respectively

crystals are close to their limit of solubility (~ $10^{19}cm^{-3}$), the masses of impurity atom are very different, (the atomic weight 14, 150, and 208 for N, Sm, and Bi, respectively), and the N and Bi impurities substitute lattice points whereas Sm occupies interstitials.

The position and the line-width of the TO line in the ordered crystals do not depend on type and concentration of impurity nor on temperature. The temperature-independent TO line-width in the aged crystals implies that the impurities do not perturb the order of the lattice, which is possible only for a very uniform environment.

It is known that the broadening and frequency shift of the phonon lines are due to anharmonicity of the lattice vibrations and disorder in the crystal. Note that the TO line-shapes in the crystals again measured after 17–19 years are more close to Lorentzian than in [41] and LO lines are narrower in the crystals with big impurity concentrations. This most likely means that the anharmonicity of the lattice vibrations in the presence of ordered impurities is less than that in the pure unordered crystals. An increase in temperature only slightly distorts the TO line-shape in these crystals implying that the anharmonicity also is weak. Considering these results, it is assumed that the anharmonicity is larger in perfect undoped crystals than in doped crystals with periodically disposed impurities. Thus, it has been observed that in long-term ordered GaP crystals different impurities presented there at high concentrations do not distort crystal lattice, lead to narrowing and increase of longitudinal optical mode LO, and do not change the symmetrical form and positions of the TO line.

After 40 years since their fabrication the impurities in these doped GaP crystals create a superlattice with a period that depends on their concentration. Absence of luminescence from bound pair excitons (which need for their creation considerably less NN spacing than it is possible at the N concentration in our GaP:N crystals) as well as the spectral position of the narrow zero phonon line A shifting along the luminescent spectrum dependently on the N impurity concentration in exact correlation with the relevant theoretical ratio (see [40] and Figure 8a) only clearly show that in the 25–40 years aged GaP:N crystals the impurity atoms are located with the equal $r_{NN}$ spacing. Thus, we get as a minimum the evidence of anti-clustering of nitrogen atoms as the crystals age. However, taking into account that the N impurity atoms substitute the host P atoms in very perfect diamond-like cubic crystal lattice (it is really perfect according to our data on the position of extremely narrow phonon replica in heavily doped aged GaP:N crystals), we may assert from very simple geometrical consideration that in the case of the equal spacing measured with high accuracy, the N impurity system also forms a cubic crystal superlattice that means not simply anti-clustering, but the ordering of impurities as well as the perfection of the host lattice modified by impurities.

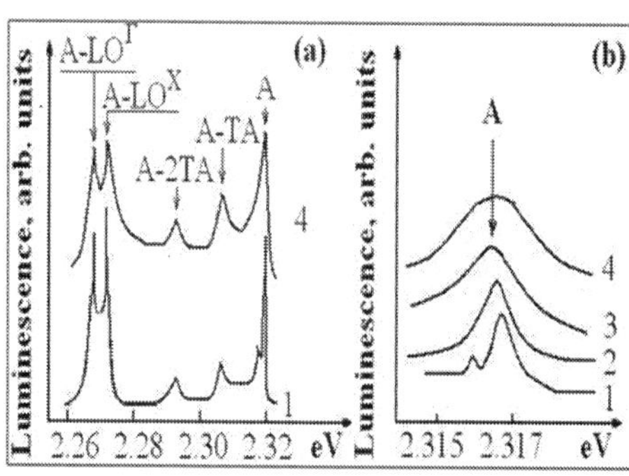


**Figure 8.** Evolution of the GaP:N luminescence with time and nitrogen concentration at a temperature of 15K. **a.** Zero-phonon line of the bound exciton A and its transversal acoustic (TA) and longitudinal optic (LO) phonon replica in as-prepared (4) and 25-year long-term ordered (1) crystals. **b.** Zero-phonon line A as a function of nitrogen (N) concentration. 1-3: 25-year-old crystals. 4: as-prepared. Curves 1 through 4 represent samples with nitrogen concentrations of $10^{17}$, $10^{18}$, $10^{19}$, and $10^{18}$ $cm^{-3}$, respectively.

Figure 8 provides a comparison of the evolution in luminescence spectra at low temperatures (80K and below) from GaP:N over a period of 25 years (original measurement in about 1963). Over this period, zero-phonon line A of single N impurity-bound excitons and their phonon replica are narrower in their line-widths when compared to the freshly prepared single crystals (Figure 8a). Further, as expected, zero phonon line and phonon replica in samples aged at room temperature for 25 years shift spectral position depending upon concentration of N impurities (Figure 8b, spectra 1-3) according to Ref. [40], while the same freshly prepared crystals exhibited broader luminescence line-widths with increasing nitrogen content (Figure 8b, spectrum 4). These, along with other half-centennial findings, including modifications of luminescence kinetics, spontaneous Raman scattering, X-ray diffraction, absorption spectra, micro-hardness, and density of dislocations, which are reported elsewhere [1, 3-24], strongly suggest that close-to-ideal GaP:N crystals are formed over time due to the equally spaced disposition of N impurities from their chaotic distribution in the same freshly prepared crystals.

As first noted in Ref. [23], these results suggest a new type of crystal lattice in which the host atoms occupy their proper (equilibrium) positions in the crystal, while the N impurities, periodically substituted into the lattice, portion it into short chains of equal length. According to the data obtained from Raman light scattering [10, 14], host atoms of this new lattice develop harmonic vibrations, and a high degree of lattice perfection leads to an abrupt decrease in the non-radiative recombination and an increase of efficiency and spectral range of luminescence.

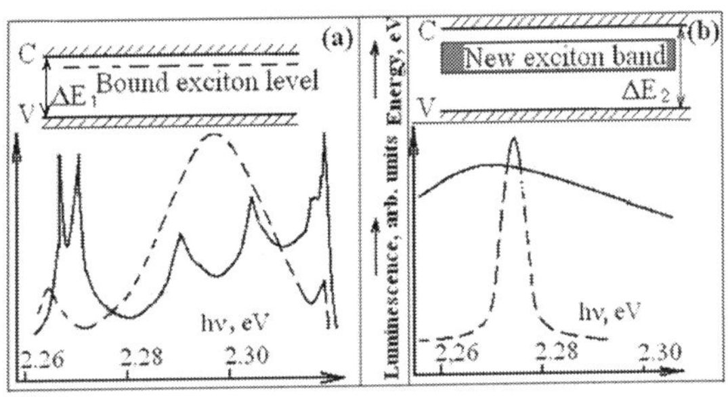


**Figure 9.** Luminescent spectra and schematic representation of the forbidden gaps ($\Delta E_1$, $\Delta E_2$) in the nitrogen-doped GaP aged for (**a**) 25 years and (**b**) 40 years. The dotted lines correspond to highly optically excited crystals. C and V represent the positions of the bottom of the conductance and the top of valence bands, respectively.

Additionally, stimulated emission of light in these temporally ordered crystals (Figure 9b) is observed. As also shown in Ref. [23], the GaP:N crystals aged for at least 40 years possess no discrete impurity level for N-bound excitons in the forbidden gap. They also demonstrated a uniform luminescence from a broad excitonic band instead of the narrow zero-phonon line and its phonon replica as observed from the less-aged 25-year-old crystals.

Thus, long-term ordered GaP:N crystals demonstrate uniform bright luminescence from a broad excitonic band instead of the narrow zero-phonon line and its phonon replica in disordered and partly ordered (25-year-old) crystals. This is due to the fact that ordered crystals have no discrete impurity level in the forbidden gap. To the best of our knowledge, such transformation of a discrete level within the forbidden gap into an excitonic band (Figure 9 a, b) is observed for the first time. In this case, the impurity atoms regularly occupy the host lattice sites and affect the band structure of the crystals, which is now a dilute solid solution of GaP-GaN with regular disposition of N atoms rather than GaP doped by occasionally located N atoms. Note that the increase of luminescence excitation in case of partly ordered GaP:N (Figure 9a, dotted line) leads to a broad luminescence band as a result of bound exciton interaction [9], while in the case of perfectly ordered crystals (Figure 9b) one can see an abrupt narrowing of the luminescence band, probably due to stimulated emission in defect-free crystals. Earlier, in freshly prepared crystals, we observed a clear stimulated emission from a GaP:N resonator at 80K [5], as well as the so-called superluminescence from GaP single crystals having natural faceting. Presently, our ordered crystals have a bright luminescence at room temperature that implies their perfection and very low light losses. In our studies [14, 23], we demonstrate that the stimulated emission in GaP also developed even at room temperature by direct electron–hole recombination of an electron at the bottom of the conduction band with a hole at the top of the valence band and the LO phonon absorption.

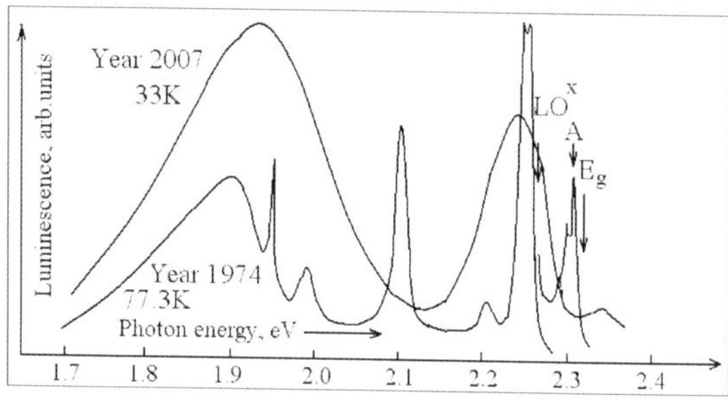


**Figure 10.** Evolution of GaP:N:Sm luminescence between 1973 and 2007. The crystals are grown in 1965.

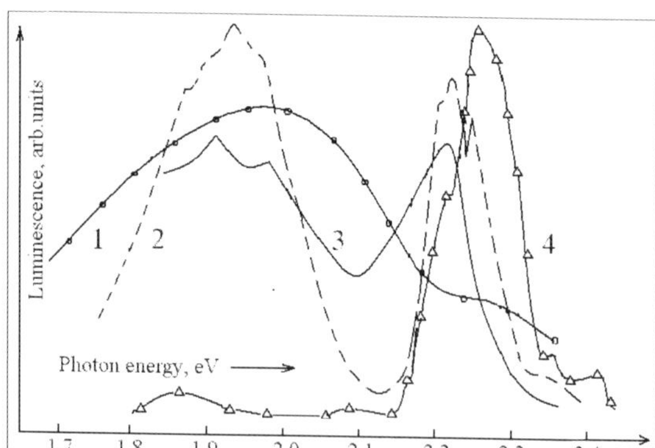


**Figure 11.** Luminescence of long-term ordered GaP:N:Sm. 1, 2—low; 3, 4—room temperatures. 1, 3—low; 2, 4—high excitation.

Investigating earlier luminescence of GaP doped by Sm and Ge, we have demonstrated bright luminescence in infrared region and opportunities to change in the wide wavelength diapason the position of luminescent maximum. For the first time, the GaP:N:Sm crystals were investigated in the year 1974 when their luminescence could be observed only at 80K and below [6]. The relevant spectrum, obtained in the year 1974, is shown in Fig. 10. Then it consisted of two parts: 1) the green part reflecting recombination of N bound excitons through the emission of the narrow zero-phonon A line and its phonon replica, arising in the emission of LO and other lattice phonons, and 2) the red and yellow parts reflecting the absorption of the bound exciton irradiation by Sm centers and its characteristic radiative recombination. In the year 2007 (Fig. 10, [13]), the spectrum of luminescence also consisted of these two parts, but instead of narrow excitonic lines we see a broad green excitonic band and also the broad red and yellow band without any fine structure.

Efficiency of Sm excitation at the N bound exciton recombination highly depends on a degree of uniformity of the N-Sm mixture. The long-term ordering gives a uniform mixture of Sm and N impurities and according to R. L. Bell [43], it provides up to 100% quantum efficiency of radiation from GaP:N:Sm system and bright luminescence at 300K. The bright emission bands at 300K, shown in Figure 11, curves 3, 4, arise as a result of formation during the period 1963–2010 of the uniform mixture of N recombination and Sm activation centers. The luminescence spectra of GaP:N:Sm at low temperatures (35K and below, Figure 11, spectra 1, 2) have the same maxima in green and red and yellow regions where the ratio of their intensities depends on the concentrations of N and Sm impurities as well as on the level of excitation. For instance, at a chosen level of excitation and an N/Sm ratio of concentration one can realize pure red or green emission of high efficiency or any combination of these colors that is an important property of a tunable light source.

Thus, doping GaP simultaneously with Sm and N, choosing necessary temperature and intensity of excitation we can get light emission from green to red. However, further expansion of the emissive spectrum in GaP can be achieved using some of the following other dopants or specific methods of crystal preparation.

Let us now discuss a possibility to converse visible emission into infrared region. Obviously, it is possible, creating in GaP an additional to the basic green another channel for light emissive electron-hole recombination in infrared region. For instance, it is known that the donor-acceptor pair $Ge^A$-$O^D$ on the base of oxygen ($O^D$) and germanium ($Ge^A$) emits the band with the maximum at 1.38 eV [44]. It was shown in Ref. [7] and in some other works fulfilled under the author's supervision, the existence of this band and its intensity, compared with the basic green emission, depends on the concentration of Ge-O pairs and intensity of excitation, obtained with the help of a Q-switched Nd glass laser supplied with the frequency doubler generating 2.34 eV photons.

Increasing the intensity of excitation, one can decrease the contribution into luminescence of light emissive recombination through the Ge-O pairs when the concentration of these pairs will be equal and less than the concentration of electron-hole pairs generated by the laser excitation source. In these conditions will be also clear detected the growing with excitation contribution into the whole light emission from the other recombination channels depressed

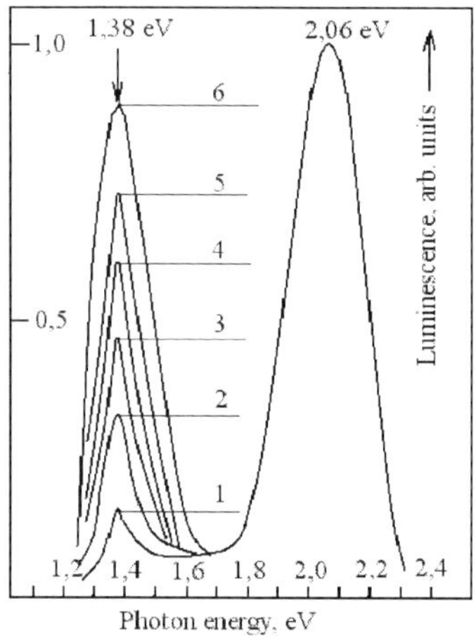


**Figure 12.** Luminescent spectra of GaP doped by Ge with a concentration of approximately $10^{18}cm^{-3}$ at 77.3K and different excitation intensity $I_{ex}$ x $10^{-24}$ $cm^{-2}sec^{-1}$: 16(1); 10(2); 4,5(3); 2,3(4); 1,2(5); and 0,46(6).

by the recombination through Ge-O pairs. Thus, the switching of light emissive recombination channels, important for application in optoelectronic devices, will be observed and demonstrated in the luminescence of GaP doped by Ge. Figure 12 demonstrates depending on excitation intensity the switching of luminescence at 77.3K from infrared with maximum at 1.38 eV to green-yellow region with maximum at 2.06 eV [7].

Moving to infrared region with the help of GaP doped by Ge, we have reached 1.2 eV in the emissive spectrum that 1 eV less than the forbidden gap of GaP crystals. Taking into account that our goal is to propose a cheap, resource-saving, and impactful way for the development of optoelectronics with the help of the special transformation of an ordinary semiconductor into the base material for various light emissive devices structures with broad and bright light emissive spectra, let us consider opportunities to expand light emission as much as possible to the ultraviolet (UV) region, using only GaP bulk and nanocrystals instead of a lot of the other semiconductors with their complicated, labor-consuming, and expensive technologies of preparation of materials with necessary reliable radiative properties.

As we have shown previously [8-17], different defects of high concentration in freshly prepared GaP single crystals completely suppress any luminescence at room temperature due to negligible small free path for non-equilibrium electrons and holes between the defects with their non-radiative recombination, while the quantum theory predicts free movement of electrons and holes in the field of an ideal crystal lattice. It was also shown that the long-term ordered and therefore close-to-ideal crystals demonstrate bright luminescence and stimulated emission repeating behavior of the best nanoparticles with pronounced quantum confinement effects. Therefore, there are only two ways of preparing a material with bright luminescence from IR to UV regions: to prepare very perfect, defectless single crystals or nanocrystals with the dimensions less than electron-hole free path in this material of standard quality. Correctness of the chosen way can be confirmed by the following comparison of optical properties of the best GaP nanocrystals and GaP perfect bulk single crystals.

## 5. Comparison of Optical Properties of GaP Nanocrystals and GaP Perfect Bulk Single Crystals

Jointly with the references [3, 7, 9-17, 21-27, 32] here we present a generalization of the results on long-term observation of luminescence, absorption, and Raman light scattering in bulk semiconductors in comparison with some properties of the best to the moment GaP nanocrystals. The combination of these characterization techniques elucidates the evolution of these crystals over the course of many years, the ordered state brought about by prolonged room-temperature thermal annealing, and the interesting optical properties that accompany such ordering. We demonstrate that long-term natural stimuli improve the perfection of our crystals, which can lead to novel heterogeneous systems and new semiconductor devices with high temporal stability. Raman light scattering confirms high quality of the long-term ordered crystals.

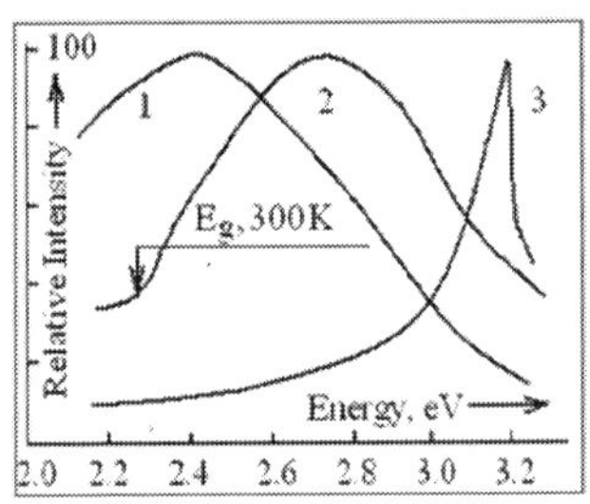


**Figure 13.** Luminescence of perfect bulk GaP single crystals (1) in comparison with the luminescence of GaP nanoparticles and GaP/polymers nanocomposites (2, 3). Nanoparticles were prepared from white P by mild aqueous or colloidal synthesis at decreased temperature, stored as the dry powder (spectrum 2) or suspension in a liquid (spectrum 3). Details: [15, 17].

We further improved upon the preparation of GaP nanocrystals using the known methods of hydrothermal and colloidal synthesis [30-32] by taking into account that the success of our activity depends on the optimal choice of the types of chemical reactions, necessary chemicals and their purity, conditions of the synthesis (control accuracy, temperature, pressure, duration, etc.), methods and quality of purification of the nanocrystals, and storage conditions for nanoparticles used in the further operations of fabrication of the GaP nanocomposites. The best quality GaP nanoparticles have been prepared by hydrothermal or colloidal synthesis from white phosphorus at decreased temperature (125°C) and intense ultrasonication; it was established that the maximum shift of their luminescence to ultraviolet and the best quality in general have the nanocomposites obtained from the nanoparticles of the same dimensions stored as a suspension in a suitable liquid.

Figure 13 compares the luminescence spectra of our long-term (up to 50 years) ordered GaP single crystals (spectrum 1) to that from high quality GaP nanoparticles and their GaP nanoparticles/polymers nanocomposites [15, 17]. Nanocrystals of the different dimensions, stored as dry powder, demonstrate rather broad luminescent band with maximum at 2.8 eV (Figure 13, spectrum 2), while the nanocrystals of about 10 nm sizes, thoroughly separated and distributed in a suspension, that prevent their coagulation, mechanical, and optical interaction, exhibit bright narrow-band luminescence with a maximum at 3.2 eV, approximately 1 eV above the position of the absorption edge in GaP at 300K (Figure 13, spectrum 3). The thoroughly washed, ultrasonicated and dried nanopowders as well as their specially prepared suspensions have been used for fabrication of blue light emissive GaP nanocomposites on the base of some optically and mechanically compatible with GaP polymers [15, 17, 30-32]. According to our measurements, the matrix polymers PGMA-co-POEGMA or BPVE used in this work provide no contribution to the spectra of luminescence of the based on these matrixes, so, the nanocomposite spectra coincide with those obtained from the relevant GaP powders or suspensions. We note that in the GaP/BPVE nanocomposite, the position of the luminescent maximum can be changed between 2.5 and 3.2 eV and the brightness is 20–30 more than in the PGMA and PGMA-co-POEGMA matrixes.

Long-term ordering leads to the creation of perfect bulk GaP crystals with considerably expanded and bright emissive band, practically the same as in the perfect GaP nanoparticles. We explain the broadening of the luminescence band and the shift of its maximum to low photon energies in luminescence of the nanocomposite based on the GaP powder by presence in the powder of the nanoparticles with the different dimensions between 10 and 100 nm. Meanwhile, the nanocomposites on the base of the suspensions containing only approximately 10 nm nanoparticles exhibit bright luminescence with a maximum at 3.2 eV due to a high transparency of 10 nm nanoparticles for these high energy emitted photons and pronounced quantum confinement effect.

In accord with our data [15, 31] the shift due to the quantum confinement effects is about a few tenths of eV and, obviously, it is impossible to explain only through this effect the dramatic 1 eV expansion of the region of luminescence at 300K to the high-energy side of the spectrum. In order to explain this interesting phenomenon, we postulate that the nanocrystals, much like the ideal long-term ordered bulk GaP single crystals, where this effect is not so strong, exhibit the huge increase in blue-shifted luminescence due to: (a) negligibly small influence of defects and non-radiative recombination of electron-hole pairs and very high efficiency of their radiative annihilation, (b) high perfection of nanocrystal lattice, and (c) high transparency of nanocrystals due to their small dimensions for the light emitted from high points of the GaP Brillouin zones, for instance, in the direct transitions $\Gamma_1^c$–$\Gamma_{15}^v$ between the conductive and valence bands with the photon energy at 300K equal to 2.8 eV [37], and (d) high efficiency of this so-called "hot" luminescence that means direct radiative recombination of electrons without their preliminary thermalization into the nearest particular point of the conductance band. Taking into account the high light absorption coefficient equal to approximately $10^5$ cm$^{-1}$ for photons with the energy in the vicinity of maximum at 3.2 eV [38], we can explain the difference in the spectrum 1 from perfect bulk GaP crystal and the spectrum 3 from its 10 nm perfect nanoparticles. Really, a big GaP single crystal, even very perfect one, in principle, cannot emit many photons in UV region, because the overwhelming majority of those photons will be immediately absorbed in the crystal; only tiny defectless 10 nm GaP spheres, transparent for this UV region, distributed in the transparent suspension or a polymer film will easily emit this UV light.

Note, our first attempts to prepare GaP nanoparticles [18, 28] yielded room temperature luminescence with the maximum shifted only to 2.4 eV in comparison with the achieved now new maximum at 3.2 eV and it confirms our significant progress in preparation of GaP nanoparticles and GaP/polymers nanocomposites. The perfect quality of the nanoparticles prepared by improved technologies is confirmed by all the used methods of characterization, while investigation of Raman light scattering evolution during 25 years (since 1989) clear confirms considerable improvement of GaP single crystal quality and the existence of new interesting phenomena characterizing only very perfect crystals.

On the base of these improved technologies for the preparation of GaP nanoparticles and GaP/polymer nanocomposites, we can change within the broad limits the main parameters of luminescence and create a framework for novel light emissive device structures using dramatic 1 eV expansion of GaP luminescence to UV region. Besides that, using all the noted in the presented review opportunities, including specially doped GaP and the necessary level of luminescence excitation, we can change the position of maximum and bandwidth of luminescence in wide, from infrared 1.2 to UV 3.2 eV, limits.

## 6. Excitonic Crystal and Its Importance in Optoelectronics

The role and application of bound excitons in nanoscience and technology are discussed in this chapter. Bound excitons are well studied in semiconductors, especially in gallium phosphide doped by nitrogen (GaP:N) [3, 4, 7, 38, 40, 45]. Doping of GaP with N leads to isoelectronic substitution of the host P atoms by N in its crystal lattice and to the creation of the electron trap with a giant capture cross-section. Therefore, any non-equilibrium electron in the vicinity of the trap will be captured by N atom, attracting a non-equilibrium hole by Coulomb interaction and creating the bound exciton — short-lived nanoparticle with the dimension of the order of 10 nm (it is the Bohr diameter of bound exciton in GaP:N). Note, that none of nanotechnology methods are used in the creation or selection of dimensions of these nanoparticles — only natural forces of electron–hole interaction and electron capture by the traps are necessary for the creation of these nanoparticles. As the result we get something like neutral short-lived atom analogue — a particle consisting of heavy negatively charged nucleus (N atom with captured electron) and a hole. So-called zero vibrations do not destroy the possible solid phase of bound excitons having these heavy nuclei that give an opportunity to reach in GaP:N a new crystal state — the short-lived excitonic crystal appeared at the necessary level of excitation, N impurity ordering and concentration, energy of photons and temperature.

Taking into account the above-mentioned preceding results, a model for the GaP:N long-term ordered crystal and its behavior at the relevant level of optical excitation for 40-year-old ordered N-doped GaP (Figure 14) can be suggested. At the relevant concentrations of N, the anion sub-lattice can be represented as a row of anions where N substitutes for P atoms with the period equal to the Bohr diameter of the bound exciton in GaP (approximately 10 nm) (Figure 14a). At some level of optical excitation, all the N sites will be filled by excitons, thereby creating an excitonic crystal (Figure 14b), which is a new phenomenon in solid-state physics and a very interesting medium for application in optoelectronics and nonlinear optics [4, 15, 17, 25, 26].

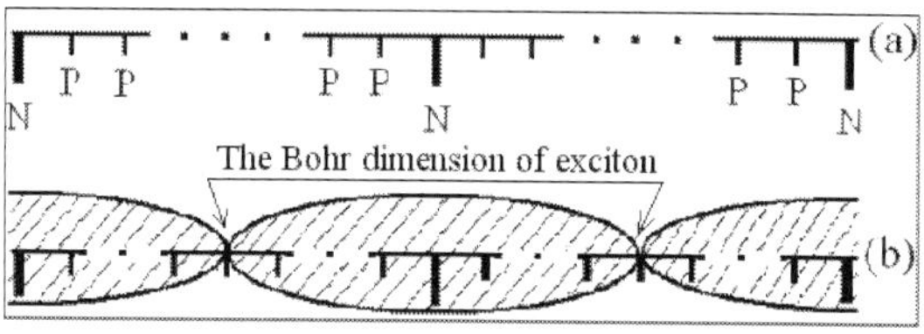


**Figure 14.** The models of 40 years ordered GaP doped by N. **a.** The new type of crystal lattice with periodic substitution of N atoms for the host P atoms. **b.** The excitonic crystal on the base of this lattice. The substitution period is equal to the Bohr diameter of exciton (~ 100Å) and optical excitation is enough for complete saturation of the N sublattice (please see details in Refs. [15, 17]).

Thus, using bound excitons as short-lived analogues of atoms and sticking to some specific rules, including the necessity to build in the GaP:N single crystal the excitonic superlattice with the identity period equal to the bound exciton Bohr dimension, we get a unique opportunity to create a new solid state media — consisting from short-lived nanoparticles excitonic crystal, obviously, with very useful and interesting properties for the application in optoelectronics, nanoscience, and technology. The following will discuss methods of preparation and possible application in optoelectronics of perfect GaP crystals, based on perfect GaP excitonic crystals and nanocrystals.

## 7. Already Discovered and Possible Nonlinear Optical Phenomena in GaP

Thus, confirmed by this semicentennial study, the impurities in doped long-term ordered GaP create a sublattice with a period that depends on their concentration. By periodically substituting for host atoms or occupying interstitial sites in the host lattice, the impurities become an intrinsic component of the modified crystal lattice and participate in the formation of a "new" phonon spectrum.

The ordered crystals with the host lattice modified by impurities could be very useful in various optoelectronic applications. Noted here are only a few potential applications in light emissive device structures. The properties of these structures will be very stable and independent on time. Uniform distribution of the recombination (N) and activator centers at the optimum concentration will yield the maximum efficiency for light emission. Further investigations of the quasi-crystalline state of excitons (or bi-excitons) bound to an impurity superlattice with a period equal to the Bohr dimension will be very interesting and useful because they should greatly strengthen nonlinear optical effects at low excitation intensities. This excitonic phase of high density of photons as an original accumulator of light also provides new opportunities for storage, transmission, and conversion of light.

With significant recent progress in semiconductor thin film deposition and growth techniques in some specific cases of device optoelectronic structure preparation, there obviously will be no longer a need to wait during years for such ordering to occur. For instance, superlattice from GaP/GaP:N with the period of the order of the bound excitons (their Bohr radius is equal to 5 nm) can be prepared by molecular beam epitaxy (MBE) or by MBE in combination with the laser-assisted epitaxy (LAE) [46-48]. Further, the preparation of a two- or three-dimensional arrangement of N impurities in a GaP film is difficult but also possible with the help of ion lithography. Of course, nowadays this technique is a frontier of our technological possibilities, but within the nearest future some very important progress likely will be obtained also in this direction. In this case, we will get a unique opportunity to design 3-dimensional impurity superlattices with configuration, symmetry, and lattice that are optimal for application in a concrete device structure or for efficient realization of specific linear or nonlinear optical phenomena. In any case, independently on the method of creation, the impurity modified crystal lattices, the excitonic (as well as bi-excitonic) phase with translational symmetry are very interesting objects, the properties and possible application of which are now under our investigation.

Nonlinear optics, starting its epoch-making development from the Nobel Prize Laureate N. Bloembergen transaction [49], with the appearance of lasers and highly supported personally by one of their inventors, the Nobel Prize Laureate A. Prokhorov, who helped the author to found in 1985 the Laser Research Laboratory in the Academy of Sciences of Moldova, since 1960s has taken its noteworthy place in investigations of GaP [5, 7, 50-52]. Here especially important for fundamentals and application in optoelectronic device structures were investigations of direct and indirect (with participation of the lattice phonons) many-quantum absorption between high symmetry points of GaP Brillouin zone (Figure 1), following the increase of photoconductivity and UV photon emission. So, exciting these high symmetry points with the help of infrared photon of a Q-switched laser, we get photoconductivity with participation of different energy bands and UV photon emission that is important equally for the investigation of band structure and for application in light frequency convertors. Addition of new opportunities due to elaboration of defectless perfect GaP bulk single crystals, its top-quality nanoparticles and multi-layered structures, discovery of the new nonlinear optic medium – excitonic crystal and its very interesting nonlinear optical phenomena [25-27] will surge of interest to this crystal, giving a new prospective industrial method of perfect crystal preparation, as well as opportunities for efficient realization in optoelectronics and electronics in general of remarkable properties of semiconductors due to a big commercial advantage from their fabrication.

## 8. Conclusions

Since the time of original preparation of gallium phosphide doped by nitrogen crystals (GaP:N) by the author in the 1960s, followed by the introduction of the excitonic crystal concept in the 1970s, the best methods of bulk, film, and nanoparticle crystal growth were elaborated. The results of semi-centennial evolution of GaP:N properties are compiled here and in the references to this paper. Novel and useful properties of GaP including an expected similarity in behavior between nanoparticles and perfect bulk crystals, as well as very bright and broadband luminescence at room temperature, are observed. These results provide a new approach to the selection and preparation of perfect materials for optoelectronics [25] and a unique opportunity to realize a new form of solid-state host — the excitonic crystal [26, 27]. In spite of the fact that the time necessary for natural long-term ordering (years) does not lead to optimism, the collected experience and results confirm expedience of the efforts directed to the formation in GaP of the N impurity superlattice having the identity period equal to the bound exciton dimension. As noted in Ref. [25], the process of preparation of top quality material for industrial electronics can be organized if the freshly grown crystals will be kept for years in a special storage and only old crystals (like to the old wine, if to use close to Moldova and other winemaking countries analogy) with the necessary properties will be annually retrieved for device fabrication while new portions of fresh crystals will be placed for long-term ordering.

Except natural aging of the relevant crystals for years, preparation of the N superlattice for excitonic crystal can be also realized by known methods of growth of multi-layer films, in particular by molecular beam and laser-assisted epitaxy [46-48].

The excitonic crystal, created by the long-term ordering or by the noted above methods of growth of multi-layer films, as well as the bulk top quality GaP crystals with the unique optic properties, obtained by the long-term ordering process of freshly prepared crystals, will be used in the new generation of optoelectronic devices, sometimes instead of nanoparticles and a lot of other materials. In particular, keeping in mind the low energy of the bound exciton creation, one can expect a low threshold for the generation of non-linear optical effects in the excitonic crystal and a good opportunity to create new and very efficient optoelectronic devices.

Note that semiconductor nanoparticles were introduced into materials science and engineering mainly in order to avoid limitations inherent to freshly grown semiconductors with a lot of different defects. However, it was shown [15] that this reason becomes unessential if, when justified, perfect long-term ordered semiconductor crystals are applied in electronics. Independently on their dimensions they demonstrate very interesting for application properties. Therefore, using the long-term ordered, perfect GaP crystals or similar on behavior and properties material in the electronic industry instead of the elaboration of very expensive and labor-consuming technologies for diverse materials and their nanoparticles with limited for application spectral region and other parameters, we get a big commercial advantage from their fabrication and application (for details please see the paper [25]).

All the obtained results presented here and included in summary reviews [15, 17, 25-27] may sufficiently change the approach to the selection of materials necessary for electronics, to make cheaper and simpler technology for the preparation of the selected materials and device structures based on them. This study of long-term convergence of bulk- and nanocrystal properties brings a novel perspective to improving the quality of semiconductor crystals. The unique collection of pure and doped crystals of semiconductors grown in the 1960s provides an opportunity to observe the long-term evolution of properties of these key electronic materials. During this half-centennial systematic investigation we have established the main trends of the evolution of their optoelectronic and mechanical properties. It was shown that these stimuli to improve the quality of the crystal lattice are the consequence of thermodynamic driving forces and prevail over tendencies that would favor disorder and destruction. Our long-term ordered and therefore close to ideal crystals repeat the behavior of the best nano-particles with pronounced quantum confinement effect.

For the first time, to the best of our knowledge, we have observed a new type of the crystal lattice where the host atoms occupy their proper (equilibrium) positions in the crystal field, while the impurities, once periodically inserted into the lattice, divide it in the short chains of equal length, where the host atoms develop harmonic vibrations. This periodic substitution of a host atom by an impurity allows the impurity to participate in the formation of the crystal's energy bands. In GaP it leads to the change in the value of the forbidden energy gap, to the appearance of a crystalline excitonic phase, and to the broad excitonic energy bands instead of the energy levels of bound excitons. The high perfection of this new lattice sharply decreases non-radiative electron-hole recombination, increases efficiency and the spectral range of luminescence, and promotes the stimulated emission of light due to its amplification inside the well-arranged, defect-free crystal. The development of techniques for the growth of thin films and bulk crystals with ordered distribution of impurities and the proper localization of host atoms inside the lattice are our high priority.

Semiconductor nanoparticles were introduced into materials science and engineering mainly to avoid limitations inherent to freshly grown semiconductors with a lot of different defects. Here and in other publications we show that this reason becomes unessential if we will apply in electronics, when it is justified, perfect long-term ordered semiconductors, which demonstrate independently on their dimensions very interesting for application properties. Especially important for application in the new generation of light emissive devices are the predicted and investigated by us crystalline state of bound excitons in GaP:N, the discovered in the framework of the STCU Project 4610 [16, 32] dramatic expansion of luminescence region in GaP perfect bulk single crystals as well as in the best prepared GaP nanocrystals and based on them composites with transparent polymers.

Using the long-term ordered GaP or similar on behavior and properties material in the electronic industry instead of elaboration of very expensive and labor-consuming technologies for diverse materials with their limited for application spectral region and other parameters, we get a big commercial advantage from their fabrication and application. So, the results of this long-term evolution of the important properties of our unique collection of semiconductor single crystals promise a novel approach to the development of a new generation of optoelectronic devices.

Besides the long-term ordering, the combined methods of laser-assisted and molecular beam epitaxies [46-48] will be applied to fabrication of device structures with artificial periodicity; together with classic methods of the perfect crystal growth, they can be employed to realize impurity ordering that would yield new types of nanostructures and enhanced optoelectronic device performance.

For the first time we demonstrate that well-aged GaP bulk crystals as well as high quality GaP nanoparticles have no essential difference in their luminescence behavior, brightness, or spectral position of the emitted light. The long-term ordered and therefore close to ideal crystals repeat the behavior of the best nanoparticles with pronounced quantum confinement effect. These perfect crystals are useful for application in top-quality optoelectronic devices and are a new object for the development of fundamentals of solid state physics.

Of course, waiting for improvement of the crystal quality for tens of years can be justified only in exceptional cases, but we propose to turn this perennial procedure of long-term ordering into the necessary one for the preparation of the top quality material for industrial electronics, which due to its unique properties will be used in electronic devices instead of a lot of various materials.

## Acknowledgements

I am glad to note that the broad discussion and dissemination of our joint results stimulate the development of our further collaboration with reliable partners from the USA, Russia, Italy, Romania, France, and other countries. I express my cordial gratitude to my teachers, the late world-known scientists: Professor Nina A. Goryunova, Nobel Prize Laureate Alexander M. Prokhorov, Professors Rem V. Khokhlov, and Sergei I. Radautsan. The author is very grateful to the US Department of State, the Institute of International Exchange, Washington, DC, the US Air Force Office for Scientific Research, the US Office of Naval Research Global, Civilian R&D Foundation, Arlington, VA, Science & Technology Center in Ukraine, to his colleagues and co-authors from Clemson University, SC, University of Central Florida, FL, Istituto di elettronica dello stato solido, CNR, Rome, Italy, Universita degli studi, Cagliari, Italy, Lomonosov Moscow State University, Ioffe Physico-Technical Institute and State Polytechnical University, St. Petersburg, Prokhorov General Physics Institute of the Russian Academy of Sciences, Moscow, Russia, the Institute of Applied Physics of the Academy of Sciences of Moldova for support and attention to this protracted (1963–present time) research.

## Author details

Sergei L. Pyshkin[1,2]

Address all correspondence to: spyshki@clemson.edu

1 Academy of Sciences of Moldova, Republic of Moldova

2 Clemson University, USA

## References

[1] Goryunova NA, Pyshkin SL, Borshchevskii AS, et al. Influence of impurities and crystallisation conditions on growth of platelet GaP crystals. Symposium on Crystal Growth at the 7th Int Crystallography Congress (Moscow, July 1966). J Growth Crystals 1969; 8 68-72, ed. N.N. Sheftal, New York.

[2] Pyshkin SL, Radautsan SI, Zenchenko VP. Processes of long-lasting ordering in crystals with a partly inverse spinel structure. J Soviet Phys – Doklady 1990; 35(4) 301-4.

[3] Pyshkin SL. Preparation and properties of gallium phosphide. Ph.D. thesis 1967, State University of Moldova, Kishinev.

[4] Pyshkin SL, Zifudin L. Excitons in highly optically excited gallium phosphide. J Luminescence 1974; 9 302-8.

[5] Pyshkin SL (presented by Nobel Prize Laureate A.M. Prokhorov). Stimulated emission in gallium phosphide. J Soviet Physics – Doklady 1975; 19 845-6.

[6] Pyshkin S. Luminescence of GaP:N:Sm crystals. J Soviet Phys Semicond 1975; 8: 912-3.

[7] Pyshkin SL. Photoconductivity and luminescence of highly optically excited semiconductors. Dr.Sc. thesis 1978, Lomonosov Moscow State University.

[8] Pyshkin S, Anedda A, Congiu F, Mura A. Luminescence of the GaP:N ordered system. J Pure Appl Optics 1993; 2 499-502.

[9] Pyshkin S, Anedda A. Time-Dependent Behaviour of Antistructural Defects and Impurities in Cd-In-S and GaP. Institute of Physics Conference Series 1998, Ternary and Multinary Compounds; 152, Section E 785-9.

[10] Pyshkin S, Ballato J, Chumanov G. Raman light scattering from long-term ordered GaP single crystals. J Optics 2007, J Opt A Pure and Appl Optics; 9 33-6.

[11] Pyshkin SL, Ballato J, Bass M, Turri G. Luminescence of Long-Term Ordered Pure and Doped Gallium Phosphide. TMS Annual Meeting, invited talk at the Symposium: Advances in Semiconductor, Electro Optic and Radio Frequency Materials, March 9–13, New Orleans, LA. J Electron Mater 2008; 37(4) 388-95.

[12] Pyshkin SL, Ballato J, Bass M, Chumanov G. Time-dependent evolution of crystal lattice, defects and impurities in CdIn2S4 and GaP. J Phys Status Sol 2009; C(6) 1112-5.

[13] Pyshkin SL, Ballato J, Bass M, Turri G. Evolution of luminescence from doped gallium phosphide over 40 Years. J Electron Mater 2009; 38(5) 640-6.

[14] Pyshkin S, Ballato J. Evolution of optical and mechanical properties of semiconductors over 40 Years. J Electron Mater 2010; 39(6) 635-41.

[15] Pyshkin SL, Ballato J. Long-term convergence of bulk- and nano-crystal properties. In: Predeep P (ed.) Optoelectronics – Materials and Techniques 2011. Rijeka: InTech; 459-76.

[16] Pyshkin S, Ballato J. Dramatic expansion of luminescence region in GaP/polymer nanocomposites: Supplemental Proceedings, vol.1: "Materials Processing and Interfaces": 353-9, TMS Annual Meetings, Orlando, FL, USA, March 11-15, 2012.

[17] Pyshkin SL, Ballato J. Advanced light emissive device structures. In: Pyshkin SL and Ballato J (ed.) Optoelectronics – Advanced Materials and Devices 2013. Rijeka: InTech; 1-24.

[18] Pyshkin S, Ballato J, Chumanov G, DiMaio J, Saha AK. Preparation and Characterization of Nanocrystalline GaP: Technical Proceedings of the 2006 NSTI Nanotech Conference; 3 194-7.

[19] Pyshkin SL, Ballato J, Chumanov G. Optical Characterization of Long -Term Ordered and Nanocrystalline GaP, (invited), 3rd Int. Conf. on Materials Science and Condensed Matter Physics (Kishinev, Moldova, Oct 2006), Abstract Book, p172; Mold. J. of the Physical Sciences 2007; 6(2) 148-159.

[20] Pyshkin SL, Ballato J, Belevschii S, Rusu E, Racu A, Van DerVeer D. Synthesis and Characterization of GaP Nanoparticles for Light Emissive Devices: The 2011 NSTI-Nanotech Conference Proceedings: www.nsti.org, ISBN 978-1 4398-7142-3 1, 327-30, Boston, MA, June 13-16.

[21] Pyshkin SL, Ballato J, Bass M, Chumanov G, Turri G. Time-dependent evolution of crystal lattice, defects and impurities in $CdIn_2S_4$ and GaP. Phys Stat Sol 2009; C(6) 1112-5.

[22] Pyshkin S, Zhitaru R, Ballato J, Chumanov G, Bass M. Structural Characterization of Long-Term Ordered Semiconductors: The 2009 MS&T Conference Proceedings, International Symposium "Fundamentals & Characterization" 698-709, Pittsburgh, October 24-29.

[23] Pyshkin S, Ballato J, Bass M, Chumanov G, Turri G. Properties of the long-term ordered semiconductors: The 2009 TMS Annual Meeting and Exhibition Suppl. Proceedings; 3 477-84, San Francisco, February 15–19.

[24] Pyshkin S, Ballato J, Chumanov G, Bass M, Turri G, Zhitaru R, et al. Optical and Mechanical Properties of Long-Term Ordered Semiconductors. Moldavian J Phys Sci 2009; 8(3-4) 287-95. The 4th Int Conference on Materials Science and Condensed Matter Physics, Kishinev, Sept 23-26, 2008.

[25] Pyshkin SL. Gallium Phosphide - New Prospect for Optoelectronics. J Adv Optoelectron Mater 2013 (ISSN Online: 2327-7661, ISSN Print: 2327-767X); 1(4) 59-66.

[26] Pyshkin SL. Excitonic crystal and nanotechnology. Symbiosis J Nanosci Technol: Open Access 2014; 1(2), http://symbiosisonlinepublishing.com/nanoscience-technology08.pdf.

[27] Pyshkin SL. Excitonic Crystal, Nanotechnology and New Prospect for Optoelectronics, J Open Optics 2015; 9 26-37.

[28] Pyshkin S, Ballato J. Advanced Light Emissive Composite Materials for Integrated Optics: The 2005 MS&T Conference Proceedings, Symposium: The Physics and Materials Challenges for Integrated Optics - A Step in the Future for Photonic Devices 3-13, Pittsburgh, PA.

[29] Pyshkin S, Ballato J, Luzinov I, Zdyrko B. Fabrication and Characterization of the GaP/Polymer Nanocomposites for Advanced Light Emissive Device Structures: The 2010 NSTI-Nanotech Conference Proceedings; 1 772-5. www.nsti.org, ISBN 978-1-4398-3401-5, Anaheim, CA, Jun 21-24, 2010.

[30] Pyshkin S, Ballato J, Chumanov G, Tsyntsaru N, Rusu E. Preparation and Characterization of Nanocrystalline GaP for Advanced Light Emissive Device Structures: The 2010 NSTI-Nanotech Conference Proceedings; 1 522-5, www.nsti.org, ISBN 978-1-4398-3401-5.

[31] Pyshkin SL, Ballato J, Luzinov I, Zdyrko B. Fabrication and characterization of GaP/polymer nanocomposites for advanced light emissive device structures. J Nanoparticle Res 2011; 13 5565-70, ISSN: 1388-0764, doi: 10.1007/s11051-011-0547-0.

[32] Pyshkin Sergei L (Project Manager). Advanced Light Emissive Device Structures. STCU (www.stcu.int) Project 4610, 2009-2012.

[33] Shanmin Gao, et al. Aqueous synthesis of III-V semiconductor GaP and InP exhibiting pronounced quantum confinement. Chem Commun 2002 3064-5.

[34] Zhengang Liu, Yujun Bai, Deliang Cui, Qilong Wang. Preliminary Insight into the formation process of InP and GaP nanocrystals. Solid State Sci 2003; 5 1037-40.

[35] Zhao-Chun Zhang, Bao-Ping Wang. Diamond and oxidized disordered graphite on the surface of gallium phosphide nanoparticles. Syst Charact 2009; 26 53–7.

[36] Paul W. Band structure of the intermetallic semiconductors from pressure experiments. J Appl Phys 1961; 32 2082-94.

[37] Zallen R, Paul W. Band structure of gallium phosphide from optical experiments at high pressure. Phys Rev 1964; 134 A1628-41.

[38] Dean PJ, Thomas DG. Intrinsic absorption-edge spectrum of gallium phosphide. Phys Rev 1966; 150(2) 690-702.

[39] Kleinman DA, Spitzer WG. Infrared lattice absorption of GaP. Phys Rev 1960; 118 (1) 110-7.

[40] Allen WJ. Energy levels of nitrogen-nitrogen pairs in gallium phosphide. J Phys C 1968, 1 1136-8.

[41] Bairamov B, Kitaev Y, Negoduiko V, Khashkhozhev Z. Light scattering by optical phonons and non-linear optical properties of GaP. J Soviet Phys Solid State 1975; 16 1323-7.

[42] Pyshkin S, Radautsan S, Zenchenko V. Raman spectra of Cd-In-S with different cation-sublattice ordering. J Soviet Phys Doklady 1990; 35(12) 1064-67.

[43] Bell RL. A proposal for a DC pumped rare-earth laser. J Appl Phys 1963; 34 1563-4.

[44] Gershenzon M, et al. Evidence for radiative recombination between deep donor-acceptor pairs in GaP at room temperature. J Appl Phys 1966; 37(2) 483-6.

[45] Thomas DG, Hopfield JJ, Frosch CJ. Isoelectronic traps due to nitrogen in gallium phosphide. Phys Rev Lett 1965; 15 857-9.

[46] Pyshkin SL, et al. Preparation and structural properties of some III-V semiconductor films grown on (100) oriented Si substrates. J Appl Surf Sci 1992; 56-58 39-43.

[47] Budyanu VA, Chechuy SN, Pyshkin SL, et al. Investigation of III-V/Si heterojunction grown by laser deposition. Phys Stat Sol (a) 1985; 91 737-44.

[48] Budyanu VA, Chechuy SN, Pyshkin SL, et al. Laser vacuum epitaxy of III-V semiconductors on silicon. Rev Roumaine de Physique 1987; 32(1-2) 216-9.

[49] Bloembergen N. Nonlinear Optics. Edition: 4th. Paperback – June, 1996.

[50] Ashkinadze BM, Pyshkin SL, Radautsan SI. Some Non-linear Optical Effects in GaP: The IX Int. Conf. on the Physics of Semiconductors Proceedings; 2 1189-93, Moscow, July 1968.

[51] Pyshkin SL, et al. Multi-quantum absorption in GaP. Opto-electronics 1970; 2 245-9.

[52] Ashkinadze BM, Kretsu IP, Pyshkin SL, et al. Luminescence of Gallium phosphide near an indirect transition in the multiphoton excitation case. Soviet Phys Semicond 1969; 2(10) 1261-3.

Chapter 2

# Efficiency Droop in III-nitride LEDs

Jie Zhang, Dongyan Zhang, Daqian Ye,
Chenke Xu and Meichun Huang

Additional information is available at the end of the chapter

http://dx.doi.org/10.5772/60454

**Abstract**

To dominate the illumination market, applications of high-power, group III-nitride light-emitting diodes (LEDs) with lower cost and higher efficiency at high injection current density must prevail. In this chapter, three possible origins of efficiency droop (including electron leakage, poor hole injection, and delocalization of carriers) in III-nitride LEDs are systematically summarized. To seek a more comprehensive understanding of the efficiency droop, experimental results based on commercialized LEDs are obtained to explain the physical mechanisms. Proposals for droop mitigation, such as (1) improving hole injection, and (2) increasing effective optical volume or reducing carrier density in the active region, are introduced. Finally, a simple expression for the effects of V-shaped pits on the droop is demonstrated.

**Keywords:** Efficiency droop, III-nitride, LEDs, Electron blocking layer (EBL), Physical mechanisms

## 1. Introduction

Lighting is, and always has been, a global market. Today, as Gallium nitride (GaN) light-emitting diodes (LED) technology gains a commercialized market, the demand for lighting continues to grow throughout the world. The transition to solid-state lighting (SSL) technology and the growth in lighting demand, coupled with the sharp growth in LED backlighting for displays, has led to a rapid expansion of LED manufacturing capacity over the last few years.

Most of people find the merits of LED engineering—high efficiency and low energy consumption—obviously attractive. A recent U.S. Department of Energy (DOE) analysis by Navigant

Consulting, Inc. (Navigant) reviewing the adoption of LED-SSL technology in the U.S. concluded that annual source energy savings from LED lighting in 2013 more than doubled from the previous year to 188 trillion British thermal units (BTUs), which is equivalent to an annual energy cost savings of about $1.8 billion [1]. In order to further match product demands for specific lighting applications and clean energy principles, one particular trend is the introduction of high-performance, low-cost, high-power LEDs. Therefore, the cost-per-lumen of packaged LEDs is estimated to be lower than 5 $/klm and efficacy projections need to improve to 220 lm/W after 2020, as shown in Fig. 1a, b [1].

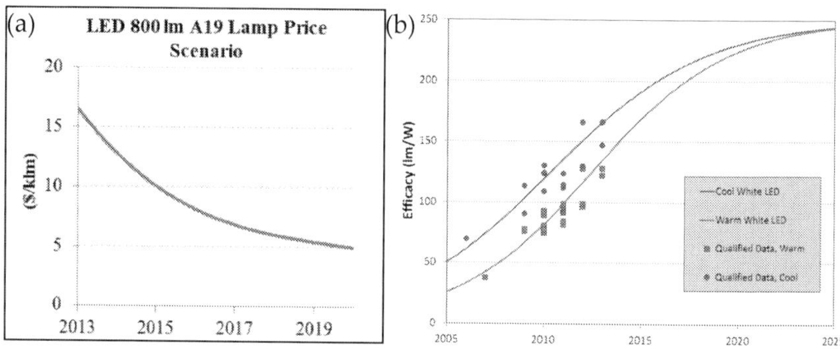


**Figure 1.** (*a*) A price scenario of 800 lm LED in the range of 2013 to 2020; (*b*) white light LED efficiency projection from 2005 to 2025.

In response to this LED-related energy saving projection, maximizing the efficiency of LED is an important innovation along the path to highly practical products. Especially for LEDs that need to be used in general lighting applications and to pave the way for high-power, high-dimension devices, it is imperative to produce high luminous fluxes with high efficiency, which necessitates high current levels. However, typical GaN LEDs are facing a substantial decrease in efficiency as injection current increases, and we call this efficiency droop effect or droop. For more than a decade, many research ideas pertaining to efficiency droop have been proposed [2]. While the real culprit is still under debate, nearly all of the proposed mechanisms stem from such origins as Auger recombination, electron leakage, delocalization of carriers and poor hole injection. Historically, phonon-assisted Auger processes were first considered to explain the droop in GaN-LEDs, because Auger recombination in InGaN would promote carriers at energies well beyond the hetero-barriers, and therefore would provide an important contribution to leakage [3–6]. In recent years, polarization fields induced band bending in active regions and electron-blocking layers were reported to enhance the leakage of injected electrons into the p-type GaN cladding layer [7–10]. At the same time, most discussions point out that the distribution and densities of carriers in the quantum wells (QWs), and the carrier injection into the active region, can make the carrier escape from localized states, forming leakage currents, which may be a key issue in identifying the origin of efficiency droop [11–13]. Furthermore, an LED is a bipolar device relying on the efficient injection of both minority carriers, and both holes and electrons need to be injected and to be distributed optimally in the active region to recombine for the effective operation of LEDs. Therefore, the low hole concentration, low hole mobility, and potential barriers for hole transport are also possible mechanisms responsible for high-current efficiency droop [14–18]. In addition to the proposals mentioned above, other models have also considered facilitating factors for electron leakage and droop, including defect-assisted tunneling effects [19, 20], junction heating [21], high plasma carrier temperatures (hot carriers) and saturation of the radiative recombination rate, as well as current crowding and related contact degradation [22–24]. In summary, the efficiency droop effect critically depends on several main mechanisms and is associated with several different promoting factors.

Based on such mechanisms of physics, many remedies to suppress the droop have been explored by many scientific researchers. Junction heating and contact degradation, both of which are the subject of vigorous research efforts, can be mitigated by increasing the efficiency and employment of packages capable of removing the dissipated heat very efficiently [25]. The intrinsic Auger losses in wide band gap semiconductors are also considered to be relatively small. Meanwhile, unfortunately, most conclusions regarding efficiency droop from Auger processes are based on theory calculations. Thus far, there is little direct experimental evidence of the Auger carrier recombination mechanism in GaN/InGaN LEDs by observing the remaining higher energy Auger electrons, which would require a spectroscopic measurement of hot electrons in the device [26, 27]. In turn, as mentioned above about the origins of droop, the radical treatments of these diseases focus on other methods in industrial mass production.

Firstly, in studies where the polarization charge has been widely proposed as the reason for electron leakage and thus efficiency droop, LEDs with an inserted electron blocking layer, generating multi-quantum barriers (QBs) and graded QBs, are expected to reduce the polarization mismatch between the QW and the barrier [28–31]. In these cases, droop is strongly attenuated in fabricated devices at the cost of reduced internal quantum efficiency (IQE) value. Interestingly, based on these conclusions, there are also reports that we can build a deeper potential well at the interface between the electron blocking layer (EBL) and the last QB, resulting in better electron confinement and improving hole injection [17, 32]. Secondly, at high driving currents, the carrier density reducing the effective active volume will get very high and lead to saturation of the radiative recombination rate, which in turn increases the carrier density. For the purpose of avoiding the droop, the ability of carriers to be captured in QWs and the mechanisms related to carrier distribution must be analyzed in terms of the quantum mechanical dwell time (the time an electron dwells over the QW) and carrier distribution. Further, increasing the QW thickness or numbers also increase the dwell time, and therefore should lead to a higher capture probability [13, 22, 33]. Thirdly, for maximum efficiency, the goal is to have equal numbers of electrons and holes injected into the active region. As reported in [18, 34–36], by employing p-type–doped barriers or other band engineering, such as using a lightly n-type-doped GaN injection layer below the InGaN multiple quantum wells (MQWs) on the n side, intended to bring electron and hole injection to comparable levels, better efficiency retention has been observed at higher current levels.

The above discussions show that the droop has been studied deeply, and some of these approaches have been successfully demonstrated in laboratory LED prototypes. However, a solution to improve efficiency droop effect requires the examination of numerical results based on commercialized LEDs. For mass production, some measures are able to inhibit the LED efficiency droop effect, yet there is still need for in-depth study of each production process, as manufacturers of market-adopted LEDs prioritize product cost reductions and quality improvements. In this chapter, we devote ourselves to finding a simple, high efficiency way to suppress the droop and trying to import it into the production flow process. First, a theoretical analysis on the physical mechanisms of efficiency droop is briefly given in Sect. 2. According to this analysis, the main factors contributing to droop are pointed out. Next, we introduce some recent reports recently to support our theoretical results. Then we present the structural design, characterization and discussion of three approaches: 1) optimization of EBL and QBs, 2) investigation of the active-volume effect in a multiple quantum well (MQW) region and 3) the use of intentionally formed V-shaped pits (V-pits) are proposed in Sect. 3. Finally, in Sect. 4, a simple summary is presented.

## 2. Investigation of physical mechanisms for efficiency droop

Internal quantum efficiency (IQE) is the ratio of photons emitted from the active region of the semiconductor to the number of electrons injected into the p-n junction of an LED. One can define the IQE as

$$IQE = \frac{I_0 / E_{ph}}{J_{in} / q} \tag{1}$$

where $I_0$ is the optical power at the central wavelength, $E_{ph}$ is the photon energy, $J_{in}$ is the current injected into the LED active region and $q$ is the electron charge.

According to the van Roosbroeck–Shockley equation model [37], the rate of radiative recombination per unit volume $R_0$ has been treated by the photon density divided by the mean lifetime of photons:

$$R_0 \approx \frac{N}{\tau} = Bn^2 \tag{2}$$

where $B$ is the radiative recombination coefficient, and $n$ is the carrier concentration. Because the $I_0 / E_{ph}$ is the number of photons emitted per unit time by the active region, we can obtain:

$$\frac{I_0}{E_{ph}} = R_0 V_{QW} = Bn^2 V_{QW} \tag{3}$$

Then the IQE can be expressed in the form:

http://dx.doi.org/10.5772/60454

$$IQE = \frac{Bn^2 V_{QW}}{J_{in}/q} \tag{4}$$

In order to better understand the physical mechanisms inside an LED, the current injected into the QW region ($J_{in}$) can be investigated by the equation:

$$\frac{J_{in}}{q} = RV_{QW} \tag{5}$$

$R$ is the carrier recombination rate and we can discuss $R$ instead of $IQE$ for the physical mechanisms of efficiency droop.

Commonly, the recombination in LEDs is described by the ABC model [38]:

$$R = An + Bn^2 + Cn^3 \tag{6}$$

This simplistic model considers $A$, $B$, and $C$ to represent the Shockley-Read-Hall (SRH), radiative, and Auger coefficients, respectively. However, this mode has a good fit only for the efficiency curve at low injected current below that of the LED peak efficiency, and the model fails to keep pace with the decline in efficiency at higher carrier densities. This suggests that there are some additional processes not included in the three conventional processes of the ABC model, such as carrier leakage or poor hole efficiency. For this reason, J. Cho et al. extended the ABC model by adding another recombination term, $f(n)$, to the model, where $f(n)$ includes carrier leakage and is allowed to contain more higher order terms of $n$ [39]. The recombination rate can be written as

$$\begin{aligned} R &= An + Bn^2 + f(n) \\ &= An + Bn^2 + Cn^3 + C_{DL} n^3 \end{aligned} \tag{7}$$

where $C_{DL}$ is a proportionality constant associated with the lowering of the injection efficiency due to drift of electrons in the p-type layer (drift leakage). The drift current of electrons injected into the p-type neutral layer, or drift-induced leakage, is given by

$$J_{drift} = qV_{QW} C_{DL} n^3 \tag{8}$$

Whereas the $J_{drift}$ at the edge of the neutral layer is

$$J_{drift} = q\mu_n \Delta n_p(0) E \tag{9}$$

where $\Delta n_p(0)$ is the injected electron concentration at the edge of the p-type neutral region of a p-n junction, and $E$ is the electric field in the p-type layer.

The current injected into the QW region can be obtained by

$$J_{in} = q\mu_p P_{p0} E \tag{10}$$

where $P_{p0}$ is the concentration of holes in the P region. In the region close to the peak-efficiency point, where radiative recombination dominates, the current depends on the carrier concentration in the QW, according to:

$$J_{in} \approx qV_{QW} Bn^2 \tag{11}$$

The $C_{DL}$ can be obtained from relationships (8), (9), (10) and (11):

$$J_{drift} = qV_{QW} \frac{\Delta n_p(0)}{n} \frac{\mu_n}{\mu_p P_{p0}} Bn^3 \tag{12}$$

$$C_{DL} = \frac{\mu_n \delta}{\mu_p P_{p0}} B \tag{13}$$

where $\delta = \Delta n_p(0)/n$. From Eqs. 8, 12 and 13, we can see that $J_{drift}$ depends on $V_{QW}$, $\delta$, $\mu_n$, $\mu_P$, and $P_{P0}$. These factors correspond to the physical mechanisms of droop, such as effective active volume, the electrons injected to the p-region, p-type carrier density, and low hole injection efficiency. Furthermore, from the Einstein relation at the edge of the neutral layer, as shown here, we can see that the drift-induced leakage current increases with the total current, and will become significant at a sufficiently large current:

$$J_{drift} = q\mu_n \Delta n_p(0) E = qD_n \frac{q}{kT} \Delta n_p(0) \frac{J_{in}}{\sigma_p} \tag{14}$$

where $D_n$ and $\sigma_P$ are the electron diffusion coefficient in the p-type GaN, and the p-type layer conductivity ($\sigma_P = qP_{p0}\mu_P$), respectively.

In the above discussions, it is noteworthy that the electron leakage into the p region, the carrier density reducing effective active volume and the poor hole injection efficiency are the three main physical mechanisms for droop. We will discuss them one by one, and introduce related pathways to overcome them.

## 2.1. Electron leakage

As we know, only the electrons captured by the MQWs in LED are able to participate in radiative recombination and contribute to the optical power that is produced. From Eqs. 12, 13 and 14, we know that the leakage current is proportional to $\delta$, $n$, and $J_{in}$, which indicates that the electrons that spill over to the p-region play a very important role in causing efficiency droop. In the process of being injected into the MQWs, the electrons face large QB barriers, and there is an EBL layer intended to confine the electrons in the active region. But, due to the mismatch polarization of InGaN and GaN, GaN and AlGaN, some sheet charges exit and attract electrons, which pull down the barrier and EBL heights [7–9]. Therefore, the QBs and the EBL layer have a triangular shape and electrons can escape to form a significant leakage current. In device simulations, J. Piprek et al. have pointed out that the band offset ratio of GaN and InGaN $(\Delta E_{C1}:\Delta E_{V1})$ and GaN and AlGaN $(\Delta E_{C2}:\Delta E_{V2})$ are important parameters associated with band bending [40, 41]. As a matter of fact, G. Verzellesi suggests that for an EBL with "nominal" electron confinement capability, the AlGaN/GaN band offset $(\Delta E_{C2}:\Delta E_{V2})$ should be kept at 70:30 [2]. In order to balance the electrons and holes in an active region, the InGaN/GaN band offsets $(\Delta E_{C1}:\Delta E_{V1})$ should be symmetric (50:50) to reduce polarization charges [2].

In recent years, many researchers have sought methods to overcome the shortcomings of polarization fields. It is possible to engineer QBs and EBL layers to achieve these objectives.

| Year | Engineered QBs/QWs | Droop* | Test current (A or A/cm²) | Chip size (mm×mm) | Experiment and/or simulation | Ref. |
|---|---|---|---|---|---|---|
| 2008 | AlInGaN QBs | Reduced droop | 300 | 1×1 | Experiment | [8] |
| 2009 | InGaN/GaN/InGaN QBs | 1.60 % | 35 | 1×1 | Experiment | [28] |
| 2010 | Insert an AlGaN spacer | 5.66 % | 521 | 0.295×0.325 | Experiment | [42] |
| 2011 | GaN/InGaN/GaN QBs | Small droop | | 0.3×000 | Simulation | [31] |
| 2011 | AlInGaN QBs | 13 % | 100 | 1×1 | Experiment and simulation | [43] |
| 2011 | Graded in composition in multiple InGaN QBs | 6 % | 200 | 0.3×0.3 | Experiment and simulation | [17] |
| 2011 | linearly graded the last InxGa1-xN barrier | 13 % | 26.7 | 0.3×0.3 | Simulation | [32] |
| 2012 | InGaN/AlGaN/InGaN QBs | Small droop | 0.3A | | Simulation | [30] |
| 2013 | Graded in content in QWs | 47 % | 160 | 0.2×0.25 | Experiment | [44] |

*The droop in Tables 1 and 2 is defined as $(\eta_{peak}-\eta_{test-current})/\eta_{peak}$, where $\eta$ are the EQE or IQE motioned in the references

**Table 1.** Development of engineered QBs/QWs for improving droop

The device designs of EBL also have a relationship with hole injection efficiency, which will be discussed later. The engineering work on QBs is summarized in Table 1, and the main strategies can be grouped as follows:

a. Use of multilayer QBs (H. Chung, 2009), for example, select InGaN/GaN/InGaN structures, as QBs. In this way, the crystal quality of epitaxy films can be ensured, and the polarization field was reduced 19 % by the time-resolved PL measurements under reverse bias. It has also been shown that the use of MLB structures increases optical power and decreases the efficiency droop [28].

b. Insertion of an AlGaN barrier between the n-type GaN layer and the MQWs (R. Lin, 2010). It was found that the EQE was improved by 5.7 % over that of a sample without an AlGaN barrier at a current density of 104.3 A/cm$^2$ [42].

c. Quaternary InAlGaN QBs (M. Schubert, P. Tu, 2008, 2011). The electroluminescence results indicated that the light performance could be effectively enhanced, and simulation results showed that the GaN LEDs with quaternary InAlGaN barrier exhibited a 62 % higher radiative recombination rate and a low efficiency droop of 13 % at a high injection current [8, 43].

d. Replacing the last GaN barrier by a linearly graded $In_xGa_{1-x}N$ barrier (C. Xia, 2011). The formation of a deep potential well in the GLB can enhance electron confinement. The forward voltage was reduced from 3.60 V to 3.25 V, and the efficiency droop was improved from 36 % to 13 % [32].

e. Use of step-stage multiple-quantum-well (MQW) structure with Si-doped hole-blocking barrier (Z. Zheng, 2013). At high injection current levels, the efficiency droop behaviour and EL wavelength stability of this structure were significantly improved. The author ascribed these improvements to the enhanced carrier injection resulting from the reduction of the polarization field in the active region by step stage QWs, as well as the hole-blocking effect by the Si-doped barriers [44].

All these methods are possible ways to achieve an improved efficiency droop effect in GaN LEDs.

### 2.2. Effects of volume and carrier density in the active region

From the calculations in the last section, we can see that the volume of the active region $V_{QW}$ is related to the drift current causing the droop. Most of time, it has been assumed that all the MQW layers act as light-emitting active regions and the carrier density in MQWs is uniform. However, actual carrier distribution in InGaN MQWs is significantly inhomogeneous and the effective light-emitting region can be greatly reduced for several reasons. N. F. Gardner and J. Son et al. have been investigated the relation of piezoelectric polarization and effect active regions in MQWs [45, 46]. The simulation results showed that the strong internal polarization fields cause the electron and hole wave functions to be mainly distributed near the edge of the QW in the opposite direction, and the small overlap of electron and hole wave functions effectively reduced the active volume. When the severe band bending in InGaN quantum-well was improved as the piezoelectric polarization was reduced, the improved overlap of electron and hole wave functions increased the internal quantum efficiency and reduced efficiency droop significantly. Another reason for the reduction of the effected active region was the strong fluctuation of In composition inside InGaN QWs. Since the recombination of electrons and holes mainly occurred in the In-rich region, A. Kaneta and J. I. Shim et al. have pointed out that the active volume acting as a light-emitting region would be much smaller than the physical volume of QWs, and the carrier density around the In rich cluster should be higher than the one in the uniformity distribution region [47, 48].

Another reason for effective volume reduction is the inefficient hole transport through QWs. Due to the low mobility and low hole density, hole carriers are mostly distributed at a few QWs closest to the p-side layers, and only a limited number of QW layers act as effective carrier recombination regions [16, 49]. Because of this aforementioned effect, the effective active volume could be greatly reduced. Hole injection efficiency will be discussed in the next section. In fact, research regarding the effected active region has long utilized two methods: optimization of the QW thickness and of the numbers. In 2007, N. F. Gardner et al. compared the LED having 9-nm thick QWs and another one with 2.5-nm thick QWs [45]. The results exhibit a significantly reduced droop in the former device, which is attributed to the reduced Auger recombination resulting from affected carrier density distribution in QW volumes. M. Maier et al. have investigated the optimal QW thickness for LEDs fabricated on sapphire substrate and free-standing GaN substrate [50]. From the electroluminescence (EL) efficiency results, LEDs on freestanding GaN with an 18-nm thick InGaN wide-well active region show the highest efficiency. In contrast, LEDs on sapphires grown with conventional low temperatures exhibit optimum well width at 3 nm. S. Tanaka et al. improved the droop property by increasing the QW number from 6 to 9 on a patterned sapphire substrate (PSS). The droop ratio was improved from 45.9 to 7.6 % [51]. At a wavelength of 447 nm, and with standard on-header packaging, the 9 QW PSS-LED had an output power of 27.6 mW and an EQE of 49.7 % at a current of 20 mA. The output power of the 9 QW PSS-LED remains linear with increasing drive current, and the EQE is almost constant, even up to a relatively high current density. X. Li et al. studied the efficiency droop of double heterostructure (DH) LEDs with different active region thicknesses separated by thin and low barriers for LEDs at high injection, and experimental results were supported by numerical simulations [52]. They concluded that the use of thin and low barriers was crucial to enhance carrier transport across the active region, and increasing active region thickness from 3 to 6 nm resulted in a decrease in IQE; however, the peak EQE increased. A further increase of the DH active region thickness to 9 nm improved EQE only at very high injection levels, while 11-nm thick DH showed significantly lower EQE.

All of this progress has provided us with QW active region design, the main physical mechanism and an estimation of conclusion. We now have a clearer physical picture, and the main experimental basis of droop improvement is considered to be the design of effective volume and optimization of carrier density in the active region.

### 2.3. Low efficiency of hole injection and transportation

Electron and hole transport characteristics in GaN-based devices are vastly different. On the one hand, electrons typically have a fairly high mobility of 200 $cm^2 V^{-1}s^{-1}$ or more, but holes in

GaN have a lower mobility with values on the order of 10 $cm^2 V^{-1}s^{-1}$, which is less than an order of magnitude than for electrons. On the other hand, due to the relatively low ionization energy of the n-type doping Si, high electron concentrations are easily achievable. By contrast, the ionization energy of the p-type dopant Mg is around 170 meV, and therefore, high hole concentrations are difficult to achieve. Such asymmetrical transportation behaviours of electrons and holes enhance electron overflow and lower the effective volume of the active region. Inefficient transportation of holes as the major reason for efficiency droop has also been identified in our calculations, as demonstrated in Eqs. 12 and 13, where the low $\mu_P$ and $P_{P0}$ can lead to the high drift leakage current mentioned in section 2.1. Approaches focused on the aim of improving hole injection into the LED active region include p-type doping in the QBs and engineering of the EBL, and some of these results will be briefly reviewed and are summarized in Table 2.

| Year | Engineered QBs/EBL | Droop | Test current (A/ cm²) | Chip size (mm×mm ) | Experiment and/or simulation | Ref |
|---|---|---|---|---|---|---|
| 2008 | p-doping QBs | efficiency peak occurs at 900 A/cm² | 900 | 0.250 mm diameter | Experiment | [53] |
| 2010 | p-doping the last QB | 24.2% | 167 | 0.3×0.3 | Simulation | [54] |
| 2010 | graded EBL | 4% | 200 | 0.3×0.3 | Experiment and simulation | [18] |
| 2010 | InAlN EBL | 18% | 350 | 0.35×0.35 | Experiment | [55] |
| 2011 | superlattice (SL) EBL | 17% | 300 | 0.2×0.5 | Experiment | [56] |
| 2012 | N-polar MQW | 7% | 192 | 0.25×0.25 | Experiment | [14] |
| 2013 | Graded SL-EBL | 8% | 28 | 0.6×0.6 | Experiment | [57] |

*The droop in Tables 1 and 2 is defined as $(\eta_{peak} - \eta_{test-current}) / \eta_{peak}$, where $\eta$ are the EQE or IQE motioned in the references

**Table 2.** Development of engineered EBLs for improving hole injection

In 2008, J. Q. Xie et al. used pulsed electroluminescence measurements to show that droop can be mitigated by p-doping all QBs, and the current density at the efficiency peak can be moved up to ~900 A/cm² [53]. Along the way, Y. K. Kuo simulated the results that only p-doped the last barrier (closest to EBL) with a doping concentration of $10^{17}$ $cm^{-3}$ compared with the LEDs with no doped barrier [54]. The simulation results show that the efficiency droop is significantly improved when the last undoped GaN barrier in a typical blue LED is replaced by a p-type GaN barrier. The results suggest that the improvement in efficiency droop is mainly due to the decrease of electron current leakage and the increase in hole injection efficiency. At the same time, C. H. Wang et al. designed a graded-composition electron blocking layer (GEBL) with aluminium composition increasing along the (0001) direction [18]. The experiments and simulation results demonstrate that such GEBL can effectively enhance the capability of hole transportation across the EBL, as well as electron confinement. Consequently, the efficiency droop is reduced from 34 % in conventional LEDs to only 4 % from the maximum value at low injection current to 200 A/cm$^2$. In order to avoid the added polarization effects caused by the AlGaN EBL, S. Choi et al. used an InAlN EBL instead of an $Al_{0.2}Ga_{0.8}N$ EBL in visible LEDs [55]. A significant enhancement of the EL intensity and light output in blue LEDs with an $In_{0.18}Al_{0.82}N$ EBL was demonstrated. Also, it has been shown that an $In_{0.18}Al_{0.82}N$ EBL is more effective than a conventional $Al_{0.2}Ga_{0.8}N$ EBL in improving quantum efficiency and reducing efficiency droop at high injection current densities. To investigate the effect of electron blocking layer (EBL) on the efficiency droop, R. B. Chung et al. studied two different types of EBLs—single AlInN:Mg layer and AlInN:Mg (2 nm)/GaN:Mg (2 nm) superlattice (SL) structure with seven periods [56]. It was found that the output power and operating voltage of a single EBL LED were sensitive to EBL thickness. On the other hand, an LED with SL EBL showed no deterioration of optical power and operating voltage, while its efficiency droop (17 % at 300A/cm$^2$) was reduced by more than one-half compared to a conventional $Al_{0.2}Ga_{0.8}N$ (20 nm) EBL LED (36% at 300A/cm$^2$). Furthermore, J. H. Park et al. introduced $Al_xGa_{1-x}N$/GaN superlattice EBLs with gradually decreasing Al composition toward the p-type GaN layer. It was experimentally demonstrated that GaInN/GaN LEDs with the GSL-EBL show lower efficiency droop and higher EQE, as well as comparable or even lower operating voltage, compared to LEDs with conventional AlGaN EBLs[57].

## 3. Experimental procedures, results and discussion — Structure design, characterizations and study of mechanisms

### 3.1. Preparation and measurements of the LEDs

A set of epitaxial structures (emitting at 455 nm) were grown on c-plane PSS in a high-speed, rotating-disk metal organic chemical vapour deposition (MOCVD) system (Veeco K465i). All structures had a similar structure, consisting of 7–11 periods of ~3-nm thick InGaN wells and ~5-nm thick GaN barriers. The underlying GaN buffer consisted of a ~1.5 μm nominally undoped GaN layer, followed by a 2-μm n-type GaN with an approximately $1 \times 10^{19}$ cm$^{-3}$ silicon doping level. The final Mg-doped p-GaN was about 100-nm thick with a nominal hole density of 3–$7 \times 10^{17}$ cm$^{-3}$. For comparison, InGaN/GaN superlattices (SLs) or AlGaN/GaN EBLs were employed to investigate the effects on droop in some samples, as shown in Fig. 2a.

The device was designed in lateral injection geometry with a chip dimension of 0.76 x 0.25 mm$^2$ with Ti/Al/Ti/Au n-type contacts and Ni/Au p-type contacts, as shown in Fig. 2b. The surface of epitaxial structures was inspected by a Dimension 3100 AFM system in tapping mode.

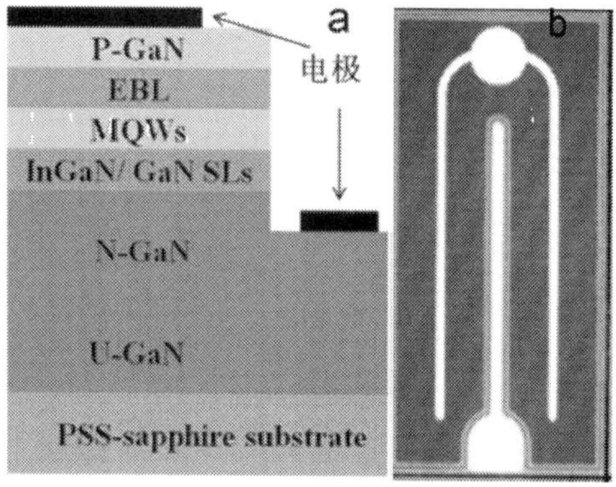


**Figure 2.** Schematic diagram of the studied epitaxial structures (*a*), and details of the graphical front surface of devices (*b*)

### 3.2. Optimization of EBL to reduce polarization charges

As we discussed in the last section, the asymmetry in carrier transport, caused by much lower concentration and mobility of holes compared to electrons, may be the dominant mechanism causing efficiency droop. Introducing a highly p-doped AlGaN electron-blocking layer (EBL) with a high Al composition may mitigate the degree of asymmetry by means of a high potential barrier for electron leakage, but a low barrier for hole injection. However, it is very difficult to realize such an ideal EBL because of the high ionization energy of the p-type dopant Mg in the AlGaN layer and the potential barrier greatly blocking the hole by the piezoelectric polarization sheet charge at the interface between the GaN spacer and the AlGaN EBL.

In this study, we present the three different EBL structures—Bulk EBL, AlGaN/GaN SL-EBLs with different loops, and graded SL-EBL (GSL-EBL), which having a graded Al mole fraction (Fig 3). For comparison, the ~20 nm-thick p-type $Al_{0.20}Ga_{0.80}N$ bulk EBL structure was used. The 6, 8 and 10-period SL-EBL consisted of $Al_xGa_{1-x}N$/GaN bi-layers with thicknesses of 1.6 nm for the AlGaN barriers and 1.8 nm for GaN wells. Likewise, six-period GSL-EBL, consisting of six periods of $Al_xGa_{1-x}N$/GaN bi-layers (x varies from 0.4 to 0.01 and from 0.01 to 0.4) with a thickness of 2 nm for both barriers and wells, were fabricated for each, respectively. Both AlGaN and GaN in the GSL-EBLs are Mg doped to realize a low-doping effect.

Fig. 4 (a) shows the representative external quantum efficiency (EQE) of LEDs, with bulk EBL and SL-EBLs increasing the loop number from 6 to 10 as a function of current density; Fig. 4b shows the I–V characteristics. From the results, we can see that, as expected, the LEDs with a reduced number of periods from the six-period GSL-EBL show the lowest operating voltage and a much reduced efficiency droop at a driving current density of 200 $A/cm^2$. However, the LEDs with an eight-period SL-EBL loop had poor efficiency droop compared to that of the bulk EBL and ten-loop SL-EBL LEDs, as well as a higher operating voltage than bulk and six-loop SL-EBL strucrures.

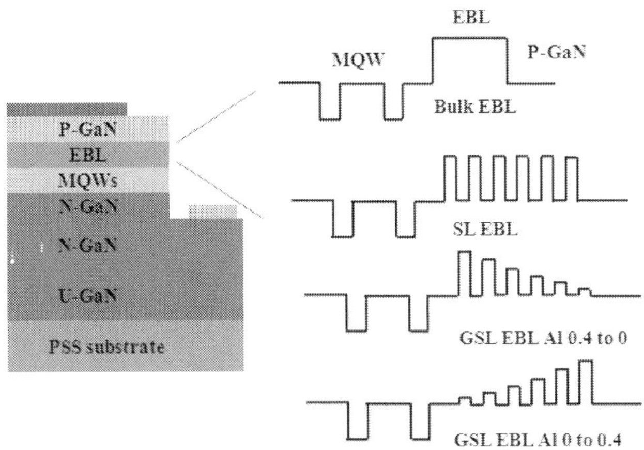


**Figure 3.** Schematic conduction band diagrams showing the three different EBL structures—the conventional bulk EBL, the supper lattice EBL and the graded supper lattice EBL

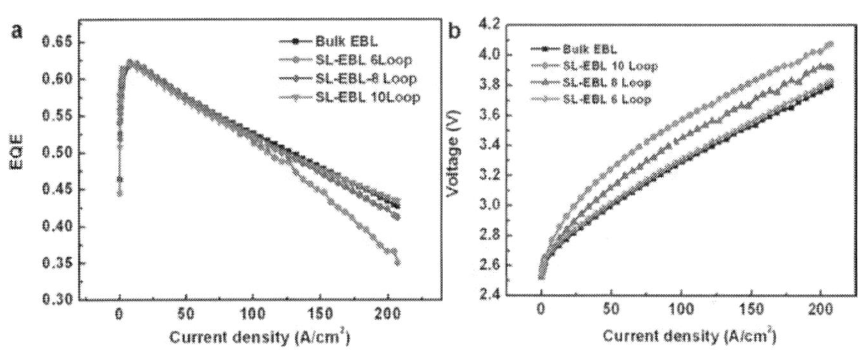


**Figure 4.** EQE (*a*) and I–V (*b*) characteristics as a function of injection current for the LEDs with Bulk EBL and SL-EBL with different loop numbers

We know that SL EBLs cause a penalty in operating voltage, due to hole transport that is hindered by the series of potential barriers at the AlGaN/GaN hetero-interfaces of the SL EBL. But in our experiments, when we reducing the loop numbers of SL-EBL to 6, the operating voltage of this LED was similar to that of bulk EBL LED, although this lower voltage is at the cost of EQE efficiency. Based on the low operating voltage results, Fig. 5a shows EQE measured as a function of current density for bulk and GSL-EBL LEDs. The LED with six-period GSL-EBL and an Al composition increasing from 0 to 0.4 shows the same EQE throughout the whole injection current density range.

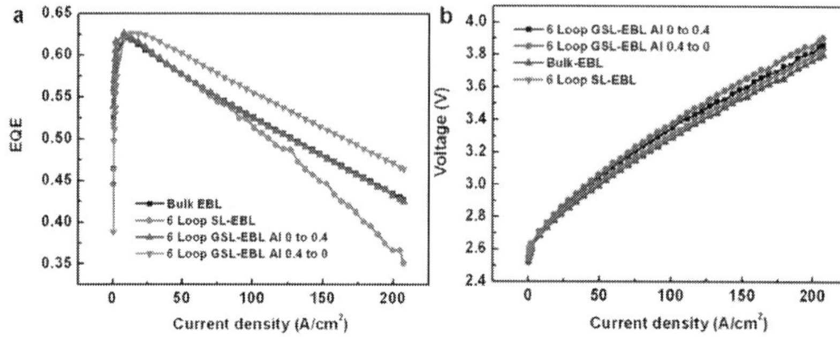


**Figure 5.** EQE (*a*) and I-V characteristics (*b*) as a function of injection current for the LEDs with Bulk EBL and SL-EBL with graded Al composition

But when we graded the Al composition from 0.4 to 0, the efficiency droop at 200 A/cm$^2$ was measured to be 25.5 %, which is higher than 32.3 % of bulk EBL LEDs and 43.8 % of 6-loop SL-EBL LED. As we discussed above, the lower Al composition on the p-type side can reduce the potential barrier for hole injection, leading to less electron leakage and a higher hole concentration at the last grown quantum well where most of the radiative recombination occurs. We think such a process clearly improved the efficiency droop. In addition, the high operating voltage of SL-EBL LED is attributed to the large lattice mismatch at the AlGaN/GaN hetero-interfaces causing large polarization-induced electric fields, as well as to the higher overall Al composition of the SL-EBL compared to the bulk EBL, which is the hard doping Mg element. By grading the Al composition, the voltage drop across the GSL-EBL becomes smaller than that for the SL-EBL, due to the smaller lattice mismatch between AlGaN and GaN layer sand lower overall Al content in the EBL. As shown in Fig. 5b, there are slight operating voltage changes in the bulk and GSL-EBL LEDs, which is as expected from our discussion.

### 3.3. Investigation of active volume effect in Multiple Quantum Well (MQW) region under high driving current

To reveal the efficiency improvement mechanism, monitoring QW with longer wavelength (480 nm) was used to detect the carrier distribution in the LEDs. We designed such a series of samples, with the MQWs consisting of seven blue wells (emitting at 455 nm) and two longer wavelength wells (emitting at 480 nm). As shown in Fig. 6, the longer wavelength wells located at different positions were introduced to experimentally clarify the carrier distribution in the MQWs. Especially at high driving currents, the carrier transport behavior in GaN/InGaN MQW LEDs can be quantitatively investigated.

As shown in Fig. 7a–f, different light output-current-voltage (L-I-V) spectra at the drive current in the range of 1 mA to 300 mA were demonstrated. It is interesting to find out that the intensity of the monitor wells changes greatly as the test current increases. The intensity of the wavelength 480 nm peak includes two parts: 1) from the recombination of electron and holes excited by the electroluminescence (EL); and 2) from the photoluminescence (PL) excited by the 455

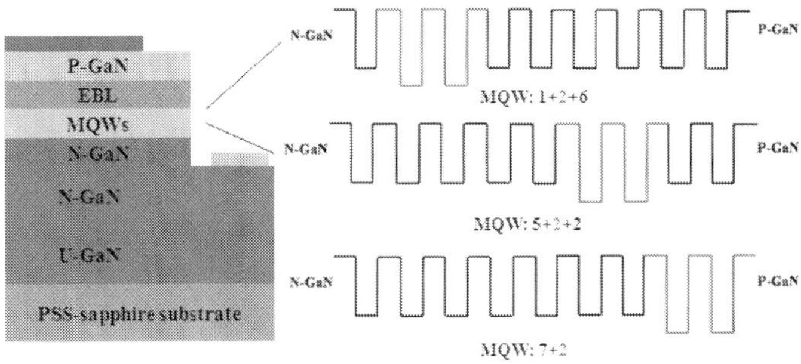


**Figure 6.** Schematic diagrams of the LEDs with different remarkable MQWs

nm light in other QWs. But we know that the blue shift comes from the polarization effect, and the PL peaks alone don't give rise to the blue shift. So, we can monitor the blue shift of the 480 nm peak to see the carrier distribution in the MQWs at different drive currents. From Fig. 7a–c, when the monitor wells are located at the positions close to the p-GaN side, the 480 nm peak shows significant blue shift behavior as the drive current increases, which indicates that these two wells play an important role in carrier recombination. Meanwhile, as we put the monitor wells at the middle of MQWs, the blue shift behavior is weakly observed. Furthermore, in the spectra from the monitor wells located at the places close to N-GaN side, we can identify that there is nearly no blue shift. Such results imply that the carriers, and especially the holes, mainly distribute at the wells close to the p-GaN side. And when the LEDs work at high drive current, the holes move the n-GaN side, but this move behavior is very limited.

As shown in Fig. 7d–f, in order to carefully study carrier transport behavior in the MQWs at high drive current, we demonstrate the I-V characteristics of LEDs with different monitor well positions at driving currents of 100 mA, 200 mA and 300 mA, respectively. For comparison, the test condition is kept at the same integrated time and external environment. At the driving current of 100 mA, in the case of monitor wells located at the middle of the MQWs, the intensity of the peak at 480 nm is clearly higher than the intensity of the 455 nm peak, which is different with the monitor wells at other places. We can conclude that at a high drive current of 100 mA, the carriers mainly distribute on these two wells, and the holes move to these two wells at the electrical driving force. When the driving current increases to 200 mA, at first the peak intensity of 455 nm increases greatly, and this implies a significant improvement in carrier distribution. Secondly, at higher driving current, the intensity at 480 nm in the monitor wells close to the p-GaN side is lower than that for monitor wells close to the n-side. When we increase the driving current to 300 mA, the intensity of the 455 nm peak increases greatly, and the variation tendency of the peak intensity of 480 nm is also apparent. We think that at this state, the wells close to the p-side show a weak contribution for lighting.

As we have investigated the active volume effect in multiple quantum well (MQW) regions under high driving current, we designed MQW structures to improve the efficiency droop. At

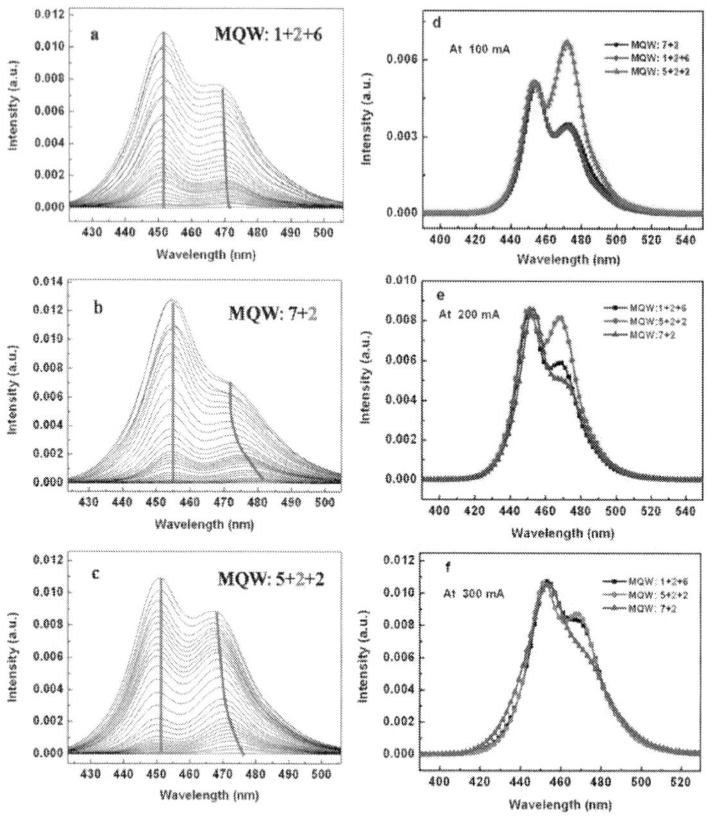

Figure 7. Light output-current-voltage (L-I-V) characteristics of the LEDs with monitor wells located at different positions. (*a–c*): driving current varied from 1 mA to 300 mA; (d–f): L-I-V characteristics at driving currents of 100 mA, 200 mA and 300 mA

first, we changed the MQW loop numbers from 7 to 11 with the QWs and QBs at the same thickness; secondly, based on the LEDs with MQW loop numbers of 11, we kept the thickness of QW at 3 nm and reduced the thickness of QBs from 5 nm to 2.5 nm. As shown in Fig. 8a, we found that when we increased the MQW loop numbers, the efficiency peak occurs at a higher current density of 20 mA/cm$^2$, rather than at 10 mA/cm$^2$. This conclusion is consistent with the previous report that increasing QW numbers can lead to more uniform electron and hole distribution across the active region and reduced peak carrier densities. When we reduce the thickness of QBs, we can see that the thinner QBs structure results in a better efficiency droop. Fig. 8b shows the enlarged efficiency peaks. We think the thinner QBs effectively reduce the irradiative recombination, leading to much more uniform hole distribution. The uniform hole distribution means a relatively lower carrier density, which is likely the reason for the improved efficiency at higher current densities.

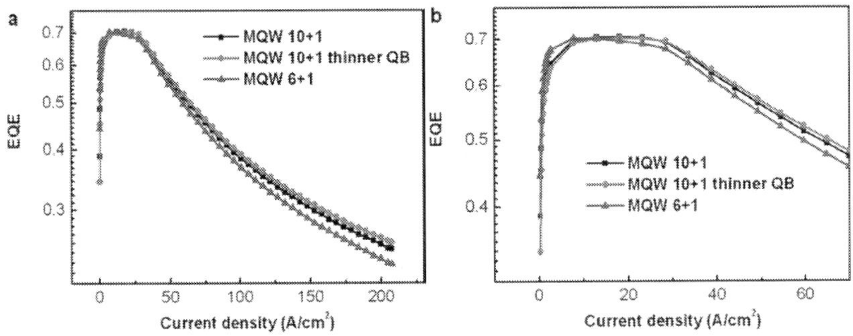


**Figure 8.** EQE as a function of injection current for the LEDs with different MQW loop numbers and QB thicknesses

### 3.4. Using Intentionally Formed V-shaped Pits (V-Pits) to mitigate efficiency droop current

Implementing a single or multiple InGaN/GaN superlattice (SL) structure formed by low content between the n-type GaN region and the MQW region could influence the distribution of strain and the morphology of V-pits. The InGaN/GaN SLs with 20 loops have a total thickness of ~50 nm, and a schematic diagram is shown in Fig. 2.

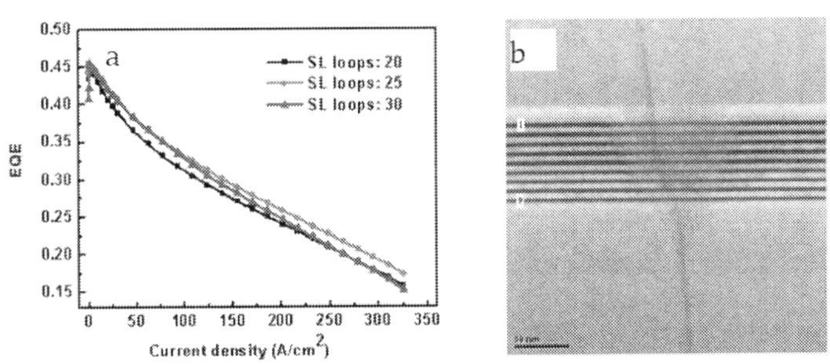


**Figure 9.** EQE as a function of the current density for three LEDs with 20-loop, 25-loop and 30-loop InGaN/GaN superlattices (*a*); transmission electron microscope (TEM) image of MQW region in LEDs with 20-loop SLs (*b*)

To clarify the effects of various InGaN/GaN SLs on the efficiency droop, the EQE of LEDs with different SL structure are calculated and plotted versus current density in Fig. 9. After a rapid increase at low driving current densities, all the LEDs show a monotonic efficiency drop with increasing current. The sharp peak EQE is about 46 % at the current density of J~2.95 A/cm$^2$. The LEDs with 20-loop and 30-loop SLs show an efficiency droop of 45.5 % at a driving current density of 250 A/cm$^2$, which is 4.4 % worse than that of LEDs with 25-loop SLs. But LEDs with 25-loop and 30-loop SLs show a 2 % higher EQE in the range of 30–150 A/cm$^2$. The results indicate that 25-loop SL LEDs show the highest EQE value in the three samples, at both high and low driving current densities.

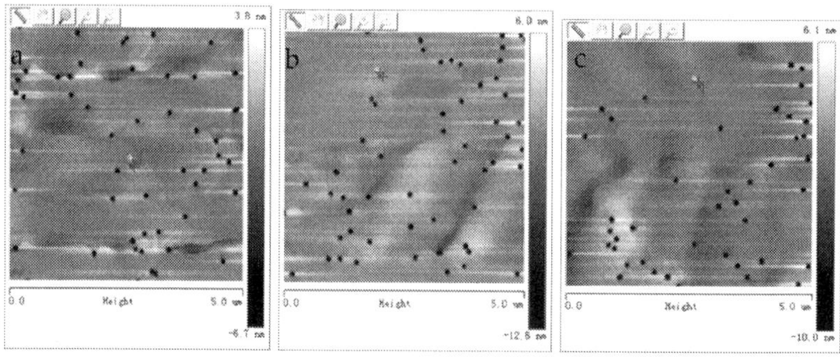


**Figure 10.** Atomic force microscope image of 9-MQW on 20-loop (*a*), 25-loop (*b*), and 30-loop (*c*) SLs

As we know, in InGaN/GaN MQW LEDs, V-shaped pits (V-pits) tend to be easily formed at the interface between InGaN and GaN layers due to lattice mismatch. As shown in Fig. 9b, insertion of InGaN/GaN SLs leads to large sized V-pits going though the MQW region. Threading dislocations penetrate through the central region of V-pits from their apices, which makes semi-polar planes surrounding the threading dislocation act as irradiative recombination centers in the MQW region. The higher energy state of thinner wells on semi-polar can screen the carrier, which diffuses from c-plane wells and enhances the IQE [58]. As shown in Fig. 10, the dimension of V-pits from AFM imagines in MQWs on 30-loop and 25-loop SLs increase to 170 and 140 nm, respectively, from 100 nm for that on 20-loop SLs. The enlarged V-pits suppress non-radiative carriers captured by threading dislocation more effectively, which leads to an increase in EQE in LEDs with 25-loop and 30-loop SLs in the range of 30–150 A/cm$^2$. However, the higher droop in LEDs with 30-loop after 150 A/cm$^2$ is attributed to the distribution of holes closer to n-type, which results from the deep tail of the Mg distribution on the SIMS spectrum (not shown here).

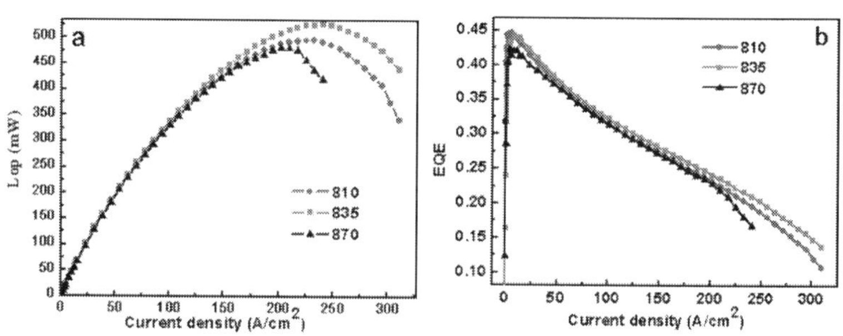


**Figure 11.** Experimental light output (*a*) and EQE (*b*) of LEDs with InGaN/GaN SLs grown at 810, 835 and 870 °C

The output power and EQE measured at current density from 0 to 300 mA are plotted in Fig. 11 for three LEDs with InGaN/GaN SLs grown at 810, 835 and 870 °C. The output power of three LEDs rises with increasing current density, rolls off as current exceeds a characteristic current density, and decreases monotonously toward higher currents. The characteristic current density increases from 200 A/cm$^2$ to 240 A/cm$^2$ when the growth temperature of InGaN/GaN SLs ramps down to 835 °C from 870 °C. However, if the growth temperature decreases lower to 810 °C, the characteristic current density drops to 225 A/cm$^2$. In fact, EQE of the LEDs with SLs grown at 835 °C reaches its peak at ~9.9 A/cm$^2$ compared to ~5 A/cm$^2$ for LEDs with SLs grown at 810 and 870 °C. At the same time, the droop effect for LEDs with SLs grown at 835 °C is ~45.2 %, which is 2.3 % better than that of LEDs with SLs grown at 810 °C. Actually, if the growth temperature of InGaN/GaN SLs decreases, V-pits with larger dimensions can be obtained in the MQW region. Similar to the effects of V-pits with different sizes mentioned above, the better droop effect is due to the larger dimension of V-pits in LEDs with InGaN/GaN SLs grown at 835 °C compared to those grown at 870 °C. But again, the degradation in LEDs with SLs grown at 810 °C is ascribed to the asymmetry in distribution of electrons and holes in the MQW region at high driving current densities.

## 4. Summary

Herein, we have presented a summary of the current state of efficiency droop research and reviewed mechanisms potentially causing the droop. At the same time, we have demonstrated three epi-layer engineered structures that offer some pathways to droop mitigation without compromising other device performance. In our study, we conclude that 1) the structure including an EBL composed of a p-doped graded-composition AlGaN/GaN superlattice can enable better hole injection and reduce electron leakage. It is experimentally shown that GaInN/GaN MQW LEDs with GSL-EBL show lower efficiency droop and higher EQE, as well as comparable or even lower operating voltage, compared to LEDs with conventional bulk AlGaN EBLs. 2) Under high driving current, we remarked on the hole shift behavior by using monitor wells at different MQW positions. Because of the asymmetry in carrier transport, caused by much lower concentration and mobility of holes even at a driving current of 300 mA, the holes move slightly to the n-GaN side and mainly concentrate at the middle well close to the p-side. Accordingly, we investigated the influence of QW numbers and the thickness of QBs for the efficiency droop. The experiments results show that increasing QW numbers and thinner QBs are helpful for carrier extending and hole mobility. 3) we used intentionally formed V-shaped pits (V-pits) to mitigate efficiency droop current. By varying the growth conditions of the SL layer, we obtained different sizes of V-pits and found that proper, larger V-pits can provide a benefit to the mitigation of droop effect.

## Acknowledgements

This work was supported by a grant from the National High Technology Research and Development Program (i.e., 863 program) (No. 2014AA032604).

## Author details

Jie Zhang[1], Dongyan Zhang[1], Daqian Ye[1], Chenke Xu[1] and Meichun Huang[2,3]

1 Xiamen San'an Optoelectronics Co., Xiamen, China

2 Department of Physics, Xiamen University, Xiamen, China

3 CCAST (World Laboratory), Beijing, China

## References

[1] U.S. Department of Energy, Solid-State Lighting Research and Development: Manufacturing Roadmap. Available online at http://www1.eere.energy.gov/buildings/ssl/techroadmaps.html. Accessed August 2014.

[2] Verzellesi G, Saguatti D, Meneghini M, Bertazzi F, Goano M, Meneghesso G, and Zanoni E. Efficiency droop in InGaN/GaN blue light-emitting diodes: Physical mechanisms and remedies. Journal of Applied Physics, 2013; 114(7) 071101.

[3] Delaney K T, Rinke P, and Van de Walle C G. Auger recombination rates in nitrides from first principles. Applied Physics Letters, 2009; 94(19) 191109.

[4] Kioupakis E, Rinke P, Delaney K T, and Van de Walle C G. Indirect Auger recombination as a cause of efficiency droop in nitride light-emitting diodes. Applied Physics Letters, 2011; 98(16) 161107.

[5] Scheibenzuber W G, Schwarz U T, Sulmoni L, Dorsaz J, Carlin J F, and Grandjean N. Recombination coefficients of GaN-based laser diodes. Journal of Applied Physics, 2011; 109(9) 093106.

[6] Shen Y C, Mueller G O, Watanabe S, Gardner N F, Munkholm A, and Krames M R. Auger recombination in InGaN measured by photoluminescence. Applied Physics Letters, 2007; 91(14) 141101.

[7] Kim M-H, Schubert M F, Dai Q, Kim J K, Schubert E F, Piprek J, and Park Y. Origin of efficiency droop in GaN-based light-emitting diodes. Applied Physics Letters, 2007; 91(18) 183507.

[8] Schubert M F, Xu J, Kim J K, Schubert E F, Kim M H, Yoon S, Lee S M, Sone C, Sakong T, and Park Y. Polarization-matched GaInN/AlGaInN multi-quantum-well light-emitting diodes with reduced efficiency droop. Applied Physics Letters, 2008; 93(4) 041102.

[9] Xu J, Schubert M F, Noemaun A N, Zhu D, Kim J K, Schubert E F, Kim M H, Chung H J, Yoon S, Sone C, and Park Y. Reduction in efficiency droop, forward voltage, ide-

ality factor, and wavelength shift in polarization-matched GaInN/GaInN multi-quantum-well light-emitting diodes. Applied Physics Letters, 2009; 94(1) 011113.

[10] Zhang L, Wei X C, Liu N X, Lu H X, Zeng J P, Wang J X, Zeng Y P, and Li J M. Improvement of efficiency of GaN-based polarization-doped light-emitting diodes grown by metalorganic chemical vapor deposition. Applied Physics Letters, 2011; 98(24) 241111.

[11] Bochkareva N I, Rebane Y T, and Shreter Y G. Efficiency droop and incomplete carrier localization in InGaN/GaN quantum well light-emitting diodes. Applied Physics Letters, 2013; 103(19) 191101.

[12] Hammersley S, Watson-Parris D, Dawson P, Godfrey M J, Badcock T J, Kappers M J, McAleese C, Oliver R A, and Humphreys C J. The consequences of high injected carrier densities on carrier localization and efficiency droop in InGaN/GaN quantum well structures. Journal of Applied Physics, 2012; 111(8) 083512.

[13] Ryu H-Y, Shin D-S, and Shim J-I. Analysis of efficiency droop in nitride light-emitting diodes by the reduced effective volume of InGaN active material. Applied Physics Letters, 2012; 100(13) 131109.

[14] Akyol F, Nath D N, Krishnamoorthy S, Park P S, and Rajan S. Suppression of electron overflow and efficiency droop in N-polar GaN green light emitting diodes. Applied Physics Letters, 2012; 100(11) 111118.

[15] Dai Q, Shan Q, Cho J, Schubert E F, Crawford M H, Koleske D D, Kim M-H, and Park Y. On the symmetry of efficiency-versus-carrier-concentration curves in GaInN/GaN light-emitting diodes and relation to droop-causing mechanisms. Applied Physics Letters, 2011; 98(3) 033506.

[16] Liu J P, Ryou J H, Dupuis R D, Han J, Shen G D, and Wang H B. Barrier effect on hole transport and carrier distribution in InGaN/GaN multiple quantum well visible light-emitting diodes. Applied Physics Letters, 2008; 93(2) 021102.

[17] Wang C H, Chang S P, Ku P H, Li J C, Lan Y P, Lin C C, Yang H C, Kuo H C, Lu T C, Wang S C, and Chang C Y. Hole transport improvement in InGaN/GaN light-emitting diodes by graded-composition multiple quantum barriers. Applied Physics Letters, 2011; 99(17) 171106.

[18] Wang C H, Ke C C, Lee C Y, Chang S P, Chang W T, Li J C, Li Z Y, Yang H C, Kuo H C, Lu T C, and Wang S C. Hole injection and efficiency droop improvement in InGaN/GaN light-emitting diodes by band-engineered electron blocking layer. Applied Physics Letters, 2010; 97(26) 261103.

[19] Akyol F, Krishnamoorthy S, and Rajan S. Tunneling-based carrier regeneration in cascaded GaN light emitting diodes to overcome efficiency droop. Applied Physics Letters, 2013; 103(8) 081107.

[20] Bochkareva N I, Voronenkov V V, Gorbunov R I, Zubrilov A S, Lelikov Y S, Latyshev P E, Rebane Y T, Tsyuk A I, and Shreter Y G. Defect-related tunneling mechanism of

efficiency droop in III-nitride light-emitting diodes. Applied Physics Letters, 2010; 96(13) 133502.

[21] Efremov A A B N I, Gorbunov R I, Larinovich D A, Rebane Y T, Tarkhin D V, and Shreter. Effect of the joule heating on the quantum efficiency and choice of thermal conditions for high power blue InGaN/GaN LEDs. Semiconductors, 2006; 40(5) 605.

[22] Hader J, Moloney J V, and Koch S W. Temperature-dependence of the internal efficiency droop in GaN-based diodes. Applied Physics Letters, 2011; 99(18) 181127.

[23] Kudryk Y Y and Zinovchuk A V. Efficiency droop in InGaN/GaN multiple quantum well light-emitting diodes with nonuniform current spreading. Semiconductor Science and Technology, 2011; 26(9) 095007.

[24] Malyutenko V K, Bolgov S S, and Podoltsev A D. Current crowding effect on the ideality factor and efficiency droop in blue lateral InGaN/GaN light emitting diodes. Applied Physics Letters, 2010; 97(25) 251110.

[25] Chu C -F C C-C, Liu W -H, Chu J -Y, Fan F -H, Cheng H -C, Doan T, and Tran C A. High-brightness GaN vertical light-emitting diodes on metal alloy for general lighting application. Proceeding of the IEEE, 2010; 98(7) 2455.

[26] Bertazzi F, Goano M, and Bellotti E. A numerical study of Auger recombination in bulk InGaN. Applied Physics Letters, 2010; 97(23) 231118.

[27] Bertazzi F, Goano M, and Bellotti E. Numerical analysis of indirect Auger transitions in InGaN. Applied Physics Letters, 2012; 101(1) 011111.

[28] Chung H J, Choi R J, Kim M H, Han J W, Park Y M, Kim Y S, Paek H S, Sone C S, Park Y J, Kim J K, and Schubert E F. Improved performance of GaN-based blue light emitting diodes with InGaN/GaN multilayer barriers. Applied Physics Letters, 2009; 95(24) 241109.

[29] Han S-H, Lee D-Y, Shim H-W, Kim G-C, Kim Y S, Kim S-T, Lee S-J, Cho C-Y, and Park S-J. Improvement of efficiency droop in InGaN/GaN multiple quantum well light-emitting diodes with trapezoidal wells. Journal of Physics D: Applied Physics, 2010; 43(35) 354004.

[30] Kuo Y-K, Wang T-H, and Chang J-Y. Advantages of blue InGaN light-emitting diodes with InGaN-AlGaN-InGaN barriers. Applied Physics Letters, 2012; 100(3) 031112.

[31] Kuo Y-K, Wang T-H, Chang J-Y, and Tsai M-C. Advantages of InGaN light-emitting diodes with GaN-InGaN-GaN barriers. Applied Physics Letters, 2011; 99(9) 091107.

[32] Sheng Xia C, Simon Li Z M, Lu W, Hua Zhang Z, Sheng Y, and Wen Cheng L. Droop improvement in blue InGaN/GaN multiple quantum well light-emitting diodes with indium graded last barrier. Applied Physics Letters, 2011; 99(23) 233501.

[33] Meyaard D S, Shan Q, Cho J, Fred Schubert E, Han S-H, Kim M-H, Sone C, Jae Oh S, and Kyu Kim J. Temperature dependent efficiency droop in GaInN light-emitting diodes with different current densities. Applied Physics Letters, 2012; 100(8) 081106.

[34] Hangleiter A, Hitzel F, Netzel C, Fuhrmann D, Rossow U, Ade G, and Hinze P. Suppression of nonradiative recombination by V-shaped pits in GaInN/GaN quantum wells produces a large increase in the light emission efficiency. Physical Review Letters, 2005; 95(12) 127420.

[35] Hwang S, Jin Ha W, Kyu Kim J, Xu J, Cho J, and Fred Schubert E. Promotion of hole injection enabled by GaInN/GaN light-emitting triodes and its effect on the efficiency droop. Applied Physics Letters, 2011; 99(18) 181115.

[36] Li Z, Lestrade M, Xiao Y, and Li Z S. Improvement of performance in p-side down InGaN/GaN light-emitting diodes with graded electron blocking layer. Japanese Journal of Applied Physics, 2011; 50(8) 080212.

[37] Shuji Nakamura S F C. Introduction to nitride semiconductor blue lasers and light emitting diodes. London: Taylor & Francis, 2000; 5–7.

[38] Dai Q S, Cho J, Schubert E F, Crawford M H, Koleske D D, Kim M-H, and Park Y. On the symmetry of efficiency-versus-carrier-concentration curves in GaInN/GaN light-emitting diodes and relation to droop-causing mechanisms. Applied Physics Letters, 2011; 98(3) 033506.

[39] Cho J, Schubert E F, and Kim J K. Efficiency droop in light-emitting diodes: Challenges and countermeasures. Laser & Photonics Reviews, 2013; 7(3) 408.

[40] Piprek J. Efficiency droop in nitride-based light-emitting diodes. Phys.Status Solidi A, 2010; 207(10) 2217.

[41] Saguatti D, Bidinelli L, Verzellesi G, Meneghini M, Meneghesso G, Zanoni E, Butendeich R, and Hahn B. Investigation of efficiency-droop mechanisms in multi-quantum-well InGaN/GaN blue light-emitting diodes. IEEE Transactions on Electron Devices, 2012; 59(5)1402.

[42] Lin R-M, Lai M-J, Chang L-B, and Huang C-H. Effect of an asymmetry AlGaN barrier on efficiency droop in wide-well InGaN double-heterostructure light-emitting diodes. Applied Physics Letters, 2010; 97(18) 181108.

[43] Tu P-M, Chang C-Y, Huang S-C, Chiu C-H, Chang J-R, Chang W-T, Wuu D-S, Zan H-W, Lin C-C, Kuo H-C, and Hsu C-P. Investigation of efficiency droop for InGaN-based UV light-emitting diodes with InAlGaN barrier. Applied Physics Letters, 2011; 98(21) 211107.

[44] Zheng Z, Chen Z, Chen Y, Wu H, Huang S, Fan B, Wu Z, Wang G, and Jiang H. Improved carrier injection and efficiency droop in InGaN/GaN light-emitting diodes with step-stage multiple-quantum-well structure and hole-blocking barriers. Applied Physics Letters, 2013; 102(24) 241108.

[45] Gardner N F, Muller G O, Shen Y C, Chen G, Watanabe S, Gotz W, and Krames M R. Blue-emitting InGaN–GaN double-heterostructure light-emitting diodes reaching maximum quantum efficiency above 200A/cm[sup 2] Applied Physics Letters, 2007; 91(24) 243506.

[46] Son J H and Lee J-L. Numerical analysis of efficiency droop induced by piezoelectric polarization in InGaN/GaN light-emitting diodes. Applied Physics Letters, 2010; 97(3) 032109.

[47] A. Kaneta T M, Y. Kawakami, S. Fujita, G. Marutsuki, Y. Naru-kawa, and T. Mukai. Discrimination of local radiative and nonradiative recombination processes in an InGaN/GaN single-quantum-well structure by a time-resolved multimode scanning near-field optical microscopy. Applied Physics Letters, 2003; 83(17) 3462.

[48] Shim J I, Kim H, Shin D S, and Ryu H Y. An explanation of efficiency droop in ingan-based light emitting diodes: saturated radiative recombination rate at randomly distributed in-rich active areas. Journal of the Korean Physical Society, 2011; 58(3) 503.

[49] Ni X, Li X, Lee J, Liu S, Avrutin V, Ozgur U, Morkoc H, and Matulionis A. Hot electron effects on efficiency degradation in InGaN light emitting diodes and designs to mitigate them. Journal of Applied Physics, 2010; 108(3) 033112.

[50] Maier M, Kohler K, Kunzer M, Pletschen W, and Wagner J. Reduced nonthermal rollover of wide-well GaInN light-emitting diodes. Applied Physics Letters, 2009; 94(4) 041103.

[51] S. Tanaka Zhao Y, Koslow I, Pan C-C, Chen H-T, Sonoda J, DenBaars S P and Nakamura S. Droop improvement in high current range on PSS-LEDs. Electronics Letters, 2011; 47(5) 335.

[52] Li X, Okur S, Zhang F, Avrutin V, Ozgur U, Morkoc H, Hong S M, Yen S H, Hsu T S, and Matulionis A. Impact of active layer design on InGaN radiative recombination coefficient and LED performance. Journal of Applied Physics, 2012; 111(6) 063112.

[53] Xie J, Ni X, Fan Q, Shimada R, Ozgur U m, and Morkoc H. On the efficiency droop in InGaN multiple quantum well blue light emitting diodes and its reduction with p-doped quantum well barriers. Applied Physics Letters, 2008; 93(12) 121107.

[54] Yen-Kuang Kuo M-C T, Sheng-Horng Yen, Ta-Cheng Hsu, and Yu-Jiun Shen. Effect of P-type last barrier on efficiency droop of blue InGaN light-emitting diodes. IEEE Journal of Quantum Electronics; 2010; 46(8) 1214.

[55] Choi S, Kim H J, Kim S-S, Liu J, Kim J, Ryou J-H, Dupuis R D, Fischer A M, and Ponce F A. Improvement of peak quantum efficiency and efficiency droop in III-nitride visible light-emitting diodes with an InAlN electron-blocking layer. Applied Physics Letters, 2010; 96(22) 221105.

[56] Chung R B, Han C, Pan C-C, Pfaff N, Speck J S, DenBaars S P, and Nakamura S. The reduction of efficiency droop by Al0.82In0.18N/GaN superlattice electron blocking

http://dx.doi.org/10.5772/60454

layer in (0001) oriented GaN-based light emitting diodes. Applied Physics Letters, 2012; 101(13) 131113.

[57] Park J H, Yeong Kim D, Hwang S, Meyaard D, Fred Schubert E, Dae Han Y, Won Choi J, Cho J, and Kyu Kim J. Enhanced overall efficiency of GaInN-based light-emitting diodes with reduced efficiency droop by Al-composition-graded AlGaN/GaN superlattice electron blocking layer. Applied Physics Letters, 2013; 103(6) 061104.

[58] Hangleiter A, Hitzel F, Netzel C, Fuhrmann D, Rossow U, Ade G, and Hinze P. Suppression of nonradiative recombination by V-shaped pits in GaInN/GaN quantum wells produces a large increase in the light emission efficiency. Physical Review Letters, 2005; 95(12) 127402.

Chapter 3

# Colorless and Transparent high – Temperature-Resistant Polymer Optical Films – Current Status and Potential Applications in Optoelectronic Fabrications

Jin-gang Liu, Hong-jiang Ni, Zhen-he Wang,
Shi-yong Yang and Wei-feng Zhou

Additional information is available at the end of the chapter

http://dx.doi.org/10.5772/60432

**Abstract**

Recent research and development of colorless and transparent high-temperature-resistant polymer optical films (CHTPFs) have been reviewed. CHTPF films possess the merits of both common polymer optical film and aromatic high-temperature-resistant polymer films and thus have been widely investigated as components for microelectronic and optoelectronic fabrications. The current paper reviews the latest research and development for CHTPF films, including their synthesis chemistry, manufacturing process, and engineering applications. Especially, this review focuses on the applications of CHTPF films as flexible substrates for optoelectrical devices, such as flexible active matrix organic light-emitting display devices (AMOLEDs), flexible printing circuit boards (FPCBs), and flexible solar cells.

**Keywords:** colorless polymer films, high temperature, synthesis, flexible substrates

## 1. Introduction

Various polymer optical films have been widely applied in the fabrication of optoelectronic devices [1]. Recently, with the ever-increasing demands of high reliability, high integration, high wiring density, and high signal transmission speed for optoelectronic fabrications, the service temperatures of polymer optical films have dramatically increased [2, 3]. For instance, in the fabrication of new-generation flexible thin-film transistor-driven active matrix liquid crystal display devices (TFT-LCDs) or active matrix organic light-emitting display devices (AMOLEDs), the processing temperature on the flexible plastic substrates might be higher than 300°C [4-6]. Most of the common polymer optical films would lose their optical and mechanical properties at such high processing temperatures. Thus, colorless and transparent high-temperature-resistant polymer optical films (CHTPFs) have attracted increasing attentions from both the academic and engineering aspects in the past decades.

According to the different servicing temperatures or glass transition temperatures ($T_g$), the polymer optical films could be roughly classified into three types, including conventional optical films ($T_g < 100°C$), common high-temperature optical films ($100 \le T_g < 200°C$), and high-temperature optical films ($T_g \ge 200°C$), as shown in Figure 1. The typical chemical structures for the polymer optical films are illustrated in Figure 2. Main physical and chemical characteristics for the typical optical polymers were tabulated in Table 1 [7, 8]. It can be clearly seen that conventional polymer optical films, such as polyethylene terephthalate (PET, $T_g$: ~78°C) or polyethylene naphthalate (PEN, $T_g$: ~123°C), possess excellent optical transparency. However, they are facing great challenges in advanced optoelectronic fabrication due to their limited service temperatures. On the other hand, high-temperature-resistant polymer films such as wholly aromatic polyimide films (PI) exhibit excellent thermal stability up to 300°C. However, they suffer from deep colors and poor optical transmittance in optoelectronic applications. Thus, achieving a compromise between the pale color and high thermal stability for the polymer optical films is one of the most challenging projects for optoelectronic polymeric films development.

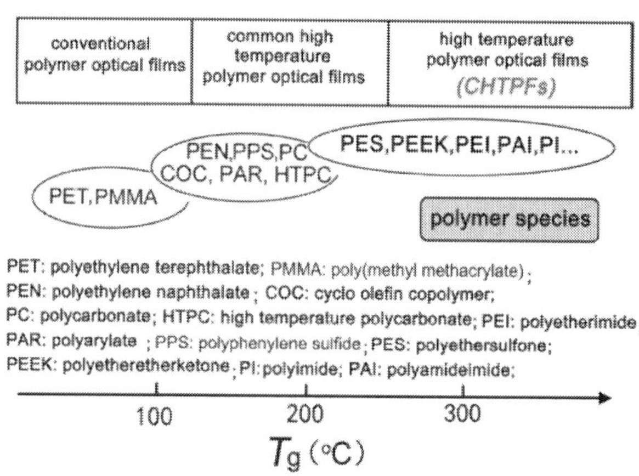


**Figure 1.** Classification of polymer optical films

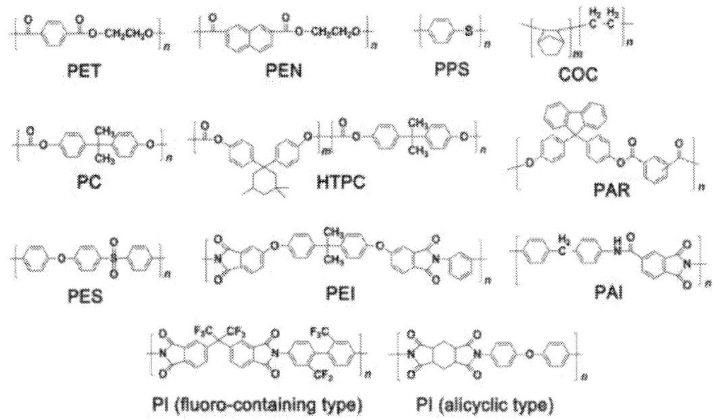


**Figure 2.** Typical chemical structures for polymer optical films

| Item[1] | Unit | PET | PEN | PC | PPS | PEI | PES | PI | CPI[2] |
|---|---|---|---|---|---|---|---|---|---|
| Density | $g/cm^3$ | 1.40 | 1.36 | 1.20 | 1.35 | 1.27 | 1.37 | 1.43 | 1.23 |
| Transmittance | % | 90 | 87 | 92 | 85 | 80 | 89 | 30-60 | 90 |
| $T_m$ | °C | 256 | 266 | 240 | 285 | 365 | 380 | NA[3] | NA |
| $T_g$ | °C | 78 | 123 | 150 | 90 | 217 | 223 | >300 | 303 |
| WVTR | $g/m^2$ day | 21 | 6.9 | 60 | 8 | 43.5 | 73 | 64 | 93 |
| OTR | $cm^3/m^2$ day | 6 | 2 | 300 | 6 | 220 | 235 | 22 | NA |
| Water uptake | % | 0.3 | 0.4 | 0.2 | 0.05 | 1 | 0.5 | 1.3 | 2.1 |
| $\sigma$ | MPa | 225 | 275 | 98 | 250 | 130 | 95 | 274 | 112 |
| $E_b$ | % | 120 | 90 | 140 | 50 | 70 | 70 | 90 | 12 |
| D.S. | V/μm | 280 | 300 | 250 | 250 | 250 | 260 | 280 | NA |
| $\varepsilon$ | - | 3.2 | 3.0 | 3.0 | 3.0 | 3.5 | 4.0 | 3.3 | 2.9 |

[1]$T_m$: melting point; $T_g$: glass transition temperature; WVTR: water vapor transmission rate; OTR: oxygen transmission rate; σ: tensile strength; $E_b$ : elongation at break; D.S.: dielectric strength; ε: dielectric constant;. [2] Data from colorless PI film Neopulim® L-3430 developed by MGC, Japan; [3] Not available.

**Table 1.** Typical properties of polymer optical films

In the past decades, considerable progress has been achieved in both the academic development and commercialization for novel CHTPFs. According to the classification in Figure 1, the leading materials, in terms of comprehensive properties and potential market volume, include colorless polyimide (PI) films, polyethersulfone (PES) films, polyetheretherketone (PEEK) films, polyamide (PA), and polyamideimide (PAI) films. They have been the main components for CHTPF families. According to the statistics from Techno Create Corp. (TCC, an authoritative consulting agency in Japan), the market of CHTPFs in 2011 has been close to 1 billion Japanese yen and the market will see a rapid increase higher than 15% per year in the following years [9].

In this review, the state of art and future development of CHTPFs in optoelectronic fabrications has been reviewed. The molecular design, synthesis chemistry, and film fabrication techniques for CHTPFs were introduced first. Then, the applications of CHTPFs in several important optoelectronic fields including flexible display, flexible printing circuit boards (FPCBs), and flexible solar cells were presented.

## 2. CHTPF manufacturing technology

Generally, the overall production process for CHTPF products consists of several steps, including monomers synthesis, polymer resin preparation, and the film preparation. Technically, these three steps all have their own core technologies and are usually interwinded and interrelated. The reactivity and purity of monomers will definitely affect the physical and chemical properties of the derived polymer resins, including molecular weights and their distribution, inherent viscosities, solubility in organic solvents, appearance, color, and so on. The features of the resins have great effects on the properties of the final polymer films, including their color, optical transparency, mechanical strength, thermal stability, and dielectric properties. Meanwhile, the preparing technologies for the films, including casting procedure, the uniaxial or biaxial stretching process, high-temperature curing program, and even the final winding and rewinding process will also affect the features of the CHTPF products. Thus, the manufacture of CHTPFs is usually a multidisciplinary technology.

The manufacturing techniques for polymer films usually include several types, such as casting, melting extrusion, and blowing procedures. Extrusion is the process of forming a film continuously through an opening. Most extruders do this by rotating a screw inside a stationary heated cylindrical barrel, to melt the polymer resins and pump the melt through a suitably shaped slit. This is used for direct manufacture of finished film products. It may also be used to feed a second process such as injection molding, blow molding, coating, laminating, or thermo-forming process. Blowing procedure is usually performed from a single-screw extruder by extruding polymer resins, cooling it with external and/or internal air streams, stretching it in the machine direction (MD) by pulling it away from the die; stretching it in the transverse direction (TD) by internal air pressure, flattening it by passing through nip rolls, and winding it onto a cylindrical roll. Optional post-stretching operations may include flame or corona surface treatment for wettability, adhesion, and sealing.

Many factors influence the choice of suitable procedures for polymer film manufacture, including physical and chemical properties of the polymer resins, color and appearance demands, and the current abilities of film-producing equipment, and so on. For example, as shown in Figure 3, for crystalline polymer resins that have clear melting points, such as PET and PEN, nonsolvent melting extrusion technique is mainly used. However, for amorphous polymers with low to moderate $T_g$ values, such as PC and PES, both melting extrusion and solution casting techniques can be used. As for high-$T_g$ amorphous polymers, such as PIs, solvent-casting procedure is usually the optimal choice.

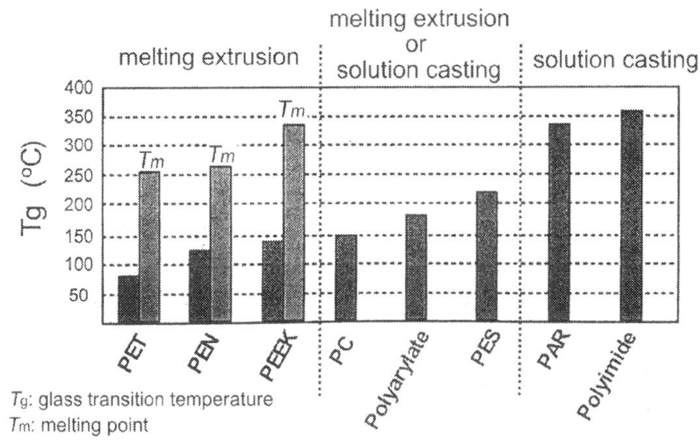


**Figure 3.** Processing methods for polymer films

The manufacturing techniques for CHTPF films have the similarities with the common optical films; however, they have their own uniqueness at the same time. This is mainly due to their relatively higher $T_g$ values (or melting points) and lower solubility in common solvents caused by the more rigid molecular skeletons compared with the common optical polymers. Thus, for CHTPF films, solvent-casting procedure is most commonly used, especially in laboratory. For the solvent-casting procedure, it can be classified into two approaches: uniaxial stretching (machine direction, MD) and biaxial stretching (transverse direction, TD, and machine direction, MD) techniques. Biaxial stretching at temperatures above the $T_g$ values of the CHTPF resins can usually improve the high-temperature dimensional stability of the obtained films.

Then, in the present paper, PI films are taken as examples to illustrate the development of lab-scale and industrial-scale manufacturing techniques for CHTPF films.

### 2.1. Laboratory preparation of PI films

Before industrial-scale manufacturing for CHTPF optical films, it is quite necessary to make a film prototype in laboratory so as to determine the optimal processing parameters. For PI films, the common fabrication techniques include two pathways: standard route via poly(amic acid) (PAA) and new route from organosoluble PIs, as illustrated in Figure 4. Both routes have their advantages and drawbacks. Generally speaking, the first standard PAA route is suitable to all kinds of PI films. In this procedure, dianhydride and diamine monomers will first polymerize in *N,N*-dimethylacetamide (DMAc) to afford PAA solution. The obtained PAA solution is sensitive to heat and moisture, which is easily degrading when stored at room temperature. Thus, it had been better using the newly synthesized PAA for preparing PI films. The PAA solution is cast onto clean glass or stainless steel substrates, followed by thermally curing from room temperature to elevated temperatures. This curing process consists of not only the physical course of solvent evaporation but also the chemical course of imidization or cyclization with the elimination of water. It has been well established that the imidization temperatures as high as 300-350°C is necessary to finish the transition from PAA to PI. Such a high temperature will definitely affect the color of the produced PI films. On the other hand, during the elimination of water from the system, microscopic defects such as pinholes and crack might occur. Thus, the imidization condition should be deliberately controlled in order to produce high-quality PI films.

Technically, the second route is only useful for PI resins which are soluble in organic solvents (mainly DMAc). As we know, the solubility of PI resins is particularly associated with its structure. Introduction of flexible linkages (-O-, -$CH_2$-, etc.), bulky substituents (alkyl groups, phenyl, etc.), and unconjugated structure (aliphatic or alicyclic groups) are all beneficial increasing the solubility of PI resins in organic solvents. From this point of view, this route is quite useful for colorless PI films production, because most of the PI resins for colorless PI films are soluble in polar solvents. In addition, the curing procedure for preimidized PI solution is nearly pure physical course of solvent evaporation. Thus, the PI films can be produced at relatively low temperature and exhibit good surface smoothness. This is undoubtedly beneficial for the production of colorless PI films.

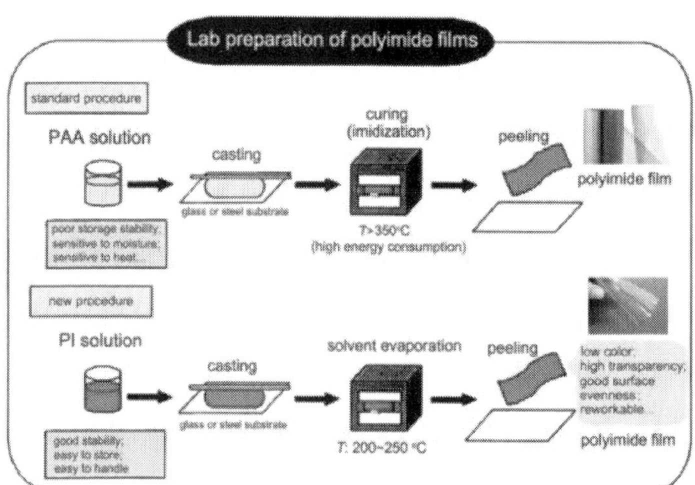


**Figure 4.** Lab-scale preparation of PI films

### 2.2. Industrial preparation of PI films

Ever since the commercialization of PI films with the trademark of Kapton® [poly(pyromellitic anhydride-oxydianiline), PMDA-ODA] in 1960s by DuPont Corporation in the USA, they have been becoming one of the most important basic materials for modern industry [10]. PI films have found various applications in civil and military high-tech fields. In 2011, the worldwide consumption for PI films is more than 8000 metric tons and this consumption is estimated to reach 13000 tons in 2016. The three major markets for PI film are flexible printed circuit substrates, high-temperature wire and cable wrapping, and magnetic wire insulation. The wide applications of PI films are mainly attributed to their excellent properties, including extreme servicing temperatures (-296-400°C for Kapton), high mechanical properties, excellent dielectric features, and good environmental stability. The superior property for PI films, on one hand, is associated with their heteroaromatic molecular structures, and on the other hand, owes to their unique producing techniques.

Compared with the lab-scale preparation, the greatest difference for industrial manufacturing of PI films is the stretching process [11]. Stretching process, either uniaxial or biaxial stretching of the gel-like PAA films, will result in the full orientation and extension for the PI molecular chains. The gelation of PAA films can be achieved either by partially evaporating the solvent or by chemical treatment with a dehydrating agent (acetic anhydride, dicyclohexylcarbodiimide, etc.) and its catalyst (pyridine). From a viewpoint of polymer physics, stretching will greatly enhance the mechanical properties of the obtained PI films. For example, the values of elongations at break for the lab-making PI films without any stretching treatment are usually below 20%. However, this value can be increased several times after stretching treatments.

A diagram of the biaxial stretching production line for PI films is shown in Figure 5. In this procedure, the monomers are first feed into the polymerization reactor containing fully dried DMAc solvent. After polycondensation, the obtained PAA solution is deaerated and cast from the slit die in the form of a continuous film onto the surface of a heated rotating stainless steel drum. The solvent in PAA is partially evaporated and a portion of imidization reaction occurs in PAA at the same time. Thus, a self-supporting PAA film is formed. Alternatively, the PAA solution on the rotating drum can pass through a bath containing dehydrating agent and cyclization catalyst to afford a gel-like PAA film. Then, the gel-like PAA film is peeled from the metal drum and first stretched in the machine direction (MD) while controlling the stretching rate using nip rolls. The stretching ratio can usually be regulated by the drive source and a speed regulator. The gel film stretched in the machine direction is subsequently introduced into a tenter frame where it is gripped at both transverse edges. Various means may be employed to grip the film, including pins, clips, clamps, and rollers. The gel film is then stretched in the transverse direction due to outward movement of the tenter clips, the volatile organic solvent is removed by evaporation, and the film is heat-treated by means of hot air or radiant heat from an electrical heater to give a biaxial oriented polyimide film (BOPI). The transverse stretching is carried out at temperatures around 350°C to facilitate the imidization of PAA. Such a procedure has been widely used for PI film production and there has been significant patent activity in the past half century since the commercialization of PI films in 1960s. Up to now, most of the commercially available wholly aromatic PI films have been produced by such kind of procedure.

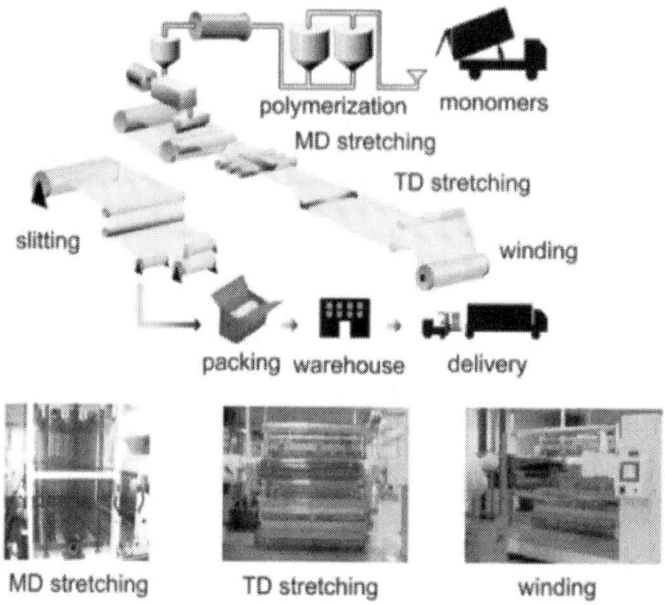


**Figure 5.** Industrial-scale manufacturing for PI films via PAA precursors

Because the full imidization temperature for PAA is usually higher than 350°C, the manufacturing procedure mentioned above might be difficult for colorless PI film production, whose color and transparency is highly sensitive to high temperatures. Thus, a new manufacturing technique has been developed in recent years [12].

As illustrated in Figure 6, the new procedure uses soluble PI resins as the starting materials instead of PAAs. The key elements for this procedure include: (a) the PI resin must be soluble in a volatile solvent; (b) a stable PI solution with a reasonable solid content and viscosity should be formed; and (c) formation of a homogeneous film and release from the casting support must be possible. In the procedure, PI resins are first dissolved in polar solvents to afford the PI solution, which are purified by filtration through screen mesh. Then, the PI solution is cast onto stainless steel belt, followed by thermally drying at high temperatures to remove the solvent. This drying procedure is only to remove the solvent in the PI solution. Thus, the temperature is usually lower than the common imidization temperature (300-350°C). Similarly, the PI films can also be stretched at an appropriate solvent content. For instance, NASA (National Aeronautics and Space Administration, USA) Langley research center investigated the molecularly oriented colorless PI films for space applications [13]. In large space structures with designed lifetimes to be 10-30 years, there exists a need for high-temperature (200-300°C) stable, flexible polymer films that have high optical transparency in the visible light region. For this purpose, a colorless and transparent PI film, LaRC-CP1, derived from 6FDA and fluoro-containing diamine, 2,2-bis[4-(4-aminophenoxy)phenyl]hexafluoropropane (BDAF), has been developed in NASA. This film is prepared from soluble PI resin. The LaRC-CP1 film was uniaxially stretched at 1.5, 1.75, and 2 times the original length of the film. Table 2 shows the influence of stretching on the physical and mechanical properties of LaRC-CP1 film. Apparently, the tensile properties of the film increased with increased stretching ratio. For example, the tensile strength of 2.0× stretched film increased from 93.0 to 145.4 MPa after stretching treatment; and the elongations increased from 16% to 65%. After stretching treatment, the dimensional stability, stiffness, elongation, and strength of the film were greatly enhanced, which are crucial for the applications in space environments.

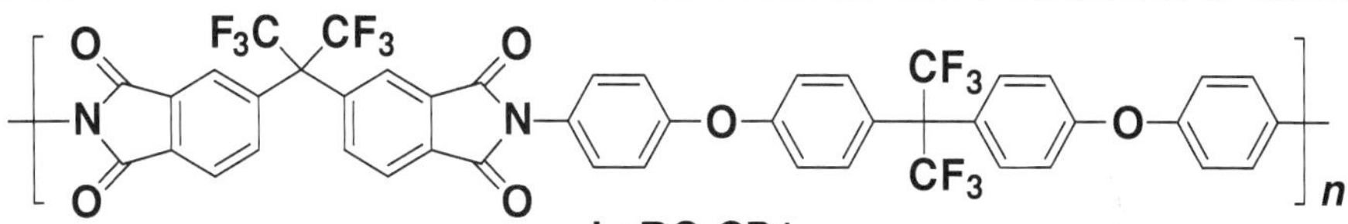


LaRC-CP1

| Stretch ratio | CTE (ppm/°C) | Tensile strength (MPa) | Modulus (GPa) | Elongation (%) |
|---|---|---|---|---|
| None | 50 | 93.0 | 2.0 | 16 |
| 1.5× | 42 | 88.9 | 1.7 | 20 |
| 1.75× | 44 | 107.5 | 1.8 | 49 |
| 2.0× | 46 | 145.4 | 2.1 | 65 |

**Table 2.** Characterization of stretched LaRC-CP1 film

US Patent 8357322 assigned to Mitsubishi Gas Chemical Company describes a method for producing colorless and transparent PI films by a solution casting procedure. The biaxially stretched colorless PI films exhibit excellent optical transparency, heat resistance, and reduced dimensional changes [14]. The films were produced with the soluble PI resin as the starting materials, which were derived from 1,2,4,5-cyclohexanetetracarboxylic dianhydride and aromatic diamines by one-step high-temperature polycondensation route. The PI film was biaxially stretched in the machine direction by 1.01 times and in the transverse direction by 1.03 times at 250°C for 11 min under a stream of nitrogen. Then, the PI film was dried by blowing nitrogen containing 1000 ppm oxygen at a flow rate of 3.3 m/sec at 280°C for 45 min. The obtained PI film had a thickness of 200 μm, a total light transmittance of 89.8%, a yellow index of 1.9, and a haze of 0.74%. The solvent residual ratio in the film was 0.5% by weight. By virtue of these properties, the colorless PI films might find various applications in optoelectronic applications, such as transparent conductive film, transparent substrates for flexible

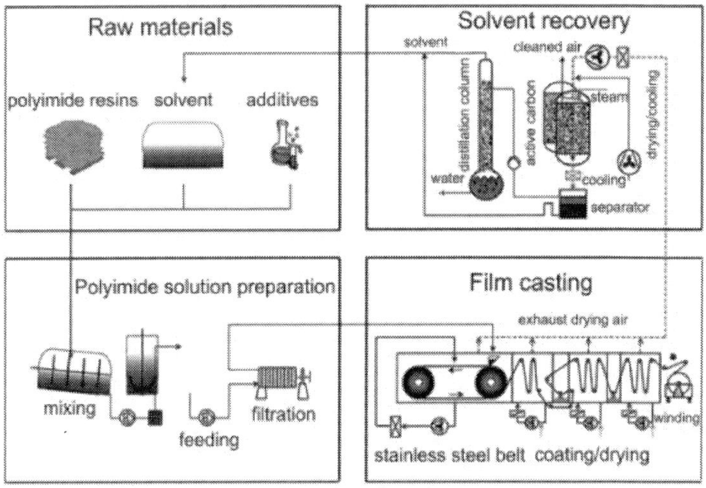


adapted, modified, and reprinted from ref. [12]

**Figure 6.** Industrial-scale manufacturing for PI films via soluble PI resins

display, flexible solar cells, and flexible printing circuit board (FPCB). Similar procedures were also reported by the company [15].

In addition, US Patent 7550194 assigned to DuPont Company [16] and US patent 8846852 to Kolon Industries [17] report the low-color PI films derived from the copolymers of fluoro-containing dianhydride, 2,2-bis(3,4-dicarboxyphenyl)hexafluoropropane dianhydride (6FDA), 3,3′,4,4′-biphenyl-tetracarboxylic dianhydride (BPDA), and fluoro-containing diamine, 2,2′-bis(trifluoromethyl)-benzidine (TFMB). The copolymers were prepared via PAA precursors, followed by chemical imidization of the PAAs to afford the gel-like PAA films or soluble PI resins. Then, the PI films were produced from these intermediums at high temperature up to 300°C. Flexible and tough PI films with low color and high transparency were obtained.

### 2.3. CHTPF films analysis and evaluation techniques

In practical applications for CTHTP optical films, various properties have usually been analyzed and evaluated. For different applications, specific properties might be specially emphasized. For example, in the fabrication of AMOLEDs, the water vapor transmission rate (WVTR) and oxygen transmission rate (OTR) of the flexible substrates are severely limited to be below $10^{-4}$ $cm^3/m^2$ day and $10^{-6}$ $g/m^2/day$, respectively, because the penetration of water and oxygen through the substrates might poison the emitting components, resulting in the reduced operating life of the devices [18]. In flexible solar cells, optical transmittance and yellowness might be the most concerned parameters because yellowness of the polymer substrates might decrease the conversion efficiency of solar light to electricity. Thus, it would be helpful to understand the analysis and evaluation techniques for CHTPF films.

### 2.3.1. *Optical properties*

Common optical properties, such as yellow index (YI), haze, optical transmittance at specific wavelength, ultraviolet-visible cutoff wavelength, refractive index are usually needed to be evaluated for CHTPFs.

Yellowness index (YI) indicates the degree of departure of an object color from colorless or from a preferred white toward yellow. Haze value indicates the degree of cloudiness in a film. The YI and haze value of a film can usually be measured by a colorimeter and can be computed by a given procedure from colorimetric or spectrophotometric data [19]. Optical transmittance of a film indicates the percentage of incident light that is transmitted by the film. The reciprocal of optical transmission is the haze value, which increases as the percent of transmission decreases. As a general rule, 0% haze relates to complete transparency, up to 30% is translucent, and more than 30% haze is considered opaque. Optical transmittance of a film can usually be measured with an ultraviolet-visible light spectrophotometer.

Refractive index of a film indicates the ratio of the velocity of light in vacuum to that in a film. It is the ratio of the sine of the angle of incidence to the sine of the angle of refraction. Refractive index values can be measured with a prism coupler. Birefringence of a film indicates the difference in the refractive indices of two perpendicular directions in a film. When the refractive indices measured along three mutually perpendicular axes are identical, they are classified as optically isotropic. When the film is stretched, providing molecular orientation, and the refractive index parallel to the direction of stretching is altered so that it is no longer identical to what is perpendicular to this direction, the film displays birefringence. The common apparatus for optical parameters measurement of CHTPF films are shown in Figure 7.

### 2.3.2. *Thermal properties*

As mentioned before, the thermal stability of optical films is becoming increasingly important for their applications in optoelectronic fabrications. The thermal properties of an optical film include thermal decomposition temperature ($T_d$), glass transition temperature ($T_g$), coefficient of thermal expansion (CTE), and high-temperature dimensional stability. Generally, thermal analysis for an optical film indicates any analysis of physical or thermodynamic properties of the film in which heat is directly involved, with the heat either being added or removed. Different methods are used with each method providing certain useful data or information.

Thermogravimetric analysis (TGA) is an analysis by the measurement of weight changes of an optical film as a function of increasing temperature with time. Properties measured include thermal decomposition temperature and relative thermal stability. Dimensional stability of a polymer optical film indicates its ability to retain the precise initial shape and size. It is the temperature above which the films lose their dimensional stability. For most films, the main determinant of dimensional stability is their glass transition temperature.

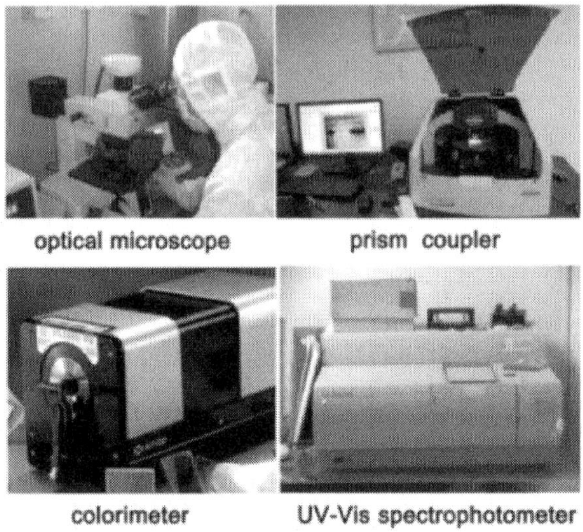


**Figure 7.** Optical properties measurement system

Below a certain temperature, polymer optical films will behave as hard glass-like substance. When heated above this temperature, individual segments of the polymer films will achieve large mobility; as a result the films become soft and elastic. The temperature at which this change happens is called the glass transition temperature ($T_g$). In other words, $T_g$ indicates the reversible change in phase of a film from a brittle glassy state to viscous or rubbery state. At $T_g$, the film's volume or length increases, and above it, the properties of the film decrease. The $T_g$ value of a film can be determined with specific equipment, such as differential scanning calorimetry (DSC), dynamic mechanical analysis (DMA), and thermal mechanical analysis (TMA) and the obtained values depend on the method used. For crystalline polymer films, such as PET and PEN, the crystalline melting points are usually above $T_g$.

In a typical DSC measurement, two pans are placed on a pair of identically positioned platforms connected to a furnace by a common heat flow path. One pan contains the polymer film, the other one is empty (reference pan). Then, the two pans are heated up at a specific rate. The computer guarantees that the two pans heat at exactly the same rate, despite the fact that one pan contains polymer and the other one is empty. The polymer film sample will take more heat to keep the temperature of the sample pan increasing at the same rate as the reference pan. A plot is created where the difference in heat flow between the sample and the reference is plotted as a function of temperature. The inflection point in the heat flow plot is recorded as the $T_g$ value for the film.

DMA indicates a technique in which either the modulus or the damping of a polymer film under oscillatory load or displacement is measured as a function of temperature, frequency, time, or other combinations. TMA indicates a test that measures the dimensional changes as a function of temperature. The dimensional behavior of a film material can be determined precisely. Measurements made include coefficient of linear thermal expansion (CTE), $T_g$, and softening characteristic. CTE value of a film reflects the change in volume per unit volume resulting from a change in temperature of the material. The mean coefficient is commonly referenced to room temperature and expressed in mm/mm °C. CTE value is quite important for polymer optical films which are used with other heterogeneous materials, such as metal, glass, or ceramic. The unmatched CTE values between the polymer films with the other materials are thought to be one of the most important reasons for delamination, cracking, and other failures in the devices.

### *2.3.3. Gas permission properties*

When a plastic substrate is used for the flexible OLED application, the water-vapor transmission rate (WVTR) and oxygen transmission rate (OTR) feature of the plastic substrate become critical because most high-performance semiconductor organic compounds show degraded performance when exposed to environmental moisture [20]. As mentioned before, WVTR and OTR of the flexible substrates are severely limited to be below $10^{-4}$ cm$^3$/m$^2$ day and $10^{-6}$ g/m$^2$ day, respectively, for AMOLED and organic solar cells [21]. Unlike glass, plastic substrates usually cannot provide sufficient protection to the permeants. For example, general PI films have WVTR values of $10^0$-$10^2$ g/m$^2$/day dependent on the aggregation structures of their molecular chains. Addition of some specific additives, such as graphene [22], might improve their moisture barrier properties to a limited extent. Thus, inorganic thin films with extremely higher barrier properties have to be used on the substrate in practical applications. In order to evaluate the WVTR and OTR features of one polymer film, it is necessary to understand these two parameters. WVTR and OTR can now be measured with water vapor or oxygen gas permeation measurement systems produced by Mocon Corp., USA. The Mocon test has a measurement limit of ~$10^{-4}$ g/m$^2$ day. A lower WVTR measurement has to be measured by calcium test, which is able to measure up to ~$10^{-6}$ g/m$^2$ day.

# 3. Applications of CHTPF films in optoelectronics

## 3.1. Commercialization of CHTPF films

It is safe to say that the commercialization of CHTPFs is highly promoted by their potential applications for flexible optoelectronic devices, such as flexible light-emitting diodes (F-LED), flexible solar cells or photovoltaic cells (PV), flexible thin-film transistors (F-TFT), flexible printing circuit boards (FPCB), and so on. Many present and future applications of optical films make greater demands for higher properties, and especially combinations of properties, than are available from the commodity materials. To satisfy these requirements, organic polymer chemists and chemical engineers have developed and commercialized many types of polymers, offering improved properties. Table 3 briefly summarizes the commercially available and R&D CHTPF optical films in the world. Optical polymers containing various thermal-stable units, such as PI, PAI, PA, PES, and PS, have been extensively investigated and commercialized. In addition, some kind of inorganic-organic hybrid optical films have also been developed. As mentioned before, a statistical date from Techno Create Corp shows that the market of CHTPF optical films in 2011 has been close to 1 billion Japanese yen and the market will see a rapid increase higher than 15% per year in the following years [9]. Some of the typical CHTPF optical films are shown in Figure 8.

Up to now, CHTPF optical films have found various applications as plastic substrate candidates for flexible optoelectronic devices, including FPCB, flexible display (TFT-LCDs or AMOLEDs, etc.), touch panel, electronic paper, and thin photovoltaic cells. Plastic substrates with both optical transparency and high-temperature resistance have great potential applications in these areas due to the superior flexibility, lightness, cost-effectiveness, and processability to their fragile and expensive glass analogs. For instance, in the fabrication of flexible bottom-emission AMOLED devices, the processing temperatures of light emitting-components on the flexible substrates usually precede 300°C. Under such processing conditions, only CTHTP optical films such as colorless PI films could meet the severe demands.

| Company | Product name | Resin | Transmission % | $T_g$ °C |
|---|---|---|---|---|
| Mitsubishi Gas Chemical | Neopulim® | PI | 89–90 | >300 |
| DuPont-Toray | Colorless Kapton® | PI | 87 | >300 |
| Kolon | NA[1] | PI | 88 | >300 |
| Japan Synthetic Rubber | Lucera® | NA | 88 | 280 |
| Toyobo | HM type | Polyamideimide (PAI) | 91 | 225 |
| Nippon Steel Chemical | Sillplus® | Resin+glass | 91–92 | NA |
| Toray | Aramid® | Polyamide (PA) | NA | 315 |
| Sumitomo Bakelite | Sumilite® FS-1300 | Polyethersulfone (PES) | 89 | 223 |
| Showa Electricity | Shorayal® | NA | 92 | 250 |
| Tosoh | OPS film | Polysulfone (PS) | 93 | 220 |
| Kurabo | Examid® | Polyamide (PA) | NA | 220 |

[1] Not available.

**Table 3.** Commercialization of CHTPF optical films in the world

### 3.2. Applications of CHTPF films

There has been growing interest in the use of plastic film substrates in the fabrication of future electronic devices, such as flexible displays, photovoltaics, batteries, sensors, and antennas [23]. This developing trend provides great opportunities for the development of CHTPF optical films. As shown in Figure 9, CHTPFs have found widespread applications in optoelectronics as various substrates for flexible display devices, flexible solar cells, FPCBs, touch panels, and so on.

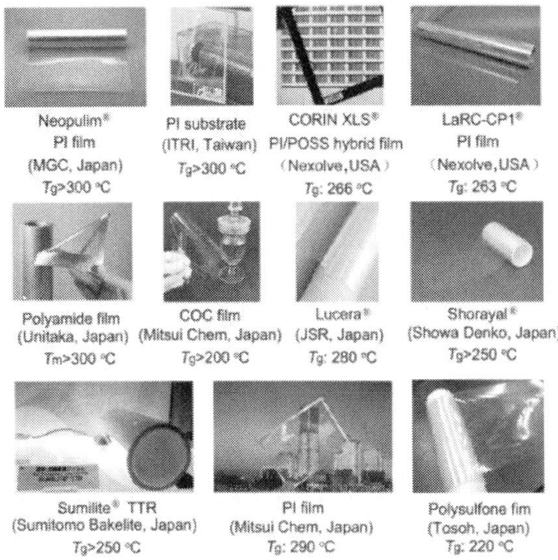


**Figure 8.** Commercially available or R&D CHTPF products in the literature

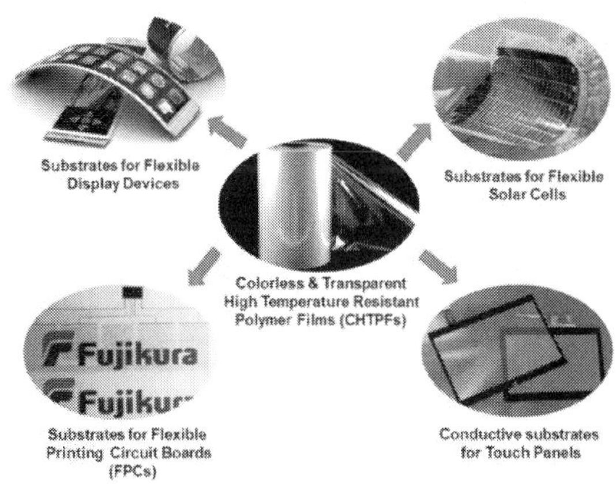


**Figure 9.** Potential applications of CHTPFs in optoelectronics

### *3.2.1. Substrates for advanced flexible display devices*

As the structural support and optical signal transmission pathway and medium, flexible substrates are playing ever-increasing important roles in advanced optoelectronic display devices [24], the characteristics and functionalities of flexible substrates have been becoming

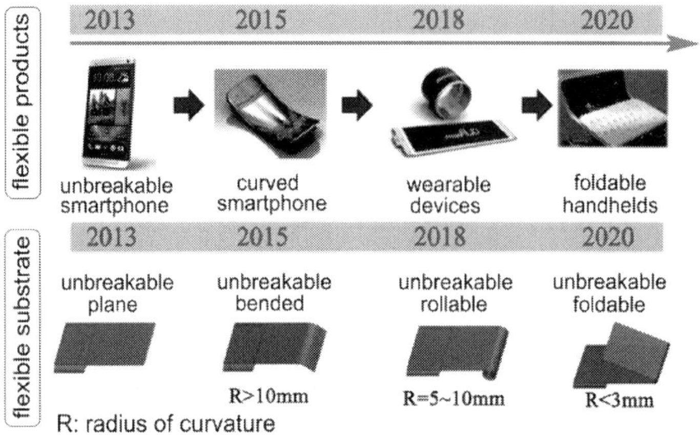


**Figure 10.** Roadmap for flexible products and substrates [24]

the important factors that affect the quality of flexible devices. Currently, there are mainly three types of substrates for flexible displays: thin glass, transparent plastic (polymer), and metal foil. Transparent plastic substrates possess good optical transmittance similar to that of thin glass; meanwhile the good flexibility and toughness comparable to those of metal foils. Thus, they are ideal for flexible display. A flexible display using a plastic substrate is considered to be one of the promising displays because of attractive features, such as thinness, lightweight, and good flexibility. For instance, as shown in Figure 10, the development of flexible substrates is experiencing a roadmap of plane (current)→ bended (2015)→ rollable (2018)→ foldable (2020) in the following years. The radius of curvature of the highly transparent flexible substrates might reach below 3 mm in the year of 2020. At that time, transparent plastic substrates might be the best candidate that can meet the demands.

However, in order to achieve a practical application for transparent plastic substrates in flexible display, several issues have to be addressed. First, currently, the performance of thin-film transistors (TFTs) built on common optical films or sheets are limited by the low-temperature process caused by the low thermal stability of current plastic substrates, typically below 250°C. For instance, for flexible display devices, such as active matrix-driven organic light-emitting diodes (AMOLED) processing, fabrication of TFTs on flexible substrates is one of the most important procedure. Up to now, there have been four types of production technologies for TFT fabrications in AMOLED, including amorphous silicon (a-Si) TFTs, low-temperature polysilicon (LTPS) TFTs, oxide TFTs, and organic TFTs (OTFTs). The key features for the current TFTs are summarized in Table 4 [25].

It can be seen that LTPS TFTs technique exhibits the highest field-effect mobility and stable electrical performance. However, the procedure requires a high process temperature of about 500°C during silicon crystallization. Conventional polymer optical film substrates cannot meet the application. For a-Si TFTs process, it has been widely used for AMOLED devices owing to uniform electrical characteristics over large areas, reasonable field-effect mobility, low-temperature process (< 300°C), and low cost compared to the other techniques.

| | a-Si TFTs | LTPS TFTs | OTFTs | IGZO TFTs |
|---|---|---|---|---|
| Field-effect mobility ($cm^2$/V-s) | < 1 | 50–100 | 0.1–1 | 10–30 |
| Process temperature (°C) | <300 | 300–500 | <300 | <300 |
| Device stability | Challenging | Good | Challenging | OK |
| uniformity | Good | Challenging | OK | OK |
| Manufacturability | Excellent | Maturing | Developing | Developing |
| cost | Low | Medium | To be determined | |

**Table 4.** Key features for a-Si TFTs, LTPS TFTs, OTFTs, and IGZO TFTs [25]

ITRI (Industrial Technology Research Institute, Taiwan) developed a unique flexible-universal-plane (FlexUP) solution for flexible display applications [26]. This new technique relies on two key innovations: flexible substrate and a debonding layer (DBL). As for the flexible substrate, ITRI developed a colorless PI substrate, which exhibits good optical transmittance (90%), high $T_g$ (>300 °C), low CTE (28 ppm/°C), and good chemical resistance. In addition, the PI substrate with barrier treatment shows a WVTR value less than $4\times10^{-5}$ g/m$^2$/day. Moreover, this barrier property suffered only to a minor drop, to $8\times10^{-5}$ g/m$^2$/day, after the flexible panel had been bent 1000 times at a radius of 5 cm. The substrate used a hybrid technique, which contains a high content (>60 wt%) of inorganic silica particles in the PI matrix. A 6-inch flexible color AMOLED display device was successfully fabricated using this substrate. By using this colorless PI substrate, flexible touch panel was also successfully prepared.

A 7-inch flexible VGA transmissive-type active matrix TFT-LCD display with a-Si TFT was successfully fabricated on the colorless PI substrate developed by ITRI [27]. The colorless PI substrate has the features of high $T_g$ (>350°C) and high light transmittance (>90%), which ensure the successful fabrication of 200°C a-Si:H TFT in the flexible device, as shown in Figure 11. The flexible panel showed resolution of 640×RGB×480, pixel pitch of 75×225 mm, and brightness of 100 nit. This technique is fully a-Si TFT backplane compatible, which makes it attractive for applications in high-performance flexible display. Similarly, a-Si TFTs deposited on clear plastic substrates (from DuPont) at 250-280°C was reported [28]. The free-standing clear plastic substrate has a $T_g$ value higher than 315°C and a CTE value below 10 ppm/°C. The maximum process temperature of 280°C has been close to the temperature used in industrial a-Si TFT production on glass substrates (300-350°C).

Toshiba Corp., Japan, successfully fabricated a flexible 10.2-inch WUXGA (1920×1200) bottom-emission AMOLED display device driven by amorphous indium gallium zinc oxide (IGZO) TFTs on a colorless and transparent PI film substrate, as shown in Figure 12 [29]. Firstly, a transparent PI film was formed on a glass substrate and then a barrier layer was deposited to prevent the permeation of water. Then, the gate insulator, IGZO thin film, source-drain metal,

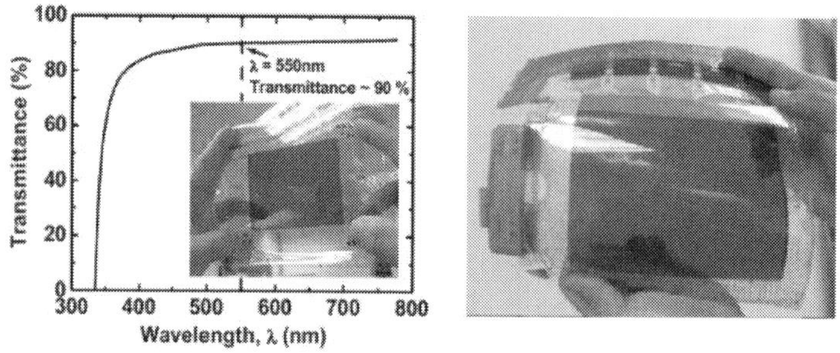


adapted, modified, and reprinted from ref. [27]

**Figure 11.** Colorless PI substrate and the color VGA flexible TFT-LCD

and passivation layer were successively deposited to afford the IGZO TFT. Secondly, the flexible AMOLED panel was fabricated using the IGZO TFT, color filter, white OLED, and encapsulation layer. Finally, the OLED panel was debonded from the glass substrate to afford the final AMOLED panel. The threshold voltage shifts of amorphous IGZO TFTs on the PI substrates under bias-temperature stress have been successfully decreased to less than 0.03 V, which is equivalent to those on glass substrates. ITRI also reported high-performance flexible amorphous IGZO TFTs on transparent PI-based nanocomposites substrates [30].

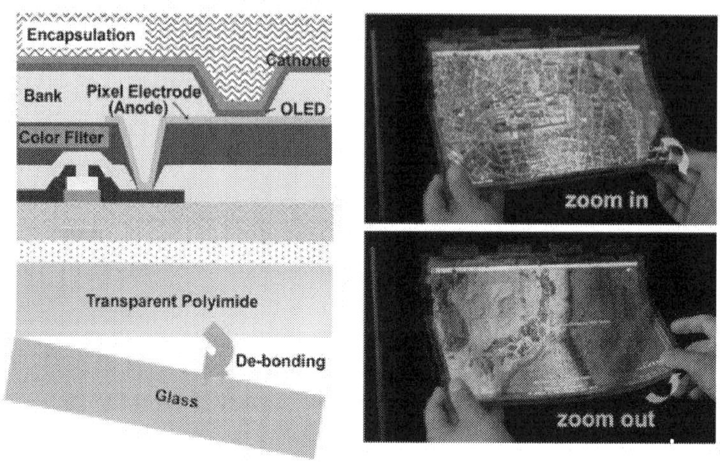


**Figure 12.** Flexible 10.2-inch AMOLED devices on transparent PI substrates [29]

Besides PI flexible substrates, other CHTPF substrates have also been developed. For instance, Teijin Ltd, Japan, developed novel high-temperature polycarbonate (PC) substrates for flexible displays [31]. The PC base film was obtained by a solvent casting process from dichloromethane solution and exhibited high optical transmittance (91%), high $T_g$ value (215 °C), ultra-low intrinsic birefringence and low retardation (1 nm), good elastic and dimensional stability, and an extremely smooth surface. The new substrates consisted of the high-temperature PC base film, silicon oxide gas barrier layer, and transparent indium zinc oxide (IZO) conductive film which showed promise in overcoming the obstacles in producing many kinds of thin, lightweight, and flexible display devices. Similarly, a high heat resistance PC film with the $T_g$ of 240°C and optical transmittance higher than 90% in the visible light region has been reported by General Electric [32]. A transparent, high barrier, and high heat substrate for organic electronics was successfully prepared by the film.

In summary, with the development of CHTPF optical films, the fabricated TFTs have showed similar characteristics to those of industry-standard a-Si TFTs fabricated on glass in the 300-350°C range. This result represents an important step toward a generic TFT backplane on flexible and optically clear film substrates.

### *3.2.2. Substrates for transparent Flexible Printing Circuit Boards (FPCBs)*

Over the years, the FPCB applications have always been the largest market for high-temperature polymer films, such as PI, polyamideimide, and polyetherimide films. The flexible nature of FPCBs allows their convenient use in compact electronic equipment such as portable computer, digital cameras, watches, and panel boards. Generally, the traditional FPCB is mainly prepared from flexible copper-clad laminates (FCCLs), as shown in Figure 13. FCCLs consist of a layer of PI film bonded to copper foil. Depending on the intended use of the laminate, copper may be applied to one (single-sided) or both sides (double-sided) of the PI film. PI film almost completely dominates the portion of FCCL market in which heat resistance is needed to withstand the soldering temperatures. Recently, with the development of flexible displays, necessity for a transparent film substrate in place of glass substrate is increasing. Correspondingly, a transparent film substrate for FCCLs is increasingly desired. However, most of the all-aromatic PI films currently used in FCCLs show colors from yellow to deep brown, and thus cannot be used in transparent FCCLs.

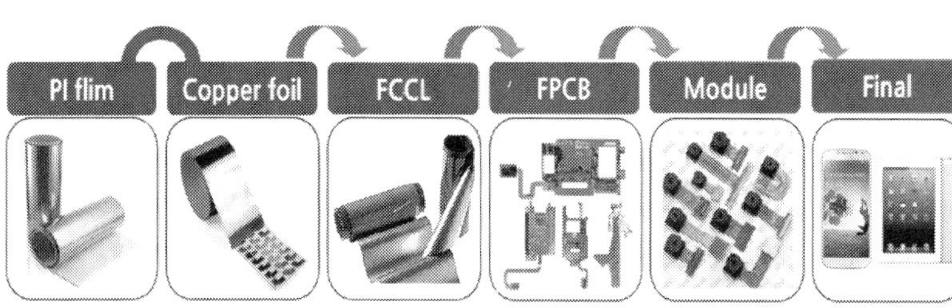


**Figure 13.** FPCB industry chains from FCCL to final products

Toyobo Corp., Japan, recently patented a colorless and transparent FCCL and the derived FPCB based on a PAI film [33]. The PAI film was synthesized from 1,2,4-cyclohexanetricarboxylic anhydride (HTA) (Table 2) and aromatic diisocyanate monomers (Figure 6) and the curing procedure was 200°C/1 h, 250°C/1 h, and 300°C/ 30 min under nitrogen. The film exhibited good thermal stability with $T_g$ of 300°C, light transmittance of 89%, good tensile properties with tensile strength of 140 MPa, elongation at break of 30%, tensile modulus of 3.9 GPa, and low CTE of 33 ppm/K. The single-side FCCL from the PAI film and copper coil showed good soldering resistance, high bonding strength (10.6 N/cm), and good dimensional stability under the condition of 150°C for 30 min. In addition, the FCCL showed good optical transparency with a transmittance of 75% at the wavelength of 500 nm.

Very recently, there has been vigorous activity in developing and commercializing transparent FPCB products in the world. This is mainly driven by the urgent needs of such products for mobile communication optoelectronics. Typical products reported by multiple manufacturers in public are summarized in Figure 14. Various optical films including PEN, PAI, and PI films have been used as the substrates in these new products. It can be anticipated that CHTPF optical films will play an increasingly important role for the future development of transparent FPCBs.

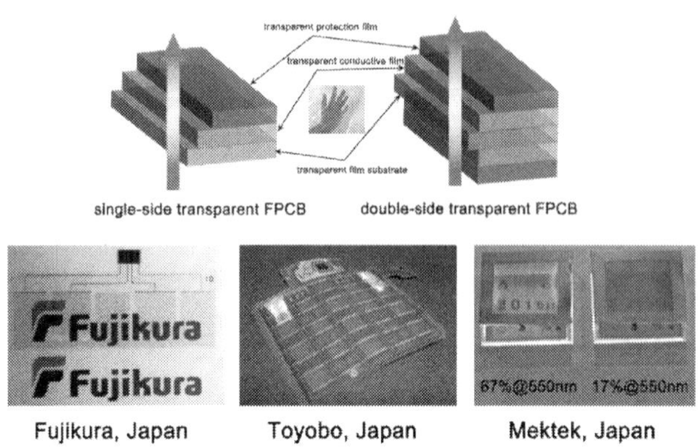


**Figure 14.** Application of CHTPF optical films in transparent FPCBs

### *3.2.3. Flexible substrates for thin-film solar cells*

Solar cells or photovoltaics (PV) have been intensively studied in energy industries due to their potential ability to reduce the cost per Watt of solar energy and improve lifetime performance of solar modules [34]. Conventional thin film solar cells are usually manufactured on transparent conducting oxide coated 3-5 mm thick soda-lime glass substrates and offer no weight advantage or shape adaptability for curved surfaces. Fabricating thin-film solar cells on flexible polymer substrates seems to offer several advantages in practical applications, such as weight saving, cost saving, and easy fabrication. The polymer substrates for thin-film solar cell fabrications should be optically transparent and should withstand the high processing temperatures. For example, for the current cadmium telluride (CdTe) cell fabrication techniques, the processing temperatures are in the range of 450-500°C. Most of the transparent polymers will degrade at such a high temperature. Undoubtedly, the lack of a transparent polymer which is stable at the high processing temperature of solar cells is one of the biggest obstacles for the applications of polymer substrates in flexible solar cells.

Wholly aromatic PI films, such as Kapton® (DuPont, USA) and Upilex® (Ube, Japan) can withstand a high temperature round 450°C. However, they exhibit deep colors and strongly absorb visible light. CdTe solar cells on such PI substrates will yield only low current due to large optical absorption [35]. The development of colorless PI film with good high-temperature stability makes it possible to produce high-efficiency solar cells. One of the most promising reports on the successful applications of colorless PI films in flexible solar cells fabrication might be the work carried out in Swiss Federal Laboratories for Materials Science and Technology (Empa) [36]. As one of the Empa's continuous work on developing high-efficiency thin-film solar cells aiming at enhancing their performance and simplifying the fabrication processes, they utilized colorless PI film (developed by DuPont) as the flexible substrate for CdTe thin-film PV modulus in 2011 (Figure 15). A conversion efficiency of 13.8% using the new substrates was achieved, which was the new record among this type of solar cells at that time.

**Figure 15.** CdTe solar cells on colorless PI substrate (Source: Empa)

## 4. Conclusions

Undoubtedly, CHTPFs represent a class of new materials with both high technological contents and high additional value. High comprehensive properties make them good candidates for advanced optoelectronic devices. It can be anticipated that, with the ever-increasing demands of optoelectronic fabrication, CHTPFs will attract more attentions from both the academia and the industry. For example, demand will continue to grow for displays of smart phones, tablet PCs, and other types of mobile electronic devices. Furthermore, these displays will be continuously improved in terms of visibility, flexibility, durability, and lightweight. In this context, CHTPF optical films are facing great developing chance. However, up to now, these still have several obstacles that should be overcome for the wide applications of CHTPFs in advanced fields. First, very limited commercially available CHTPF products greatly increase their cost, which lead to a very limited application only in high-end optoelectronic products. Low-cost CHTPFs are highly desired for their wide applications. Secondly, the combined properties of current CHTPFs should be further enhanced, such as further improving their optical transmittances at elevated temperatures, improving their mechanical and gas barrier properties. Thirdly, the manufacturing technology for CHTPFs should be further perfected in order to increase their uniformity, colorlessness, and dimensional stability at high temperatures.

## Acknowledgements

Financial support from National Basic Research Program (973 Program) of China (2014CB643605), National Natural Science Foundation of China (51173188) and Beijing Science and Technology Project (D141100003314002) are gratefully acknowledged.

## Author details

Jin-gang Liu[1*], Hong-jiang Ni[1], Zhen-he Wang[1], Shi-yong Yang[1] and Wei-feng Zhou[2]

*Address all correspondence to: liujg@iccas.ac.cn

1 Laboratory of Advanced Polymer Materials, Institute of Chemistry, Chinese Academy of Sciences, Beijing, People's Republic of China

2 BOE Technology Group Co. Ltd., Beijing, People's Republic of China

## References

[1] MacDonald WA. Engineered films for display technologies. J Mater Chem 2004;14:4-10.

[2] Choi MC, Kim YK, Ha CS. Polymers for flexible displays: from material selection to device applications. Prog Polymer Sci 2008;33:581-630.

[3] Logothetidis S. Flexible organic electronic devices: materials, process and applications. Mater Sci Eng B 2008;152:96-104.

[4] Huang JJ, Chen YP, Lien SY, Weng KW, Chao CH. High mechanical and electrical reliability of bottom-gate microcrystalline silicon thin film transistors on polyimide substrate. Curr Appl Phys 2011;11:S266-70.

[5] Nakano S, Saito N, Miura K, Sakano T, Ueda T, Sugi K, Yamaguchi H, Amemiya I, Hiramatsu M, Ishida A. Highly reliable a-IGZO TFTs on a plastic substrate for flexible AMOLED displays. J Soc Inform Display 2012;20(9):493-8.

[6] Yamaguchi H, Ueda T, Miura K, Saito N, Nakano S, Sakano T, Sugi K, Amemiya I, Hiramatsu H, Ishida A. 11.7-inch flexible AMOLED display driven by a-IGZO TFTs on plastic substrate. SID Sympos Digest Tech Pap 2012;43(1):1002-5.

[7] Ei-chi S. The thermal resistant films for electronics. Kogyo Chosakai Publishing: Tokyo, Japan; 2006 (in Japanese).

[8] Ito H, Oka W, Goto H, Umeda H. Plastic substrates for flexible display. Japan J Appl Phys 2006;45(5B):4325-9.

[9] Tsukidate K. Current Status and Prospect of Heat-resistant transparent film market. Techno-Create Month J 2012;2:1-3.

[10] Ghosh MK, Mittal KL. Polyimides. New York: Marcel Dekker; 1996.

[11] DeMeuse MT. ed. Biaxial stretching of film. Principles and applications. Woodhead: Cambridge; 2011.

[12] Siemann U. Solvent cast technology - a versatile tool for thin film production. Prog Colloid Polymer Sci 2005;130:1-14.

[13] Fay CC, Stoakley DM, St Clair AK. Molecularly oriented polymeric thin films for space applications. High Perform Polymers 1999;11(1):145-56.

[14] Oishi J, Hiramatsu S, Kihara S. Process and apparatus for production of colorless transparent resin film. US Patent 8357322, Jan 22, 2013.

[15] Oguro H, Kihara S, Bito T. Process for producing solvent-soluble polyimide. US Patent 7078477, Jan 18, 2006.

[16] Simone CD, Auman BC, Carcia PF. Low color polyimide compositions useful in optical type applications and methods and compositions relating thereto. US Patent 7550194, Jun 23, 2009.

[17] Jeong Y H, Cho HM, Park H J. Polyimide film, US Patent 8846852, Sep. 30, 2014.

[18] Lewis J. Material challenge for flexible organic devices [J]. Mater Today 2006;9(4): 38-45.

[19] Rosato DV. Concise Encyclopedia of Plastics. New York: Springer; 2000.

[20] Erlat AG, Yan M, Duggal AR. Substrates and thin-film barrier technology for flexible electronics. In: Wong WS, Salleo A (eds.) Flexible Electronics: Materials and Applications. New York: Springer Science + Business Media; 2009.

[21] Park JS, Chae H, Chung HK, Lee S I. Thin film encapsulation for flexible AM-OLED: a review [J]. Semiconduct Sci Technol 2011;26:034001.

[22] Tsai MH, Tseng IH, Liao YF, Chiang JC. Transparent polyimide nanocomposites with improved moisture barrier using graphene. Polymer Int 2013;62(9):1302-9.

[23] Crawford GP. Flexible Flat Panel Displays. Chichester: John Wiley & Sons Ltd; 2005.

[24] Chen JL, Liu CT. Technology advances in flexible displays and substrates. IEEE Access 2013;1:150-8.

[25] Pang HQ, Tajan K, Silvernail J. Recent progress of flexible AMOLED displays. Proc SPIE 2011;7956:79560J-1-10.

[26] Chen J L, Ho J C. A flexible universal plane for displays. J Soc Inform Display 2011;2(11):6-9.

[27] Yeh YH, Cheng CC, Ho KY, Chen PC, Lee MH, Huang JJ, Tyan HL, Leu CM, Chang CS, Lee KC, Fang Sy, Chen TH, Pan CY. 7-inch color VGA flexible TFT LCD on colorless polyimide substrate with 200°C a-Si:H TFTs. SID Sympos Digest Tech Pap 2007;38(1):1677-9.

[28] Long K, Kattamis AZ, Cheng IC, Gleskova H, Wagner S, Sturm JC. Stability of amorphous-silicon TFTs deposited on clear plastic substrates at 250°C to 280°C. IEEE Electron Device Lett 2006;27(2):111-3.

[29] Miura K, Ueda T, Saito N, Nakano S, Sakano T, Sugi K, Yamaguchi H, Amemiya I. Flexible AMOLED display driven by amorphous InGaZnO TFTs. The 20th International Workshop on Active-matrix Flat panel Displays and Device: Conference proceedings, July 2-5, 2013, Kyoto, Japan.

[30] Chien CW, Wu CH, Tsai YT, Kung YC, Lin CY, Hsu PC, Hsieh HH, Wu CC, Yeh YH, Leu CM, Lee TM. High-performance flexible a-IGZO TFTs adopting stacked electrodes and transparent polyimide-based nanocomposites substrates. IEEE Trans Electronic Devices 2011;58(5):1440-6.

[31] Hanada T, Shiroshi I, Negishi T, Shiro T. Plastic substrate technologies for flexible displays. Proc SPIE 2010;7618:76180Q-1~8.

[32] Yan M, Kim T W, Erlat AG, Pellow M, Foust DF, Liu J, Schaepkens M, Heller CM, McConnelee PA, Feist TP, Duggal AR. A transparent, high barrier, and high heat substrate for organic electronics. Proc IEEE 2005;93(8):1468-77.

[33] Shimeno K, Ito T, Aoyama T, Nishimoto A, Nagata S, Kurita T. Polyamideimide resin, as well as a colorless and transparent flexible metal-clad laminate and circuit board obtained therefrom. US Patent 82222365, Jul. 17, 2012.

[34] Fraas I, Partain L. Solar Cells and their Applications, Second Edition. New Jersey: Wiley; 2010.

[35] Mathew X, Enriquez JP, Romeo A, Tiwari AN. CdTe/CdS solar cells on flexible substrates. Solar Energy 2004;77:831-8.

[36] Aliyu MM, Islam MA, Hamzah NR, Karin MR, Martin MA, Sopain K, Amin N. Recent developments of flexible CdTe solar cells on metallic substrates: Issues and prospects. Int J Photoenergy 2012; 2012:1-10, doi:10.1155/2012/351381.

Chapter 4

# InP-Based Antimony-Free MQW Lasers in 2-3 μm Band

Yi Gu and Yong-Gang Zhang

Additional information is available at the end of the chapter

http://dx.doi.org/10.5772/60705

**Abstract**

Mid-infrared semiconductor lasers in the wavelength range of 2-3 μm have aroused increasing interests as they are highly desired for a wide range of applications ranging from medical diagnostics to environmental sensing. Access to this wavelength range was mainly achieved by antimony-containing compound semiconductor structures on GaSb substrates. Besides, InP-based $In_xGa_{1-x}As$ (x>0.53) type-I multiple quantum well laser is a promising antimony-free approach in this band. The emission wavelength can be tailored to the 2-3 μm band by increasing the indium composition in the quantum wells. During the demonstration of this kind of lasers, controlling the strain and keeping fair structural quality is the main obstacle.

In this chapter, the route for developing this kind of lasers is reviewed. The schemes of pseudomorphic and metamorphic structures are discussed for the 2-2.5 μm and 2.5-3 μm range, respectively. In the pseudomorphic scheme, triangular quantum wells grown by digital alloy technology are applied to restrict the strain and increase the lasing wavelength. Lasers at 2.43 μm were demonstrated under continuous wave operation at room temperature. To extend the emission wavelength longer, an InP-based metamorphic template with larger lattice constant was produced and InAs quantum wells were then grown. The lasing wavelength was further increased up to 2.71 μm. The details on the gas source molecular beam epitaxial growth, device processing as well as performance characterization are presented.

**Keywords:** Lasers, Quantum well, InP, 2-3 μm, Antimony-free

## 1. Introduction

The 2-3 μm wavelength range covers both the short side of the mid-infrared (2.5-25 μm) band and the long side of the near-infrared (0.7-2.5 μm) band. Semiconductor laser diodes emitting in this spectral range are very attractive for free space communications, ultra-low loss fluoride fiber communication, solid or fiber lasers (amplifiers) pumping and seeding, light detection and ranging (LIDAR), tunable diode laser absorption spectroscopy (TDLAS), etc., due to the abundant spectroscopic features in this wavelength range. For example, the 2.1-2.4 μm band is free of water and carbon dioxide absorption as shown in Figure 1, which forms a clear atmosphere window; on the other hand, the strong and wide absorption band of water (mainly around 2.7 μm) and the moderate absorption band of carbon dioxide (around 2.0-2.1μm and 2.65-2.85 μm) forms a high contrast area, which is usable for humidity and carbon dioxide monitoring, active spectroscopic imaging of water containing subjects, etc.

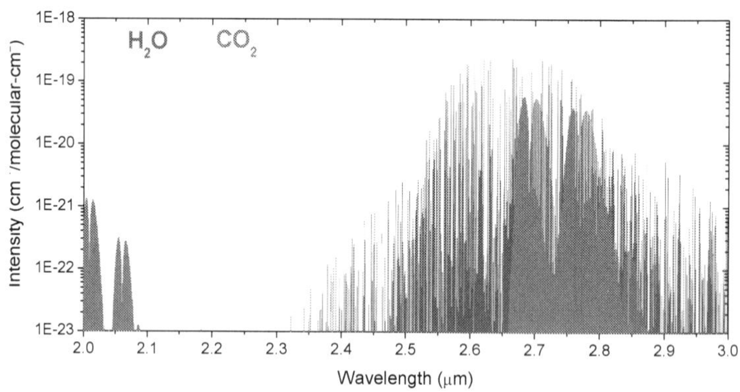


**Figure 1.** Absorption intensity of $H_2O$ and $CO_2$ in the 2-3 μm band.

Furthermore, as shown in Figure 2, the fingerprint absorption lines of several molecules also exist in this band, and therefore, semiconductor lasers in this band are of great interest in atmospheric pollution monitoring and medical diagnostics by using TDLAS [1, 2]. As a monochromatic light source, the laser diode in this wavelength band is also desired for the evaluation of optoelectronic materials and devices such as photodetectors and focal plane arrays operated in this wavelength range, because in the measurement schemes wavelength match is very important to acquire their actual features [3, 4].

The choices and developments of semiconductor lasers in the 2-3 μm band will be reviewed at first. Then we will introduce the recent progress for the design, growth and demonstration of InP-based antimony-free multiple quantum well (MQW) lasers in this band. Schemes of pseudomorphic and metamorphic structures will be discussed for the 2-2.5 μm and 2.5-3 μm ranges, respectively.

http://dx.doi.org/10.5772/60705

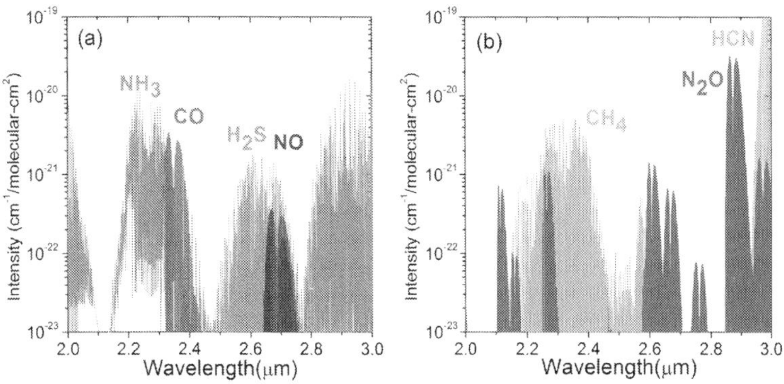


**Figure 2.** Absorption intensity of (a) $NH_3$, CO, $H_2S$, NO molecules and (b) $CH_4$, $N_2O$, HCN molecules in the 2-3 μm band.

## 2. Development of semiconductor lasers in the 2-3 μm band

There are several options of semiconductor lasers in the 2-3 μm wavelength band. Traditionally, such semiconductor lasers are developed in an antimony (Sb)-containing material system on GaSb or InAs substrates. InGaAsSb/AlGa(In)AsSb type-I quantum well (QW) lasers are the primary approach, and high power and single-mode lasers have been achieved and the wavelength range has covered 2-3.8 μm [5-10], and continuous wave (CW) operation above 80°C has been reached for 2.1 μm lasers [10]. The laser diodes have been applied in TDLAS demonstrations [11, 12]. There were also some sporadic reports on InGaAsSb/GaSb type-II QW lasers covering this band [13, 14]. GaSb-based interband cascade lasers combining type-II heterostructure and cascade concept have been developed adequately, but it is still a big challenge to shift the lasing wavelength below 3 μm, which requires new material systems [15].

Compared to GaSb, InP substrates display superior quality and are easier to acquire. Sb-containing InGaAs/GaAsSb type-II structures on InP substrates have been researched while room temperature (RT) photoluminescence (PL) and sporadic device results have been reported [16, 17]. InAsSb quantum dots (QDs) on InP are another potential scheme to demonstrate light sources in 2-3 μm, but the improvements of material quality are still needed to achieve more experimental progress [18]. On the other hand, the growth and processing technologies of the Sb-containing materials are still less mature than those of InP-based Sb-free materials. Besides, the thermal characteristics of antimonide are much inferior to the phosphides and arsenide [19, 20]. Many impressive results on mid-infrared intersubband emission of InP-based Sb-free InGaAs/InAlAs quantum cascade lasers have been reported in the recent two decades, whereas the emission wavelengths are mainly longer than 3 μm [21].

Besides the aforementioned options, InP-based Sb-free type-I $In_xGa_{1-x}As$ ($x>0.53$) QW laser is a promising approach in this band. As the bandgap wavelength of InAs at RT is as long as 3.5 μm, the emission wavelength of In(Ga)As QW can be tailored to the 2-3 μm range [22]. For this InP-based Sb-free In(Ga)As system, mature growth and processing technologies can be relied on, and a simple structure can be applied for the type-I QW lasers using interband emission, so the laser diodes with better performance could be expected. The QW lasers applying InP-lattice-matched InGaAsP as QW layers have been well developed with the aim of telecom lasers [23]. However, the lasing wavelength is shorter than 1.7 μm. To extend the wavelength beyond 2 μm, a higher InAs composition should be applied and thus strain would be introduced into the QW layers.

The InP-based InGaAs strained QW lasers were first demonstrated by Forouhar et al. in the early 1990s [24]. The lasing wavelengths are around 2 μm at RT, and $In_{0.75}Ga_{0.25}As$ layers grown by metal-organic vapor phase epitaxy (MOVPE) were applied as the QW layers and other layers were all lattice matched to InP [25-27]. As shown in Figure 3(a), in these pseudomorphic structures, the InGaAs (x>0.53) QW layers are compressively strained whereas the other layers, such as cladding, waveguide and barrier layers, are all nearly lattice-matched to InP. The Fabry-Perot (FP) and distributed feedback buried (DFB) devices around 2.07 μm were then demonstrated by NTT's group [28-31]. After the year 2000, FP and vertical-cavity surface-emitting lasers (VCSELs) at 2.3 μm were demonstrated by Amann et al. using molecular beam epitaxy (MBE) grown InAs-containing triangular QW active region [32-34]. Using MOVPE grown 5 nm pure InAs as the QW layer, Mitsuhara et al. in NTT's group reported FP and DFB lasers with lasing wavelength at 2.33 μm [35-37]. Lasers up to 2.4 μm were recently demonstrated by using triangular QWs grown by MBE digital alloy technology [38].

As mentioned earlier, further increase of well width is restrained by the strain between the QW part and InP substrate, and thus the longest emission wavelength is limited. To extend the emission wavelength to longer wavelengths, a metamorphic "virtual substrate" can be produced with a larger lattice constant than InP, e.g. metamorphic $In_{0.8}Ga_{0.2}As$ layer as shown in Figure 3(b). In this case, the critical thicknesses can be larger than those of pseudomorphic structures, therefore the QW width can be increased, which would increase the emission wavelength. Recently, step-graded $InAs_xP_{1-x}$ buffers on InP have been applied to achieve 2.8-3 μm PL at RT [39-41], where 16.5 nm InAs or 22 nm $InAs_{0.94}P_{0.06}$ were used as the well layers. By using 15 nm InAs QWs on continuously-graded metamorphic $In_xAl_{1-x}As$ buffers, RT PL of QWs at 3.05 μm [42] and low temperature lasing of laser diodes at 2.7 μm have been achieved [43]. This scheme can even be explored to mid-infrared metamorphic InAs QWs on GaAs substrate, which is even more attractive than on InP substrate [44].

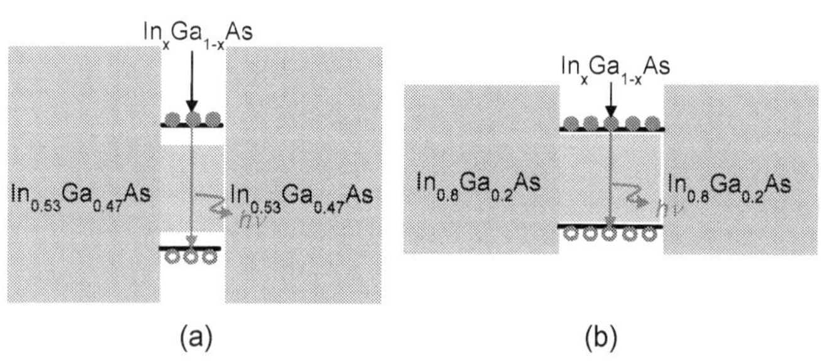


**Figure 3.** Schematic diagrams of the transition for the InP-based (a) pseudomorphic and (b) metamorphic $In_xGa_{1-x}As$ QWs.

## 3. Theoretical evaluation of InP-based Sb-free type-I QW lasers

At first, we theoretically evaluate the type-I transition energy and emission wavelength of rectangular pseudomorphic $In_xGa_{1-x}As/In_{0.53}Ga_{0.47}As$ (x>0.53) and metamorphic $In_xGa_{1-x}As/In_{0.8}Ga_{0.2}As$ (x>0.8) QWs. The subband energies of the QWs were calculated by numerically solving the Schrödinger equation with strain-contained 4×4 Luttinger-Kohn Hamiltonian, and the transition energy from the first electron energy level to the first heavy hole energy level was calculated. The emission wavelength versus the well width of $In_xGa_{1-x}As/In_{0.53}Ga_{0.47}As$ (x>0.53) and $In_xGa_{1-x}As/In_{0.8}Ga_{0.2}As$ (x>0.8) QWs are shown in Figure 4(a) and Figure 4(b), respectively. The well width is limited by the critical thickness calculated from the force balance model as shown in Figure 4. It is seen that the critical thickness can be extended by using the metamorphic scheme, and therefore, a thicker well width can be used, leading to a longer wavelength.

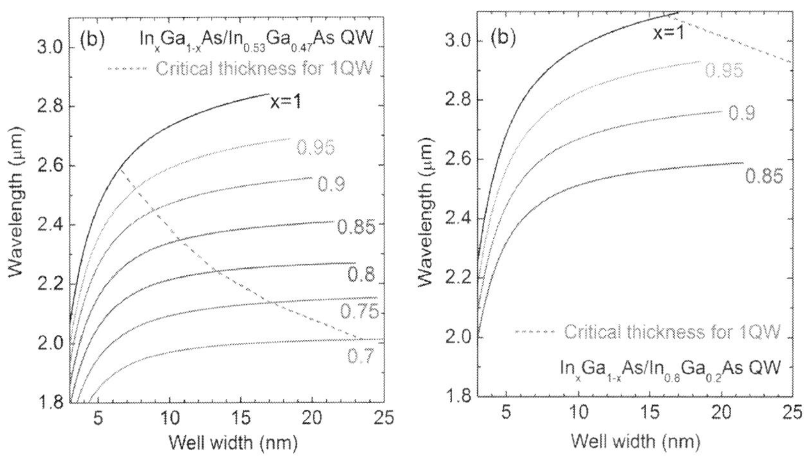


**Figure 4.** Calculated transition wavelength versus well width of (a) pseudomorphic $In_xGa_{1-x}As/In_{0.53}Ga_{0.47}As$ QW (x>0.53) and (b) metamorphic $In_xGa_{1-x}As/In_{0.8}Ga_{0.2}As$ QW (x>0.8) with rectangular QW shape. The critical thickness for a single QW is indicated as the dotted line.

Furthermore, calculation shows that the lasing wavelength can be extended dramatically if the energy band shape is changed from a rectangular to a triangular one [45, 46]. Figure 5(a) shows the schematic band structure of the $InAs/In_{0.53}Ga_{0.47}As$ triangular QW (TQW) and $In_{0.765}Ga_{0.235}As/In_{0.53}Ga_{0.47}As$ rectangular QW (RQW). The two QW structures are considered to have the same strain extent in the QW part. From Figure 5(b), we can see that changing the energy band from the rectangular shape to a triangular one redshifts the lasing wavelength markedly. For the well width of 16 nm, the calculated wavelength of RQW is only about 2.1 μm, but it is beyond 2.5 μm for TQW. The theoretical estimations show the promising potential of the TQW to extend the wavelength and improve the performances.

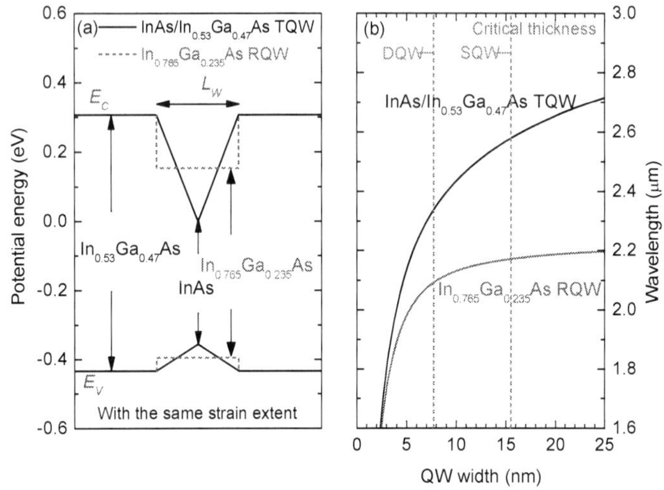


**Figure 5.** (a) Schematic band structures and (b) calculated transition wavelength as a function of well width of InAs/$In_{0.53}Ga_{0.47}As$ TQW and $In_{0.765}Ga_{0.235}As/In_{0.53}Ga_{0.47}As$ RQW with the same strain extent. The critical thicknesses for single QW and double QWs are indicated.

# 4. Pseudomorphic scheme for lasing wavelength below 2.5 μm

The pseudomorphic scheme is based on the lattice-mismatched material system except the active QW region. Therefore, the main concern of such lasers is the control of strain in the QW and to keep fair structural quality. The gas source molecular beam epitaxy (GSMBE) growth parameters and the structures of the QW parts need to be optimized at first [47, 48]. Then the whole laser structures were grown and the lasers were demonstrated and analyzed.

## 4.1. GSMBE growth and optimization of pseudomorphic triangular QWs

To construct the QWs with triangular shape in practice, usually two methods can be applied. The first one uses analogue alloy where the emission source flux changes gradually with an extremely low rate [49], and the other one uses the so-called digital alloy, where short period superlattice composed of two materials are grown at digitally setting thicknesses [50]. The growth rate is normal for the latter one, and this method is especially much more convenient for molecular beam epitaxy (MBE). In MBE, the time for opening and closing a shutter is far below 1 s, so the control precision of the short period superlattice can be high enough. It is also found that digital alloy technology can extend the critical thickness and restrain the three-dimensional material growth [51, 52].

A VG Semicon V80H GSMBE system was applied for the growth. The best background vacuum achieved in this system was about $1\times10^{-11}$ Torr. The elemental indium and gallium cells with two heaters as well as aluminum standard cell were used as group III sources. The fluxes of these group III emission sources were controlled by the cell temperatures. Group V sources were $As_2$ and $P_2$ cracked from Arsine ($AsH_3$) and phosphine ($PH_3$) cracking cells at around 1000 °C. The fluxes of $As_2$ and $P_2$ were controlled by pressure. Standard beryllium and silicon cells were used as *p*- and *n*-type doping sources, respectively. The doping levels were controlled by the temperatures of the emission cells. The growth rates of InP-lattice-matched InGaAs(P) and InP layers were all controlled to be around 1 μm/h.

| Sample | Designed | | | $T_g$ (°C) | Parameters from HRXRD simulation | | PL wavelengths (μm) | | PL FWHM (meV) | |
|---|---|---|---|---|---|---|---|---|---|---|
| | $\lambda_{300}$ (μm) | $d_W$ (nm) | $d_B$ (nm) | | $d_W$ (nm) | $d_B$ (nm) | 300 K | 12 K | 300 K | 12 K |
| 1 | 2.1 | 9 | 15 | 500 | 9.1 | 16.2 | 2.14 | 1.94 | 35 | 24 |
| 2 | 2.1 | 9 | 15 | 530 | 9.1 | 16.4 | 2.11 | 1.93 | 35 | 19 |
| 3 | 2.3 | 13 | 15 | 530 | 14.0 | 16.4 | 2.25 | 2.07 | 31 | 20 |
| 4 | 2.5 | 19 | 15 | 530 | 19.2 | 16.4 | 2.38 | 2.16 | 36 | 24 |
| 5 | 2.5 | 19 | 20 | 530 | 19.2 | 20.6 | 2.38 | 2.16 | 33 | 17 |
| 6 | 2.6 | 25 | 20 | 530 | N/A | N/A | N/A | 2.26 & 2.31 | N/A | 24 & 31 |

**Table 1.** The structure, growth parameters, HRXRD and PL results of the samples. $\lambda_{300}$ is the wavelength at 300 K, $T_g$ is the growth temperature, $d_W$ and $d_B$ is well width and barrier width, respectively.

The grown QW samples consisted of a 200-nm-thick InP buffer layer, two triangular QWs and a 100-nm-thick InP cap layer. The triangular QWs were constructed by growing InGaAs/InAs digital alloy with very short periods. In each period the thicknesses of $In_{0.53}Ga_{0.47}As$ ($d_1$) and InAs ($d_2$) is designed from the following equation:

$$\begin{cases} (d_1 \times 0.53 + d_2) / d = \alpha \\ d_1 + d_2 = d \end{cases} \quad (1)$$

where $\alpha$ is the expected average indium composition in each short period, and $d$ is the period of the short period setting to be 1 nm here. The barrier layer is AlInGaAs grown using the growth parameters of InP-lattice-matched $In_{0.52}Al_{0.48}As$ and $In_{0.53}Ga_{0.47}As$, thus the composition is around $Al_{0.33}In_{0.36}Ga_{0.31}As$.

Table 1 lists the structure parameters, growth conditions, high resolution X-ray diffraction (HRXRD) and PL results of the grown QW samples. The HRXRD (004) scanning curves of all samples are shown in Figure 6. For all the samples, the upper and lower curves show the measured and simulated results, respectively. The good agreement between the measurement and simulation in samples 1-5 indicates the pseudomorphic growth. By simulation, the average indium composition in the well and barrier layers were about 0.765 and 0.355 in samples 1-5, the deduced well widths $d_w$ and barrier widths $d_B$ for all the samples are also listed in Table 1, agreeing well with the designed values. Figure 7 presents the PL spectra of samples 1-6 at RT and 12 K, respectively. The peak wavelengths and full-width at half-maximum (FWHM) of the PL spectra are listed in Table 1.

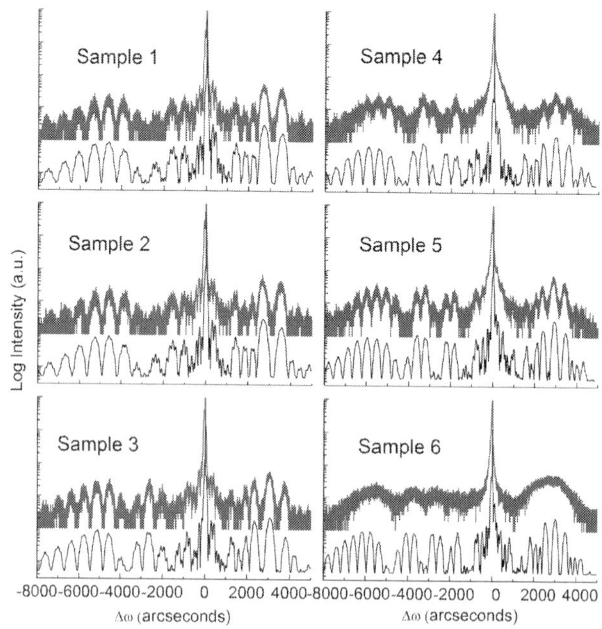


**Figure 6.** HRXRD (004) scanning curves of the TQWs with various structures and growth parameters [48]. (Reprinted with permission from Elsevier)

The growth temperature $T_g$ is a crucial growth parameter for high crystalline and optical quality during the MBE growth. For the growth of high indium materials, generally $T_g$ needs to be lowered. However, for the $In_{0.53}Ga_{0.47}As$ barriers and the Al-containing AlInGaAs barriers, a relatively higher growth temperature is preferred to improve the optical properties. It is really unpractical to increase and decrease $T_g$ in the interface of well and barrier layers, because the interface quality would be significantly deteriorated by the growth interruption. Therefore, $T_g$ is needed to be optimized, especially for the QW growth. Samples 1 and 2 are with different growth temperatures at 500 °C and 530 °C, respectively. Figure 6 shows the sharp satellite peaks in the HRXRD scanning curves of samples 1 and 2, denoting the nice crystalline quality of both samples. Similarly, PL spectra of the two samples have been observed at 300 K as shown in Figure 7, whereas at 12 K the PL intensity of sample 2 is stronger than that of sample 1. The FWHM at 12 K is 19 meV for sample 2, and 24 meV for sample 1. It is shown that the sample grown at a relatively higher temperature shows better optical quality.

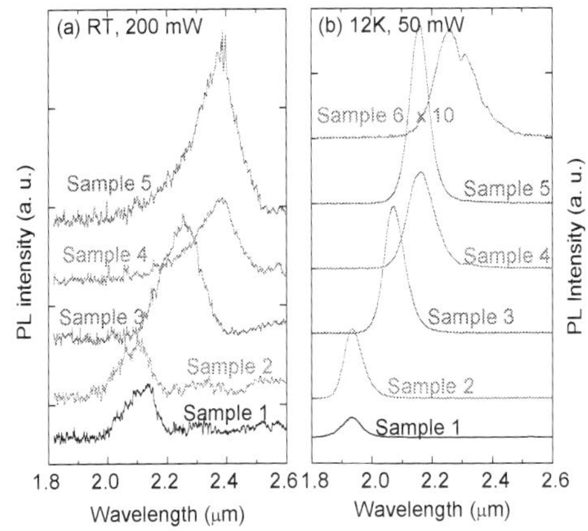


**Figure 7.** PL of triangular QWs with various structures and growth parameters at RT and 12 K [48]. (Reprinted with permission from Elsevier)

By increasing the well width from 9 nm of sample 2 to 13 nm of sample 3 and 19 nm of sample 4 while keeping the same barrier width, the HRXRD curve becomes indistinct as shown in Figure 6, indicating the deteriorated crystalline quality of sample 4. As the QW width increases, more defects may be generated due to the larger lattice mismatch. The wider QW also weakens the quantum restriction, and the carriers have more probability to occupy the second subband, which induces the larger PL FWHM and the much shorter peak wavelength than designed for sample 4. For sample 5, by increasing the barrier width on the basis of sample 4, the satellite peaks of HRXRD curves recover to be distinct and the PL intensity is enhanced. The PL FWHM is reduced to 17 meV at 12 K and 33 meV at 300 K for sample 5, comparing to 24 meV at 12 K and 36 meV at 300 K for sample 4. The well width of sample 6 is further enlarged to 25 nm in order to extend the wavelength further. It can be seen from Figure 6 that the satellite peaks of the HRXRD curve become indistinct, and two PL peaks at 12 K exist as shown in Figure 6(b), whereas no PL signal is observed at 300 K. It suggests that 2.4 µm is around the ceiling wavelength for the In(Ga)As strained QW.

### 4.2. Demonstration of CW-operated lasers above room temperature

In our preliminary work, InP layers were applied as waveguides for the convenience of the growth, but the lasing could only be achieved below 150 K [53, 54]. Subsequently, InGaAsP quaternary alloys with bandgap around 1.1 eV and lattice-matched to InP were used as the waveguide layers, and the laser performances were dramatically improved. In this section, the performances of RQW and TQW lasers at 2.2 µm will be presented. And the demonstration of TQW lasers beyond 2.4 µm with increased QW width will also be introduced.

For the lasers with designed lasing wavelength at 2.2 µm, two structures with RQW and TQW were grown [52]. The active QW regions of both samples were sandwiched between 120-nm-thick InGaAsP waveguide layers, as well as 1000-nm-thick *n*-type bottom and 1700-nm-thick *p*-type upper InP cladding layers, and 300-nm-thick p-type $In_{0.53}Ga_{0.47}As$ layers were grown as a top contact layer. The active QW region was formed by four QW layers sandwiched between 20-nm-thick $In_{0.53}Ga_{0.47}As$ barriers for both samples. The QW layers were 3-nm-thick compressive InAs layers in the RQW structure, and 10-nm-thick triangular $In_{0.53}Ga_{0.47}As$/InAs/$In_{0.53}Ga_{0.47}As$ layers in the TQW structure. The total strain content of the TQW was still larger than that of the RQW due to the much thicker well width.

Ridge waveguide lasers with a strip width of 6 μm were fabricated by standard lithography and wet chemical etching process. Then, a 300-nm-thick $Si_3N_4$ layer was deposited by using plasma enhanced chemical vapor deposition (PECVD). On the top of the ridge a 4-μm-wide window was opened. The top and bottom metallic contacts were sputtered Ti/Pt/Au and evaporated Ge/Au/Ni/Au, respectively. The chips were cleaved into 0.8-mm-cavity-length bars with uncoated facet, soldered on copper heat sinks and wire bonded.

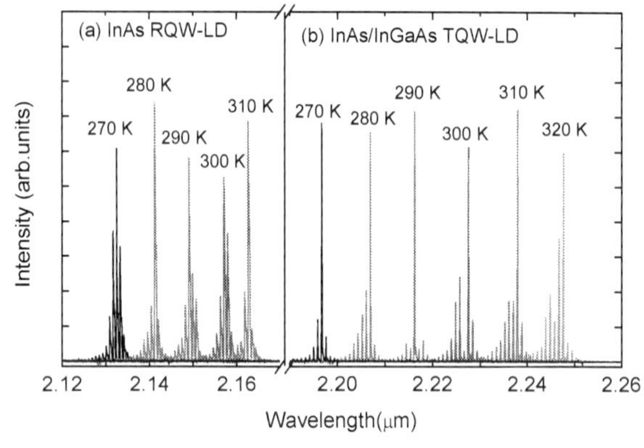


**Figure 8.** Lasing spectra of (a) rectangular QW laser and (b) triangular QW laser around 2.2 μm at various temperatures under CW driving condition.

Figure 8 shows the CW lasing spectra at a driving current of 1.2 times the threshold current in a temperature step of 10 K. The maximum operation temperature of the RQW laser is about 310 K and the lasing peak is about 2.16 μm at 310 K, whereas the TQW laser shows an increased maximum operation temperature up to 330 K. Over the whole temperature range, the lasing peaks of the TQW laser are about 60 nm longer than those of the RQW laser, probably due to the model error in theoretical design as well as due to the growth error. These results show that the TQW laser can operate at higher temperatures even with a longer lasing wavelength, proving the benefit of TQW. The average temperature coefficient of the lasing wavelength $\Delta\lambda/\Delta T$ is about 0.8 nm/K for the RQW laser and 1 nm/K for the TQW laser.

The temperature-dependent I-P characteristics and the I-V curve at 300 K of the lasers are shown in Figure 9. The threshold current of the RQW laser is 124 mA at 300 K, corresponding to a threshold current density of 2.58 kA/cm², and the output power is 3.6 mW/facet at an injection current of 400 mA. In contrast, the TQW laser shows a much lower threshold current of 68 mA (1.42 kA/cm²) at 300 K. Also, a much higher output power of 10.4 mW/facet is obtained at 300 K, almost three times that of the RQW laser. The turn-on voltage and the differential resistance for both lasers are almost the same at about 1.25 V and 1.87 Ω at 300 K.

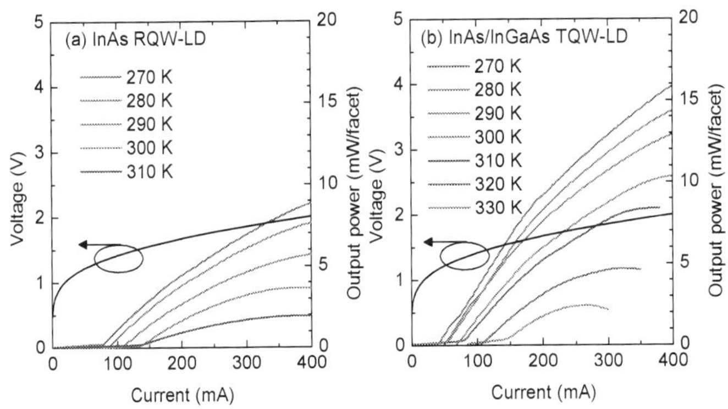


**Figure 9.** I-P and I-V characteristics of (a) rectangular QW laser and (b) triangular QW laser around 2.2 μm at various temperatures under CW driving condition.

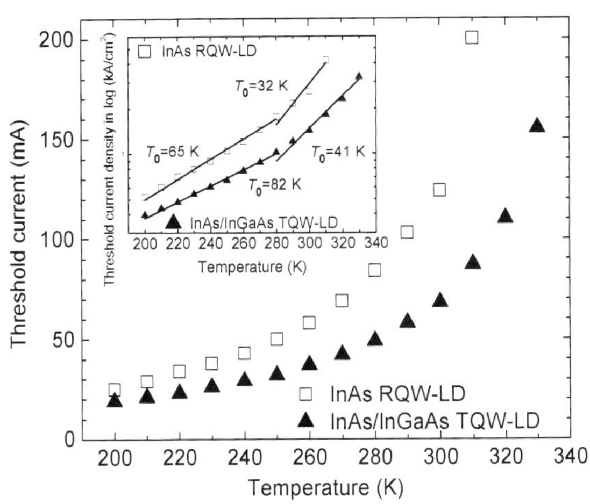


**Figure 10.** Temperature-dependent threshold current of rectangular QW laser and triangular QW laser around 2.2 μm at various temperatures under CW driving condition. The inset shows the temperature-dependent threshold current densities and characteristics temperatures.

The threshold current as a function of temperature is shown in Figure 10, covering the temperature range from 200 K to 310 K and 330 K for the RQW and TQW lasers, respectively. The characteristic temperatures were achieved from the temperature-dependent threshold current density as shown in the inset of Figure 10. In the low temperature range 200-280 K, a characteristic temperature of $T_0$=82.1 K was obtained for the TQW laser, higher than the value of $T_0$=65.2 K for the RQW laser. In the high temperature range beyond 290 K, $T_0$ is decreased for both lasers, but the TQW laser also has a higher value of $T_0$=40.5 K than the $T_0$=32.1 K for the RQW laser.

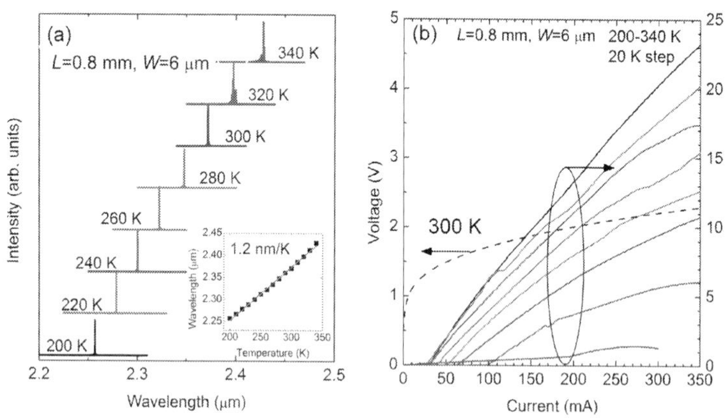


**Figure 11.** (a) The CW lasing spectra of the around 2.4 μm TQW laser at various temperatures. The inset shows the temperature-dependent lasing wavelength. (b) Output power at different temperatures and voltage at 300 K as a function of injected current of the around 2.4 μm TQW laser [38]. (Copyright 2014 The Japan Society of Applied Physics)

By further increasing the well width of TQW laser to 16 nm, the lasing wavelength of the device has been extended. Figure 11(a) shows the CW lasing spectra in the temperature range of 200 K to 340 K. The lasing wavelength is 2.37 μm at 300 K and red-shifts to 2.43 μm at 340 K. The average temperature coefficient of the wavelength is 1.2 nm/K in 200-340 K as shown in the inset of Figure 10(a). As shown in Figure 11(b), the output power at 200 K in CW mode is 23 mW/facet injected by a current of 350 mA and at 300 K the power dropped to around 11 mW/facet. At 200 K the threshold current is 24 mA and at 300 K the threshold current is increased to 62 mA. The corresponding threshold current density is 1.3 kA/cm$^2$ for the laser with four QWs and 325 mA/cm$^2$ for each QW, respectively. At 340 K the maximum output power is decreased to 1.4 mW/facet with a threshold current of 186 mA.

The laser threshold current density and external differential quantum efficiency $\eta_d$ as a function of temperature are shown in Figure 12(a). The characteristic temperature $T_0$ is derived to be about 99 K in the temperature range of 200-300 K, and decreases to 35 K in 300-340 K range. The external differential quantum efficiency at 200 K is 41% and is decreased to 23% at 300 K and is further dropped to 7% at 340 K. The quantum efficiency characteristic temperature of $T_1$=142 K is obtained by fitting the external differential quantum efficiency in 200-320 K temperature range.

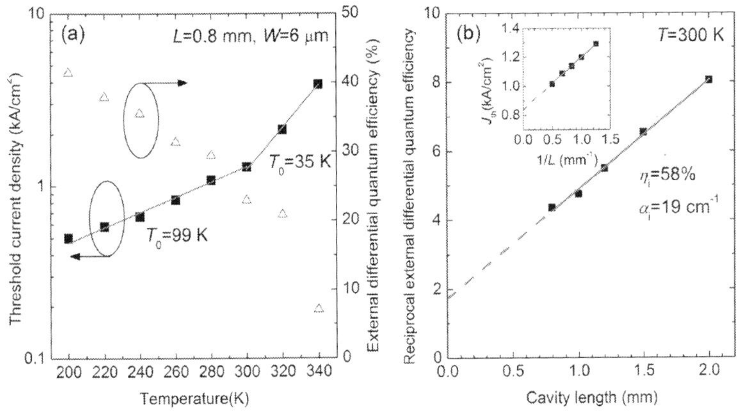


**Figure 12.** (a) Temperature dependence of the threshold current density and external differential quantum efficiency of the 2.4 µm TQW laser. The lines are the fitting lines of threshold current density in two temperature ranges. (b) Reciprocal external differential quantum efficiency of the TQW lasers as a function of the cavity length at 300 K. The inset shows the threshold current density ($J_{th}$) versus the reciprocal cavity length ($1/L$). The solid lines are the fitting of the measured data, and the dashed lines are the extrapolated data [38]. (Copyright 2014 The Japan Society of Applied Physics)

The reciprocal external differential quantum efficiency $\eta_d$ as a function of the cavity length is shown in figure 12(b). The internal quantum efficiency $\eta_i$ and internal loss $\alpha_i$ are calculated by fitting the data following $1/\eta_d=(1/\eta_i)[1+\alpha_i L/\ln(1+R)]$, where $L$ is the cavity length and $R$ is the facet reflectivity. Taking $R$ as 0.35 for the as-cleaved facet in the calculation, the internal quantum efficiency and internal loss is calculated to be 58% and 19 cm$^{-1}$, respectively. Note that the internal quantum efficiency value is favorable but the loss is relatively high, which limits the laser performances. The intervalence band absorption in the *p*-type InP up cladding is one main cause of the large internal loss. The inset of Figure 11(b) illustrates the threshold current density versus the reciprocal cavity length. By extrapolation, the threshold current density at infinite cavity length is 841 A/cm$^2$ (210 A/cm$^2$ per QW).

## 5. Metamorphic scheme for lasing wavelength beyond 2.5 µm

For the metamorphic scheme, InAlAs graded buffers were grown on InP substrate to form the metamorphic template with larger lattice constant than InP. The InAs QW structures were then grown on the template with thicker QW width, and the wavelength could be beyond 2.5 µm. For this kind of metamorphic laser structures, the material quality is very crucial for the device performance, therefore both of the buffer layers and growth parameters were optimized to improve the buffer quality. On the basis of the buffer optimization, the lasers were grown, demonstrated and characterized.

### 5.1. GSMBE growth and optimization of metamorphic buffers and QWs

As a preliminary experiment, the metamorphic buffer composed of a 2.5-μm-thick $In_xAl_{1-x}As$ continuously graded buffer and a 1-μm-thick $In_{0.8}Ga_{0.2}As$ virtual substrate layer were grown on InP substrate after the growth of a 100-nm-thick InP buffer layer. In the graded buffer, the indium composition x was graded from 0.52 to 0.8 through the simultaneously linear increase of indium source temperature and decrease of aluminum source temperature. After that, two periods of $InAs/In_{0.53}Ga_{0.47}As$ strained QWs were grown. The $In_{0.53}Ga_{0.47}As$ barrier is used to compensate the compressive strain of InAs well comparing to the virtual substrate layer. The lattice mismatch of InAs well and $In_{0.53}Ga_{0.47}As$ barrier with respect to $In_{0.8}Ga_{0.2}As$ virtual substrate layer is +1.3% and -1.8%, respectively. Two samples were grown and the widths of InAs well and $In_{0.53}Ga_{0.47}As$ barrier were 10 nm/7 nm and 15 nm/10 nm, respectively to form a strain compensated QW structure with minimal residual strain. The $In_{0.53}Ga_{0.47}As$ barrier can also stop the diffusion of exited carriers between the QW region and the rest part of the sample.

The grown samples show mirror-like surface morphology without haziness under optical microscopy. Regular cross-hatch patterns are observed on the surfaces with ridges and troughs along the [110] and [1-10] crystal directions. The root mean square (RMS) roughness measured by atomic force microscope (AFM) over 40×40 μm$^2$ is 3.9 nm and 6.1 nm for the 10 nm and 15 nm QW lasers, respectively. From the cross-sectional transmission electron microscope (XTEM) images of 15 nm QW laser measured at 160 kV as shown in Figure 13, the dislocations are mainly localized within the $In_xAl_{1-x}As$ graded buffer and do not propagate into the upper structures. The $In_{0.8}Ga_{0.2}As$ virtual substrate layer is free of dislocations in the XTEM measurements, which means the threading dislocation density is under the detection limit (about $10^7$ cm$^{-2}$). In the amplificatory image of the $InAs/In_{0.53}Ga_{0.47}As$ QWs region, distinct interfaces between the well and barrier layers are observed. The thickness fluctuations of wells and barriers are revealed around ±1 nm by a careful inspection.

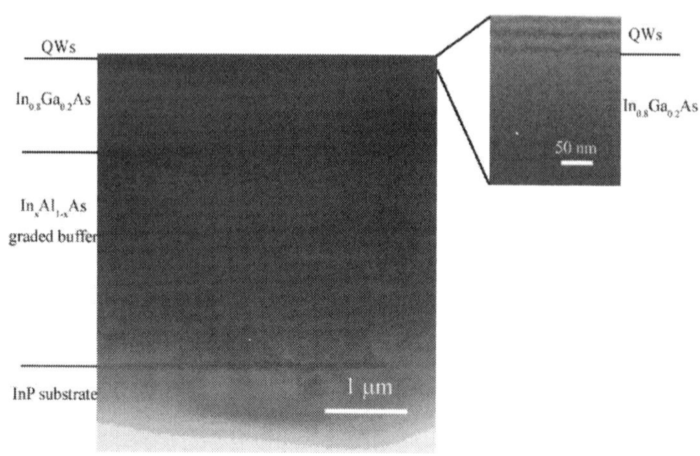


**Figure 13.** Bright-field XTEM images of the InP-based metamorphic QW epitaxy structure [42]. (Reprinted with permission from AIP)

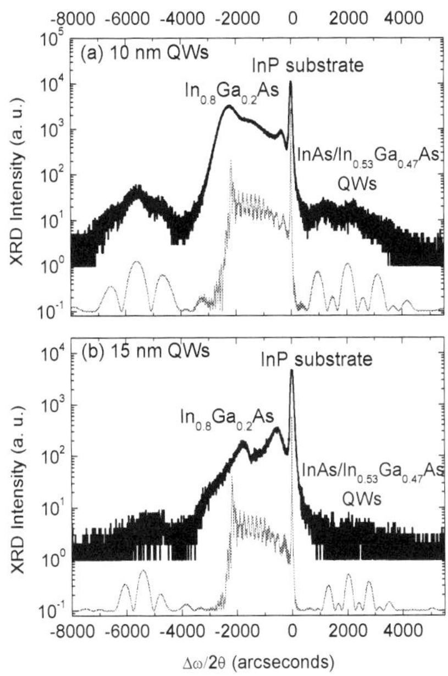


**Figure 14.** HRXRD (004) scanning curves of InP-based metamorphic (a) 10 nm QWs and (b) 15 nm QWs. The thick and thin lines show the measured and simulated results, respectively [42]. (Reprinted with permission from AIP)

The HRXRD ω/2θ (004) scanning curves were measured and shown in Figure 14 (thick lines). The peak with the largest intensity corresponds to InP substrate, and the wide peak with a relaxed mismatch of $1.8\times10^{-2}$ with respect to InP corresponds to $In_{0.8}Ga_{0.2}As$ virtual substrate layer. The HRXRD signals from the virtual substrate of the two samples show some difference, but become the same after etching away the QW regions at surface. Rough satellite peaks, related to the $InAs/In_{0.53}Ga_{0.47}As$ QWs, are observed symmetrically on the both sides of the $In_{0.8}Ga_{0.2}As$ peak, indicating the strain compensation effect with respect to the $In_{0.8}Ga_{0.2}As$ virtual substrate layer. The lattice dynamical simulations were also performed and the simulated curves are shown in Figure 14 (thin lines). In the simulation, the thicknesses of the wells/barriers were set the same as the design. The $In_xAl_{1-x}As$ buffer was assumed graded relaxed and the $In_{0.8}Ga_{0.2}As$ virtual substrate layer fully relaxed with respect to InP substrate, meanwhile InAs wells and $In_{0.53}Ga_{0.47}As$ barriers were fully strained with respect to $In_{0.8}Ga_{0.2}As$ virtual substrate layer.

The PL measurements were performed as shown in Figure 15. The photo-excited carriers are very sensitive to threading dislocations due to the diffusion length of photo-excited carriers in micrometers. Therefore, the PL intensity can reflect the material quality. The strong PL intensity at 300 K in Figure 15(a) indicates few threading dislocations in QW structures. For each sample, two emission peaks can be observed at 300 K, where the peak at about 2.4 μm corresponds to $In_{0.8}Ga_{0.2}As$ virtual substrate layer, the peaks at 2.90 μm and 3.05 μm for the

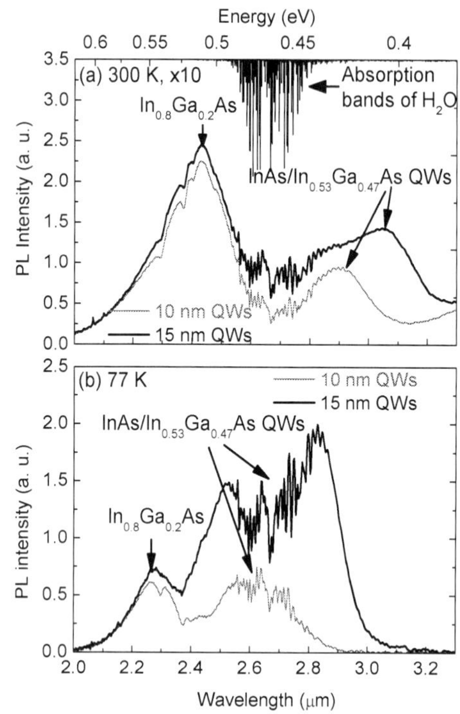


**Figure 15.** PL results of the metamorphic QW samples at (a) 300 K and (b) 77 K. The thin and thick lines show the results of 10 nm and 15 nm QWs, respectively. The absorption bands of water are also indicated [42]. (Reprinted with permission from AIP)

two samples correspond to the active QWs. In Figure 15(b) the peak corresponds to $In_{0.8}Ga_{0.2}As$ buffer blue-shifts to about 2.3 μm at 77 K. The PL signals of active QWs blue-shift to about 2.6 μm and 2.7 μm at 77 K for the 10 nm and 15 nm QW lasers, respectively, and the peak shapes are affected significantly by the water absorption. The relatively wide PL peaks even at low temperatures are possibly due to the presents of lattice defects and the poor carrier confinement on the top side of the QW. The metamorphic QWs with 15 nm thick InAs layer still show moderate optical quality. This structure is promising for the lasers around 3 μm.

### 5.2. Demonstration of metamorphic CW-operated lasers at low temperatures

To demonstrate the metamorphic lasers, the effects of strain-compensated QWs and separate confinement heterostructure (SCH) were studied. The growth of the samples started with a 200-nm-thick $n^+$ InP buffer layer grown, followed by a 1800-nm $n^+$ $In_xAl_{1-x}As$ continuously graded buffer layer, both highly doped with Si to about $2\times10^{18}$ $cm^{-3}$. In the graded buffer the indium composition was graded from 0.52 to 0.84 introducing an indium composition overshoot of 0.04 to achieve the full relaxation of the metamorphic layer. Then an 800-nm $n^+$ $In_{0.8}Al_{0.2}As$ cladding layer with carrier density of about $2\times10^{18}$ $cm^{-3}$ was grown to form a template. Then a 150-nm-thick $n^-$ waveguide layer, undoped active double QWs and a 150-nm-thick $p^-$ waveguide layer were grown [55]. The schematic band energy of the active regions for samples A, B and C can be found in Figure 16, where the solid lines indicate samples A and B whereas the dotted lines indicate sample C. For sample A $In_{0.8}Al_{0.2}As$ was used as the waveguide layers, which was the same as the cladding layers. For samples B and C, $In_{0.8}Ga_{0.2}As$ was used as the waveguide layers so that the SCH consisting of $In_{0.8}Ga_{0.2}As$ waveguide and $In_{0.8}Al_{0.2}As$ cladding layers were constructed. In the double QW region, a strain-compensated structure was used for sample A, formed by two 12-nm InAs wells sandwiched by $In_{0.6}Ga_{0.4}As$ barriers with the strains of +1.4% and -1.4% with respect to the $In_{0.8}Al_{0.2}As$ template. On the other hand, $In_{0.8}Ga_{0.2}As$ was used as the barrier layers in sample B, thus no strain compensation was applied in the QW region. Sample C combined the SCH and strain-compensated QWs, where 15-nm-thick InAs well and 15-nm-thick $In_{0.6}Ga_{0.4}As$ barrier layers were applied. At last, a 1700-nm-thick $p^+$ $In_{0.8}Al_{0.2}As$ cladding layer and a 300-nm-thick $p^+$ $In_{0.8}Ga_{0.2}As$ contact layer were grown, heavily doped with Be to higher than $5\times10^{18}$ $cm^{-3}$.

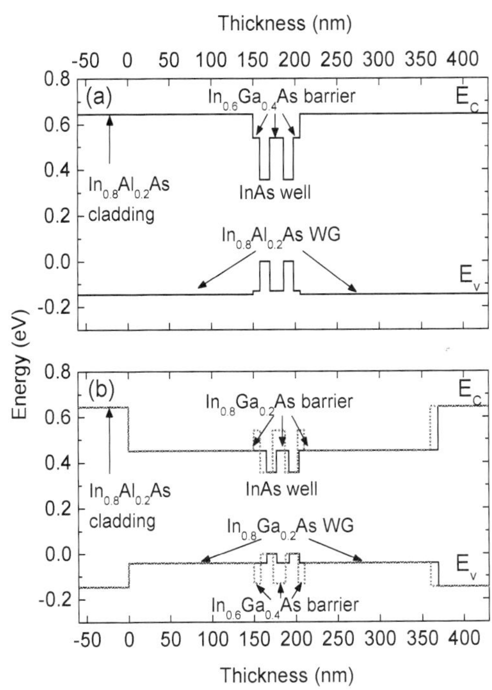


**Figure 16.** Schematic energy-band diagrams of the active regions of (a) sample A, (b) sample B (solid line) and sample C (dotted line) [55]. (Reprinted with permission from IOP)

Figure 17 shows the HRXRD scanning curves and PL spectra of the samples after etching away the p-type contact layer and p+ $In_{0.8}Al_{0.2}As$ cladding layer. Rough satellite peaks, related to the QWs, can be observed on the left side of the $In_{0.8}Al_{0.2}As/In_{0.8}Ga_{0.2}As$ peaks for the samples, although they are not so distinct. Comparing the intensities of $In_{0.8}Al_{0.2}As/In_{0.8}Ga_{0.2}As$ and QW peaks for the three samples, it is observed that the layer peak intensities of samples A and C with strain-compensated QWs are stronger than those of sample B without strain compensation. The lattice dynamical simulations were also performed and the HRXRD intensities of layer peaks on the left side of InP substrate were very close for all the samples in the simulations. Therefore, the measured stronger layer peak intensities of samples A and C indicate their better material quality than sample B due to the strain compensation of QW regions.

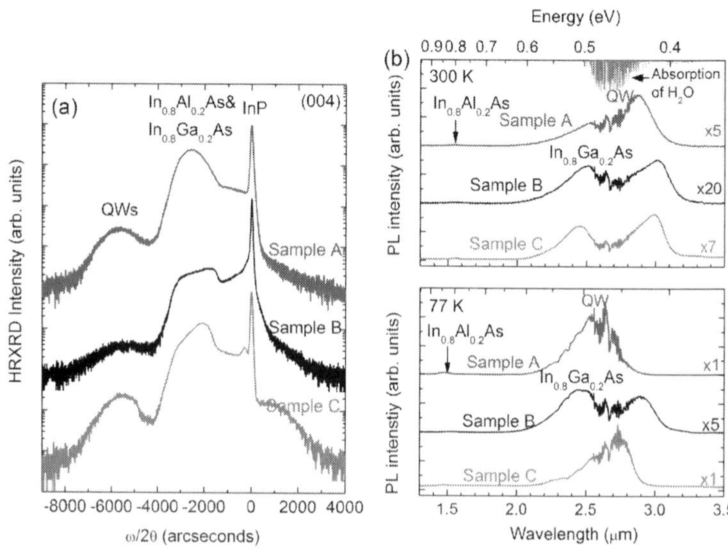


**Figure 17.** (a) HRXRD ω/2θ scanning curves of the QW lasers on InP-based metamorphic InAlAs buffers. (b) PL emissions of the QW structures on InP-based metamorphic InAlAs buffers at 300 K and 77 K [55]. (Reprinted with permission from IOP)

As shown in Figure 17(b), the PL signals between 2.6-2.8 μm are affected by the absorption of water vapor. Two PL peaks are observed for sample A, the relatively weak peak at 1.55 μm corresponds to the $In_{0.8}Al_{0.2}As$ cladding and waveguide layers, and the peak at 2.88 μm corresponds to the QWs. For sample B, there are also two PL peaks, and the PL signal of $In_{0.8}Al_{0.2}As$ is not clear in linear scale. The two PL peaks at 2.51 μm and 3.02 μm correspond to $In_{0.8}Ga_{0.2}As$ waveguide layers and the QWs, respectively. The PL intensities of the QWs for samples A and C are several times of that of sample B at both 300 K and 77 K, confirming the better material quality of samples A and C. The phenomenon of PL measurements is consistent with the results of HRXRD.

The ridge waveguide lasers were then demonstrated and the cross-section of the ridge waveguide structure was shown in the upper inset of Figure 18(a), where the upside of the ridge was narrowed due to the lateral etching. For the devices fabricated by sample A, no lasing was observed even at 77 K and injected by current up to 1 A. For sample B, around 2.3 μm lasing was observed at low temperatures. This implies that the SCH is very crucial for the device performances, even more important than the crystalline quality of the QWs. For sample B the optical confinement factor is calculated about 0.22, but for sample A the confinement factor is only 0.12 because only the QWs provide a very limited optical confinement in sample A. Figure 18 shows the typical 77 K CW spectrum of the device fabricated by sample B injected by the current of 195 mA. The lasing was declined markedly as the temperature increases. The maximum operation temperature was about 100 K in CW mode and 170 K in pulsed mode with a 5% duty cycle. The lasing transition energy is close to the transition from $In_{0.8}Ga_{0.2}As$ waveguide layers, indicating that excessive band filling exists and higher gain from the QW transition is required. Quite a few carriers may escape from the QWs to the $In_{0.8}Ga_{0.2}As$ layers, due to the small offsets of conductive and valence bands between InAs and $In_{0.8}Ga_{0.2}As$ layers, especially when injected by relatively high currents. The 77 K I-V-P curves of sample B are shown in the lower inset of Figure 18(a). The output power was smaller than 1 mW/facet and the turn-on voltage was about 0.5 V.

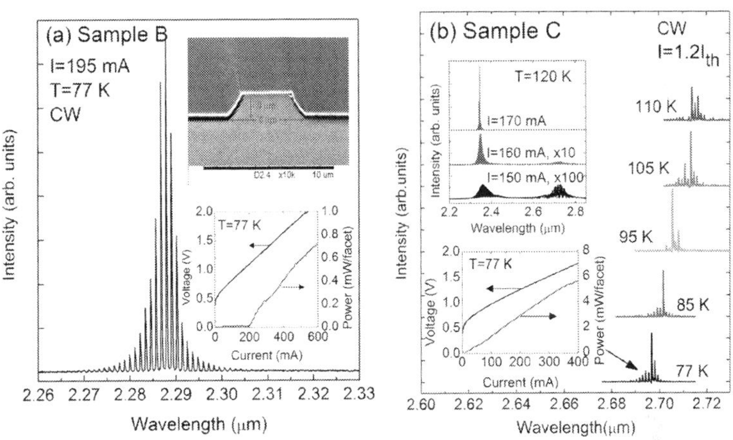


**Figure 18.** (a) CW lasing spectrum of sample B at 77 K. The upper inset shows the cross-section image of the ridge waveguide structure and the lower inset shows the I-V-P spectra of sample B at 77 K. (b) The temperature-dependent spectra of sample C. The upper inset shows the spectra at different injection currents at 120 K. The lower inset shows the I-V-P spectra at 77 K [55]. (Reprinted with permission from IOP)

For sample C, the CW lasing as long as 2.7 μm was observed, and the temperature-dependent lasing spectra are shown in Figure 18(b). The laser was operated in CW mode and injected by the current of 1.2 times the threshold current. As shown in the lower inset of Figure 18(b), the output power at 77 K was 5.7 mW/facet injected by the current of 400 mA and the turn-on voltage was about 0.65 V. The lasing wavelength moved from 2.70 μm at 77 K to above 2.71 μm at 110 K. When the temperature increased to 120 K, the lasing bounded to around 2.34 μm. The luminescence spectra at different injection currents at 120 K are shown in the upper inset of Figure 18(b). Two luminescence envelops at about 2.34 μm and 2.7 μm can be observed under the injection current of 150 mA. As the injection current increased the 2.34 μm luminescence was enhanced whereas the 2.7 μm luminescence was declined, and only the lasing at 2.34 μm could be observed when the injection current was increased to 170 mA.

Sample C combines with SCH and strain-compensated QWs and shows the best performances in these three metamorphic QW laser structures. The calculated optical confinement factor of sample C is about 0.21, close to that of sample B with SCH. The larger bandgap of $In_{0.6}Ga_{0.4}As$ barrier layers compared to that of $In_{0.8}Ga_{0.2}As$ barriers in sample B can also enhance the carrier confinement and gain of QWs. On the other hand, the strain-compensated QWs ensure the feasible material quality. Nevertheless, the confinements in this laser structure are still unfavorable, as the bandgap of $In_{0.8}Ga_{0.2}As$ waveguide layers is smaller than that of $In_{0.6}Ga_{0.4}As$ barriers, and the refraction index is larger in $In_{0.8}Ga_{0.2}As$ waveguide. The excessive band filling carrier overflow still exists and the lasing at 2.34 μm is possibly due to the transition from the waveguide layers. Generally, increasing the QW number is an effective approach to improve the optical gain and confinement, but more QWs may introduce more strain in the QW region. The growth process needs to be optimized and a tradeoff needs to be considered between the material quality and the QW number.

## 6. Conclusion

In conclusion, we have demonstrated InP-based Sb-free QW lasers in 2-3 μm wavelength band by optimizing the structural design, GSMBE growth and device processing. To control the strain in the pseudomorphic QW region, tne QWs with triangular shape were grown by using digital alloy technology. RT-CW lasers up to 2.43 μm have been successfully achieved by this approach and the performances are competitive to those well-developed GaSb-based QW lasers. Metamorphic InAs QWs were grown on InP-based $In_{0.8}Al_{0.2}As$ template to extend the lasing wavelength even longer. The lasers with CW lasing up to 2.71 μm have been demonstrated by applying 15 nm thick InAs QWs, although they still can only work at a low temperature of 110 K. Our efforts make clear that the extending of the lasing wavelength of InP-based Sb-free QW lasers into 2-3 μm band is not only attractive, but also practical. For CW operation of the laser at RT, the lasing wavelength up to about 2.5 μm should be possible through the fine design of the epitaxial structure, as well as optimization of the growth and processing technologies, adopting pseudomorphic or metamorphic schemes. For longer wavelengths, limited by the strain in pseudomorphic scheme or poor material quality and confinements of both photons and carriers in metamorphic scheme, the reachable performances of the lasers are degraded remarkably.

## Acknowledgements

This work is supported by the National Basic Research Program of China under grant No. 2012CB619200 and the National Natural Science Foundation of China under grant Nos. 61275113, 61204133 and 61405232.

http://dx.doi.org/10.5772/60705

## Author details

Yi Gu* and Yong-Gang Zhang

*Address all correspondence to: ygu@mail.sim.ac.cn

State Key Laboratory of Functional Materials for Informatics, Shanghai Institute of Microsystem and Information Technology, Chinese Academy of Sciences, China

## References

[1] Lei W, Jagadish C. Lasers and photodetectors for mid-infrared 2-3 μm applications. J. Appl. Phys. 2008; 104(9), 091101.

[2] Tittel F K, Richter D, Fried A. Mid-infrared laser applications in spectroscopy. Topics Appl. Phys. 2003; 89, 445–516.

[3] Zhang Y G, Gu Y. Gas source MBE grown wavelength extending InGaAs photodetectors. In: Gian-Franco Dalla Betta (ed.) Advances in photodiodes. Rijeka: InTech; 2011. pp. 349–376.

[4] Dhar N K, Dat R, Sood A K. Advances in infrared detector array technology. In: Pyshkin S L, Ballato J M (ed.) Optoelectronics – Advanced materials and devices. Rijeka: InTech; 2013. pp. 149–190.

[5] Bleuel T, Müller M, Forchel A. 2-μm GaInSb-AlGaAsSb distributed-feedback lasers. IEEE Photon. Technol. Letts. 2001; 13(6), 553.

[6] Belenky G, Shterengas L, Kipshidze G, Hosoda T. Type-I diode lasers for spectral region above 3 μm. IEEE J. Select Top. Qutuantum Electron. 2011; 17, 1426.

[7] Sanchez D, Cerutti L, Tournie E. Single-mode monolithic GaSb vertical-cavity surface-emitting laser. Optics Express 2012; 20, 15540.

[8] Vizbaras K, Amann M C. Room-temperature 3.73 μm GaSb-based type-I quantum-well lasers with quinternary barriers. Semicond. Sci. Technol. 2012; 27, 032001.

[9] Zhang Y G, Li A Z, Zheng Y L, Lin C, Jian G Z. MBE grown 2.0 μm InGaAsSb/AlGaAsSb MQW ridge waveguide laser diodes. J. Cryst. Growth 2001; 227–228, 582–585.

[10] Zhang Y G, Zheng Y L, Lin C, Li A Z, Liu S. Continuous wave performance and tunability of MBE grown 2.1 μm InGaAsSb/AlGaAsSb MQW lasers. Chin. Phys. Lett. 2006; 23(8), 2262.

[11] Zhang Y G, Zhang X J, Zhu X R, Li A Z, Liu S. Tunable diode laser absorption spectroscopy detection of $N_2O$ at 2.1 μm using antimonide laser and InGaAs photodiode. Chin. Phys. Lett. 2007; 24(8), 230.

[12] Zhang Y G, Gu Y, Zhang X J, Li A Z, Tian Z B. Gas sensor using a robust approach under time multiplexing scheme with a twin laser chip for absorption and reference. Chin. Phys. Lett. 2008; 25(9), 3246–3249.

[13] Baranov A N, Cuminal Y, Boissier G, Alibert C, Joullié A. Low-threshold laser diodes based on type-II GaInAsSb/GaSb quantum-wells operating at 2.36 μm at room temperature. Electron. Lett. 1996; 32, 2279.

[14] Rossner K, Hummer M, Lehnhardt T, Muller M, Forchel A, Fischer M, Koeth J. Continuous-wave operation of GaInAsSb/GaSb type-II ridge waveguide lasers emitting at 2.8 μm. IEEE Photon. Technol. Lett. 2006; 18(13), 1424.

[15] Yang R Q. Infrared laser based on intersubband transitions in quantum wells. Superlattices Microstruct. 1995; 17(1), 77.

[16] Sprengel S, Andrejew A, Vizbaras K, Gruendl T, Geiger K, Boehm G, Grasse C, Amann M C. Type-II InP-based lasers emitting at 2.55 μm. Appl. Phys. Lett. 2012; 100(4), 041109.

[17] Chang C H, Li Z L, Pan C H, Lu H T, Lee C P, Lin S D. Room-temperature mid-infrared "M"-type GaAsSb/InGaAs quantum well lasers on InP substrate. J. Appl. Phys. 2014; 115(6), 063104.

[18] Cornet C, Doré F, Ballestar A, Even J, Bertru N, Corre A, Loualiche S. InAsSb/InP quantum dots for midwave infrared emitters: A theoretical study. J. Appl. Phys. 2005; 98, 126105.

[19] Zhu C, Zhang Y G, Li A Z, Zheng Y L. Comparison of thermal characteristics of antimonide and phosphide MQW lasers. Semicond. Sci. Technol. 2005; 20, 563.

[20] Zhu C, Zhang Y G, Li A Z, Zheng Y L, Tang T. Heat management of MBE-grown antimonide lasers. J. Cryst. Growth 2005; 278, 173.

[21] Razeghi M, Bandyopadhyay N, Bai Y B, Lu Q Y, Slivken S. Recent advances in mid infrared (3–5 μm) quantum cascade lasers. Opt. Mater. Express 2013; 3(11), 1872.

[22] Vurgaftman I, Meyer J R, Ram-Mohan L R. Band parameters for III–V compound semiconductors and their alloys. J. Appl. Phys. 2001; 89, 5815.

[23] Zhang Y G, Chen J X, Chen Y Q, Qi M, Li A Z, Frojdh K, Stoltz B. Characteristics of strain compensated 1.3 μm InAsP/InGaAsP ridge waveguide laser diodes grown by gas source MBE. J. Cryst. Growth 2001; 227–228, 329–333.

[24] Forouhar S, Ksendzov A, Larsson A, Temkin H. InGaAs/InGaAsP/InP strained layer quantum well lasers at ~2 μm. Electron. Lett. 1992; 28(15), 1431.

[25] Forouhar S, Keo S, Larsson A, Ksendzov A, Temkin H. Low threshold continuous operation of InGaAs/InGaAsP quantum well lasers at ~2.0 μm. Electron. Lett. 1993; 29(7), 574.

[26] Major J S, Nam D W, Osinski J S, Welch D F. High-power 2.0 μm InGaAsP laser diodes. IEEE Photon. Technol. Lett. 1993; 5(6), 594.

[27] Martinelli R U, Menna R J, Triano A, Harvey M G, Olsen G H. Temperature dependence of 2 μm strained-quantum-well InGaAs/InGaAsP/InP diode lasers. Electron. Lett. 1994; 30(4), 324.

[28] Ochiai M, Temkin H, Forouhar S, Logan R A. InGaAs-InGaAsP buried heterostructure lasers operating at 2.0 μm. IEEE Photon. Technol. Lett. 1995; 7(8), 825.

[29] Oishi M, Yamamoto M, Kasaya K. 2.0-μm single-mode operation of InGaAs-InGaAsP distributed-feedback buried-heterostructure quantum-well lasers. IEEE Photon. Technol. Lett. 1997; 9(4), 431.

[30] Mitsuhara M, Ogasawara M, Oishi M, Sugiura H. 2.05-μm wavelength InGaAs-InGaAs distributed-feedback multiquantum-well lasers with 10-mW output power. IEEE Photon. Technol. Lett. 1999; 11(1), 33.

[31] Mitsuhara M, Ogasawara M, Oishi M, Sugiura H. Metalorganic molecular-beam-epitaxy-grown $In_{0.77}Ga_{0.23}As$/InGaAs multiple quantum well lasers emitting at 2.07 μm wavelength. Appl. Phys. Lett. 1998; 72(24), 3106.

[32] Kuang G K, Böhm G, Grau M, Rösel G, Meyer R, Amann M C. 2.12 μm InGaAs-InGaAlAs-InP diode lasers grown in solid-source-molecular-beam epitaxy. Appl. Phys. Lett. 2000; 77(8), 1091.

[33] Ortsiefer M, Böhm G, Grau M, Windhorn K, Rönneberg E, Rosskopf J, Shau R, Dier O, Amann M C. Electrically pumped room temperature CW VCSELs with 2.3 μm emission wavelength. Electron. Lett. 2006; 42(11), 640.

[34] Boehm G, Grau M, Dier O, Windhorn K, Roenneberg E, Rosskopf J, Shau R, Meyer R, Ortsiefer M, Amann M C, Growth of InAs-containing quantum wells for InP-based VCSELs emitting at 2.3 μm. J. Cryst. Growth. 2007; 301–302, 941–944.

[35] Sato T, Mitsuhara M, Kondo Y. 2.33 μm-wavelength InAs/InGaAs multiple-quantum-well lasers grown by MOVPE. Electron. Lett. 2007; 43(21), 1143.

[36] Sato T, Mitsuhara M, Nunoya N, Kasaya K, Kano F, Kondo Y. 2.33-μm-wavelength distributed feedback lasers with InAs-$In_{0.53}Ga_{0.47}As$ multiple-quantum wells on InP substrates. IEEE Photon. Technol. Lett. 2008; 20(12), 1045.

[37] Sato T, Mitsuhara M, Kakitsuka T, Fujisawa T, Kondo Y. Metalorganic vapor phase epitaxial growth of InAs/InGaAs multiple quantum well structures on InP substrates. IEEE J. Sel. Topics Quan. Electron. 2008; 14(4), 992.

[38] Gu Y, Zhang Y G, Cao Y Y, Zhou L, Chen X Y, Li Hsby, Xi S P. 2.4-μm InP-based antimony-free triangular quantum well lasers in continuous-wave operation above room temperature. Appl. Phys. Express 2014; 7, 032701.

[39] Krier A, Chubb D, Krier S E, Hopkinson M, Hill G. Light sources for wavelengths >2 μm grown by MBE on InP using a strain relaxed buffer. IEE Proc. 1998; 145(5), 292.

[40] Kirch J, Garrod T, Kim S, Park J H, Shin J C, Mawst L J, Kuech T F, Song X, Babcock S E, Vurgaftman I, Meyer J R, Kuan T S. $InAs_yP_{1-y}$ metamorphic buffer layers on InP substrates for mid-IR diode lasers. J. Cryst. Growth 2010; 312, 1165–1169.

[41] Jung D, Song Y, Yu L, Wasserman D, Lee M L. 2.8 μm emission from type-I quantum wells grown on InAsP/InP metamorphic graded buffer. Appl. Phys. Lett. 2012; 101, 251107.

[42] Gu Y, Zhang Y G, Wang K, Fang X, Li C, Cao Y Y, Li A Z, Li Y Y. InP-based InAs/InGaAs quantum wells with type-I emission beyond 3 μm. Appl. Phys. Lett. 2011; 99, 081914.

[43] Cao Y Y, Zhang Y G, Gu Y, Chen X Y, Zhou L, Li Hsby. 2.7 μm InAs quantum well lasers on InP-based InAlAs metamorphic buffer layers. Appl. Phys. Lett. 2013; 102, 201111.

[44] Gu Y, Zhang Y G, Chen X Y, Cao Y Y, Fang X, Ding G Q, Zhou L. $InAs/In_{0.83}Al_{0.17}As$ quantum wells on GaAs substrate with type-I emission at 2.9 μm. Appl. Phys. Lett. 2013; 102, 121110.

[45] Gu Y, Zhang Y G, Liu S. Strain compensated AlInGaAs/InGaAs/InAs triangular quantum wells for lasing wavelength beyond 2 μm. Chin. Phys. Lett. 2007; 24, 3237–3240.

[46] Gu Y, Zhang Y G. Properties of strain compensated symmetrical triangular quantum wells composed of InGaAs/InAs chirped superlattice grown using gas source molecular beam epitaxy. Chin. Phys. Lett. 2008; 25, 726–729.

[47] Chen J X, Li A Z, Chen Y Q, Guo F M, Lin C, Zhang Y G, Qi M. Quasi RT-CW operation of InGaAs/InGaAsP strained quantum well lasers. J. Cryst. Growth 2001; 227–228, 338–342.

[48] Gu Y, Zhang Y G, Wang K, Li A Z, Li Y Y. Optimization of AlInGaAs/InGaAs/InAs strain compensated triangular quantum wells grown by gas source molecular beam epitaxy for laser applications in 2.1–2.4 μm range. J. Cryst. Growth 2009; 311, 1935–1938.

[49] Giugni S, Tansley T L, Green F, Shwe C, Gal M. Optical transitions in symmetric, compositionally graded triangular AlGaAs quantum wells grown by molecular beam epitaxy. J. Appl. Phys. 1992; 71, 3486–3491.

[50] Lin D Y, Lin F C, Huang Y S, Qiang H, Pollak Fred H, Mathine D L, Maracas G N. Piezoreflectance and photoreflectance study of GaAs/AlGaAs digital alloy compositional graded structures. J. Appl. Phys. 1996; 79, 460–466.

[51] Jourba S, Gendry M, Marty O, Pitaval M, Hollinger G. High-quality highly strained InGaAs quantum wells grown on InP using $(InAs)_n(GaAs)_{0.25}$ fractional monolayer superlattices. Appl. Phys. Lett. 1999; 75, 220.

[52] Cao Y Y, Zhang Y G, Gu Y, Chen X Y, Zhou L, Li Hsby. Improved performance of 2.2 μm InAs/InGaAs QW lasers on InP by using triangular wells. IEEE Photon. Technol. Lett. 2014; 26(6), 571–574.

[53] Gu Y, Wang K, Li Y Y, Li C, Zhang Y G. InP-based InGaAs/InAlGaAs digital alloy quantum well laser structure at 2 μm. Chin. Phys. B 2010; 19, 077304.

[54] Cao Y Y, Gu Y, Zhang Y G, Li Y Y, Fang X, Li A Z, Zhou L, Li Hsby. InAs/InGaAs digital alloy strain-compensated quantum well lasers. J. Infrared Millim. Waves 2014; 33(3), 213–217.

[55] Gu Y, Chen X Y, Zhang Y G, Cao Y Y, Fang X, Li Hsby. Type-I mid-infrared InAs/InGaAs quantum well lasers on InP-based metamorphic InAlAs buffers. J. Phys. D: Appl. Phys. 2013; 46, 505103.

Chapter 5

# Dye-Sensitized Solar Cells with Graphene Electron Extraction Layer

Lung-Chien Chen

Additional information is available at the end of the chapter

http://dx.doi.org/10.5772/60644

**Abstract**

This work aims to improve the conversion efficiency of dye-sensitized solar cells (DSSCs) by introducing a new material, graphene, into the DSSC structure. Graphene is a potential material for many applications due to their high electron mobility, outstanding optical properties, and thermal, chemical, and mechanical stability. Therefore, this study changes several parameters, structures, and methods to optimize and compare with the traditional DSSCs. There are three major respects about with or without graphene, the method of plating or sputtering, and the structure of graphene/$TiO_2$ or $TiO_2$/graphene/$TiO_2$ in DSSCs solar cells. Finally, this research knows that the method of sputtering is much better than plating; the conversion efficiency of solar energy with graphene/$TiO_2$ was increased from 1.45 % to 3.98 %, and the conversion efficiency with $TiO_2$/graphene/$TiO_2$ sandwich structure was increased from 1.38 % to 3.93 %. It means that the new material, graphene, works in enhancing the conversion efficiency of DSSCs.

**Keywords:** DSSCs, Solar cell, Graphene, Sandwich structure, $TiO_2$

## 1. Introduction

This chapter aims to review the dye-sensitized solar cells (DSSCs) with graphene structure. DSSCs have been under extensive research. Since the color of the device can be easily varied by choosing different dyes and cells on flexible substrates have been already demonstrated, DSSCs are especially attractive for building integrated photovoltaics. The cell concept can reduce the production costs and energy payback time significantly compared to standard silicon cells or other thin film cells.

However, one of the major issues hindering the rapid commercialization of DSSCs is their lower conversion efficiency compared to conventional p-n junction solar cells [1]. That may be attributed to poor charge separation in DSSC structure. Therefore, charge transfer structure, such as Au nanoparticles and quantum dots, has been employed in a DSSC to improve the device performance through charge separation in the photoelectrodes [2-5].

Graphene is a potential material for many applications due to their high electron mobility, outstanding optical properties, and thermal, chemical, and mechanical stability [6-10]. Therefore, the second section in this article illustrates the principle of electron extraction layer. $TiO_2$ plays an important role on the electron-extraction layer. We will discuss the electron transmission on dye-sensitized solar cells. The effect of the electron-transporting layer of the solar cell is very important. Therefore, we show the I-V characteristics of the DSSCs. The cell performance was measured, which had different electron-extraction layer structures.

The third section discusses the preparation method of the graphene. Graphene is a potential material for many applications due to their high electron mobility, outstanding optical properties, and thermal, chemical, and mechanical stability.

In the fourth section, the graphene was introduced into the DSSC structure to improve electron conversion efficiency. This study investigates the effect on the graphene layer as electron transport layer in the DSSC structure deposited by the magnetron sputtering method; in particular, it examines the performance of the DSSCs with the graphene electron transport layer.

The fifth section reveals a new DSSC structure. The structure has provided excellent performance and higher photoelectric conversion efficiency by DSSC with the $TiO_2$/graphene/$TiO_2$ sandwich structure. This section focuses on the improvement that is associated with the increase in electron transport efficiency and the absorption of light in the visible range.

The concluding paragraphs will summarize some parameters of DSSC with or without a graphene layer which was prepared by sputtering and then discuss the DSSC's parameters and reasons which have different preparation methods of graphene layer. Finally, there are some concluding remarks.

## 2. Principle of electron transport layer

Figure 1 sketches the structure of the basic DSSC which can be divided into several parts. They have a basic structure that comprises two conductive substrates (one is photoelectrode and the other is counter electrode), an absorbing layer of semiconductor materials, dye molecules, and a redox electrolyte. The basic principle of operation of DSSCs includes the following: (1) the light irradiates on the DSSC and the photons will pass through the photoelectrode to the dye layer which is absorbed by the photosensitizer dye molecule. (2) The photosensitizers are

http://dx.doi.org/10.5772/60644

excited from the ground state (S) to the excited state (S*). The excited electrons are injected into the conduction band of the $TiO_2$ electrode. This results in the oxidation of the photosensitizer ($S^+$). (3) Electrons are injected from the photoexcited dye into the conductive band of the semiconductor. The electrons will pass from the electric transport layer to the external circuit. (4) The oxidized photosensitizer ($S^+$) accepts electrons from the $I^-$ ion redox mediator, leading to regeneration of the ground state (S), and the $I^-$ is oxidized to the oxidized state, $I_3^-$, and transports the positive charges to the counter electrode. (5) The oxidized redox mediator, $I_3^-$, diffuses toward the counter electrode and then it is reduced to $I^-$ ions.

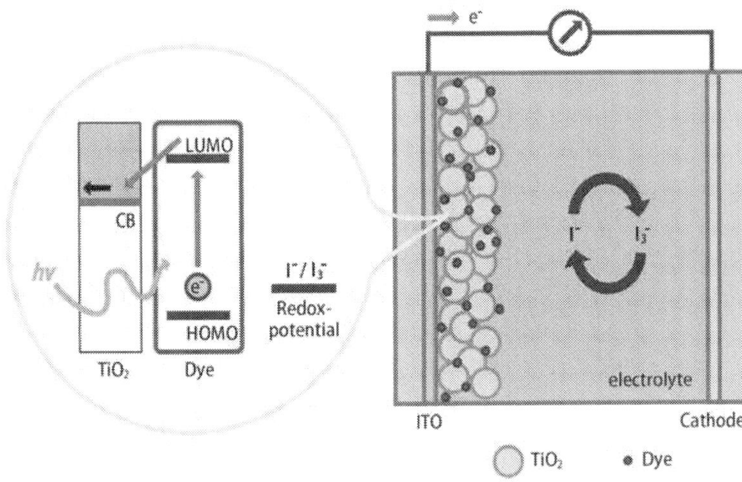


**Figure 1.** Schematic cross section of the completed structure

The principle of electron transport (or extraction) layer inserted in the traditional DSSC structure had been reported [11-17]. Figure 2 shows the energy level diagram and mechanism of photocurrent generation in $TiO_2$ DSSCs with the graphene layer. The work function of the graphene layer is around 4.5 eV [18,19]. Graphene has a work function similar to that of the indium tin oxide (ITO) (4.8 eV) electrode. The graphene layer does not prevent the flow of injected electrons down to the ITO electrode because its work function exceeds that of the ITO electrode [20-22]. Therefore, the brief operating process is as follows. Dye N719 was excited by incident light, and electrons transit from HOMO to LUMO. The LUMO and HOMO are the lowest unoccupied molecular orbit and highest occupied molecular orbit, respectively. Electrons are injected into the graphene electron transport layer via the $TiO_2$ photoelectrode. The electrons transferred to the graphene electron transport layer were collected at the back contact to generate a photocurrent. Therefore, the inserted graphene layer collects electrons and acts as a transporter in the effective separation of charge and rapid transport of the photogenerated electrons.

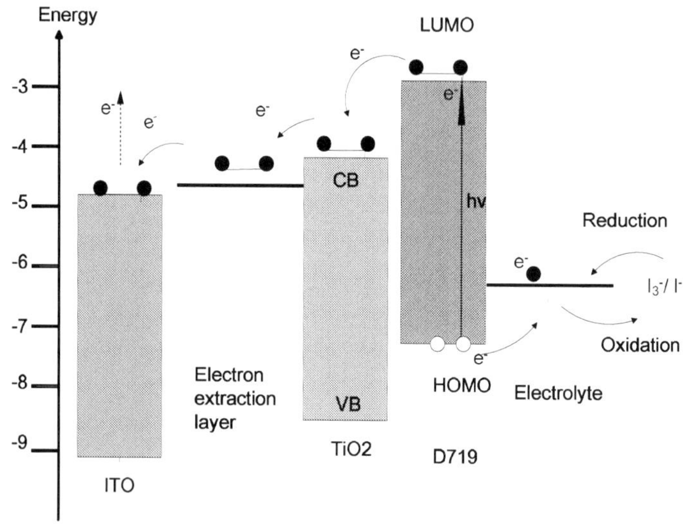


**Figure 2.** Energy level diagram and mechanism of photocurrent generation in the DSSCs with the graphene electron transfer layer

## 3. Preparation and characteristics of graphene

Graphene is a potential material for many applications such as extensively utilized in organic photovoltaic (PV) cells. It has excellent optical and electrical characteristics which are exploited in transparent conductive films or electrodes by their high electron mobility [6-8,10,17], outstanding optical properties, and thermal, chemical, and mechanical stability [6–10,12]. However, it is hard to produce high-quality graphene to use in the sputter deposition method. Therefore, this study uses plating method to plate graphene and compare with the traditional method.

First, acetylacetone and Triton X-100 were added into 10 ml water by using syringe. The $TiO_2$ compound solution was stirred for 24 h using a magnetic stirrer. After mixing the $TiO_2$ compound solution, the $TiO_2$ colloid is obtained. The graphene was stacked on the ITO substrate by electroplating process. The plating solution is graphene dispersion. The plating solution was injected into the beaker and stirred with an air pump. Figure 3 shows the electroplating process. The anode connected to the graphite, and the cathode connected to the ITO substrate. The speeds of the coating process were 500 rpm for 20 s and 2,000 rpm for 60 s. The thickness of $TiO_2$ is about 13 μm. After the annealing process by using RTA (rapid thermal annealing) at 450 °C for 30 min, the strength of the anatase structure would be enhanced. When the samples cool down to room temperature, they were soaked into the N719 solution; the

N719 solution is mixed ethanol and N719 powder. The samples will produce electrons when they are illuminated with light after they are soaked into the N719 solution.

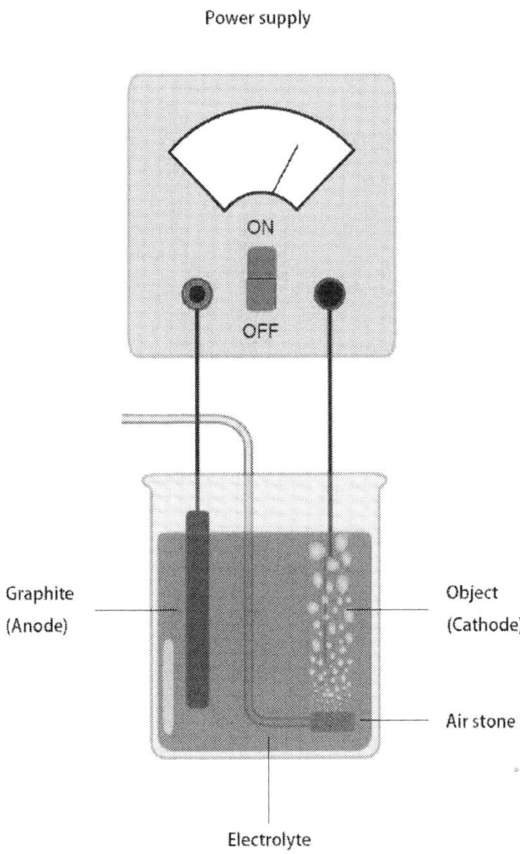


**Figure 3.** The graphene electroplating methods

A graphene layer was sputtered on indium tin oxide (ITO) conductive glass substrate by radio-frequency magnetron sputtering with a graphite target. It is the electron transport layer that improves the electron transfer in the DSSC structure.

First, the solution consisting of $TiO_2$ nanocrystalline powder, Triton X-100, acetic acid, and deionized water was mixed as a colloidal solution, and the colloidal solutions were daubed uniformly onto the graphene electron transfer layer to form a thick film. After annealing, the photoelectrode with the graphene layer was immersed in N719 dye absorption ($(Bu4N)_2$-[Ru(dcbpyH)2$(NCS)_2$] complex) in ethanol for 24 h. To increase its anatase content, the samples were sintered at 450 for 30 min. The electrolyte was composed of iodide and lithium iodide with and without 4-tertbutylpyridine (TBP) in propylene carbonate. Then a thick layer of platinum was sputtered onto ITO substrate as a counter electrode. Cells were fabricated by placing sealing films (SX1170-60, Solaronix) between the two electrodes and leaving just two via-holes for injection of electrolyte. The sealing process was carried out on a hot plate. Then the electrolyte was injected into the space between the two electrodes through the via-holes. Finally, the via-holes were sealed using epoxy with low vapor transmission rate.

This study fabricated three different samples: samples A and B are plated with graphene for 20 min and 30 min, respectively, and sample C is a normal DSSC. Figure 4 and Table 1 show the I-V curves and the measurement values. The cell is measured under AM 1.5 illumination at 25 °C. The active area is 0.3×0.3 cm$^2$. The short-circuit current densities of the samples are 4.97 mA/cm$^2$ (electroplated with graphene for 20 min), 5.42 mA/cm$^2$ (electroplated with graphene for 30 min), and 11.2 mA/cm$^2$ (normal DSSC), respectively. The value of open-circuit voltage between the samples only has a slight difference. The efficiency of the samples are 0.796 % (electroplated with graphene for 20 min), 0.844 % (electroplated with graphene for 30 min) and 3.93 % (normal DSSC), respectively.

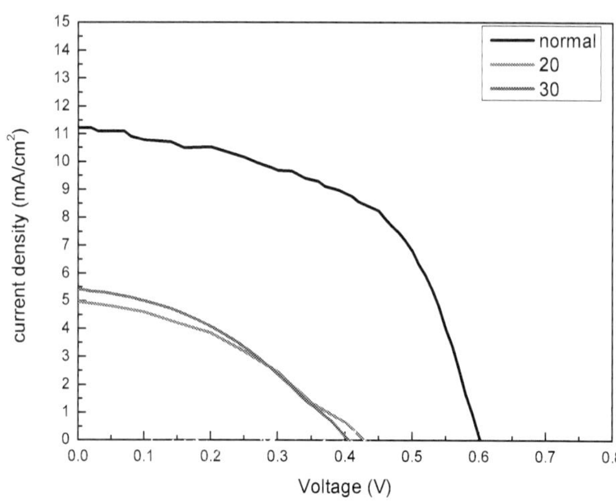


**Figure 4.** I-V curves of three samples

| Sample | Jsc (mA/cm$^2$) | Voc (mV) | FF% | Efficiency (%) |
|---|---|---|---|---|
| A(20 min) | 4.97 | 0.4 | 0.401 % | 0.796 % |
| B(30 min) | 5.42 | 0.4 | 0.389 % | 0.844 % |
| C (normal DSSC) | 11.2 | 0.6 | 0.585 % | 3.93 % |

**Table 1.** The measurement values of this study

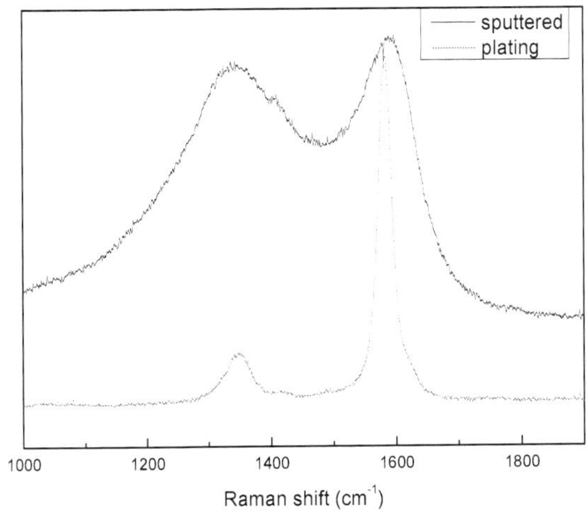


**Figure 5.** The Raman spectra of plated graphene and sputtered graphene in G-band and D-band

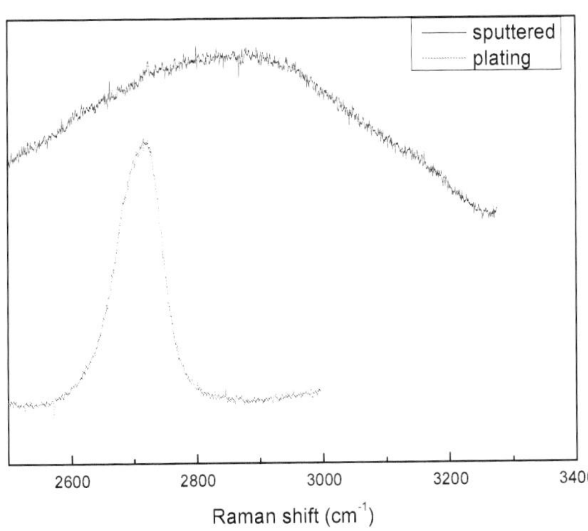


**Figure 6.** The Raman spectra of plated graphene and sputtered graphene in 2D-band

Figures 5 and 6 show the Raman spectra of electroplated graphene and sputtered graphene (normal DSSC). As shown in Figure 5, electroplated graphene and sputtered graphene have Raman peaks at 1,350 $cm^{-1}$ and 1,580 $cm^{-1}$ [23–25]. The ID/IG of plating graphene is 0.52, and the sputtered graphene is 0.96. The higher value of ID/IG shows good preservation of the highly crystalline structure of graphene. Figure 5 compares the D-band of two different procedures [26]; it shows that using the plating process is much better than using the sputtering process. The G-band is a doubly degenerate phonon mode at the Brillouin zone center; the D-band is a defect and the phonon branches around K point [27]. As shown in Figure 6, the 2D-band is a two-phonon double-resonance process [28] and is similar to the G-band but has a more complicated peak structure [29-31]. It depends on the photon energy and polarization.

Figure 7 shows the top-view SEM image of electroplating graphene on ITO glass. As shown in Figure 7, the graphene flakes with 10 μm width were stacked on the ITO glasses. Figure 8 shows the cross-sectional SEM image of plating graphene on the ITO glass. The graphene is successfully plated on the ITO glasses, and the thickness is around 8 μm.

**Figure 7.** Top view of electroplating graphene on ITO glass surface

## 4. DSSCs with graphene/TiO2 active layer

Because graphene has high electron mobility, we use graphene as an electron transport layer to improve the electron transfer in the DSSC. That is DSSCs with graphene/$TiO_2$ active layer. The graphene flakes prepared by using the electroplating method have demonstrated a superior graphene property by Raman scattering. However, the DSSCs with graphene flakes exhibited poor power conversion efficiency, owing to the high series resistance caused by the discontinue graphene flakes. Therefore, sputtered graphene was employed to replace the graphene flakes prepared by electroplating to improve the electrical properties of the DSSCs even the sputtered graphene including graphene oxide.

**Figure 8.** Cross-section SEM image of the plated graphene

First of all, a 60-nm-thick graphene layer was sputtered on indium tin oxide (ITO) conductive glass substrate by radio-frequency magnetron sputtering as an electron transport layer. Next, the solution consisting of $TiO_2$ was mixed as a colloidal solution which was daubed uniformly onto the graphene electron transfer layer to form a thick film. Then a 100-nm-thick layer of platinum was sputtered onto ITO substrate as a counter electrode. Cells were fabricated by placing sealing films between the two electrodes and leaving just two via-holes for injection of electrolyte. Then, the electrolyte was injected into the space between the two electrodes through the via-holes. Finally, the via-holes were sealed using epoxy with low vapor transmission rate. Figure 9 shows the cross section of the completed structure.

Afterward, we began examining its results by comparing the 60-nm-thick graphene electron transport layer with the 100-nm-thick graphene layer. Figure 10 shows the absorption of $TiO_2$ DSSCs with and without the graphene electron transfer layer in visible range. As shown in Figure 10, the graphene electron transport layer has an increased absorption coefficient in the range of 310–400 nm. Therefore, the graphene electron transport layer is also an absorption layer to improve the absorption of the solar cells.

Figure 11 shows the *I-V* characteristics of the DSSCs. This figure shows cell performance between $TiO_2$ DSSCs and $TiO_2$/graphene under AM 1.5 illumination with a solar intensity of 100 mW/cm$^2$ at 25°C. The cell has an active area of 3×3 mm$^2$ and no antireflective coating.

Finally, we examine its result by measuring the cell parameters, open-circuit voltage ($V_{oc}$), short-circuit current ($J_{sc}$), fill factor (FF), and energy conversion efficiency ($E_{ff}$) which are summarized in Table 2 [12].

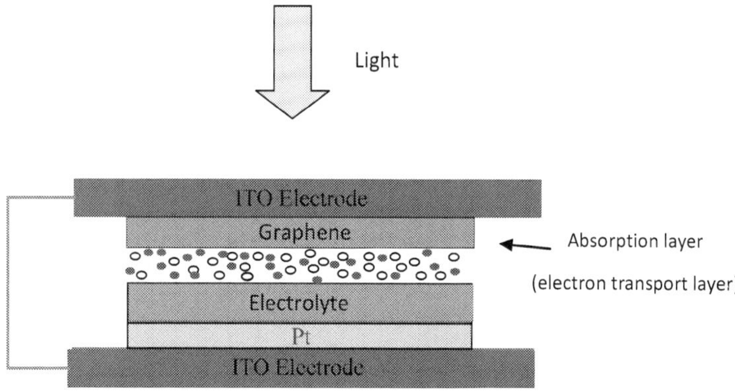


**Figure 9.** Schematic cross section of the completed structure

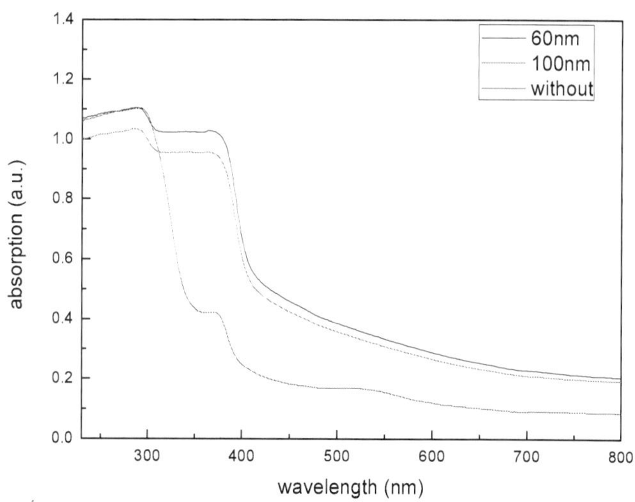


**Figure 10.** Absorption spectra of the DSSCs with and without the graphene electron transfer layer [12]

| | $TiO_2$ | Graphene+$TiO_2$ |
|---|---|---|
| Jsc ($mA/cm^2$) | 6.9 | 17.5 |
| $V_{oc}$ (V) | 0.5 | 0.5 |
| FF | 0.419 | 0.456 |
| η (%) | 1.45 | 3.98 |

**Table 2.** The parameters of TiO2 DSSCs with and without graphene electron transport layer [12]

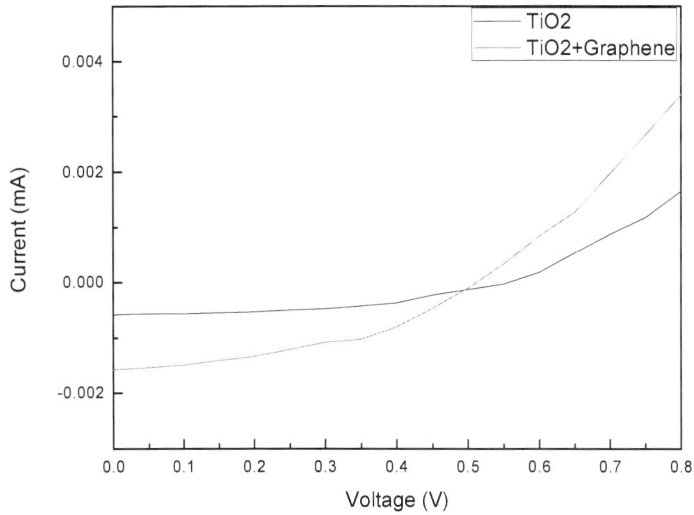


**Figure 11.** *I-V* curves of the DSSCs with and without the graphene electron transfer layer under illumination [12]

According to Figures 10 and 11 and Table 2, the short-circuit current rises up to 17.5, fill factor to 0.456, and energy conversion efficiency to 3.98 %. The enhanced performance of DSSCs with a graphene was attributed to the increase in electron transport efficiency and light absorption in visible range.

# 5. DSSCs with $TiO_2$/Graphene/$TiO_2$ sandwich structure

Because of the $TiO_2$/graphene sandwich structure, the efficiency on traditional DSSCs improved. As a result, we use three sandwich structures to achieve the desired outcomes of the following experiment. The enhanced performance of DSSCs with the sandwich structure can be attributed to an increase in electron transport efficiency and in the absorption of light in the visible range. The preparation of $TiO_2$ photoelectrodes is done by the following: the $TiO_2$ slurry was prepared by mixing 6 g of nanocrystalline powder, 0.1 mL Triton X-100, and 0.2 mL acetylacetone. The graphene film is deposited on the surface of the first photoelectrode layer, a single $TiO_2$ photoelectrode layer. This is spin-coated with the rate of rotation of 2,000 rpm, a sandwich structure with three rotational speed to 4,000 rpm, in the present experiment for comparison.

In summary, the DSSC with the sandwich structure in this study exhibited a Voc of 0.6 V, a high Jsc of 11.22 mA $cm^{-2}$, a fill factor (FF) of 0.58, and a calculated $\eta$ of 3.93 %, which is 60 % higher than that of a DSSC with the traditional structure.

Figure 12(a) shows the top-view SEM image of the $TiO_2$ nanoparticles with mean diameter of 50 nm. Figure 12(b) shows the cross-sectional SEM image of a $TiO_2$/graphene/$TiO_2$ sandwich structure. The thickness of the graphene electron extraction layer is around 60 nm.

**Figure 12.** SEM images of (a) TiO2 nanoparticles and (b) $TiO_2$/graphene/$TiO_2$ sandwich structure [17]

Figures 13(a) and 13(b) present the Raman scattering spectra of the graphene film that was deposited on the glass substrate using the process that was described in the section on the preparation of graphene. The spectra include important peaks that correspond to the D-band (approximately 1,350 $cm^{-1}$), the G-band (approximately 1,580 $cm^{-1}$), and the 2D-band (approximately 2,700 $cm^{-1}$).

Figure 14 displays the UV-vis spectra of photoelectrodes with different structures before and after they were loaded with dye. Clearly, the photoelectrode with the $TiO_2$/graphene/$TiO_2$ sandwich structure has a higher absorption than those with the traditional structure both before and after loading with dye.

Figure 15 presents the energy level diagram of the DSSC with the $TiO_2$/graphene/$TiO_2$ sandwich structure. Under illumination, electrons from the photoexcited dye are transported to the conduction band (CB) of $TiO_2$ via the CB of the graphene and $TiO_2$. The transportation path via the CB of graphene is in addition to the traditional path. Owing to the excellent electrical conduction of the graphene, the graphene layer bridges behave as a channel for transferring electrons and rapidly transport the photoexcited electrons. The graphene is homogeneous throughout the system, and the excited electrons are captured by the graphene without any obstruction. The collected electrons can be rapidly and effectively transported to the CB of $TiO_2$ through graphene bridges. In the interface of graphene and $TiO_2$, the resistance through which charges are transported is reduced relative to the DSSC without graphene electron transport layer, and the recombination and back-reaction processes are suppressed.

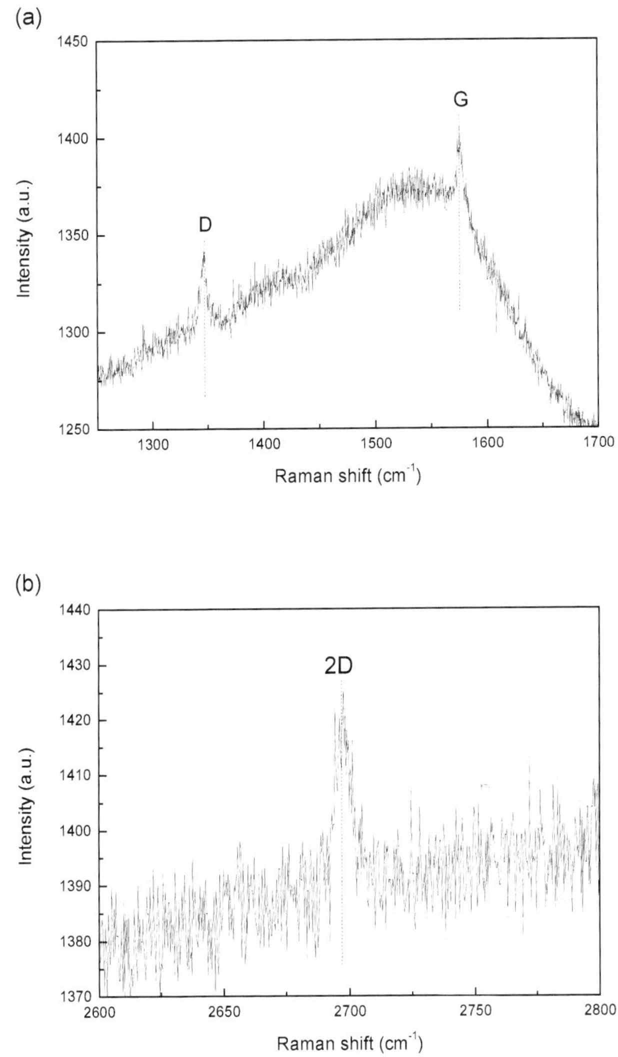


**Figure 13.** Raman scattering spectra of graphene film deposited on glass substrate. The spectra include important peaks that correspond to the D-band (1,350 $cm^{-1}$), the G-band (1,580 $cm^{-1}$), and the 2D-band (2,700 $cm^{-1}$) [17]

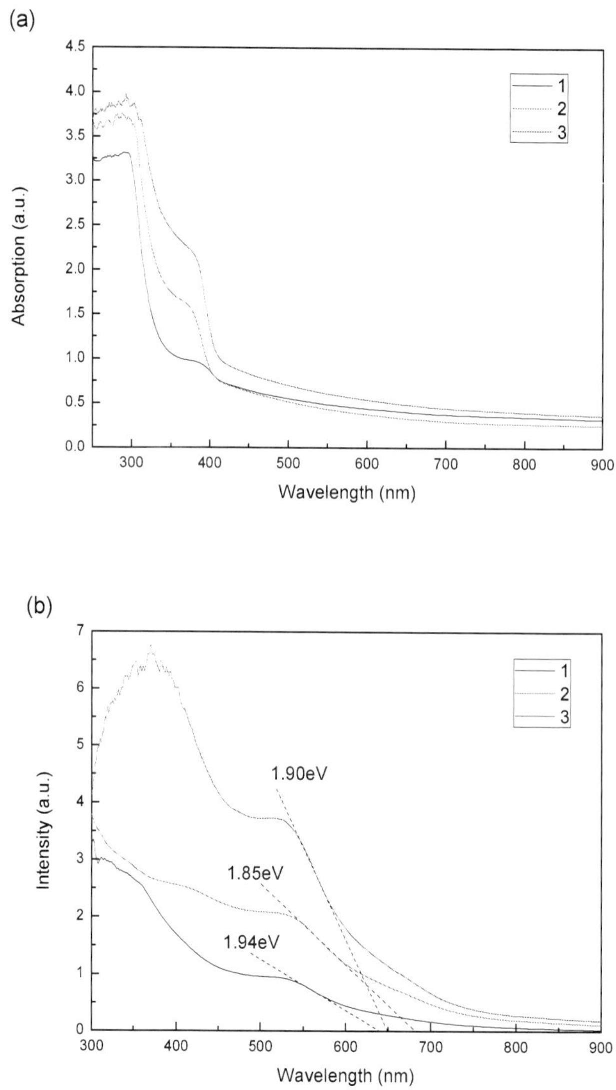


**Figure 14.** UV-vis absorption spectra of DSSCs with different structures (a) before and (b) after dye loading [17]

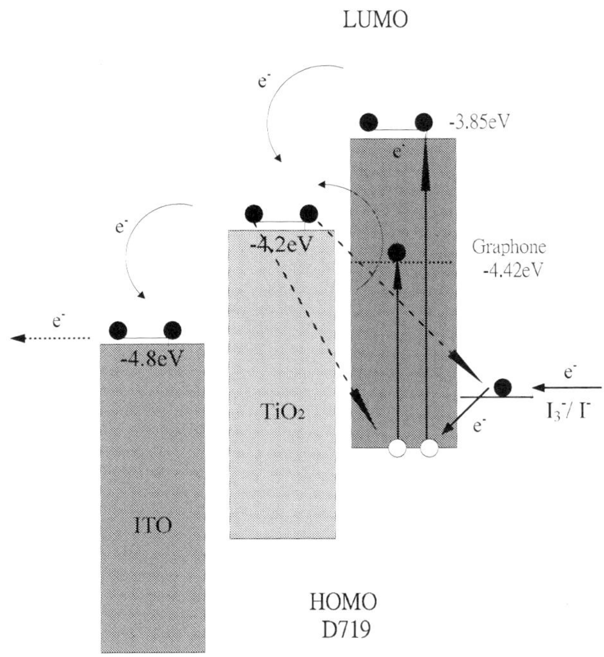


**Figure 15.** Energy level diagram and mechanism of photocurrent generation in DSSCs with $TiO_2$ /graphene/$TiO_2$ sandwich structure [17]

## 6. Conclusion

To enhance the performance of DSSCs, this work used nanostructure graphene electron transfer layer by plating or sputtering and compared the difference between the DSSC structure with graphene/$TiO_2$ and with $TiO_2$/graphene/$TiO_2$. From the I-V curves, sputtered graphene is much better than plated graphene because the plated graphene has a scattered distribution of ITO. The enhanced performance of DSSCs with a graphene may be attributed to the increase in electron transport efficiency and light absorption in visible range, especially in the range of 310–400 nm. Therefore, the efficiency of conversion of solar energy with graphene+$TiO_2$ to electricity were increased from 1.45 % to 3.98 %, and the efficiency of conversion of solar energy with $TiO_2$/graphene/$TiO_2$ sandwich structure to electricity was increased from 1.38 % to 3.93 %, respectively, under simulated full-sun illumination. This improvement in performance is associated with an increase in the absorption of light, a wide range of absorption wavelengths, shorter charge transportation distances, and the suppression of charge recombination when the graphene is applied.

## Author details

Lung-Chien Chen

Address all correspondence to: ocean@ntut.edu.tw

Department of Electro-optical Engineering, National Taipei University of Technology, Taipei, Taiwan

## References

[1] T. Bora, H. H. Kyaw, S. Sarkar, S. K. Pal, and J. Dutta, "Highly efficient ZnO/Au Schottky barrier dye-sensitized solar cells: role of gold nanoparticles on the charge-transfer process," Beilstein J Nanotech, vol. 2, no. 1, 681–690, 2011.

[2] C. Hagglund, M. Zach, and B. Kasemo, "Enhanced charge carrier generation in dye sensitized solar cells by nanoparticle plasmons," Appl. Phys. Lett., vol. 92, no. 1, 2008.

[3] S. Barazzouk and S. Hotchandani, "Enhanced charge separation in chlorophyll a solar cell by gold nanoparticles," J. Appl. Phys., vol. 96, no. 12, 7744–7746, 2004.

[4] S. W. Tong, C. F. Zhang, C. Y. Jiang et al., "Improvement in the hole collection of polymer solar cells by utilizing gold nanoparticle buffer layer," Chem. Phys. Lett., vol. 453, no. 1–3, 73–76, 2008.

[5] L. C. Chen, C. C. Wang, and B. S. Tseng, "Enhancement in nano-crystalline TiO2 solar cells sensitized with ZnPc by nanoparticles, " J Optoelectron Biome.,vol.1, no.3, 249–254, 2009.

[6] X. Du, I. Skachko, A. Barker, and E. Y. Andrei, "Approaching ballistic transport in suspended grapheme," Nat Nanotechnol., vol. 3, no. 8, 491–495, 2008.

[7] R. R. Nair, P. Blake, A. N. Grigorenko et al., "Fine structure constant defines visual transparency of graphene," Science, vol. 320, no. 5881, 1308, 2008.

[8] X. Wang, L. Zhi, N. Tsao, Z. Tomovic, J. Li, and K. M. Mullen, "Transparent carbon films as electrodes in organic solar cells," Angew Chem. Int. Edit., vol. 47, no. 16, 2990–2992, 2008.

[9] X. Wang, L. Zhi, and K. Mullen, "Transparent, conductive graphene electrodes for dye-sensitized solar cells," Nano Lett., vol. 8, no. 1, 323–327, 2008.

[10] S. Bae, H. Kim, Y. Lee et al., "Roll-to-roll production of 30-inch graphene films for transparent electrodes," Nat. Nanotechnol, vol. 5, no. 8, 574–578, 2010.

[11] D. W. Chang, H. J. Choi, A. Filer, and J. B. Baek, "Graphene in photovoltaic applications: Organic photovoltaic cells (OPVs) and dye-sensitized solar cells (DSSCs)", J. Mater. Chem. A, vol. 2,12136–12149, 2014.

[12] C. H. Hsu, J. R. Wu, L. C. Chen, P. S. Chan, and C. C. Chen, "Enhanced performance of dye-sensitized solar cells with nanostructure graphene electron transfer layer," Adv. Mater. Sci. Eng, 107352, 2014.

[13] H. Zhang, W. Wang, H. Liu, R. Wang, Y. Chen, and Z. Wang, "Effects of TiO2 film thickness on photovoltaic properties of dye-sensitized solar cell and its enhanced performance by graphene combination," Mater. Res. Bull., vol. 49, 126–131, 2014.

[14] H. N. Kim, H. Yoo, and J. H. Moon, "Graphene-embedded 3D TiO2 inverse opal electrodes for highly efficient dye-sensitized solar cells: Morphological characteristics and photocurrent enhancement," Nanoscale, vol. 5, 4200–4204, 2013.

[15] C. Yang, H. Bi, D. Wan, F. Huang, X. Xie, and M. Jiang, "Direct PECVD growth of vertically erected graphene walls on dielectric substrates as excellent multifunctional electrodes," J. Mater. Chem. A, vol. 1, 770–775, 2013.

[16] B. Tang and G. Hu, "Two kinds of graphene-based composites for photoanode applying in dye-sensitized solar cell," J. Power Source, vol. 220,95–102, 2012.

[17] L. C. Chen, C. H, Hsu, P. S. Chan, X. Zhang, and C. J. Huang, "Improving the performance of dye-sensitized solar cells with TiO2/graphene/TiO2 sandwich structure," Nanoscale Res. Lett., vol. 9, 380, 2014.

[18] M. Koshino and T. Ando, "Electronic structures and optical absorption of multilayer graphenes," Solid State Commun., vol. 149, no. 27-28, 1123–1127, 2009.

[19] G. Giovannetti, P. A. Khomyakov, G. Brocks, V. M. Karpan, J. van den Brink, and P. J. Kelly, "Doping graphene with metal contacts," Phys. Rev. Lett., vol. 101, no. 2, 2008.

[20] R. Czerw, B. Foley, D. Tekleab, A. Rubio, P. M. Ajayan, and D. L. Carroll, "Substrate-interface interactions between carbon nanotubes and the supporting substrate," Phys. Rev. Lett. B, vol. 66, 2002.

[21] A. Kongkanand, R. Martinez-Dominguez, and P. V. Kamat, "Single wall carbon nanotube scaffolds for photoelectron chemical solar cells. Capture and transport of photogenerated electrons," Nano. Lett., vol. 7, no. 3, 676–680, 2007.

[22] F. Xu, J. Chen, and X. Wuetal, "Graphene scaffolds enhanced photogenerated electron transport in ZnO photoanodes for high- efficiency dye-sensitized solar cells," J Phys. Chem. C, vol. 117, 8619–8627, 2013.

[23] A. C. Ferrari, J. C. Meyer, V. Scardaci, C. Casiraghi, M Lazzeri, F. Mauri, S. Piscanec, D. Jiang, K. S. Novoselov, S. Roth, and A. K. Geim, "Raman spectrum of graphene and graphene layers," Phys. Rev. Lett., vol. 97, 187401, 2006.

[24] X. L. Fang, M. Y. Li, K. M. Guo, Y. D. Zhu, Z. Q. Hu, X. L. Liu, B. L. Chen, and X. Z. Zhao, "Photoelectrodes modification by N doping for dye-sensitized solar cells," Electrochim. Acta 65, 174–178, 2012.

[25] X. L. Fang, M. Y. Li, K. M. Guo, X. L. Liu, Y. Zhu, B. Sebo, and X. H. Zhao, "Graphene-compositing optimization of the properties of dye-sensitized solar cells Solar," Energy, vol. 101, 176–181, March 2014.

[26] Y. Y. Wang, Z. H. Ni, T. Yu, Z. X. Shen, H. M. Wang, Y. H. Wu, W. Chen, and A. T. S. Wee, "Raman studies of monolayer graphene," The Substrate Effect J. Phys. Chem. C, vol. 112, no. 29, 10637–10640, 2008.

[27] J. N. Zheng, S. S. Li, F. Y. Chen, N. Bao, A. J. Wang, J. R. Hen, and J. J. Feng, "Facile synthesis of platinum–ruthenium nanodendrites supported on reduced graphene oxide with enhanced electrocatalytic properties," J Power Sources, vol. 266, 259–267, 2014.

[28] AZoM, "Characterization of graphene using Raman spectroscopy," http://www.azom.com/article.aspx?ArticleID=6271.

[29] M. Huang, H. Yan, T. F. Heinz, and J. Hone, "Probing strain-induced electronic structure change in graphene by Raman spectroscopy," Nano Lett., vol. 10, 4074–4079, 2010.

[30] O. Frank, M. Mohr, J. Maultzsch, C. Thomsen, I. Riaz, R. Jalil et al. "Raman 2D-band splitting in graphene," Theory Expt ACS Nano., vol. 5, no. 3, 2231–2239, 2011.

[31] D. Yoon, Y. W. Son, and H. Cheong, "Strain-dependent splitting of the double-resonance Raman scattering band in graphene," Phys. Rev. Lett., vol. 106, no. 15, 155502-1–4, 2011.

Chapter 6

# Long Wavelength VCSELs and VCSEL-Based Processing of Microwave Signals

M. E. Belkin, L. Belkin, A. Loparev, A. S. Sigov and V. Iakovlev

Additional information is available at the end of the chapter

http://dx.doi.org/10.5772/60480

**Abstract**

We address the challenge of decreasing the size, cost and power consumption for practical applications of next generation microwave photonics systems by using long-wavelength vertical cavity surface emitting lasers. Several demonstrations of new concepts of microwave photonics devices are presented and discussed.

**Keywords:** Vertical cavity surface emitting laser, microwave photonics, optoelectronic processing

## 1. Introduction

Microwave photonics (MWP) is a relatively new branch of science and technology gradually penetrating into telecom (ultra-high-speed optical fiber systems, fiber-to-wireless (FiWi) access networks) and defense (phased-array antenna radars, electronic warfare (EW) systems) industries. Progress in developing MWP solutions is mainly following the progress in low cost and high power, efficient optical sources [1]. Among other semiconductor laser technologies, long-wavelength vertical cavity surface emitting lasers (LW-VCSELs) are regarded as an enabler for superior features in high-performance photonic circuits [2], and have the potential to do so for MWP as well [3]. In fact, the application of LW-VCSELs emitting in the telecom wavelength range is expected to benefit from the availability of low cost components already developed in the framework of information and communication technology (ICT) [2]. However, as compared with other applications of VCSELs (datacom, sensing, etc.), their use in

MWP requires paying further attention to their spectral purity, noise, single-mode operation, and linearity features [4]. Specifically, the possibility of wavelength division multiplexing (WDM) and the low sensitivity to electromagnetic perturbations of optical fiber-based transmission allow the introduction of novel concepts into on-board or ground-based modules for antenna's remote control, signal distribution and processing of broad-bandwidth analogue signals. For example, the development of electrical-to-optical converters is in fact a continuous challenge for lower noise and higher linearity, which would be fully compatible with the required dynamics in phased-array antennas. Therefore, for the next generations of telecom and radar microwave-to-fiber systems, it is imperative to find solutions for increasing the spurious-free dynamic range (SFDR) of photonic devices and links. In addition, the penetration of optoelectronic technologies into modern radar systems has also been realized through the optical implementation of a large number of vital functions like optical beam-forming, adaptive filtering or microwave signal analog-to-digital conversion [5, 6].

In view of these trends, the engineering and cost benefits of VCSELs make them increasingly attractive for optical fiber–based applications [7] in general and for microwave photonics in particular. That is why there is an increasing interest to evaluate the potential of LW-VCSELs for microwave photonics application areas [4].

Starting from 1990s, the development of LW-VCSEL technology was benefitting from what is known today as the 'telecom bubble', with the main motivation coming from high-bandwidth (mainly metropolitan) optical networks.

It was expected that LW-VCSEL technology will follow the success story of short-wavelength spectral range ($\lambda$ ~ 800–1000 nm) VCSEL technology. It turns out that the main difficulty of LW-VCSELs technology is the need of having in a monolithic semiconductor device the active materials providing high optical gain and mirrors with high-reflectivity, low optical absorption and high thermal conductivity. The best semiconductor materials for these tasks are, from one side, InP/InAlGaAs quantum wells (InP based) and, from another side, GaAs/AlGaAs DBRs (GaAs based). Unfortunately, these semiconductor structures are mutually incompatible from the epitaxial growth point of view, even virtually impossible to grow such VCSELs structure in all-epitaxial growth fabrication approach. This difficulty has triggered many attempts to develop LW-VCSELs to overcome this materials problem without yielding the expected results. As a result, the progress in development of long-wavelength VCSEL technology was much slower than initially expected.

As of today, we see progress in LW-VCSELs development and industrialization towards meeting the requirements of previously identified applications as well as toward the new emerging application areas [8]. This progress is supported only by a few technologies, which have proven to yield LW-VCSEL with acceptable performance [2]. It turns out, that so called wafer-fused LW-VCSEL technology, that is employing strained InP/InAlGaAs quantum well active regions, tunnel junction and GaAs/AlGaAs distributed Bragg reflectors for generation of photons, carrier and optical confinement and for optical feedback respectively is the only technology that have reached the industrial production stage and proven reliability [2, 8]. A particular advantage of these industrially fabricated LW-VCSELs is in covering the full ITU-T spectral range from O-band to U-band. Concerning MWP, an important feature of LW-

VCSELs is their adaptability with future silicon-based photonic integrated circuits, which should provide many advantages when implementing microwave photonic devices and techniques. Investigations of VCSEL-based MWP devices enabled to find and exploit unique features of LW-VCSELs, for example, excellent compatibility with optical injection locking techniques to enhance key dynamic features (modulation bandwidth, extinction ratio, SFDR) [9]. A specific benefit of LW-VCSELs is the suitability for silicon large-scale integrated technology where it could effectively be used as source and modulator for optical interconnects [10].

In this chapter we will describe benefits of long wavelength wafer-fused VCSEL compared with other types of optical sources, double fused LW-VCSEL design and fabrication, procedure and results of Telcordia-grade reliability testing, metrology and characteristics of the devices, VCSEL-based devices processing of microwave signals and future trends in long wavelength VCSEL photonics.

## 2. Benefits of Long Wavelength Wafer-Fused VCSEL

The ongoing success of VCSELs is particularly due to the intrinsic advantages of this laser type as compared to the edge-emitting laser diodes. The most important features of VCSELs are their low beam divergence leading to relaxed fiber alignment tolerances, the small threshold currents and high slope efficiencies leading to low electrical power consumption, and their potential for integration to 1D and 2D laser arrays. Additionally, VCSELs are usually longitudinal single-mode [11].

All the short wavelength VCSEL features hold for the VCSELs operating at long wavelength range beyond 1.3 μm. In addition, for geometrical reasons, the longer wavelengths make the transverse mode and polarization control easier so that true single-mode devices with a stable polarization may be achieved even for rather large current aperture diameters of around 5–7 μm [12, 13]. Accordingly, long-wavelength VCSELs have a tremendous meaning for a wide variety of applications, ranging from short to long range optical communications, parallel data transmission as well as optical measurement and gas sensing. In many of these applications, therefore, long-wavelength VCSELs turn out to be cost-effective and superior performance substitutes for conventional Fabry-Perot or distributed feedback (DFB) lasers [2, 10]. As compared with edge emitting technology, the main advantage is the fact that devices can be fully tested on wafer, without the need to form laser resonator by cleaving. In fact VCSEL technology is similar to LED technology that is yielding light emitting devices with the quality of beam much superior as compared with the beam of edge emitting laser.

For ICT, for example, in [14] it was demonstrated an order of magnitude better efficiency for a VCSEL in comparison to that for a DFB laser for radio frequency (RF) to optical power conversion. The performance of optical link comprising wired and wireless circuits with direct modulation have confirmed the viability of VCSELs as power-efficient optical sources. According to the results of detailed investigation of VCSEL-based optoelectronic frequency converter and novel sub-harmonic frequency multiplicator for the circuitry of RoF's systems [15], the potential of long wavelength VCSEL based MWP devices for the application in future equipment for ultra-wide band telecom and radar systems and in measurement techniques is demonstrated.

## 3. Design and fabrication.

As stated above, the development of LW-VCSELs has converged to only a few technology solutions for fabrication of devices with performances matching and even surpassing state of the art short-wavelength VCSEL. Three main common elements of these few technology solutions should be outlined. First, the strained InP/InAlGaAs quantum wells for high optical gain at high temperatures. Second, GaAs/AlGaAs or dielectric distributed Bragg reflectors (DBR) for high mirror reflectivity. Third, the tunnel junctions for intra-cavity contacting to reduce mirror optical loss. In Table 1 [2] it is presented a summary of the three main approaches that are currently being used in LW-VCSEL design, as well as the resulting device performance.

| Aperture type | Mirror design | SM $P_{max}$, 20 °C | SM $P_{max}$, 80 °C |
|---|---|---|---|
| Undercut QWs | Top and bottom as-grown DBRs: InAlGaAs/InAlAs, or AlGaAsSb | 1.6 mW | 0.5 mW @ 70 |
| Buried TJ | One as-grown InAlGaAs /InAlAs DBR; one dielectric DBR | 4.3 mW | 1.4 mW |
| Regrown TJ | Both wafer-fused AlGaAs/GaAs DBRs | 6 mW | 2.5 mW |
| SM $P_{max}$ 20 (80 )°C, single-mode maximum power output at 20 °C (80 °C); TJ- tunnel junction; QW-quantum well | | | |

**Table 1.** Performances of LW-VCSELs fabricated by three main approaches.

In the approach based on undercut quantum wells (QW) all laser cavity parts (DBRs and QWs) are fabricated epitaxially in InAlGaAs/InAlAs(InP) material system, and are grown in a single epitaxial run. In this material system it is possible to define device aperture by undercut selective chemical etching of a part of the QWs. The best devices fabricated following this approach have demonstrated quite low single-mode output of 0.5 mW at 70°C, that is rather low, mainly because of too poor thermal dissipation from the active region both in the lateral direction (through air gap) and in the vertical direction (through quaternary layers of DBR), as thermal conductivity of the InAlGaAs layers is much lower as compared with AlGaAs layers in GaAs based DBRs, normally used in the well-established short-wavelength VCSEL technology.

Further progress was made possible by introducing buried tunnel junction (TJ) for current and optical localization and a dielectric DBR. Even though the thermal dissipation is still inefficient in the vertical direction, it is considerably improved in the lateral direction because of the relatively good heat transfer through the regrown InP. This improvement enables the device to reach a single-mode output power of about 1.4 mW at 80 °C.

It turns out that the above mentioned two approaches do not show good performance devices in O-band (1300 nm wavelength range), especially at the shorter wavelengths of 1270 and 1290

http://dx.doi.org/10.5772/60480

nm, as compared with the performances in C-band (1550 nm waveband). On top of poor thermal characteristics, the optical properties of InP based DBRs for this wavelength suffer from the low refractive index-contrast of the 1310-nm wavelength DBR that needs to be set lower to exclude the edge-band absorption in InAlGaAs quarter-wavelength layers.

The best performances of long-wavelength VCSELs (single-mode output powers as high as 6 mW at room temperature and 2.5 mW at 80°C) are demonstrated with devices that have a regrown TJ aperture on the InAlGaAs/InP active cavity and AlGaAs/GaAs DBRs attached to this cavity using wafer fusion (Figure 1a). The results comes with enhanced thermal dissipation from the active region in both lateral (through InP spacers) and vertical directions (through GaAs based DBRs) as well as with high reflectivity and thermal conductivity of the AlGaAs/GaAs DBRs. With such performances VCSELs may be used in both single- and multi-channel applications. Wafer fusion for LW-VCSEL fabrication was pioneered by the group at the University of California in Santa Barbara in 1995, but without regrown tunnel junction (current localization were performed by introducing an oxide aperture in the same way as for 850-nm wavelength VCSELs).

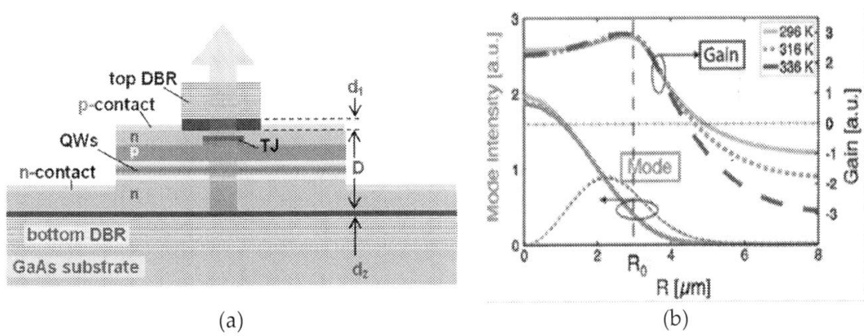


**Figure 1.** (a) Schematic sectional presentation of the wafer-fused VCSEL; (b) Radial gain and optical field distributions of a device with 3um TJ mesa radius ($R_0$)

VCSEL's wafer fabrication is schematically depicted in Figure 2. Below is the description of the details of double fused LW-VCSEL design and fabrication process as described in [8]. The VCSEL has an InP-based 5/2λ-active cavity. Un-doped AlGaAs/GaAs DBRs are wafer fused on both side of this cavity, as schematically depicted in Figure 1a.

Four to six compressively strained quantum wells are inserted in this InAlGaAs/InP active region, as well as a p++/n++ InAlGaAs tunnel junction. All epitaxial material for the double fused wafers is grown by low pressure metal-organic vapor phase deposition (MOVPE) on 2-inch (100) wafers. The mesa-structures of 3-3.5 μm radius is formed in the tunnel junction layer after first growth run, and then is re-grown with n-type InP layer. This regrown structure serves for carrier and photon confinement. The top and bottom intra-cavity n-InP layers are used for electrical contacting that allows using un-doped top and bottom DBR mirrors. It is important to note the possibility to introduce InGaAsP cavity adjustment layers in the InP based epi structure. Such layers on both sides of the active cavity help for precise adjustment of the emission wavelength. Numerical simulations of such a structure [16] indicates (Figure 1b), that in normal operation conditions in the device, operating predominately in the fundamental mode, the gain and mode profiles spread outside the active region defined by the tunnel junction mesa by about 1–1.5 μm.

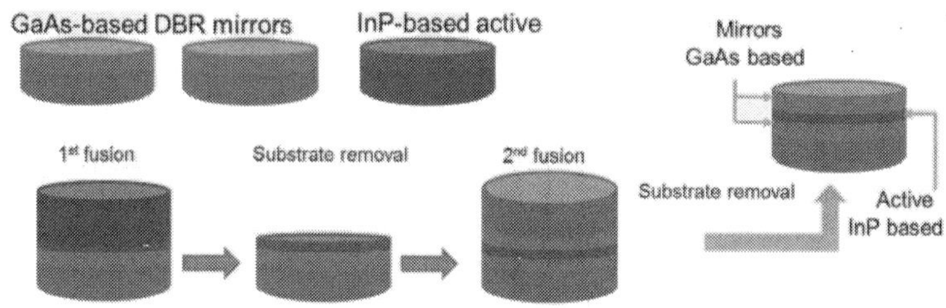


**Figure 2.** Fabrication of wafer-fused VCSEL

The details of fabrication process are presented in Figure 3, starting from mesas etched in the tunnel junction (a), then after re-growth (b). One can see that before first fusion step (c), and after the first fusion step (d) the surface is not flat. After re-growth the mesas have elliptical shape and a size that exceeds 2–2.5 times the initially round shaped mesas. This slight elongation of the mesas and planarization is occurring during re-growth by MOVPE. The elliptical shape of the re-grown mesa is a strong factor for discrimination of polarization modes.

The oblique line in Figure 3c depicts the border between two regions on the wafer with different cavity lengths. In wafer fusion process it is possible to have different cavity lengths on the same wafer. For this purpose, before the first fusion process, the InGaAsP cavity adjustment layer is selectively etched.

For bonding, the 2-inch wafers of InP-based active region and GaAs-based DBR are brought into contact. The parameters of the process are: vacuum better than $10^{-5}$ mBars, 600°C and a force of 7000 N for 30 min. The process is performed in an industrial custom-built wafer-bonding machine. At these process conditions, both wafers undergo a slight plastic deformation and a uniform contact on a nanometer scale is achieved and the strong covalent bonds are formed between atoms at interface of fused wafers. During cooling to room temperature, the wafer is bowed with a radius of curvature of about 1m due to thermal expansion coefficients mismatch between GaAs (lattice parameter of 0.5653 nm, thermal expansion coefficient $5.8\times10^{-6}$ $K^{-1}$) and InP (lattice parameter of 0.5868 nm and thermal expansion coefficient $4.8\times10^{-6}$ $K^{-1}$).

The selective etching the InP substrate is the releasing the strain that was bowing the fused stack, the remaining GaAs substrate with thin InP based layered active region re-gains its planarity. The second DBR is then bonded to the InP-based active cavity in the second fusion process under the same conditions as during the first fusion. After the second fusion the fused stack is no longer bowed. SEM and TEM micrographs of the double-fused VCSEL heterostructure are depicted in Figure 4. High resolution SEM and TEM images indicate that the

http://dx.doi.org/10.5772/60480

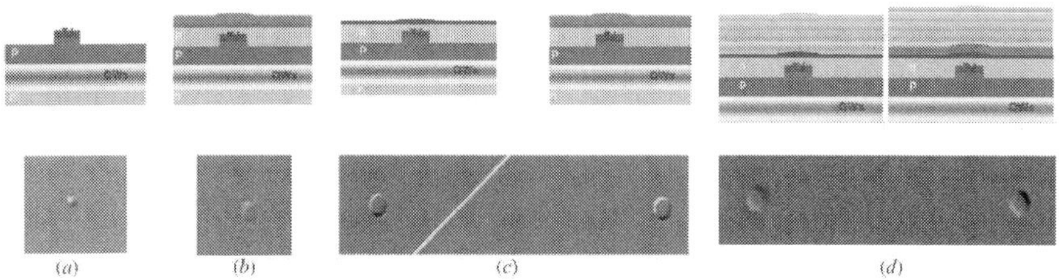


**Figure 3.** Images of TJ mesas: (a) initially etched 7μm mesas, (b) re-grown mesas, (c) before fusion, (d) after fusion.

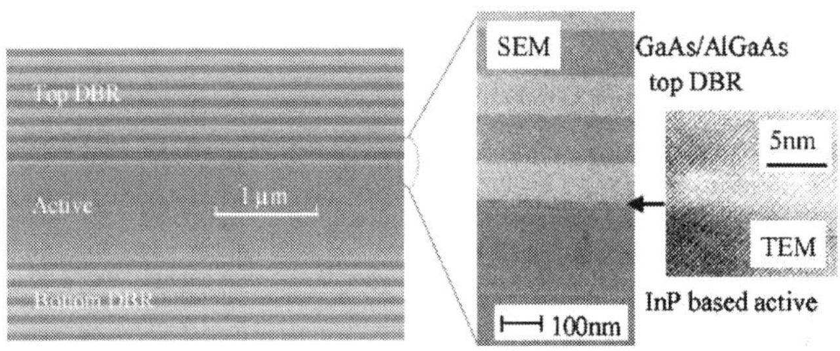


**Figure 4.** SEM and TEM micrographs of the double-fused VCSEL wafer.

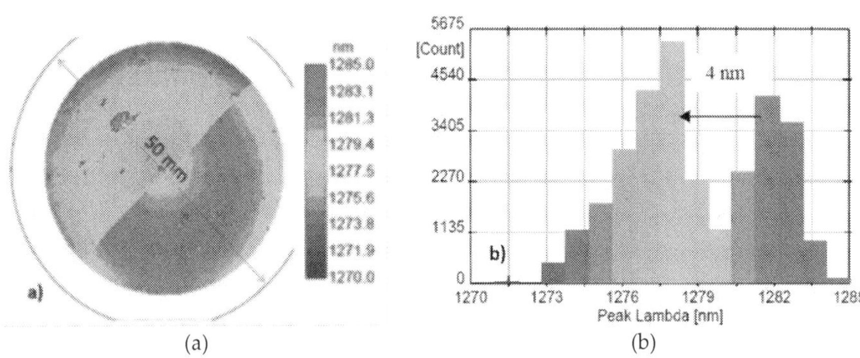


**Figure 5.** a) Map of a wafer. Cavity adjustment was performed on half of the wafer; b) Histograms of the cavity wavelength across the wafer shown in Figure 2. a)

misfit dislocations resulting from the lattice mismatch of GaAs and InP are well confined only at the fused interface and do not propagate inside the VCSEL structure. As presented in [17], the high quality of bonding process was attested by reliability tests of fabricated devices. Figure 5, a, depicts the cavity length distribution obtained from photoluminescence emission spectra mapping on the double-fused wafer under optical pumping with a 980-nm laser pump spot of 5 μm diameter with 0.1 mm step. One can observe two regions with different emission wavelengths resulting from the cavity engineering performed before the first fusion, as described above. In this way one can perform wavelength control in fabrication of wafer fused VCSELs by applying a nanometer precision cavity adjustment. In more details the procedure and results of cavity length adjustment is presented in [18] and is described shortly below.

The strategy for increasing the yield of wavelength setting in wafer-fused VCSELs consists in working with batches of identical wafers (at least 3 batches, see Figure 2) of InP-based active cavity material and DBR. After fabricating the first double-fused VCSEL wafer and measuring the emission wavelength distribution, it is possible to introduce corrective actions when fabricating the next wafers for improving the emission wavelength setting yield. For the situations when the emission wavelength distribution is shifted to longer wavelengths relative to the requested wavelength slot, a nanometer precision cavity trimming technique [18] was developed. According to this technique, several oxidation-desoxidation calibrated steps (that removes one nanometer at one step) were performed in order to decrease the effective cavity length. Figure 5a depicts the cavity mode wavelength mapping of a VCSEL wafer on which half wafer has undergone 4 of such cavity etching steps before fusing to the bottom DBR. As one can see from Figure 5b, the distribution peak of cavity wavelength on the adjusted part of the wafer is shifted by 4 nm relative to the cavity distribution on the non-etched part.

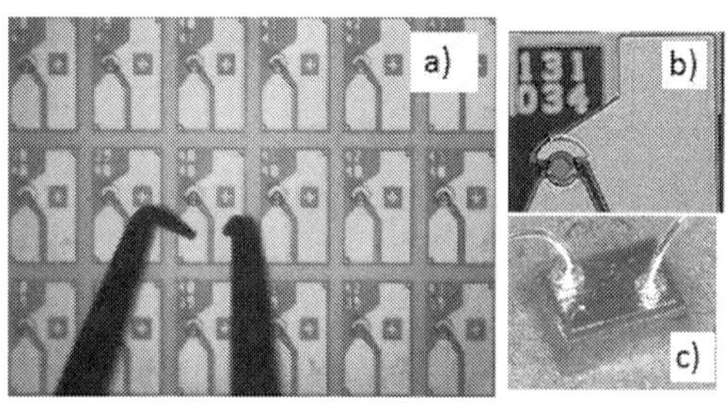


**Figure 6.** a) Microscope picture of the VCSEL wafer under test, b) VCSEL chip with electroplated contact pads, c) 250x350μm VCSEL chip mounted on header.

The processing of the double-fused VCSEL wafer includes (i) reactive ion etching of the top DBR, (ii) selective chemical etching steps in the InAlGaAs/InP active cavity region, (iii) dielectric deposition, (iv) dry etching steps and (v) e-beam deposition of metals for contacts. The processed VCSEL wafer (Figure 6) offers on-wafer testing possibility that decrease manufacturing cost as compared to edge emitting lasers. Full wafer electroplating for bond pads allows depositing thick metal that is a benefit for wire bonding.

The industrialization of above presented VCSEL technology was started for wavelength division multiplied (WDM) optical fiber communication applications. In ref. [19], it was demonstrated that devices with single-mode emission power in excess of 1.2mW in the temperature range of 20-80°C at operating voltage below 2.5 V, single-mode emission with 40-

http://dx.doi.org/10.5772/60480

dB side mode suppression ratio are in compliance with the requirements of the 10 GBASE-LX4 IEEE.802.3ae standards. The progress in 1300 nm wavelength wafer-fused VCSELs was backed by the progress in 1500 nm wafer-fused VCSELs [20]. The recent status of wafer-fusion VCSEL technology [21], as compared with its status presented earlier [2], is a result of the dramatic improvement of the quality of the final double fused-wafer as well as of the processing steps that allows a successful demonstration of their reliability.

In Figure 7 [22] are presented examples of the VCSEL chip SEM cross section overview (a) and high resolution SEM of TJ region.

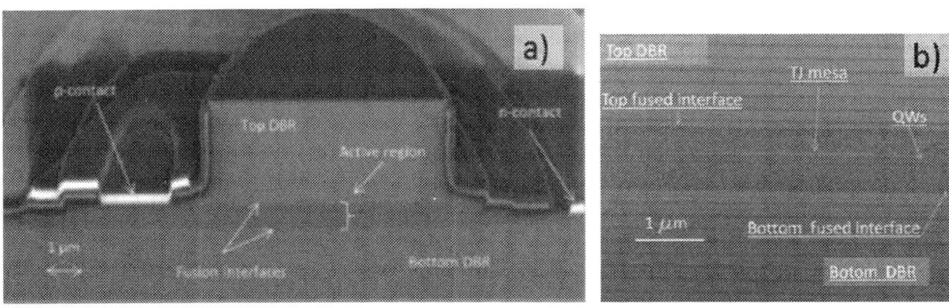


**Figure 7.** SEM of VCSEL chip cross section overview (a) and high resolution SEM of TJ region (b)

## 4. Telcordia-grade reliability testing

For the reliability tests, VCSELs were assembled onto hermetically sealed standard headers. A burn-in (BI) procedure has been designed and applied in order to exclude any early period failure. Reliability tests were performed according to the GR-468-CORE Telcordia Generic Reliability Assurance Requirements for Optoelectronic Devices [23]. Overall, the tests have demonstrated that wafer-fused VCSELs behave like any other all-grown laser devices. The emission power in time of a qualification lot of 11 devices at 10 mA and ambient temperature of 90°C that correspond to a junction temperature close to 120°C and to a current density through the tunnel junction of 26 kA/cm$^2$ is depicted in Figure 8. One cannot see any visible degradation at these operating conditions during 5000 hours of operation. Four more groups of devices from the same VCSEL wafer were tested for accelerated wear-out at higher values of junction temperature and current densities, as shown in Figure 8. The performance of devices was periodically tested at room temperature. Figure 9 depicts evolution in time of emission power at 9 mA (left graph) and threshold current (right graph) of a group of 8 devices under aging test at 10 mA and temperature 150°C. Under these test conditions one can observe devices with small changes (for example, devices 32, 35, 37) in output power and threshold current and devices that show more pronounced gradual degradation, like devices 30, 31, 33 and 36. Based on the statistical distributions of the totality of accelerated lifetime test (ALT) data, VCSEL aging parameters were calculated: activation energy value of $E_a$=0.67 eV and current exponent factor N=3.93. The value of activation energy of 0.67 eV is close to 0.79 eV - the activation energy of 1550 nm VCSELs with active region in the same (InAlGaAs/InP) material system grown by MOVPE [24], even though devices with undercut apertures investigated in [24] do not exhibit gradual wear-out in time during ALT tests. Predicted operation lifetimes are calculated based on the accelerated life-time data and applying the acceleration factors and the pass-fail condition of 2-dB change. The maximum driving current for VCSELs is set 9 mA which is quite sufficient for high-speed modulation at 8-9 mA bias current and the maximum ambient operating temperature is set to 70°C according to telecom industry requirements.

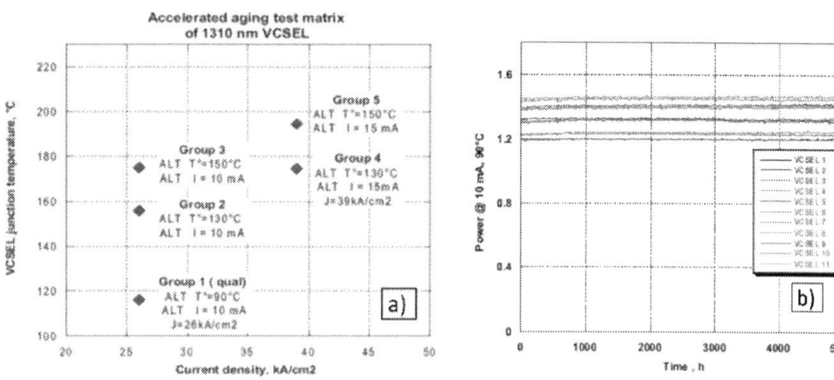


**Figure 8.** a) Accelerated aging test matrix and b) Emission power in time of 11 devices operating at 10 mA driving current and ambient temperature of 90°C

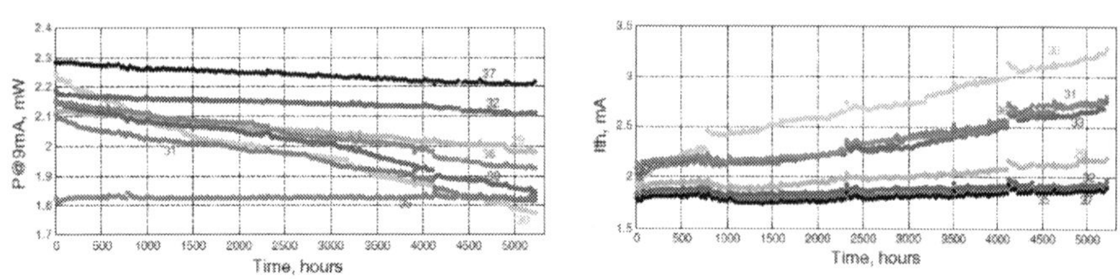


**Figure 9.** left: Evolution in time of emission power at 9 mA ; right: Threshold current of 8 devices under ALT test at 10 mA and temperature 150°C.

Figure 10 shows the scaled data from all ALT conditions on lognormal plots. The extracted time to 1%-failure is of 291 years at 25°C and 18 years at 70°C. With decreasing the driving currents to 8 mA and 7 mA the time to 1%-failure at 70°C increases to 30 years and 50 years respectively. These lifetime values meet the telecom industry requirements for the time to 1%-failure of more than 10 years at 70°C. In real-life applications actual lifetimes are expected to be longer since the devices spend most of the time in less demanding operating conditions. In addition, the wafer fused VCSELs have successfully passed all mechanical and electrical Telcordia qualification tests.

http://dx.doi.org/10.5772/60480

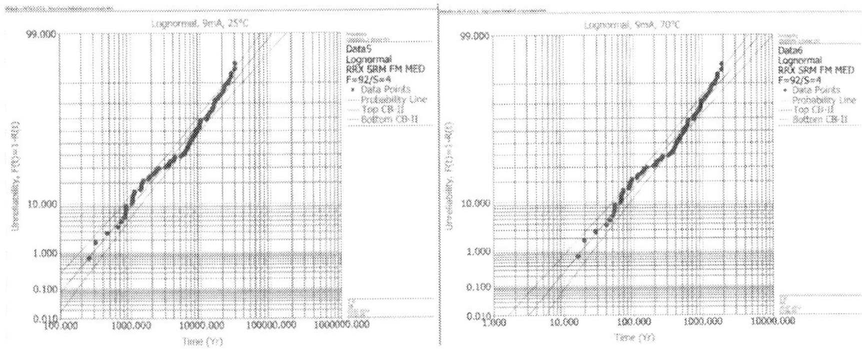


Figure 10. Lognormal plots of lifetimes (left: 25°C, right: 70°C) from the totality of test conditions scaled to an operating condition of 9 mA.

## 5. LW-VCSEL characterization

Below we will present typical DC and dynamic characteristics of the wafer-fused LW-VCSELs used further for processing of microwave signals. The VCSELs considered here have a device structure and performances similar to those reported in [8, 25, 26] and references therein.

### 5.1. Light – current characteristics

Figure 11 shows the typical light-current characteristics of the wafer fused LW-VCSELs emitting in the fundamental mode in the temperature range from 0 to 110°C. For the devices under test, the room-temperature threshold current varied from near 2 mA at 1.3 μm to less than 4 mA in 1.55-μm spectral band, and a quasi-linear dependence of optical power vs. current was observed up to 5 mW at 20°C and up to 3 mW at 70°C. Besides, room-temperature electric power consumption in quasi-linear mode is as low as 20 mW at 1.3 μm and 15 mW in 1.55-μm spectral band.

### 5.2. Spectral characteristics

In Figure 12 it is presented the spectral evolution of the 1560-nm VCSEL emission with current and temperature. A remarkable single-mode behavior at any of these operation conditions is observed (see also the inset of Figure 11b showing the emission spectra at room temperature and current 15 mA). As it can be easily observed, right combinations of temperature and driving current allow adjustment of the emission wavelength at any value between 1562 nm and 1572 nm. On a full wafer, VCSEL inventories of specified wavelengths can thus be selected in a quite wide spectral range of 40 nm (1550 nm to 1590 nm). A typical VCSEL spectral line shape for a 1565 nm, at 20°C and emission power 2mW is depicted in Figure 12b. As one can see, a full width at half maximum (FWHM) is as narrow as 4.5 MHz that allows VCSEL to be used effectively even in coherent fiber-optic systems.

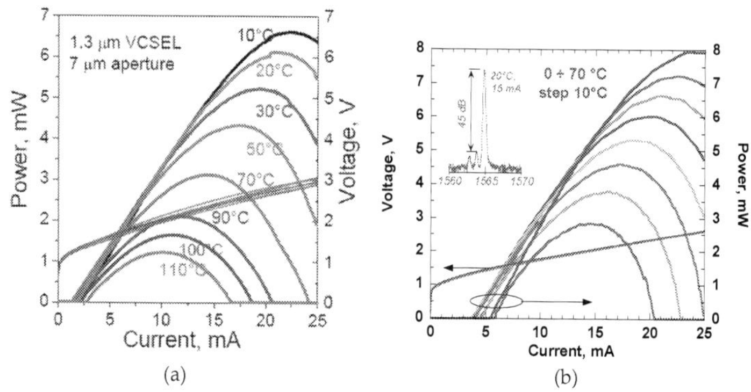


Figure 11. Light-current and volt-current characteristics of the VCSELs emitting at 1.3 μm (a) and at 1.55 μm (b)

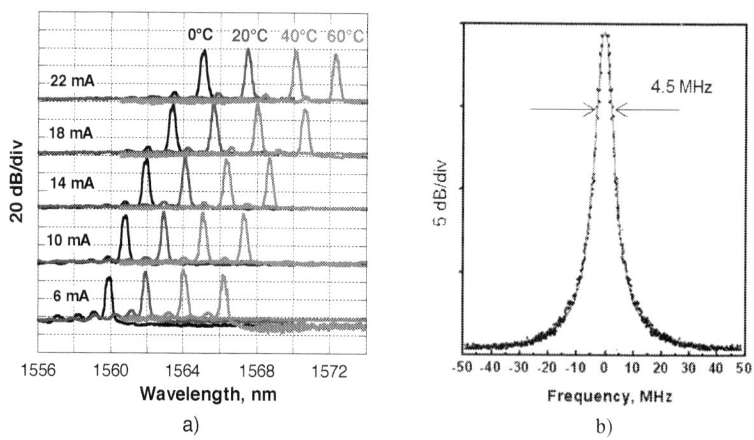


Figure 12. (a) Emission spectra of a 1560 nm VCSEL; (b) line shape of a 1565 nm VCSEL at 20°C and 2 mW emission power.

### 5.3. Relative intensity noise (RIN)

Figure 13 summarize the measured RIN-current characteristics of the LW-VCSELs tested at 1 and 6 GHz for the ambient temperatures of 20 and 70°C. Clearly, the RIN values decrease with increasing current and increase with increasing temperature and modulation frequency, as expected. Besides, at room temperature the RIN is as low as -160 dB/Hz (noise floor of the test setup employed) when the DC current of the laser under test is only 4.7 mA, at a modulation frequency of 1 GHz.

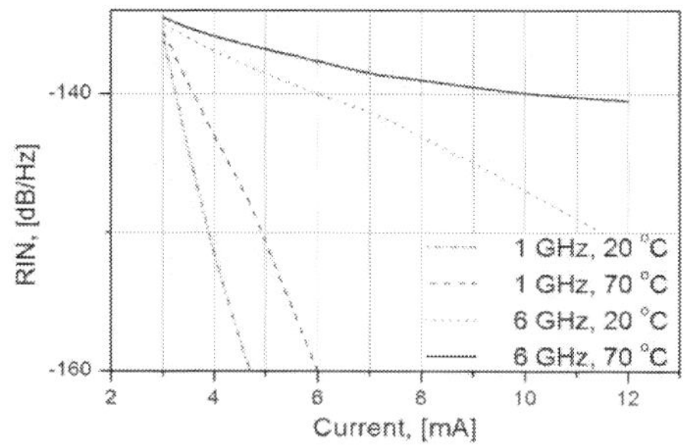


**Figure 13.** RIN characteristics of the wafer-fused LW-VCSELs

## 5.4. Small signal modulation

Figure 14 presents the small-signal transmission gain (TG) of an optoelectronic pair comprising the LW-VCSEL under test and a reference photodiode (40 GHz bandwidth, 0.6 A/W responsivity). As one can see, at lower modulation frequencies the TG value is 30 dB and the -3dB bandwidth of the LW-VCSEL under test is more than 9 GHz at 10-mA bias current.

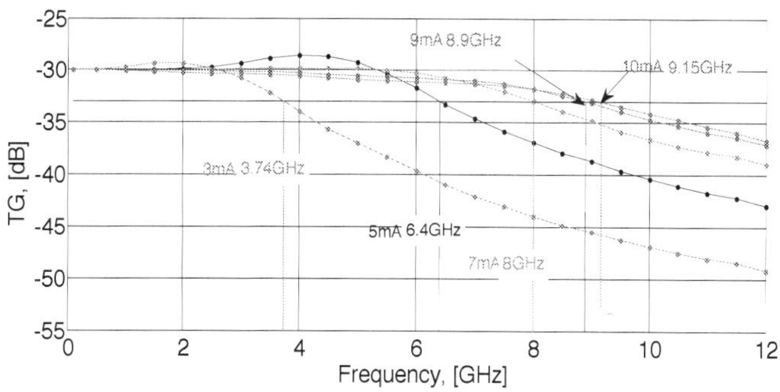


**Figure 14.** Small-signal transmission gain characteristics of wafer-fused LW-VCSELs

## 5.5. Large signal modulation

As well known, the most obvious way to assess the linear properties of an active device, in this case a semiconductor laser, is to determine its input intercept point (IIP) [27]. This parameter is most accurately determined graphically by comparing the amplitude characteristics of the laser fundamental tone and the intermodulation distortion (IMD) of corresponding order. The advantage of this parameter is the ability to compare different devices, regardless of the power of the modulating signal $P_i$. Simplified assessment of the third-order and fifth-order IIP (IIP3 and IIP5) values can be carried out using the following formulas:

$$IIP3 = P_i + IMD3/2 \tag{1}$$

$$IIP5 = P_i + IMD5/4 \tag{2}$$

Figure 15 depicts a typical example of IMD measurements of the tested LW-VCSELs obtained by the procedure described in [28]. Based on the measurement data and the above formulas, calculated data of *IIP3* and *IIP5* for the LW-VCSEL under test in the modulation frequency slots of 1 GHz and 6 GHz are listed in Table 2. The power of each of the input signals is $P_i$ = 8dBm in the first frequency slot and $P_i$ = 6dBm in the latter one.

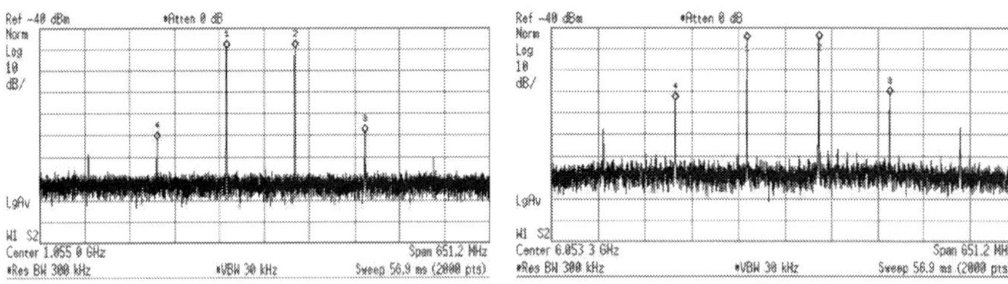


**Figure 15.** Measured intermodulation distortions of the wafer-fused LW-VCSELs at bias current of 9 mA in the vicinity of 1 GHz (a) and 6 GHz (b). Markers 1, 2 are the fundamental tones; markers 3, 4 are the corresponding IMD3 tones; the two extreme right and left peaks are the IMD5 tones.

| Frequency [GHz] | IIP3 [dBm] | IIP5 [dBm] |
|---|---|---|
| 1 | 26 | 20 |
| 6 | 20 | 17 |

**Table 2.** Results of calculation of IIP3 and IIP5

It follows from Table 1, that the level of *IIP*, and, therefore, the linearity of the LW-VCSELs under test, decreases with increasing modulation frequency, which agrees with other reported simulations and experimental results [28].

As mentioned above, a general figure of merit for multichannel analog optical fiber links is the SFDR that quantifies the combined effects of noise and nonlinear distortions. As shown in [29], third-order SFDR

$$SFDR = \frac{2}{3}\left[IIP3 - EIN - 10\log\left(BW\right)\right] \tag{3}$$

where *BW* is the system bandwidth and *EIN* is the equivalent input noise. Considering *RIN* as a predominant noise source,

http://dx.doi.org/10.5772/60480

$$EIN = RIN - TG \tag{4}$$

where $TG$ is the transmission gain of the laser-photodiode pair (see Figure 14). The third-order SFDR is usually expressed in dB•Hz$^{2/3}$. In accordance with Figure 13, for the bias current of 10 mA that is still inside the quasi-linear regime of the light-current characteristic (Figure 11), the room temperature RIN values of the LW-VCSEL under test are below -160 dB/Hz at 1 GHz, and near -148 dB/Hz at 6 GHz. Referring to Figure 14, we find TG=-30 dB for both modulation frequencies. Using the above data and formulas, we calculate the third-order SFDR as 104 dB•Hz$^{2/3}$ for 1 GHz and 93 dB•Hz$^{2/3}$ for 6 GHz. These values are within the requirements of radio-over-fiber applications [29] in the bandwidth of 1-6 GHz, in accordance with 3GPP standard (GSM, WiFi, Bluetooth, WiMAX).

## 6. VCSEL-based microwave signal processing

In this section we demonstrate several examples of the application of wafer-fused LW-VCSELs for microwave analog signal processing that we have explored using VCSELs from a batch with parameters and characteristics similar to those presented in previous section.

### 6.1. Optoelectronic microwave signal oscillator

An optoelectronic oscillator (OEO) studied in this section is the most valuable example of microwave photonics breakthrough. As a matter of fact, OEO opens up new horizons of highly stable RF oscillators in a frequency range from 100s MHz to 100s GHz [30]. OEO has a key advantage of higher spectral purity in comparison with traditional RF and microwave oscillators [31].

The state of the art for OEOs is currently considered to be a single-frequency (within the X-band) OEO product from OEwaves Inc. [32] with a class leading phase noise level of -163 dBc/Hz at a 10 kHz offset from the carrier. The enhanced spectral purity of an OEO's central mode occurs due to its well-known feature of improving with delay time, and very long delays can be easily achieved with a nearly lossless optical fiber line extending for several kilometers. At the other hand, OEO is able to combine successfully higher spectral purity and extremely wide frequency range of carrier tuning within the limits of some octaves [33] that is impossible for traditional microwave transistor oscillators.

Up-to-date in the most publications devoted to OEO a circuit arrangement with unmodulated laser source and feedback on Mach-Zehnder external modulator has been studied. But recently a version combined the functions of optical emitter and feedback control in one vertical cavity surface emitting laser (VCSEL) preferred by potential low cost, power consumption, and integrability has been also investigated [34]. Thus, there exists a need for detailed comparison of the above OEO design approaches. Following this, below spectral and phase noise characteristics of two microwave-band OEO layouts based on an unmodulated distributed feedback (DFB) laser and an external Mach-Zehnder intensity modulator (MZM), or based on a direct modulated VCSEL will be researched and compared.

### 6.1.1. OE-CAD modeling

The typical layout of an OEO scheme with external modulator [35] is presented in Figure 16a. The OEO consists of two principal sections: optical and electrical. The optical section includes semiconductor laser module (SLM), electro-optical modulator (EOM), optical fiber line (OFL) and photodetector module (PDM). To ensure a reliable operation of the OEO with low laser noise levels and low loss for the optical section an optical isolator (OI) and polarization controller (PC) are introduced into the scheme. The electrical section includes a low-noise RF amplifier (LNA), tunable band-pass filter (BPF), power amplifier (PA), and electrical coupler (EC). The operating principle of the OEO is based on the conversion of the continuous optical radiation energy emitted from the SLM into periodical discrete energy bursts in the RF-band. To achieve this, the EOM is controlled by a positive-feedback optoelectronic loop (see Figure 16a). A current OEO carrier frequency in the RF band is determined by the BPF's band pass, while the overall energy storage time of OEO depends on the fiber delay. The phase and amplitude balance for self-sustained oscillations can be ensured by managing the fiber length and gain of the LNA and PA for a given circuit. The electrical section of the VCSEL-based OEO being simulated (Figure 16b) is similar to the same one of OEO with external modulator. But in optical section VCSEL is modulated directly by injection current from EC so the layout is far simpler.

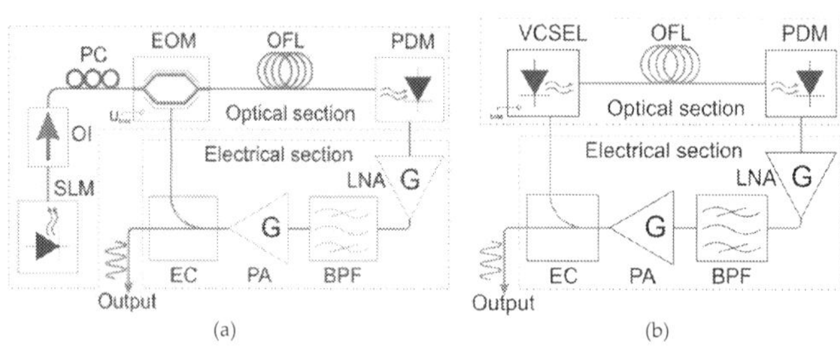


**Figure 16.** External EOM-based (a) and LW-VCSEL-based (b) OEO layouts

Previously we worked out in detail optoelectronic computer-added design (OE-CAD) based OEO model [33] by VPI System's VPI Transmission Maker software tool [36], which due to its self-excitation and large-signal steady-state operation modes allows simulating spectral and phase-noise characteristics. A single-loop EOM-based and VCSEL-based OEO computerized models are shown in Figure 17a and 17b respectively. In both cases the simulation is performed with aperiodic boundary conditions that allow concurrent simulation of the RF and optical elements of the OEO in object-oriented environment. Note that the library VCSEL model based on the rate equations does not work correctly in OEO's divergent oscillations regime, so we substituted it for a combination of equivalent laser model and ideal optical intensity modulator.

In the layouts of Figure 17 all the elements of Figure 16 have a specific interpretation. For example, the SLM is represented by single-mode rate equations-based DFB laser model, the EOM is based on a differential Mach-Zehnder interferometer model, OFL is constructed by a combination of an optical attenuator and a delay line. The spectrum of the OEO output signal

http://dx.doi.org/10.5772/60480

extracts by a spectrum analyzer library model. The specific schematic realization for simulating in VPI Transmission Maker OEO's phase noise characteristics in the bandwidth of 100 Hz at offsets of 10 kHz, 100 kHz and 1 MHz from the carrier, is presented in Figure 18.

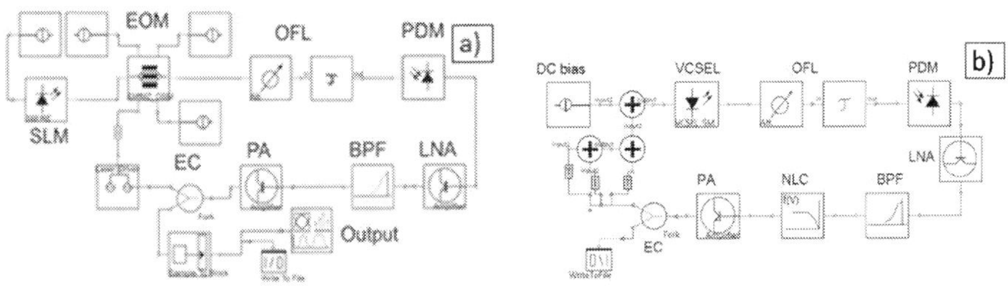


**Figure 17.** EOM-based (a) and VCSEL-based (b) OEO models in VPI Transmission Maker OE-CAD tool

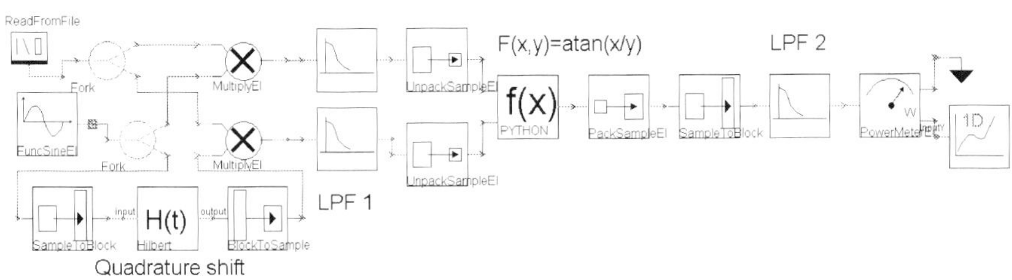


**Figure 18.** Arrangement for simulating OEO's phase noise measurement in VPI Transmission Maker

The simulation results of spectral and phase noise (in the bandwidth of 100 Hz) characteristics of EOM-based (a, b) and VCSEL-based (c, d) OEOs at operation frequency near 3 GHz and 65 m of OFL length are showed in Figure 19. As one can see, the both versions of OEO show comparable output powers and side-mode suppression ratios but the values of phase noise spectral density of the VCSEL-based OEO are somewhat higher.

### *6.1.2. Experimental verification*

Experimental verification of the simulated results was performed using a prototype of the layout presented in Figure 16b. The elements employed in the optical section were: the tested wafer-fused LW-VCSEL and a pin-photodiode XPDV3120R (wavelength range 1300-1620 nm, responsivity 0.6 A/W, 3-dB bandwidth 75 GHz) from u2t Photonics, Inc. as the PDM. For the electrical section, a tunable band pass YIG-filter (tuning range 2.5-15 GHz, insertion loss 5 dB, 3-dB bandwidth 11 MHz at 2.5 GHz), from Magneton, Inc., was used as the BPF. Besides, a set of two microwave amplifiers (total gain of near 50 dB, noise figure 3.5 dB, frequency band 2.5-8 GHz) was employed.

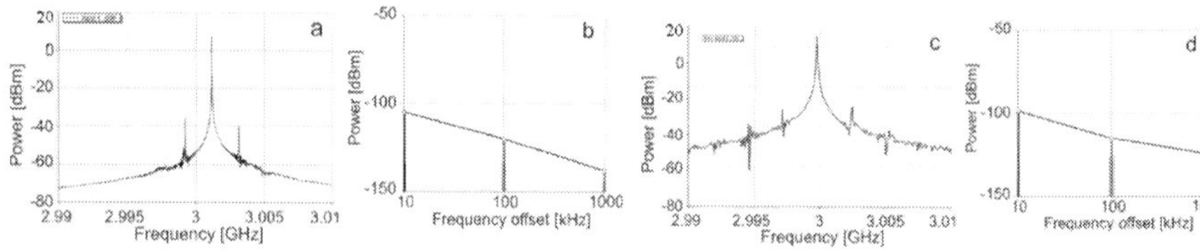


Figure 19. EOM-based OEO spectral (a) and phase noise (b) characteristics and VCSEL-based OEO spectral (c) and phase noise (d) characteristics simulated in VPI Transmission Maker OE-CAD

The examples of the experimental results of EOM-based OEO spectrum and phase noise characteristics at the oscillation frequency of 3 GHz are presented in Figures 20a and 20b [37]. In addition, the VCSEL-based OEO spectrum and phase noise characteristics at the same oscillation frequency are showed in Figure 20c and 20d.

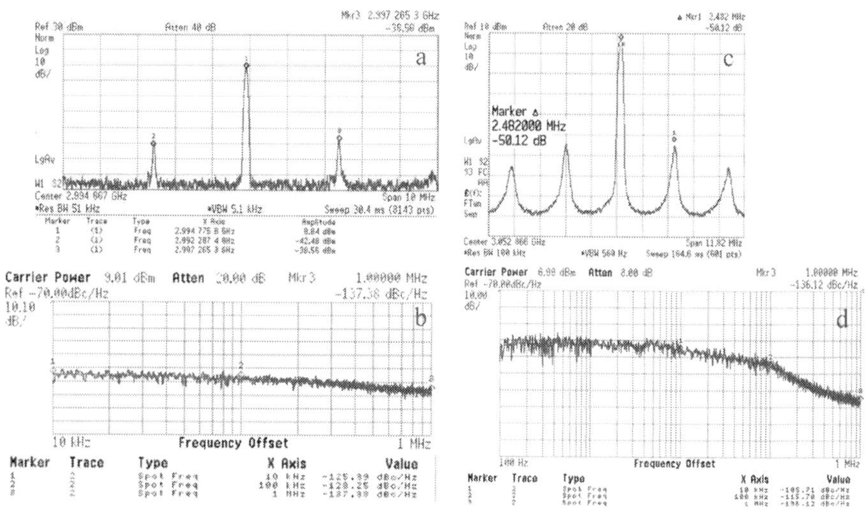


Figure 20. Measured EOM-based OEO spectrum (a) and phase noise (b) characteristics and VCSEL-based OEO spectrum (c) and phase noise (d) characteristics.

Experimental results represented in Figure 20 are closely matched with the simulation data of Figure 19 and confirm the feasibility of the proposed OEO models in VPI Transmission Maker software tool.

Table 3 presents a comparison of the VCSEL-based OEO and the one based on a distributed feedback (DFB) laser and external Mach-Zehnder intensity modulator [34, 37]. As one can see, the LW-VCSEL-based OEO exhibits similar parameters as compared to a typical EOM-based layout (except the phase noise near the carrier frequency), but also offers advantages in terms of potential integrability, low cost and low power consumption. We believe that the most probable cause of higher phase noise of the VCSEL-based OEO is the increased relative intensity noise of the directly modulated VCSEL versus the externally modulated DFB laser. Note that a fully electronic microwave counterpart to the proposed optoelectronic solution is a phase lock loop (PLL) synthesizer. To validate the benefits of the OEO design presented here, Table 2 also lists a brief technical comparison of the OEO prototypes with IC ADF4350, an advanced wideband synthesizer with integrated voltage-controlled oscillator from Analog Devices, Inc. It is clear from the table that the LW-VCSEL-based OEO prototype provides a remarkably wide tuning range (1.5 octaves) with a comparable output power, much stronger parasitic suppression and a 14 dB lower phase noise at 10-kHz offset from the carrier.

Thus, simulating and experimental investigation of two versions of the microwave-band optoelectronic oscillator reveals that directly modulated VCSEL-based OEO, which offers a number of obvious advantages such as integrability, low cost and power consumption, has commensurable parameters compared to a typical external modulator-based layout excluding near-to-carrier phase noise. We predict that most probable cause of this defect lies in an increased radiation noise of VCSEL vs. DFB used in typical OEO layout. Another bottleneck of a VCSEL-based OEO might be the relatively limited tuning range that is referred to more narrow modulation bandwidth of the VCSEL (see Figure 14) in comparison with the same of the Mach-Zehnder modulator.

| | MZM – based OEO | **VCSEL - based OEO** | ADF350 (fundamental mode) |
|---|---|---|---|
| *Frequency, [GHz]* | 2.5-15 | 2.5 - 6 | 2.2 – 4.4 |
| *Output power, [dBm]* | 9 | 7 | 5 |
| *Side-mode suppression, [dB]* | 46.5 | 50 | 13 |
| *Phase noise at 3 GHz at Offsets, [dBc/Hz]* | | | |
| *10 kHz* | -125 | -106 | -92 |
| *100 kHz* | -128 | -116 | -111 |
| *1 MHz* | -138 | -136 | -134 |

**Table 3.** EOM-based vs. VCSEL-based OEO comparison

## 6.2. Optoelectronic frequency converter and multiplicator

The microwave-band frequency conversion is usually based on the non-linear conversion features of optoelectronic devices [38], employing direct modulation of a laser [39], external modulation of passive electro-optic modulator [40, 41], semiconductor optical amplifier [38], p-i-n photodetector [42], and a combination of these components [43]. High cost and low conversion efficiency are the common drawbacks of these devices as compared with electronic counterparts. We have analyzed the above publications and came to the conclusion that the most parametric/cost-effective technique in the case of fiber-wireless applications is direct intensity modulation of semiconductor laser, low cost emitter as LW-VCSEL, being the best candidate.

In this section, first, the details of simulation and test of VCSEL-based optoelectronic frequency mixing process are presented. Then, results of simulation and experiments of novel version of long wavelength VCSEL-based microwave frequency converter for the fiber-wireless (FiWi) base stations such as an optoelectronic sub-harmonic frequency multiplicator recently proposed [15] are presented as well.

### 6.2.1. *Optoelectronic microwave frequency mixer*

The layout of the optoelectronic frequency mixer (OE-FMX) under test realized by optical fiber means is drawn in Figure 21. Input signal (RF) and local oscillator (LO) signal are mixed in passive power combiner (PC) and directly modulate emission power of the VCSEL. An optical isolator (OI) in the layout was essential for the reliable operation of the OE-FMX with low laser noise level. Short fiber-optic patchcords, p-i-n photodiode (PD) and band-pass filter (BPF) are the main elements of optical path for optical emission (modulated by the complex RF+LO signals) propagation, detection and converted frequency output selection, respectively.

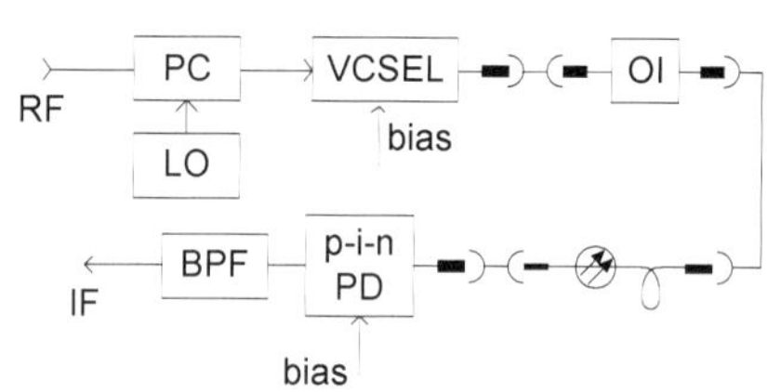


**Figure 21.** Layout of the long wavelength VCSEL-based optoelectronic frequency mixer under test

A modeling of the parameters of both devices characterized in the section was also performed by the same OE-CAD as for OEO. In this tool the library VCSEL model is addressed to single-mode rate equations and calibrated based on the measured light-current and small-signal modulating plots (see Figures 11, 14). A schematic layout of the computerized model is depicted in Figure 22 for the OE-FMX block diagram shown in Figure 21. For better visualization, the circuitry does not include a model of BPF with a central frequency of 2.5 GHz. Simulation details are reported in [15, 44].

The conversion efficiency (CE) is studied by simulation. In this case this is the difference (in dB) between IF signal power on the BPF (or PD) output and the RF signal power on the VCSEL's modulating input. Figure 23 presents the simulated (a) and PSA-measured (b) OE-FMX output spectra with up-conversion of L to S microwave bands. In both cases the input RF signal had a power of -20 dBm at 1 GHz and LO signal had a power of 6 dBm at 1.5 GHz.

At VCSEL DC bias current of 7 mA an output (IF) signal power near -55 dBm at a frequency of 2.5 GHz, i. e. CE of -35 dB was measured. The good fit of the experimental and simulation results confirm the feasibility of the proposed OE-FMX model.

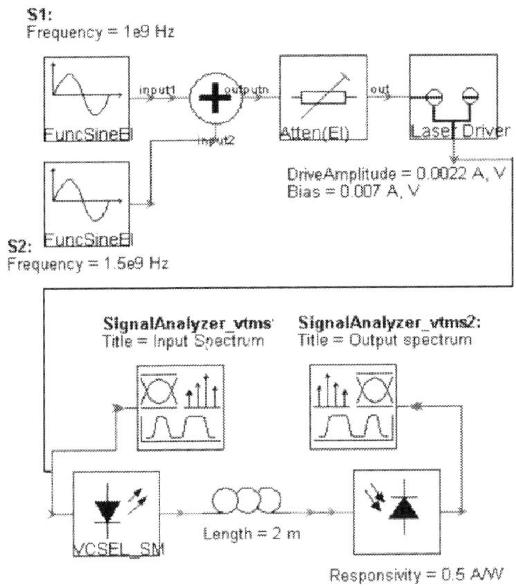


Figure 22. OE-CAD model of the OE-FMX under test

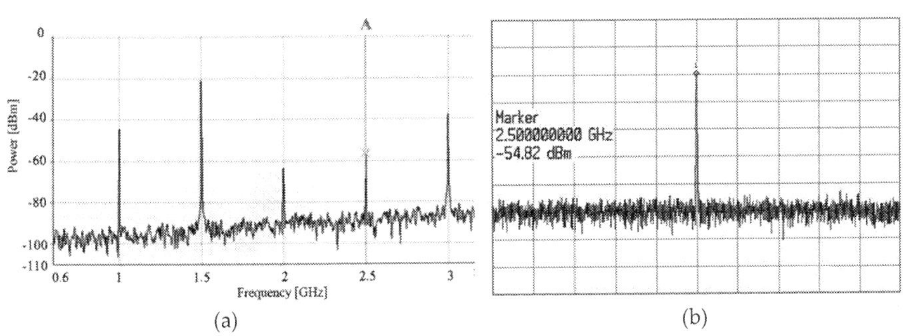


Figure 23. Simulated (a) and PSA-measured (b) OE-FMX output spectra with up-conversion of L to S microwave bands

It is clear that CE vs. frequency plot of this device is in line with the modulation characteristic of the tested VCSEL (see Figure 14). The results of prototyping the rest inherent characteristics of long wavelength VCSEL-based OE-FMX under test such as CE vs. input and local oscillator power, second-order and third-order IIP, and VSWR, are depicted in Figures 24-27.

The outcome of Figure 24 is that the input linearity threshold ($P_{-1dB}$) is near 2 dBm, that is comparable with standard transistor microwave mixer. Besides, Figure 25 shows the conversion efficiency to be remarkably local power-selective and its maximum value is near -33 dB at power as low as 3 dBm. To complete checking in accordance to traditional microwave mixers the tests of input second-, third-order intercept points (IIP2, IIP3) and voltage standing wave ratios (VSWR) for the OE-FMX prototypes are demonstrated (Figure 26). At the result, IMD values coincide exactly with the same for intrinsic VCSEL measurement (see Figure 15, Table 2), that makes clear that VCSEL is the most nonlinear element of OE-FMX layout. Additionally, VSWR values for RF and LO inputs and IF outputs (Figure 27) are lower than 1.7 within the bandwidth of 0.5-8 GHz.

### 6.2.2. *Optoelectronic microwave frequency multiplicator*

In ref. [15] we have proposed and investigated a simplified version of OE-FMX, the optoelectronic frequency multiplicator (OE-FMP) with the goal of achieving higher cost-efficiency of the VCSEL-based OE-FMX. Its block diagram is similar to that of Figure 16, but without power combiner and local oscillator. The principle of the approach is based on using the effect of a period doubling [45] under modulation of a semiconductor laser by a power RF sinusoidal signal. For this purpose, the laser should be modulated in super large-signal mode with an injection current cutoff similar to a class C regime in electronic amplifier. To reach this regime, the laser bias current has to be set in a near-threshold (but higher) area or in the onset of quasi-linear zone of the laser's light-current plot. In this regime inside the laser optical emission spectrum some sub-harmonics and the products of their mixing with the above tones are generated, in addition to fundamental frequency and its higher harmonic tones. In simulation we succeeded in identification of such a regime of semiconductor laser, where the signal levels of a fundamental modulation frequency F, a sub-harmonic tone 0.5F, and a component at 1.5F would be near equal and enough to secure the signal-to-noise ratio needed for wireless communication systems.

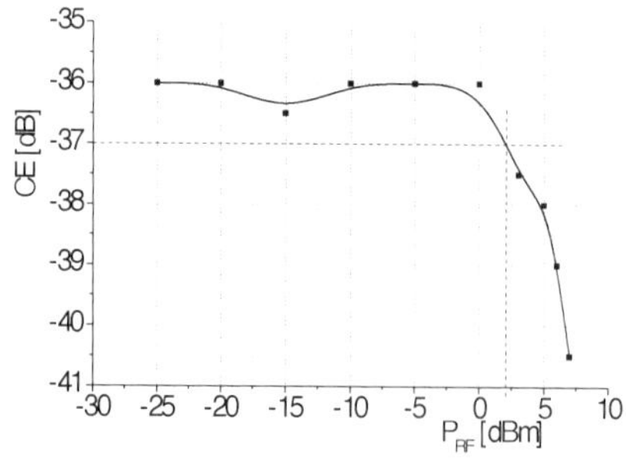


**Figure 24.** Conversion effectiveness vs. RF signal power ($P_{LO}$=6 dBm)

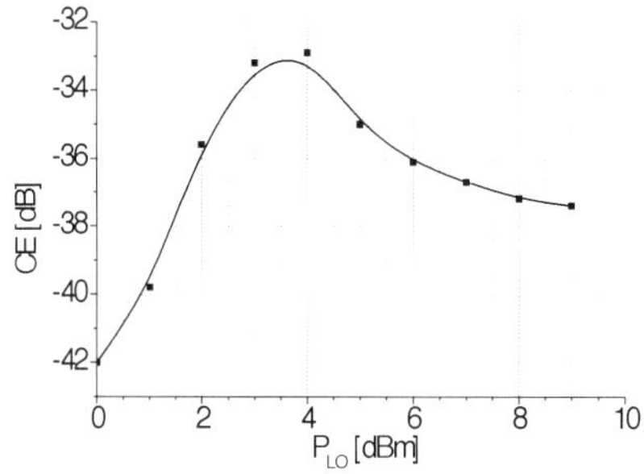


**Figure 25.** Conversion effectiveness vs. local oscillator power (Ps=-20 dBm)

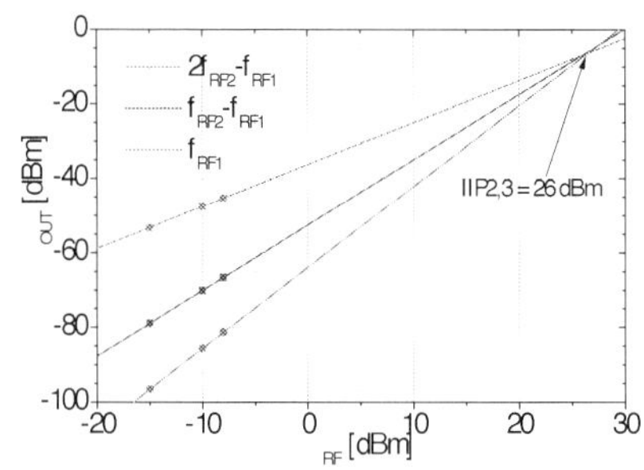


**Figure 26.** Second (IIP2) and third (IIP3) order input intercept points

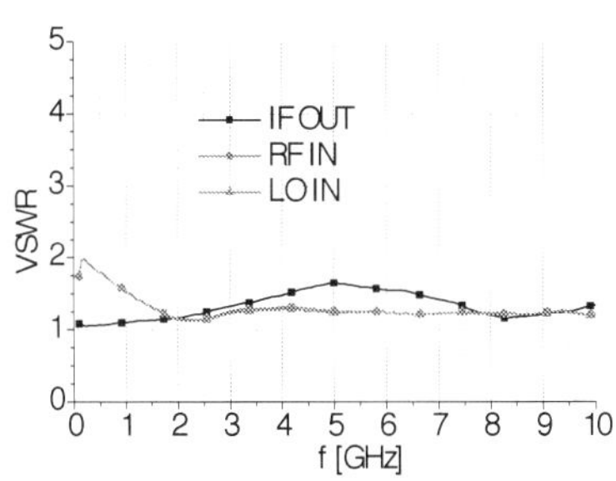


**Figure 27.** Input and output voltage standing-wave ratio (VSWR)

### *6.2.2.1. Formal-oriented VCSEL model in super large-signal mode*

As it is well-known, for example, see [46], a system of rate equations is used to address the dynamic operation of a single-mode MQW semiconductor laser, in our case, of a VCSEL. The main issue is in identification of the necessary operating range for the VCSEL under investigation, because the nonlinear rate equations do not always converge to the desired solution. To address this issue, a preliminary computation by MathCAD tool was performed, with a restrictive condition that the difference of fundamental and nearest sub-harmonic or mixed tones must be below 10 dB. The results are depicted in Figure 28. The area with horizontal hatching is of the operation area of VCSEL's effective functioning in the sub-harmonic multiplication regime.

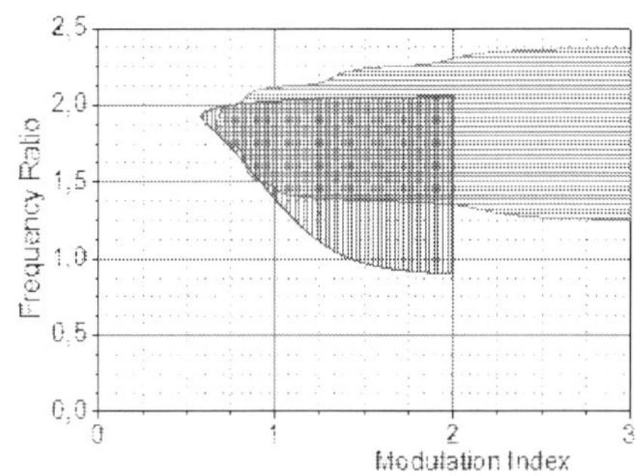


**Figure 28.** Diagram for the validating the VCSEL's effective functioning in the sub-harmonic multiplication regime

The terms "Frequency ratio" and "Modulation Index" (Figure 28) mean an input RF frequency normalized by the laser's small-signal resonance one, and AC modulation current amplitude normalized by the laser's DC bias current, respectively. As one can see, effective process of generating nearest to fundamental frequency sub-harmonic or mixed tones is obtained when a modulation frequency is higher than a laser's relaxation frequency (RFQ) and even higher than its 3-dB direct modulation bandwidth as well as when the current modulation index is more than 0.7; in good agreement with known data. The simulation results of the period doubling effect for an edge-emitted laser [45] (Figure 28, vertical hatching) show large overlapping of the areas with horizontal hatching, that confirms validity of the developed model.

*6.2.2.2. OE-FMP's object-oriented modeling, verification and example of realization in FiWi network*

Simulation data presented in Figure 28 became the basis for the modeling of optoelectronic microwave frequency multiplicator by the same OE-CAD tool as for OE-FMX. The developed object-oriented model is presented in Figure 29. All labels are the same as for Figure 22. A number of simulation experiments were performed on this model with a goal of choosing the optimal regime for each given device. The simulated (a) and PSA-measured (b) OE-FMP output spectra together with input microwave-band frequency are presented on Figure 30. For both procedures the input RF signal was kept at 0 dBm power level at a frequency F=3 GHz and VCSEL DC bias current was set at 3.2 mA. As one can see, the experimental results fit well with the simulation results thus validating the proposed OE-FMP model. The rest of the parameters of the OE-FMP prototype coincide with the same ones for the OE-FMX.

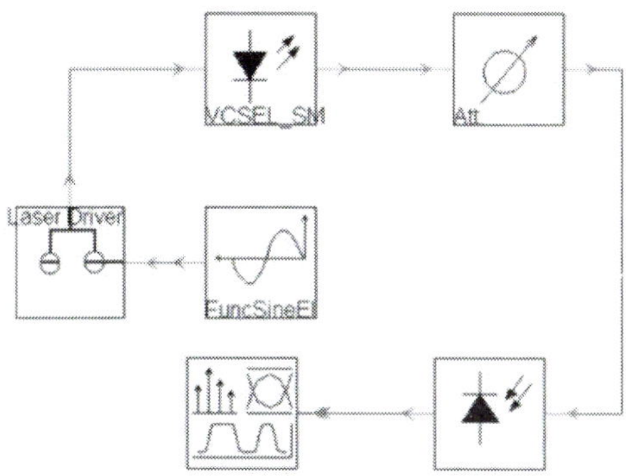


Figure 29. VPI transmission Maker™ object-oriented model of the OE-FMP

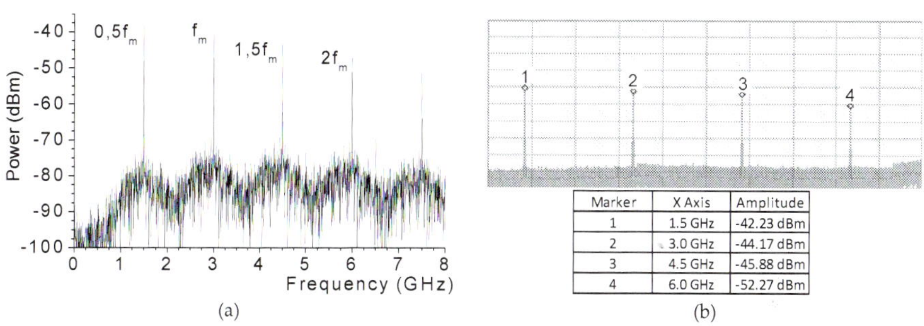


Figure 30. Simulated (a) and measured (b) output spectra of VCSEL-based OE-FMP.

Since a practical application of the developed OE-FMP is not as clear as a microwave-band optoelectronic frequency mixer, below a version of its effective realization in an advanced telecom system based on Radio-over-Fiber (RoF) technology is exemplified [47]. Hybrid Fiber-Wireless (FiWi) access network is a novel backhaul network architecture that integrates the next-generation WLAN-based wireless mesh network and Ethernet passive optical network (EPON). Its major feature lies in squeezing the cell diameters up to picocell (10s - 200 m) or femtocell (10s cm – 10s m) [48]. As a result the base station equipment must satisfy extremely rigid requirements related to the cost-effectiveness. To meet this issue we propose to use OE-FMP described beyond in the base station's uplink channel that is much simpler than the known methods using a microwave local oscillator or remote delivery of a reference oscillator signal [49]. Figure 31 shows an example of uplink circuitry realization by the OE-FMP in the popular X-band. The rest elements of the layout are: 1 - receiving antenna, 2 – sub-harmonically pumped mixer, 3 – IF filter, 4, 7 – amplifier, 5 – coupler, 6 – uplink laser, 8 – carrier recovery unit. The main benefits of the layout proposed are: (i) in realistic simplification of the base station arrangement without the concurrent complication of a central office architecture or remote delivery of oscillator signal as in known approaches; (ii) in improvement of a base station's cost features; (iii) in coarser demands to local oscillator frequency stability.

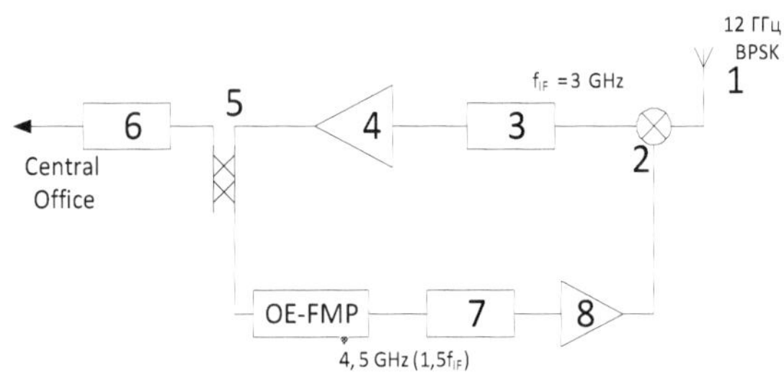


**Figure 31.** Example of OE-FMP application in uplink channel of FiWi network

Above, a detailed investigation of two low cost VCSEL-based microwave photonics devices: known optoelectronic frequency mixer and novel sub-harmonic frequency multiplicator for the circuitry of FiWi's base stations is presented. The main advantage of using optoelectronic approach in general and VCSELs in particular is in:

- Large bandwidth, that is limited by the bandwidths of laser and photodiode solely,
- Application versatility, the same block diagram for up-converter and down-converters
- Losses do not depend on a position of microwave frequencies inside the operation band, as compared with the inherent effect of increasing conversion loss with a super wide frequency of a microwave transistor mixer.
- Design simplicity (practically decoupled input and output),
- Lower power consumption (5-10-fold lower than edge-emitting laser), and
- Lower required output power of the local oscillator (only some dBm instead of 15-20 dBm for edge-emitting laser).

Table 4 lists a summary of the results referred to demonstrated LW-VCSEL-based devices for frequency converting microwave signals [44]. As seen, simulation and experimental values are rather close.

Altogether, the microwave photonics frequency conversion devices based on LW-VCSEL demonstrate a remarkable potential for the application in future equipment for ultra-wide band telecom and radar systems. (in phased-array antenna systems, measurement techniques, to name only a few). The work is in progress for decreasing conversion losses in these devices.

| | | Operating bandwidth [GHz] | LO power [dBm] | Conversion efficiency [dB] | P-1 dB [dBm] | **IIP2 [dB]** | **IIP3 [dB]** | **IIP5 [dB]** |
|---|---|---|---|---|---|---|---|---|
| OE-FMX | Theory | 6 | 6 | -36 | - | - | 25, 5 | 19, 2 |
| | Experiment | 5, 5 | 4 | -33, 2 | 2 | 26 | 26 | 20 |
| OE-FMP | Theory | 1, 2...2, 5 | - | -42, 0 | - | - | 25, 5 | 19, 2 |
| | Experiment | 1, 4...2, 4 | - | -45, 0 | - | 26 | 26 | 20 |
| | | VSWR | | | Input/output isolation [dB] | | | NF [dB] |
| | | RF input | LO input | IF output | LO-IF | 2RF-IF | RF-IF | LO-RF | |
| OE-FMX | Theory | 1* | 1* | 1* | 26 | 44 | 25 | ∞* | - |
| | Experiment | 1, 5 | 1, 3 | 1, 3 | 33 | 39 | 32 | 15 | 34 |
| OE-FMP | Theory | 1* | - | 1* | - | 48 | 41 | - | - |
| | Experiment | 1, 7 | - | 1, 3 | - | 52, 3 | 44, 2 | - | 44 |

*Ideal elements were used

**Table 4.** Summary of the results referred to developed LW-VCSEL-based devices for frequency converting microwave signals

# 7. Challenges in LW-VCSEL photonics

In the sections above we have demonstrated the potential of the long-wavelength low power consumption VCSELs for microwave photonics applications. Further development will make available such VCSELs with increased performances and lower power consumption thus allowing implementation of VCSELs in optical integrated circuits [10] and realization of new functionalities as VCSEL-based optical frequency combs [51]. Altogether such development will pave the way to photonic microwave transistor [52]. At the same time, a major shortcoming of VCSEL compared with DFB laser is its relatively low emitting power that constrains its penetration, for example, into telecom multi-branch passive optical networks (PON), and into transmission units of processing circuitry with external modulation by higher loss MZM. One of possible solution to overcome it is using vertical external cavity surface emitting lasers (VECSEL). Besides obvious fiber-optics applications in optical communications and optical clocking in supercomputers, VECSELs emitting in this wavelength range presents interest also in MWP [53]. Very recently, two-color tunable VECSEL with various gain chip configuration have demonstrated a wavelength separations ranging from nearly zero to more than two hundred nanometers [54]. Watt-level semiconductor disk lasers (SDL) emitting in the 1300 nm band were demonstrated using the wafer fusion fabrication technique [55, 56]. In [58] it is reported the state of the art in design and performance of electrically pumped VECSEL in 14XX nm band produced by the wafer fusion technique that exhibits 6.2 mW output power in continuous wave operation.

## 8. Conclusion

In this chapter a number of author's original works in the area of long-wavelength wafer-fused vertical channel surface emitting laser design, fabrication, and application is reviewed. Specifically, design, fabrication, DC and dynamic performances and a special benefit of wafer-fused LW-VCSELs for microwave photonic devices are highlighted. As described, wafer fused LW-VCSELs exhibit the above-listed unique features that might pave the way for their exploitation in future telecom and radar modules based on microwave photonics technology. The validity of this statement is supported both through the proper transmission characteristics of the wafer-fused LW-VCSEL itself and by simulation and experimental results of a number of basic microwave photonics devices presented here: an optoelectronic microwave-band oscillator, two versions of optoelectronic microwave-band frequency converters. Besides the above-described applications, some other examples of important microwave photonic devices based on LW-VCSELs have been published, for example in heterogeneous silicon photonic integrated circuits [10], coherent transmission module of the PON's upstream channel [50], and optical frequency comb generators [51]. Also, as a newer trend of VCSEL photonics, some recent author's investments referred to vertical external cavity surface emitting laser (VECSEL) are discussed. All these examples clearly testify that LW-VCSELs themselves and microwave photonics devices based on them have a great potential in future telecom and radar systems both for civil and for military applications.

## Author details

M. E. Belkin[1*], L. Belkin[1], A. Loparev[1], A. S. Sigov[1] and V. Iakovlev[2]

*Address all correspondence to: belkin@mirea.ru

1 Moscow State Technical University of Radio-Engineering, Electronics and Automation (MIREA), Moscow, Russia

2 Ecole Polytechnique Fédérale de Lausanne (EPFL), Switzerland

## References

[1] Seeds A. J., Williams K. J. Microwave Photonics. IEEE Journal of Lightwave Technology 2006; 24(12) 4628-4641.

[2] Kapon E., Sirbu A. Long-wavelength VCSELs: Power-efficient answer. Nature Photonics 2009; 3, 27-29 doi:10.1038/nphoton.2008.266.

[3] Belkin M. E., Belkin L., Loparev A.,. Sigov A. S, Suruceanu G., Sirbu A., Mereuta A., Caliman A., Ellafi D., Iakovlev V., Kapon E. VCSEL-based Processing of Microwave Signals. 2014; DOI: 10.1109/MWP.2014.6994554 ;

[4] Charlier J. C., and Krüger S. Long-wavelength VCSELs ready to benefit 40/100-GbE modules. Lightwave, 2012; http://www.lightwaveonline.com/articles/print/volume-28/issue-6/technology/longwavelength-vcsel-technology-improves.html.

[5] Yao J. Microwave Photonics. IEEE Journal of Lightwave Technology 2009. 27(3) 314.

[6] Capmany J., Novak D. Microwave photonics combines two worlds. Nature Photonics 1, 319-330 doi:10.1038/nphoton.2007.89

[7] Koyama A. F. Recent advances of VCSEL photonics. IEEE Journal of Lightwave Technology 2006.24(12) 4502-4513.

[8] Sirbu A., Iakovlev V., Mereuta A., Caliman A., Suruceanu G. and Kapon E.Wafer-fused heterostructures: application to vertical cavity surface-emitting lasers emitting in the 1310 nm band. Semicond. Sci. Technol. 2011.26(1) 014016.

[9] Chrostowski L., Chang C-H., and Chang-Hasnain C. J. Enhancement of dynamic range in 1.55-um VCSELs using injection locking. IEEE Photon. Technol.Lett. 2003.15(4) 498-500.

[10] Miller D. A. B. Device Requirements for Optical Interconnects to Silicon Chips. Proceedings of the IEEE 2009.97(7) 1166-1185

[11] Michalzik R. Fundamentals, Technology and Applications of Vertical-Cavity Surface-Emitting Lasers. Springer Series in Optical Sciences 166;2013.

[12] Shau R., Ortsiefer M., Rosskopf J., Böhm G., Lauer C., Maute M., Amann M.-C. Long-wavelength InP-based VCSELs with buried tunnel junction: properties and applications. Proceedings of SPIE 2008; 5364, 1–15.

[13] Caliman A., Mereuta A., Suruceanu G., Iakovlev V., A. Sirbu et al. 8 mW fundamental mode output of wafer-fused VCSELs emitting in the 1550-nm band. in Optics Express 2011;19(18) 16996.

[14] Gamage P. F., Nirmalathas A. Performance comparison of directly modulated VCSEL and DFB lasers in wired-wireless networks. IEEE Phot. Technol. Lett. 2008;20(24) 2102-2104.

[15] Belkin M. E., Belkin L. M. Microwave-Band Optoelectronic Frequency Converters Based on Long Wavelength VCSELs. IEEE COMCAS 2011: conference proceedings, 7-9 November, 2011, Tel Aviv.

[16] Baecker A. Transverse optical mode analysis of long-wavelength VCSEL in high single-mode operation. Proc. 8th Int. Conf. of Numerical Simulation of Optoelectronic Devices (NUSOD) 2008:conference proceeding, p. 87.

[17] Sirbu, A., Suruceanu, G., Iakovlev, V., Mereuta, A., Mickovic, Z., Caliman, A., Kapon, E. Reliability of 1310 nm Wafer Fused VCSELs. IEEE Phot. Technol. Let. 2013; 25(16)1555-1558.

[18] Sirbu A., Mereuta A., Caliman A., Iakovlev V., Suruceanu G., Ellafi D., Mickovic Z., and Kapon E. Wavelength controlled VCSELs emitting in the 1310-nm waveband. SPIE Photonic Europe 2014: conference proceedings, April, 14-17 2014, Brussels.

[19] Iakovlev V., Suruceanu G., Caliman A., Mereuta A., Mircea A., Berseth C.-A., Syrbu A., Rudra A., and Kapon E. High-performance Single Mode VCSELs in the 1310-nm Waveband. IEEE Photonics Technol. Lett. 2005; 17(5) 947 – 949.

[20] Mereuta A., Syrbu A., Iakovlev V., Rudra A., Caliman A.et al. 1.5μm VCSEL Structure Optimization for High Power and High Temperature Operation. J. Crystal Growth 2004; doi:10.1016/j.jcrysgro.2004.08.034

[21] Mereuta A., Sirbu A., Caliman A., Suruceanu G., Iakovlev V., Mickovic Z., Kapon E. Fabrication and performance of 1.3-μm 10-Gb/s CWDM wafer-fused VCSELs grown by MOVPE. Journal of Crystal Growth 2014; doi:10.1016/j.jcrysgro.2014.11.012.

[22] Sirbu A. and Kapon E. Ultra-Low Power VCSELs for optical network. Compound semiconductors 2013; June, p. 41-44.

[23] Information SuperStore : http:.telecominfo.telcordia.com/ido/AUX/GR_468_TOC.i02.pdf. (accessed 25 /01/ 2015).

[24] Rhew K. H., Jeon S. C., Lee D. H., Yoo B.-S., and Yun I. Reliability assessment of 1.55-μm vertical cavity surface emitting lasers with tunnel junction using high-temperature aging tests. Microelectron. Rel. 2009; 49(1) 42–50.

[25] Iakovlev V., Sirbu A., Mickovic Z. et al. Progress and challenges in industrial fabrication of wafer-fused VCSELs emitting in the 1310 nm band for high speed wavelength division multiplexing applications. Proc. of SPIE 2013; 8639, 863904-1 - 863904-7.

[26] Caliman A., Mereuta A., Suruceanu G., Iakovlev V., Sirbu A. et al. 8 mW fundamental mode output of wafer-fused VCSELs emitting in the 1550-nm band. Optics Express 2011;19(18)16996.

[27] Pedro J. C., Carvalho N. B. Intermodulation Distortion in Microwave and Wireless Circuits. Artech House, Inc., London; 2003.

[28] Belkin M. E., Belkin L. et al. Performances of Microwave-Band Analog Signal Transmission using Wafer-Fused Long Wavelength VCSELs. IEEE Phot. Technol. Lett. 2011; 23(20)1463-1465.

[29] H. Al-Raweshidy, S. Komaki, Radio over fiber technologies for mobile communication networks. Artech House, 2002.

[30] Yao X. S. Opto-electronic Oscillators. In: W. S. C. Chang (ed.) RF Photonic Technology in Optical Fiber Links. Cambridge University Press;2002. 255-292.

[31] Maleki L. Recent Progress in Opto-Electronic Oscillator, in conference proceedings: Microwave Photonics International Topical Meeting 2005, DOI: 10.1109/MWP.2005.203545

[32] http:.www.oewaves.com

[33] Belkin M.E., Loparev A.V., Semenova Y. et al. A Tunable RF-Band Optoelectronic Oscillator and OE-CAD Model for its Simulation. Microwave and Optical Technology Letters 2011;53(11)2474-2477.

[34] M. E. Belkin, and A.V. Loparev. "A Microwave Optoelectronic Oscillator: Mach-Zehnder Modulator or VCSEL Based Layout Comparison", *PIERS Proceeding*, pp. 1138-1142, Moscow, 2012.

[35] Yao X.S., Maleki L. Optoelectronic oscillator for photonic systems. IEEE Journal of Quantum Electronics 1996; 32(7) 1141-1149.

[36] http:.www.vpisystems.com

[37] Loparev A. V. Optoelectronic and Microelectronic Building Principles of Solid-State Microwave Oscillators, PhD thesis. MSTU MIREA; 2011.

[38] Cabon B., Le Guennec Y., Lourdiane M., and Maury G. Photonic mixing in RF modulated optical link, in LEOS 2006, Oct. 2006, pp. 408-409. DOI: 10.1109/LEOS.2006.279173

[39] Constant S. B., Le Guennec Y., Maury G., Corrao N., and Cabon B. Low-cost all-optical up-conversion of digital radio signals using directly modulated 1550-nm emitting VCSEL. IEEE Photonics Technol. Lett. 2008; 20(2)120-122.

[40] Lindsay A. C., Knight G. A., and Winnall S. T. Photonic Mixers for Wide Bandwidth RF Receiver Applications. IEEE Trans. on MTT 1995;43(9)2311-2317.

[41] Polo V., Marti J., Ramos F., Fuster J. M. Millimetre-wave Optical Harmonic Mixer Employing a Sigle Mach-Zehnder Electro-Optic Modulator. 28th European Microwave Conference;1998 DOI: 10.1109/EUMA.1998.338096

[42] Malyshev S. A., Chizh A. L. P-i-n Photodiodes for Frequency Mixing in Radio-over-Fiber Systems. IEEE J of Lightwave Technol. 2007; 25(11) 3236-3243.

[43] Cabon B. Microwave Photonics Mixing. Transactions D: Computer Science & Engineering and Electrical Engineering 2010;17(2) 149-162

[44] Belkin L. Microelectronic and optoelectronic design principles of microwave semiconductor frequency converters. PhD Thesis. MIREA, Moscow; 2012.

[45] Hemery E., Chusseau L., and Lourtioz J.-M. Dynamic Behaviors of Semiconductor Lasers under Strong Sinusoidal Current Modulation: Modeling and Experiments at 1.3 μm. IEEE J of Quantum Electron.1990; 26 (4) 633-641.

[46] Piprek J., Bowers J. E. Analog modulation of semiconductor lasers. In: Chang W. S. C. (ed.) RF Photonic Technology in Optical Fiber Links. Cambridge University Press; 2002.p 57-79.

[47] Belkin M. E.and Belkin L. M. Method and apparatus of base station building for RoF-structured telecom system. RF Patent No 2472290, priority from July, 11, 2011.

[48] Sauer M., Kobyakov A., George J. Radio over Fiber for Picocellular Network Architectures. IEEE J of Lightwave Technol. 2007; 25(11) 3301-3320

[49] Kaszubowska A., Hu L. Barry L. P. Remote Downconversion with Wavelength Reuse for the Radio/Fiber Uplink Connection. IEEE Photonics Technol. Lett. 2006;18(4)562-564.

[50] Olmos J. J. V., Rodes R., Monroy I. T. Low power consumption O-band VCSEL sources for upstream channels in PON systems. in *Proc. OECC*, pp. 130-131, Jul. 2012; DOI: 10.1109/OECC.2012.6276405

[51] Serrano A. C., Fernandez C. de Dios et al. VCSEL-Based Optical Frequency Combs: Toward Efficient Single-Device Comb Generation. IEEE Photonics Technol. Lett. 2013;25(20)1981-1984.

[52] Pérez D., Gasulla I. and Capmany J., Microwave Photonics Transistor Design Equations, 2014 International Topical Meeting on Microwave Photonics (MWP) and the 2014 9th Asia-Pacific Microwave Photonics Conference (APMP), p.82, DOI: 10.1109/MWP.2014.6994496.

[53] Kuznetsov M.VECSEL Semiconductor Lasers: A Path to High-Power, Quality Beam and UV to IR Wavelength by Design. In Semiconductor Disk Lasers: Physics and Technology, Prof. Dr. Oleg G. Okhotnikov. Published Online: 16 JUL 2010, DOI: 10.1002/9783527630394.ch1

[54] Lukowski M., Hessenius C., Fallahi M.. Widely Tunable High-Power Two-Color VECSELs for New Wavelength Generation. IEEE J. OF SELECTED TOPICS IN QE 2015;21(1), 1700208

[55] Rantamäki A., Sirbu A., Saarinen E., Lyytikäinen J., Mereuta A., Iakovlev V., Kapon E., and Okhotnikov O., High-power flip-chip semiconductor disk laser in the 1.3um wavelength band. Opt. Lett.2014; 39(16), 4855-4858.

[56] Sirbu A., Rantamäki A., Saarinen E. J., Iakovlev V., Mereuta A. et al. High performance wafer-fused semiconductor disk lasers emitting in the 1300 nm waveband. Optics Express 2014;22(24) 29398

[57] Iakovlev V., Walczak J., Gebski M., Sokol A. K., Wasiak M. et al. Double-diamond high-contrast-gratings vertical external cavity surface emitting laser. Journal Of Physics D-Applied Physics 2014;47(6).

[58] Caliman A., Sirbu A., Mereuta A., Pierściński K., Iakovlev V. et al. 14XX nm-wavelength electrically-pumped VECSELs fabricated by wafer fusion. Optics Express 2013; 21(11)13668.

Chapter 7

# Determination of the Bulk and Local Diffusion-Length Values of Charge Carriers in MCT Films and in the Absorber Layers of MCT-Based Photovoltaic IR FPA Detectors

S.A. Dvoretsky, V.V. Vasil'ev, A.V. Predein, A.V. Vishnyakov, V.A. Stuchinsky, D.V. Brunev and A.V. Zverev

Additional information is available at the end of the chapter

http://dx.doi.org/10.5772/60717

**Abstract**

In the present chapter, we describe two new photoelectric-measurement-based methods that can be used for characterization of the diffusion process of charge carriers either in mercury-cadmium-tellurium (MCT) films intended for fabrication of infrared focal plane array (IR FPA) detectors or in the absorber layers of ready MCT-based photovoltaic IR FPA detectors. First-type measurements are photocurrent measurements to be performed on special test structures involving round photodiodes provided with coaxial light-shielding cap metal contacts. Second-type measurements are spot-scan measurements of MCT photovoltaic 2D IR FPA detectors traditionally used for measuring the crosstalk value of such detectors yet implemented at low and high levels of registered diode photocurrents. Both methods permit the determination of the bulk diffusion length of minority charge carriers in MCT material. The second method, in addition, permits the determination of the local effective diffusion length of minority charge carriers in the absorber-layer region under FPA diodes. The values of the bulk diffusion length of minority carriers obtained in MCT films and in the MCT absorber material of the examined middle-wave and long-wave IR FPA detectors were found to be in good agreement with previously reported relevant data. The estimated value of the local effective diffusion length of minority carriers in the film region under back-biased FPA diodes proved to be consistent with a theoretical estimate of this length.

**Keywords:** infrared focal plane array detectors, mercury-cadmium-tellurium, diffusion, diffusion length, charge carriers, test structures, spot-scan measurements

## 1. Introduction

In infrared (IR) imaging and detection, cooled hybrid mercury-cadmium-tellurium (MCT) photovoltaic IR focal plane array (FPA) detectors have found widespread use [1, 2]. Collection of photogenerated charge carriers by FPA diodes in the photodiode arrays of such detectors is a process essentially mediated by the diffusion of minority carriers in the MCT film. That is why the performance characteristics of MCT IR FPA detectors and their figures of merit depend on the distances the excess carriers diffuse over in the photosensitive film (PF) of such detectors [3, 4]. Among the parameters that govern the diffusion process, of primary interest is the bulk diffusion length of minority charge carriers $l_d$ in the absorber material of the detector as this length characterizes the properties of the material irrespective of a particular FPA design, including the pixel architecture and array structure. Knowledge of $l_d$ is significant for quantitative analysis and numerical modeling of FPA detectors. With known mobility of minority carriers, this knowledge also permits evaluation of the lifetime of excess carriers in MCT, which is often hard to measure because of its small value. Finally, the length $l_d$ is generally regarded as a key parameter that defines the crosstalk value of FPA detectors [5-9].

In view of the aforesaid, of considerable interest are new experimental techniques and analysis methods that can be used for determining the length $l_d$ in the MCT material of MCT-based films and diode hetero-epitaxial structures.

In the present chapter, we describe two new photoelectric-measurement-based methods that can be used for characterization of the diffusion process of charge carriers either in MCT films intended for fabrication of IR FPA detectors or in the absorber layers of ready MCT-based photovoltaic IR FPA detectors. Both methods permit the determination of the bulk diffusion length of minority charge carriers $l_d$ in MCT material.

First-type photoelectric measurements can be performed on special test structures involving round photodiodes of radius $r_0$ provided with coaxial light-shielding cap metal contacts of different radii $R_i > r_0$. The contacts shield the vicinity of the photodiodes from the excitation radiation falling onto the front surface of the test structure. In the shielded region, the optical generation of excess charge carriers is nearly zero, and the diode photocurrents are therefore formed by those minority carriers that reach the diode via diffusion across the shielded annular region of width $R_i - r_0$. Numerically solving an appropriate 2D diffusion problem, we can preliminarily calculate the diode photocurrents $I_i$ as a function of $l_d$ for several experimentally implemented values of $R_i - r_0$. To avoid the necessity of performing absolute measurements and calculations, photocurrent ratios $I_1/I_2$, $I_1/I_3$, $I_1/I_4$, etc. for diode pairs with different contact-to-diode size ratios can be calculated. Then, the calculated curves $I_1/I_i$ $(l_d)$ can be used as reference dependences allowing the determination of the length $l_d$ in the MCT film under study.

Second-type measurements are spot-scan measurements of MCT photovoltaic 2D IR FPA detectors traditionally used for measuring the crosstalk value of such detectors, yet implemented at low levels of registered diode photocurrents $j_{ph}$. Low photocurrent levels can be achieved by lowering the gate potential $V_g$ of input field effect transistors (FETs) in the readout circuits of the photoelectric cells of the detector. The requirement for low values of the photocurrents $j_{ph}$ extracted by the readout circuits out of the photosensitive film of the detector arises as a condition for minimization of the influence of readout circuits on the lifetime and the effective (at $j_{ph} \neq 0$) diffusion length of photogenerated minority carriers in the photosensitive film $l_{d\ eff}$. The effective diffusion lengths $l_{d\ eff}$ can be extracted from broadened spatial diode photoresponse profiles $S(x)$ registered on successive reduction of the level of registered diode photocurrents. Then, the bulk diffusion length of minority charge carriers in the photosensitive film $l_d$ can be found as the extreme value of $l_{d\ eff}$ as $j_{ph} \rightarrow 0$. For revealing the asymptotic behavior of $l_{d\,eff} \rightarrow l_d$ for $j_{ph} \rightarrow 0$, a special diffusion model with zero PF thickness was developed. This model was subsequently used to analyze the charge-carrier diffusion process at normal operational regime of the detector with large diode photocurrents $j_{ph}$. The latter analysis has allowed us to evaluate the local diffusion length of minority charge carriers in the PF region under the back-biased FPA diodes $l_{d\ eff}$. The obtained value of this length proved to be in good agreement with a theoretical estimate of this length. As a result, an experimental approach enabling a comprehensive spot-scan characterization of the charge-carrier diffusion process in the absorber layers of MCT IR FPA detectors at arbitrary levels of diode photocurrents has been developed.

## 2. A method for determining the diffusion length of minority charge carriers in MCT films with band-graded surface layers

The method for determining the diffusion length of minority carriers in MCT films $l_d$ (or $L_n$, for electrons in *p*-type semiconductor) is based on the previously mentioned fact that this length presents an important parameter that defines, among other characteristics, the quantum efficiency of lateral diode structures prepared on such films [4].

### 2.1. Essence of the method

Methods for determining the lateral optical collection length of charge carriers $l_{opt}$ in MCT films based on quantum-efficiency measurements of lateral diode structures comprising variable-area diode arrays have previously been reported in the literature (see, for instance, [10, 11]). The procedure for determining the length $l_{opt}$ in those methods is based on the fact that measured quantum efficiency of back-biased photodiodes in such structures is essentially defined, first, by the diffusion length of minority charge carriers $l_{opt}$ and, second, by the diode radius. However, measured length $l_{opt}$ presents only a rough estimate to the diffusion length of minority carriers in the film.

In [12], we have proposed another method for determining the length $l_d$ in photosensitive semiconductors. That method was subsequently implemented to perform measurements of the electron diffusion length $L_n$ in $p$-type MCT-film-based hetero-epitaxial structures [13]. The proposed method uses the fact that measured photocurrent produced by an illuminated $p$-$n$ junction capped with a coaxial light-shielding metal contact varies in magnitude as we change the radius of the $n$-type region of the diode $r_0$ and/or the radius of the cap contact $R$. Of course, the radius $r_0$ is always smaller than $R$. A schematic of such a photodiode is shown in Fig. 1. The method works as follows. First, we measure the ratio between the photocurrents produced by two photodiodes having different values of $R$-$r_0$. Preliminarily, the photocurrent ratio is theoretically calculated to be plotted versus the electron diffusion length $L_n$. Then, with the measured photocurrent ratios at hand, we can use the theoretically calculated curves as reference dependences from which the length $L_n$ can be determined.

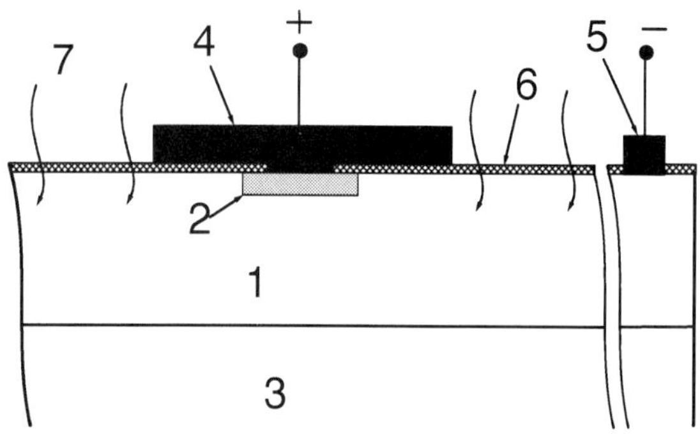


**Figure 1.** Schematic of a photodiode provided with a coaxial light-shielding cap contact. 1 – $p$-type MCT film, 2 – $n$-type diode region of radius $r_0$, 3 – substrate, 4 – shielding contact of radius $R$ to the $n$-type region, 5 – peripheral contact to the $p$-type MCT film, 6 – dielectric, 7 – illumination.

### 2.2. Calculations

The photocurrent produced by a $p$-$n$ photodiode was calculated by numerically solving the 2D stationary continuity equation for excess electron concentration in an axisymmetric quasi-neutral $p$-type region in the vicinity of the $p$-$n$ diode illuminated, from the side of the cap contact, with an IR radiation flux that modeled the radiation flux used in the experiment. With the electron mobility and the electron lifetime assumed constant, the continuity equation has the form

$$\frac{\partial^2 n'}{\partial r^2} + \frac{1}{r}\frac{\partial n'}{\partial r} + \frac{\partial^2 n'}{\partial z^2} - \frac{1}{L_n^2} \cdot n' = -\frac{g(z,r)}{D_n} \tag{1}$$

In Eq. (1), the $z$ axis is directed along the symmetry axis of the photodiode normally to the $p$-type absorber layer, and the $r$ axis is directed along radius; $n' = n - n_{p0}$ is the concentration of minority carriers in excess of their equilibrium concentration $n_{p0}$, $D_n$ is the electron diffusion constant in the $p$-type absorber material, and $g(z,r)$ is a function that describes the rate of photogeneration of charge carriers in the illuminated film with allowance for the shielding action of the cap contact:

$$g(z,\ r)=F(r)\int_{0}^{\lambda_1}\alpha(\lambda)Q(\lambda)\exp(-\alpha z)d\lambda$$

Here, $\alpha(\lambda,z)$ is the absorption coefficient, $Q(\lambda,\ T_b)=\dfrac{2\pi c}{\lambda^4[\exp(hc/\lambda kT_b)-1]}$ is the Planck distribution function for the blackbody emissivity at temperature $T_b$ (see, for instance, [1]), $\lambda_1\approx 20$ μm is a wavelength far exceeding the long-wave cutoff wavelength $\lambda_c$ of the $p$-type MCT absorber layer, and $F(r)$ is the shielding function (in a simplest model, $F(r)$=0 for $r < R$ and $F(r)$=1 for $r > R$). Since, in what follows, ratios of diode-produced photocurrents will be calculated, then the knowledge of the absolute value of the radiation flux (which depends on the aperture angle and on the coefficient of radiation reflection from the surface of the test structure) is not necessary. The absorption coefficient $\alpha$ depends on the coordinate across the photosensitive film since, starting from a distance 0.4 μm from either the lower or upper surface of the film, the cadmium content of the MCT material and, hence, its bandgap energy in the test structure gradually increase in value toward the surface. The graded-band MCT surface layers were used to suppress the surface recombination of excess carriers at the surface boundaries of the $p$-type MCT film [14]. The computational domain in which Eq. (1) was solved is shown in Fig. 2. This domain is the region referring to the coordinate ranges r=0÷$R_{ph}$ and z=0÷H; the $Cd_xHg_{1-x}Te$ material in this domain has fixed composition $x$.

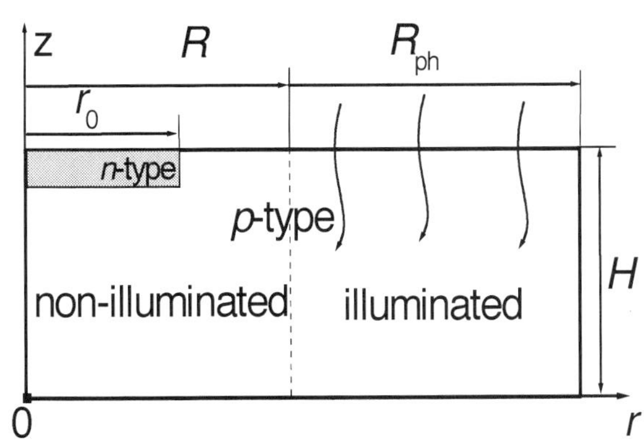


**Figure 2.** Computational domain.

In solving the continuity equation, the following boundary conditions were adopted [15]:

a. $\frac{\partial n'}{\partial r}=0$ for $r = 0$ and on the lateral cylindrical surface of the computational domain at $r = R_{ph} = r_0 + 5L_{n,\,max}$, where $L_{n,\,max}$ is the maximum value of $L_n$ for which the computations were performed, and $r_0$ is the radius of the *p-n* junction.

b. $n' = n_{p0}(\exp(qV/kT)-1)$ – on the lateral and planar surfaces of the *p-n* junction; here, $V$ is the voltage bias across the diode, $q$ is the electron charge, $k$ is the Boltzmann constant, and $T$ is temperature. In calculating the photocurrent, the bias voltage was assumed to be zero.

c. $D_n\frac{\partial n'}{\partial z} = S_r n'$ – on the upper and lower boundaries of the base layer at $z = 0$ and $z = H$, where $S_r$ is the effective rate of recombination of minority carriers at those surfaces.

With regard to the data of [16], the electron mobility value $\mu_n=6\times10^4$ cm$^2$/V×s for *p*-type $Cd_xHg_{1-x}Te$ with $x=0.223$ was adopted in the calculations. The bandgap energy $E_g$ and the intrinsic concentration of charge carriers $n_i$ in the MCT material were calculated as functions of MCT-material stoichiometry and temperature by the formulas borrowed from the monograph [1]. The MCT absorption coefficient was calculated using the formulas reported in [17]. In the calculations, the length $L_n$ was varied by changing the electron lifetime $\tau_n$ according to the formula $L_n=(D_n\,\tau_n)^{1/2}$. Equation (1) was transformed in a finite-difference equation to be solved, by the sweep method, on a grid with variable step over the *p*-type region. The total number of applied iterations reached 15000. The photocurrents across the planar and lateral surfaces of the *p-n* junction, $j_N = qD_n\frac{\partial n}{\partial z}$ and $j_L = qD_n\frac{\partial n}{\partial r}$, were calculated via numerical differentiation of the obtained solution $n'(r,z)$. The integral diode photocurrent $I$ was calculated by numerical integration of photocurrent density over the *p-n* junction surface.

### 2.3. Experimental

For experimental implementation of the method, a test photodiode structure whose fragment is shown in Fig. 3 was fabricated.

In the fabricated structure, the radius of the *p-n* junctions was fixed, equal to $r_0\approx6.5$ μm, whereas the radius of the shielding contacts varied in the range from 10 to 100 μm (see Fig. 3 and Table 1). The distance between neighbor diodes was chosen to be sufficiently large, so that their contacts exerted no influence on the photocurrents produced by the neighbor diodes. At the periphery of the lateral diode structure, a second contact (to the *p*-type absorber layer) was provided. Diode photocurrents were determined from measured current-voltage characteristics of illuminated and nonilluminated diodes. The characteristics were measured with the help of a thin probe brought in contact to the diode caps. The structure of each individual lateral photodiode involved a peripheral metal-insulator-semiconductor (MIS) region formed by the cap metal contact, by the dielectric, and by the *p*-type MCT material (see Fig. 1). For avoiding the formation of surface inversion layers in the vicinity of the diodes, a sufficiently thick dielectric layer without a noticea-

http://dx.doi.org/10.5772/60717

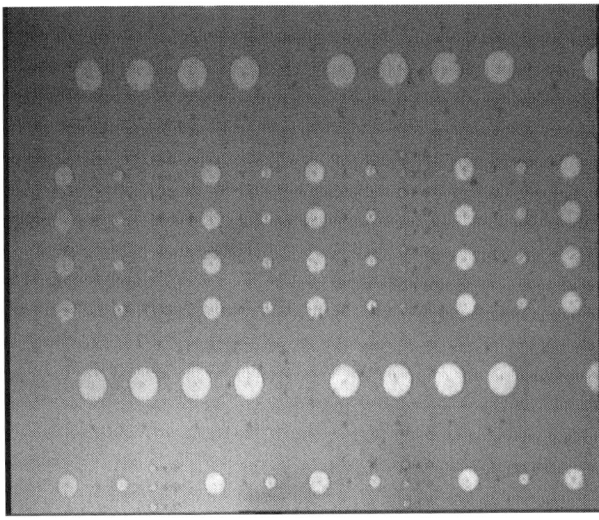

**Figure 3.** Fragment of the test structure. The test structure comprises photodiode sets with different diameters of cap contacts. In the photo, the structure is shown as viewed from the side of the cap contacts, which appear as light round regions.

ble positive built-in charge and, also, a metal with a work function value close to that of the semiconductor need to be used, and measurements are to be performed on diodes biased with a sufficiently low bias voltage. In the test structures used in our experiments, a 0.15-μm thick double-layer $SiO_2/Si_3N_4$ dielectric and indium metallization, allowing one to meet the above-formulated requirements, were used. In the experiments, a test structure fabricated on a variband MCT film grown on a GaAs substrate was placed on the surface of a corundum ceramic plate. The plate was cooled down to $T$=78 K, via a copper cylinder, by liquid nitrogen into which the copper cylinder was immersed. The ceramic plate was necessary for preventing the reflection of IR radiation having passed through the thin MCT layer and through the transparent GaAs plate, from the metal surface of the cylinder. Such reflection could otherwise entail additional absorption of reflected radiation in the MCT material under the shielding contact, which was neglected in our calculations, and, hence, it could result in an increased value of measured photocurrent. The latter in turn could induce an additional inaccuracy in the found value of $L_n$. The illumination of the test structure was performed from the side of the cap contacts by exposing the sample to 293-K thermal background radiation passing through an aperture angle of 32°. The test structure was fabricated on sample 1-MCT130607 with the following characteristics of the $p$-type MCT film: thickness 8 μm, stoichiometric coefficient $x$=0.223, hole mobility $\mu_p$= 540 cm²/V×s, hole concentration $p$=6.7×10$^{15}$ cm$^{-3}$. As illustrated by Fig. 4, thin graded-band MCT layers were provided on both sides of the grown MCT film to suppress the surface recombination of charge carriers at its boundaries.

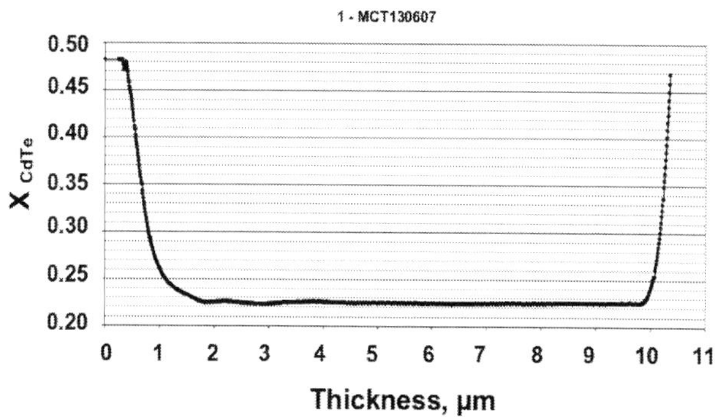


**Figure 4.** Composition of MCT material over the film thickness.

### 2.4. Experimental results

Current-voltage characteristics of the diode pairs involving neighbor diodes with different radii of cap contacts were measured, and the photocurrent values were determined from measured data. Results obtained for a diode pair with cap-contact radii $R_1$ = 10 µm and $R_5$ =70 µm are illustrated in Fig. 5.

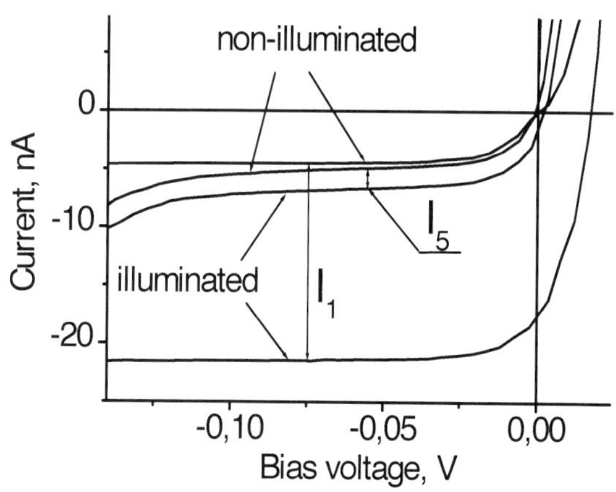


**Figure 5.** Current-voltage characteristics of a diode pair with cap-contact radii $R_1$= 10 µm and $R_5$ =70 µm.

Evidently, the photocurrents $I_1$ and $I_5$ remained roughly constant throughout the interval of low bias voltages till the bias voltage -150 mV. In addition, we have $I_1 >> I_5$, which relation proves that the peripheral MIS regions of the photodiodes induced no substantial contribution to measured photocurrent values. Typically, photocurrents were calculated at a bias voltage of

-50 mV. A bar chart of photocurrents $I_1$ and $I_5$ for seven diode pairs with cap-contact radii 10 and 70 μm is shown in Fig. 6. Here, the average photocurrent ratio $I_1/I_5$ equals 9.84.

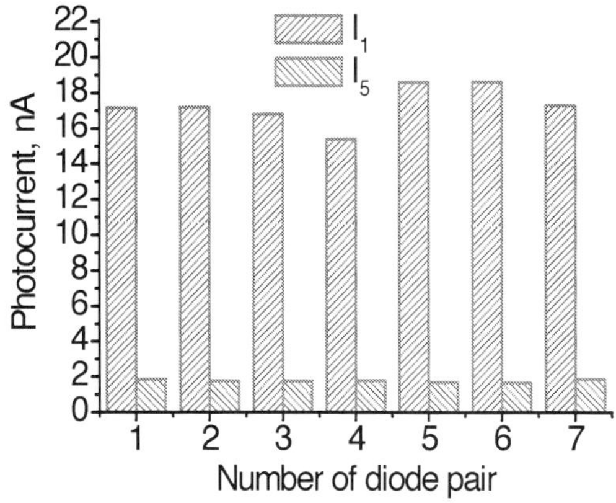


**Figure 6.** A bar chart of measured photocurrents for diode pairs comprising photodiodes with cap-contact radii 10 and 70 μm.

The theoretically calculated ratios of the photocurrents produced by a diode with the $k$-th shield ($k$ = 2÷6) (see Table 1) and by a diode with shield 1 are shown in Fig. 7; as explained above, the photocurrent ratios were calculated by solving Eq. (1). For the dimensions of the $p$-$n$ junctions, the following values were adopted: radius $r_0$=6.5 μm, inward extension of the $p$-$n$ junction in depth of the MCT film $z_0$=2 μm. The thickness of the $p$-type absorber film was assumed to be $H$=8 μm. The effective rate of surface recombination of excess carriers on both surfaces of the MCT film was assumed to be zero due to the presence of band-graded MCT layers on those surfaces (see Fig. 4). The absorption of IR radiation in the band-graded MCT layers was neglected.

| $k$ | $R_k$, $\mu m$ | $I_1/I_k$ | $L_n$ , $\mu m$ |
|---|---|---|---|
| 1 | 10 | 1 | |
| 2 | 17 | 1.2 | 22.7±5.3 |
| 3 | 24 | 1.55 | 20.4±2.1 |
| 4 | 40 | 2.88 | 20.2±0.9 |
| 5 | 70 | 9.84 | 20.5±0.8 |
| 6 | 100 | 23.8 | 22.8±0.3 |

**Table 1.** Measured photocurrent ratios for diode pairs comprising diodes with different cap-contact radii and the determined diffusion-length values $L_n$.

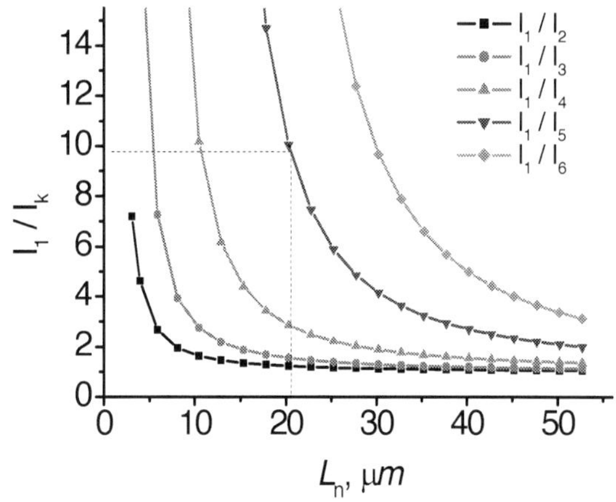


**Figure 7.** Calculated photocurrent ratios $I_1/I_k$ versus the diffusion length $L_n$.

Each of the curves in Fig. 7 has an abruptly descending portion appropriate for determining diffusion-length values in a particular interval of $L_n$.

## 2.5. Determination of the diffusion length $L_n$

With reference to the properly chosen curve in Fig. 7, the experimentally measured photocurrent ratio $I_1/I_5$ = 9.84 translates into a diffusion-length value 20.5 μm. Photocurrent ratios $I_1/I_k$ for the diode pairs which comprised diodes with other values of cap-contact radii have also been measured. The values of $L_n$ obtained from the comparison of measured with calculated photocurrent ratios for those diodes are listed in Table 1. The inaccuracy in determining the length $L_n$ can be evaluated considering the scatter of measured ratios $I_1/I_k$ for used photodiodes, whose total number for each shield radius $R_k$ typically amounted to 7. It is seen from Table 1 that very close data with average values of $L_n$ ranging from 20.2 μm to 20.5 μm were obtained for the diode pairs with cap-contact radii 10 and 24, 10 and 40, and 10 and 70 μm. Here, the data obtained from the photocurrent ratios $I_1/I_2$ and $I_1/I_3$ are the least accurate results since the curves in Fig. 4 for those diode pairs exhibit a gently sloping behavior in the range of $L_n$-value around 20 μm.

Application of the largest shield with $R_6$ =100 μm yields somewhat overestimated values of $L_n$ due to the absorption, in the shielded region, of the radiation having passed through the MCT film and having been partially reflected from the lower surface of the GaAs substrate. Let us evaluate the effect due to the latter reflection. In the case of an 8-μm thick MCT film, one can readily obtain that, for $\alpha$=2200 cm$^{-1}$, the radiation flux reaching the MCT film–substrate interface at wavelength $\lambda$=0.9$\lambda_c$ will amount to 17 % of the incident radiation flux. For reflection-coefficient value 0.27 at the air–GaAs interface, for the reflected radiation we obtain a value amounting to 3.8 % of the total flux initially incident onto the structure. Evaluation yields an additive to the photocurrent $I_k$ approximately equal to 0.05 $I_1$. The larger value of $I_k$

will entail a smaller calculated value of $I_1/I_k$ and, hence, a larger value of $L_n$ extracted from the graph of Fig. 7. Clearly, the contribution due to the mentioned additive to the photocurrent produced by the diode with the $k$-th shield can be neglected providing that $0.05\, I_1 << I_k$, or $I_1/I_k << 20$. From Table 1, we see that the above condition was obviously violated for $k$=6.

To summarize, our measurements of the electron diffusion length in the fabricated test structure performed on the diodes with cap contacts considerably differing in size have yielded an average length value $\langle L_n \rangle$=20.2–22.8 μm. A similar analysis performed for the optimum diode pairs with the photocurrent ratios $I_1/I_3$ and $I_1/I_4$, whose dependences on $L_n$ as revealed by calculations proved to be much more steeply sloping in the interval of $L_n$-values around $L_n \approx 20$ μm, has yielded for the length $L_n$ values ranging in the interval from 19.3 to 21.3 μm (with allowance for the scatter of photocurrent-ratio values obtained for 14 measured diode pairs).

## 3. Determination of the bulk and local diffusion-length values of charge carriers in the photosensing film of ready MCT IR FPA detectors

Below, we present an analysis of the charge-carrier diffusion process in a photosensing film of MCT 2D IR FPA detectors based on spot-scan measurements of such detectors. The analysis was performed with due consideration given to the fact that the extraction of minority carriers by FPA diodes out of the absorber layer, generally speaking, reduces the lifetime of excess carriers in the photosensitive film of the detector in comparison with the "bare" (without FPA diodes) MCT film, making the diffusion length of those carriers in the film regions with photocurrent suction a shorter *effective* diffusion length $l_{d\,eff}$. If the length $l_{d\,eff}$, whose magnitude depends on the level of diode photocurrents, does not exceed the lateral sizes of the FPA diodes, we may speak of a *local* effective diffusion length of photogenerated charge carriers in the film region under the photodiodes. Below, we show how the bulk and local diffusion-length values of charge carriers in different parts of the FPA structure can be evaluated using the spot-scan data measured for FPA detectors at different levels of diode photocurrents.

### 3.1. Essence of the method

For evaluating the diffusion length of charge carriers in the photosensitive film of FPA detectors, the spot-scan technique was often used [5–9]. In a spot-scan procedure, the charge-carrier diffusion-length value is deduced from spatial diode photoresponses $S(x)$ measured in the vicinity of a local illumination spot shined onto the FPA (see Fig. 8). However, the important fact that the spot-scan procedure itself largely modifies the distribution of minority carriers in the photosensitive film under measurement conditions with relatively high diode photocurrents has not been given due consideration in previous implementations of spot-scan measurements. It should be additionally mentioned here that not only the scanning diode but also its neighbor FPA diodes, which also produce their own photoresponses during scanning, seriously affect the distribution of photogenerated charge carriers in the scanned region. Like the scanning diode does, in performing scans, the neighbor diodes also continuously change their position with respect to the excitation spot. The latter circumstance hampers the treatment of spot-scan data if one attempts introducing corrections to gained data for photocurrent suction.

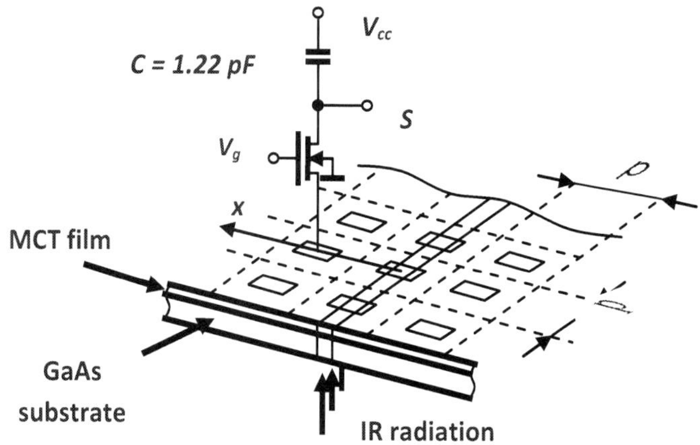


**Figure 8.** Scanning of a thin-line illumination spot with an FPA photodiode ($V_g$ is the gate potential of the photoelectric-cell input FETs with respect to the frame electrode of the diode array, $S$ is the registered photosignal, $C$ is the storage capacitor, and $V_{cc}$ is the readout integrated circuit (ROIC) supply voltage).

A good strategy toward diminishing the influence of measuring circuits on the distribution of charge carriers in the absorber layer consists in decreasing the level of registered diode photocurrents $j_{ph}$. By following this strategy, we have recently proposed a new method for determining the bulk diffusion length of charge carriers $l_d$ in the absorber layer of MCT IR FPA detectors [18]. In the new method, the spot-scan technique was implemented at low levels of diode photocurrents $j_{ph}$, and the bulk diffusion length of charge carriers in the photosensitive film $l_d$ was proposed to be determined as the extreme value of their "spatially averaged" effective diffusion length $\overline{l_{d\ eff}}$ for $j_{ph} \rightarrow 0$. The length $\overline{l_{d\ eff}}$ at low diode photocurrents $j_{ph}$ having been sufficiently large, the diffusion of photogenerated charge carriers in the photosensitive film of examined detectors was analyzed in [18] within a continuum approach. At a high level of diode photocurrents $j_{ph}$, however, the latter approach becomes invalid because of a decreased value of $\overline{l_{d\ eff}}$. In the present paper, we extend the analysis [18] of the diffusion process of charge carriers in the photosensitive film of FPA detectors to the case of large diode photocurrents $j_{ph}$. We show that, following the determination of the length $l_d$ by the method [18], treatment of spot-scan profiles $S(x)$ measured at large diode photocurrents within a simple 2D diffusion model makes it possible to determine the local effective diffusion length $l_{d\ eff}$ of minority carriers in the film region under FPA diodes in normal operational regime of the photodetector. In this way, we for the first time give a complete analysis of the charge-carrier diffusion process in a photosensitive film of MCT IR FPA detectors at arbitrary diode photocurrents.

### 3.2. Experimental

In our experiments, two 320x256 $Cd_xHg_{1-x}Te$ *n*-on-*p* photovoltaic detectors with long-wave cutoff wavelengths ~5.4 ($x \approx 0.30$) and ~10 μm ($x \approx 0.225$), hereinafter to be referred to as detectors D-1 and D-2, were used. In both detectors, the diode pitch was $p = 30$ μm, the nominal size of FPA diodes (as defined by the implantation window size in the dielectric) was 10x10 μm, and the thickness of the continuous *p*-type Hg vacancy-doped MCT photosensing layer was about 10 μm. The concentration of acceptors in the absorber layer was $N_a = (7\div10) \times 10^{15}$ $cm^{-3}$. The diode arrays in our detectors were fabricated using boron implantation. Both detectors contained graded-gap wideband MCT layers grown on both sides of the MCT film to ensure surface insulation of FPA diodes and suppress the recombination of charge carriers at the film boundaries [14]. The MCT film of the D-2 detector additionally contained a heavily doped *n*-type MCT layer with a larger bandgap energy provided to it for suppressing the debiasing effect in the diode array [19, 20].

The scanning procedure and the experimental setup used in our experiments are schematically illustrated in Figs. 8 and 9. Different decreased levels of diode photocurrents were implemented by varying the gate-potential value $V_g$ of input field effect transistors (FETs) in the photoelectric cells of the detectors (see Fig. 8). The scans were performed at a 1-μm step by translational displacements of the objective lens installed on a high-precision PC-controlled motorized XYZ-stage. The photoresponse $S$ was registered with the help of a shutter controlled by the same personal computer; usually, the photoresponse values were obtained by subtracting the average of typically 10–20 frames taken with closed shutter from the average of the same number of frames taken with open shutter.

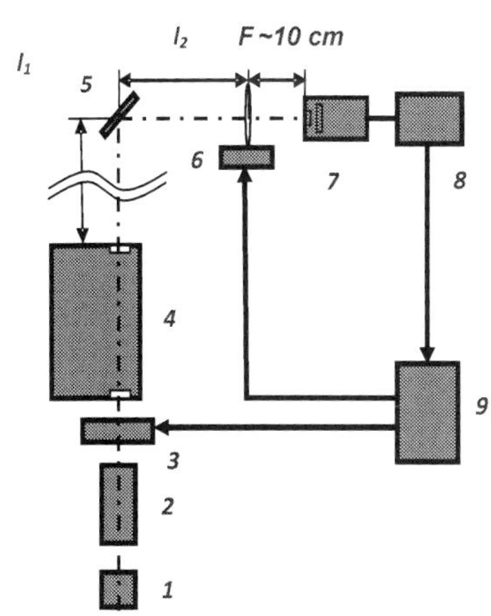


**Figure 9.** Experimental setup: 1 – globar, 2 – collimator, 3 – PC-controlled shutter, 4 – monochromator, 5 – deflecting mirror, 6 – focusing lens mounted on a motorized XYZ stage, 7 – cryostat with the mounted FPA detector, 8 – electronic equipment controlling ROIC operation and measuring the photoelectric signals, 9 – personal computer.

For diminishing the chromatic aberrations in the objective lens, as the secondary source of IR radiation, we used the exit slit of a monochromator through which IR radiation around wavelengths $\lambda$ = 4.5 or 8 μm (respectively, in the case of the D-1 and D-2 detectors) cut off from the globar emission spectrum, passed. Those wavelengths were roughly at the maximum of the spectral sensitivity of the detectors. According to [21], the IR radiation at the above-mentioned wavelengths was absorbed in the absorber layer of the D-1 and D-2 detectors over the characteristic optical lengths $l_{absorp} \approx 0.2$ and 2.4 μm. In the chosen optical configuration, for typical slit-width values of 0.1–0.2 mm, the width of the focused image of the slit at the FPA location in the geometrical optics approximation was smaller than ~8 μm. The diffraction of IR radiation caused some smearing of the slit image on the FPA. The distributions of the illumination intensity $I(x)$ across the excitation spot as implemented in the experiments with the D-1 and D-2 detectors (see Fig. 10, curves 1) were calculated under the assumption that the optical system used in our experiments was a diffraction-limited one, by summing the Airy disc distributions centered at each point of the geometric image of the slit [22]. The manner in which the radiation intensity $I$ diffracted in the vicinity of a strip-like illumination spot of width $b$ was calculated is illustrated by Fig. 11. The calculations of $I$ in the geometrically illuminated region of FPA ($x \leq b$) and outside this region ($0 \leq x \leq b$) were performed by formulas

$$I(x \geq b) = \int_{x+b}^{b} 2r \times Airy(r,\lambda) \times \left[ \mathrm{acos}(\frac{x}{r}) - \mathrm{acos}(\frac{x+b}{r}) \right] dr + \int_{x}^{x+b} 2r \times Airy(r,\lambda) \times \mathrm{acos}(\frac{x}{r}) dr,$$

$$I(0 \leq x \leq b) = \int_{x+b}^{\infty} 2r \times Airy(r,\lambda) \times \left[ \frac{\pi}{2} - \mathrm{acos}(\frac{b-x}{r}) \right] dr + \int_{0}^{b-x} \pi \times r \times Airy(r,\lambda) dr +$$

$$+ \int_{x+b}^{\infty} 2r \times Airy(r,\lambda) \times \left[ \frac{\pi}{2} - \mathrm{acos}(\frac{b+x}{r}) \right] dr + \int_{0}^{b+x} \pi \times r \times Airy(r,\lambda) dr,$$

where $Airy(r,\lambda) = \left( \frac{2J_1(\xi)}{\xi} \right)^2$ is the Airy function ($\xi = 2\pi \frac{D}{\lambda} \theta$, $J_1$ is the first-kind Bessel function, $D$ is the lens diameter, $\lambda$ is the radiation wavelength, $\theta$=atan($r/R$) is the diffraction angle, $r$ is the radius-vector in the plane of FPA, and $R$ is the lens-to-FPA separation).

The experimental conditions that were adopted in measuring the diode photoresponse profiles $S(x)$ in the D-1 and D-2 detectors are summarized in Table 2. For extending, at low and high photocurrent values, the range of photocurrents $j_{ph}$ implemented in our experiments, we used such means as increasing the integration time $t_{int}$ (e.g., from 150 μs to 2 ms in the case of the D-2 photodetector while performing measurements at low currents $j_{ph}$) or attenuating the illumination intensity in the radiation spot via narrowing the entrance slit of the monochromator (in cases where off-scale swings of the analog-to-digital converter output signal at high photocurrents $j_{ph}$ had to be avoided).

### 3.3. Results and discussion

Figure 10 shows the wings of the profiles $S(x)$ that were measured at various values of $V_g$ in experiments with the D-1 (curve 2 and symbols 3–5) and D-2 (curve 2 and symbols 3–6)

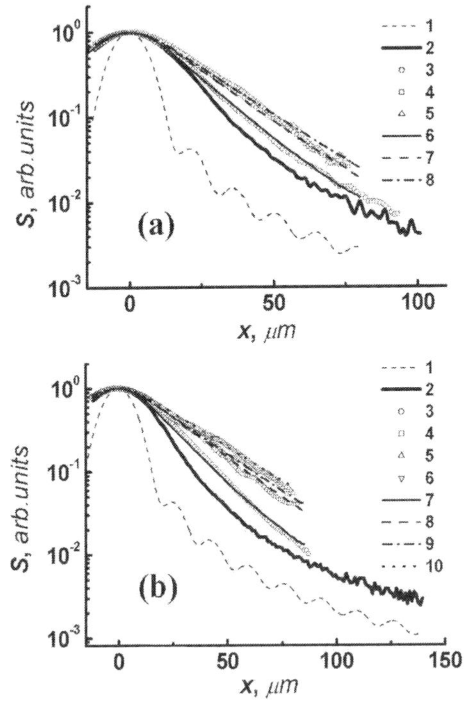


**Figure 10.** Distribution of the IR radiation intensity in the vicinity of the illumination spot (curve 1) and spot-scan profiles $S(x)$ registered in the D-1 (*a*) and D-2 (*b*) detectors at several values of $V_g$ (the rest data). (*a*) Input-FET gate potential: $V_g$=1.068 V (curve 2), and 0.96, 0.88, and 0.82 V (symbols 3, 4, and 5 and curves 6, 7, and 8, respectively). Symbols 3–5 and curve 2 show the measured spot-scan profiles $S(x)$; curves 6–8 show the profiles $S(x)$ calculated for small values of $V_g$ by the 1D diffusion model. (*b*) Input-FET gate potential: $V_g$=0.95 V (curve 2), and 0.90, 0.85, 0.74, and 0.73 V (symbols 3, 4, 5, and 6, and curves 7, 8, 9, and 10, respectively). Symbols 3–6 and curve 2 show the measured spot-scan profiles $S(x)$; curves 7–10 are the profiles $S(x)$ calculated for small values of $V_g$ by the 1D diffusion model.

detectors. As the right and left wings in the measured profiles were slightly asymmetrical, presumably because of some nonuniformity of the recombination properties of the absorber material and because of a slight difference in the electrical characteristics of the neighbor FPA diodes, Fig. 10 shows only the right wings of the profiles, from which data used in subsequent calculations were taken. For clarity, the curves $S(x)$ measured for the D-1 detector in operation with $V_g \leq 0.96$ V and for the D-2 detector in operation with $V_g \leq 0.90$ V are shown in Fig. 10 in a smoothed form. The relation between the as-measured and smoothed profiles $S(x)$ for the noisiest experimental curves (data for D-2 measured at $V_g$=0.73 and 0.74 V) is illustrated in Fig. 12. It is seen that the case of $V_g$=0.73 V was the noisiest experimental situation, while already at $V_g$= 0.74 V the spot-scan profile S(x) was measured quite reliably. The profiles $S(x)$ measured in both detectors at $V_g \leq 0.9$ V are noticeably widened in comparison with the profiles taken from the D-1 and D-2 detectors, respectively, at $V_g$ = 1.086 V and $V_g$ = 0.95 V, that is, at gate-potential values close to the optimum ones in normal operational regime of the detectors (see Fig. 10). On increasing the gate-potential value in excess of 0.95–1.00 V, the shape of the registered spatial photoresponses exhibited saturation, and the normalized profiles $S(x)$, therefore, remained identical. Wishing to avoid dull data repetition, in Fig. 10 we do not illustrate this fact. Below, we give an analysis to measured profiles $S(x)$ separately for the range of $V_g$-values in which distinct widening of the bell-shaped profiles $S(x)$ with decreasing the level of diode photocurrents $j_{ph}$ was observed ($V_g$<0.95–1.00 V) and for the range of $V_g$-values in which the shape of the profiles $S(x)$ became independent of $V_g$ ($V_g$>0.95–1.00 V) (conventionally, the cases of small and large diode photocurrents).

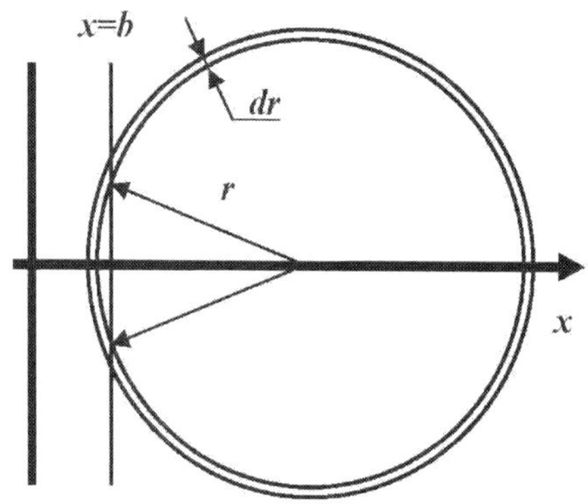


**Figure 11.** On the calculation of the distribution of the radiation intensity diffracted in the vicinity of an illumination spot shaped as a strip of width $b$ ($0 \le x \le b$), in the region outside the spot ($x \ge b$). The radiation intensity in the region $0 \le y \le b$ was calculated similarly, by summing the contributions due to circular rings and arcs.

| | | | | | **D-1** (λ = 4.5 μm) | | | | | | |
|---|---|---|---|---|---|---|---|---|---|---|---|
| $V_g$, V | $t_{int}$, μs | $I_0$, nA | $I$, arb. units | $s$, mm | $l$, cm | $D$, Cm | $F$, cm | $\Delta$, Mm | $l_{d\ eff}$, μm | $\bar{k}$ | $k_{ph\ opt}$ |
| 0.82 | 1300 | 0.14 | 1 | 0.1 | 214 | 3.33 | 10 | 4.9 | 17.5 | 0.24 | ≈1 |
| 0.88 | 1300 | 0.2 | 1 | 0.1 | 214 | 3.33 | 10 | 4.9 | 16 | 0.485 | ≈2 |
| 0.96 | 1300 | 0.5 | 1 | 0.1 | 214 | 3.33 | 10 | 4.9 | 13 | 1.25 | ≈7 |
| 1.068 | 1300 | 0.29 | ~0.4 | 0.1 | 214 | 3.33 | 10 | 4.9 | - | - | ≈20-40 |
| | | | | | **D-2** (λ = 8 μm) | | | | | | |
| $V_g$, V | $t_{int}$, μs | $I_0$, nA | $I$, arb. units | $s$, mm | $l$, cm | $D$, Cm | $F$, cm | $\Delta$, Mm | $l_{d\ eff}$, μm | $\bar{k}$ | $k_{ph\ opt}$ |
| 0.73 | 2000 | 0.00026 | 1 | 0.2 | 274 | 5 | 10 | 7.6 | 24 | <0.1 | <0.8 |
| 0.74 | 2000 | 0.0026 | 1 | 0.2 | 274 | 5 | 10 | 7.6 | 22.5 | 0.14 | ≈0.9 |

| | | | | | | | | | | | |
|---|---|---|---|---|---|---|---|---|---|---|---|
| 0.85 | 250 | 0.49 | 1 | 0.2 | 274 | 5 | 10 | 7.6 | 20 | 0.44 | ≈2.5 |
| 0.90 | 150 | 1.46 | 1 | 0.2 | 274 | 5 | 10 | 7.6 | 14.5 | 1.74 | ≈9 |
| 0.95 | 150 | 1.68 | ~0.5 | 0.2 | 274 | 5 | 10 | 7.6 | - | - | ≈20-40 |

**Notation**: $V_g$ is the gate potential of photoelectric-cell input FETs with respect to the frame electrode of the diode array, $t_{int}$ is the integration time, $I_0$ is the photocurrent produced by a photodiode centered at the axis of the illumination spot, $I$ is the IR radiation intensity in the illuminated spot (in arbitrary units different for D-1 and D-2), $s$ is the width of the monochromator exit slit, $l=l_1+l_2$ is the distance from the monochromator exit slit to the lens (see Fig. 9), $D$ and $F$ are the lens diameter and focal length, $\Delta$ is the width of the geometric image of the monochromator exit slit at the FPA location, $\overline{l_{d\ eff}}$ is the "spatially averaged" effective diffusion length of minority carriers in the PF of the FPA detector in the model with uniform photocurrent suction, $\bar{k}$ is the value of the coefficient $k$ found in the 1D diffusion model with uniform photocurrent suction, and $k_{ph\ opt}$ is the suction coefficient in the region under the photodiodes in the 2D diffusion model with structured photocurrent suction.

**Table 2.** Conditions of the spot-scan tests of the D-1 and D-2 detectors and the obtained values of $\overline{l_{d\ eff}}$, $\bar{k}$, and $k_{ph\ opt}$.

### 3.3.1. *The case of small diode photocurrents $j_{ph}$*

As the IR radiation intensity reaching the $n$-regions of the FPA diodes was negligible, the measured diode photoresponses were primarily due to photoelectrons excited by IR radiation in the $p$-type absorber layer and sinking to the $n$-regions of the photodiodes. The spot-scan data obtained at small values of $V_g$ were analyzed within a continuum diffusion model that was developed to clarify the expected asymptotic behavior of $\overline{l_{d\ eff}}$ $(j_{ph})$ for $j_{ph} \rightarrow 0$. This model could be used because, at the low levels of $j_{ph}$, we had $\overline{l_{d\ eff}} \geq l_{SCR}/2$ and $\overline{l_{d\ eff}} \geq d$ (here $l_{SCR}$ is the separation between the depletion edges of adjacent FPA diodes and $d$ is the PF thickness) [18].

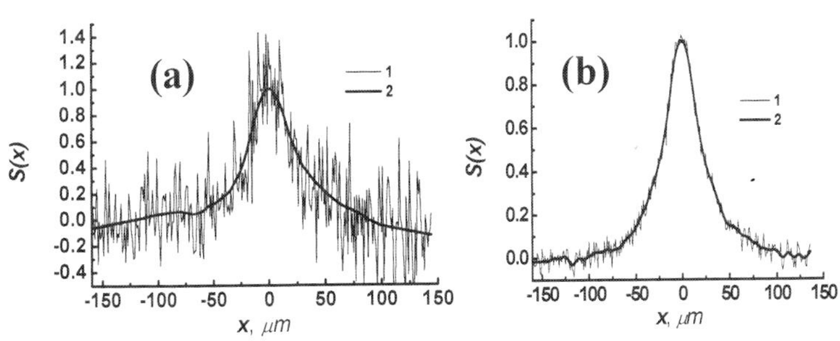


**Figure 12.** Relation between the most noisy spot-scan curves measured for the D-2 detector at $V_g$=0.73 (*a*) and 0.74 V (*b*) (curves 1) and their averaged counterparts $S(x)$ (curves 2).

#### 3.3.1.1. *Continuum diffusion model. Derivation of the asymptotic $l_{d\ eff} \rightarrow l_d$ for $j_{ph} \rightarrow 0$*

The model is based on the following assumptions: (i) the lateral diffusion of the photogenerated charge carriers proceeds in an infinitely thin film; (ii) the photoelectrons are extracted out of the photosensitive film, instead of a discrete diode array, by a large-area electrode continuously covering the entire surface of the film; (iii) the local photocurrent density $j_{ph}$ varies over the diode area in proportion to the local sheet density of excess minority carriers $n_s$ with a coefficient whose magnitude depends on $V_g$; (iv) the total photocurrent across the diode can be calculated by integrating the local photocurrent density $j_{ph}$ over the diode area, which has an efficient size $t$. For our detectors, a value $t$=14 μm was adopted in the calculations described below (the nominal window size plus 2 μm on either side of the photodiode) [23, 24]. In the formulated model, the spot-scan profiles $S(x)$ can be found by solving the 1D diffusion equation

$$l_d^2 \times \frac{\partial^2 n_s}{\partial x^2} - (k+1) \times n_s + g \times \tau = 0,$$

which can also be written as

$$l_{deff}^2 \times \frac{\partial^2 n_s}{\partial x^2} - n_s + g \times \tau_{eff} = 0. \tag{2}$$

Here, $l_{d\,eff} = l_d / \sqrt{1+k}$ and $\tau_{eff} = \tau / (1+k)$ are the effective diffusion length and the effective lifetime of excess carriers in the photosensitive film ($\tau$ is the lifetime of excess carriers in the bulk absorber material, $k$ is the coefficient that relates the local photocurrent density $j_{ph}$ to $n_s$, $j_{ph}=k \times n / \tau$), and $g$ is the rate of sheet photogeneration of charge carriers in the photosensitive film. For the normalized profiles $S(x)$, we then have [18]:

$$S(x) = \frac{\displaystyle\int_{x-t/2}^{x+t/2} \left( \int_{-\infty}^{\infty} g(x') \times \exp\left( -\frac{|x'-\xi|}{l_{d\,eff}} \right) d\xi \right) dx'}{\displaystyle\int_{-t/2}^{t/2} \left( \int_{-\infty}^{\infty} g(x') \times \exp\left( -\frac{|x'-\xi|}{l_{d\,eff}} \right) d\xi \right) dx'}. \tag{3}$$

The curves $S(x)$ calculated by formula (3) with the values of $l_{d\,eff}$ providing the best fit of experimental with calculated data for the central part of the bell-shaped spatial photoresponses for $V_g$=0.73, 0.74, 0.85, and 0.90 V (D-1) and $V_g$=0.82, 0.86, and 0.96 V (D-2) are shown in Figs. 10(a) (curves 6–8) and 10(b) (curves 7–10). The obtained values of $l_{d\,eff}$, $l_{d\,eff}$, are listed in Table 2; in Fig. 13, they are plotted (for fixed illumination intensity in the excitation spot) against the photocurrent $I_0$ that was produced by the scanning diode when the diode was centered at the spot axis (the current $I_0$ was evaluated from the maximum photosignal $\Delta V_0$ in the registered scans by the formula $I_0 = C \times \Delta V_0 / t_{int}$, where $C$=1.22 pF is the photoelectric-cell charge storage capacity and $t_{int}$ is the integration time).

The curves in Fig. 13 show the dependences $l_{d\,eff}(I_0)$ as predicted by the model; those dependences are defined by a nondimensional parameter $\alpha = (j_0 \times l_d)/G$ that varies in proportion to $I_0$

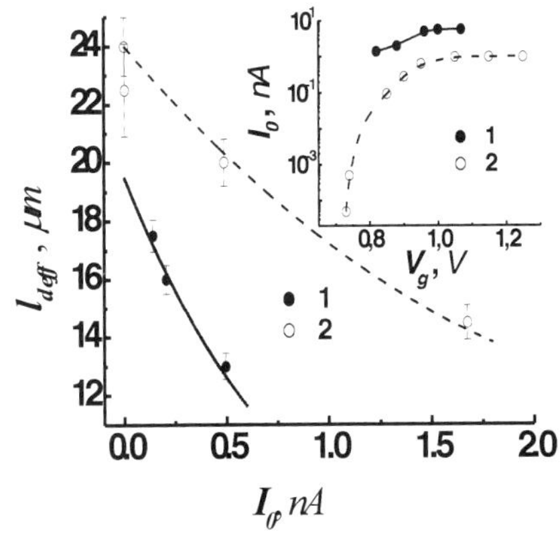


**Figure 13.** "Spatially averaged" effective minority-carrier diffusion length $l_{deff}$ versus the photocurrent $I_0$ registered by the measuring diode at the maximum of the spot-scan profiles *S(x)*. Symbols 1 and 2 are the data for the D-1 and D-2 detectors; the curves are the dependences calculated by formula (4) with $\alpha/I_0$=0.9 and 0.33 nA$^{-1}$ for D-1 and D-2, respectively. The inset shows the dependence of $I_0$ on $V_g$ for the D-1 and D-2 detectors at fixed (yet different for D-1 and D-2) levels of the IR radiation intensity in the illumination spot. Symbols 1 and 2 are the data for the D-1 and D-2 detectors, respectively.

(here $j_0=I_0/(q\times p^2)$ is the average flux of photoelectrons through the spot-centered pixel area ($q$ is the electron charge) and $G$ is the rate of the sheet photogeneration of charge carriers in the illumination spot per unit length of the spot):

$$l_{d\,eff}=l_d\times\left(\sqrt{\alpha^2+1}-\alpha\right)=l_d\cdot\left(1-\alpha+\frac{\alpha^2}{2}+\ldots\right). \quad (4)$$

Formula (4) can be derived as follows. Assuming that $n(x)\propto\exp(\pm x/l_{d\,eff})$, putting $g\approx 0$ outside the spot, and writing the coefficient $k$ as $j_0/(n_0\times\tau)$ (here, $n_0$ is the sheet density of photogenerated charge carriers in the spot), from Eq. (2) we obtain:

$$\frac{l_d^2}{l_{d\,eff}^2}=1+\frac{j_0\times\tau}{n_0}. \quad (5)$$

Considering now the problem over the scale ~ $l_{d\,eff}$ and treating the distribution of illumination intensity in the spot as a delta-function, from the condition of matching the lateral photocurrents that spread from the spot to the left and to the right, for the photoelectron density $n_0$ we obtain:

$$n_0=\frac{G\times l_{d\,eff}}{2D}. \quad (6)$$

On substitution of expression (6) into relation (5), for $x = \frac{l_{d\,eff}}{l_d}$ we then derive a quadratic equation whose solution yields formula (4).

According to the asymptotic of (4), at low currents $I_0$ the effective diffusion length $l_{d\,eff}$ decreases quasi-linearly with increasing current $I_0$ in such a way that the slope $\alpha/I_0$ defines simultaneously the weakly nonlinear behavior of $l_{d\,eff}(I_0)$. As it is seen from Fig. 13, both conclusions fairly well agree with the experimental data. Interestingly, from the found value of $\alpha/I_0$ one can readily obtain an estimate of $G$.

The 1D continuum model described above possesses the following advantageous features: (i) in treating the withdrawal of charge carriers out of the photosensitive film, this model permits leaving aside such things as boundary conditions for the minority carrier concentration $n$, which would be dealt with in the 3D diffusion model; (ii) the model explicitly involves a quantity $l_{d\,eff}$ whose asymptotic value at low photocurrents gives the bulk diffusion length of minority charge carriers $l_d$ in the photosensitive film; (ii) the model admits derivation of an analytical expression for the asymptotic behavior of $l_{d\,eff}\,(j_{ph})$ as $j_{ph} \to 0$.

*3.3.1.2. Evaluation of the length $l_d$ in the absorber layer of the examined IR FPA detectors*

Approximation of the curve $l_{d\,eff}\,(I_0)$ to $I_0$=0 according to asymptotic (4) yields for $l_d$ a value of 19.5 μm in the absorber material of the D-1 detector and a value of 24 μm in the absorber material of the D-2 detector. Those values are in good agreement with relevant data that were previously reported in the literature (see, e.g. [5, 13, 25], where the values of 19 μm [25], 20–23 μm [13], and 25–35 μm [5] were reported for the carrier diffusion length $l_d$ in $p$-type MCT at 78 K), and also with the diffusion-length values calculated from the lifetime and mobility data for $p$-type MCT of similar stoichiometry (see, e.g. [26], where the lifetime and mobility values $\tau$=10–15 ns and $\mu_e$=6.8×10$^4$ cm$^2$/V×s, translating into $l_d$=21.4–26.2 μm, were obtained for a $Cd_xHg_{1-x}Te$ material with $x$=0.20–0.23).

It would also be instructive to correlate the diffusion-length data obtained in the present study with the available minority-carrier lifetime data for Hg-vacancy-doped MCT. According to [27], the lifetime of excess carriers in the $p$-type $Cd_xHg_{1-x}Te$ material with $x$=0.225 and 0.30 doped with Hg vacancies to a concentration of (7–10)x10$^{15}$ cm$^{-3}$ is approximately $\tau$=2-4 and 5.5–10 ns. Evaluating the electron mobility in MCT by traditional formulas [21], we obtain the mobility values $\mu_e$=1.06x10$^5$ and 4.46×10$^4$ cm$^2$/V×s for such materials. Then, for the electron diffusion length, we obtain the values of 12–17 μm and 13–18 μm, which fairly well compare with the values of 19.5–24 μm obtained in our detectors. Thus, the diffusion-length values for MWIR and LWIR detectors differ little, in line with the results reported in [16]. Thus, here again we have a good agreement with the literature.

It should be additionally noted here that, as a result of the neglect of the film thickness in the applied approach, our analysis of the profiles $S(x)$ yields somewhat underestimated values of $l_d$ because, for reaching a photodiode, the photogenerated charge carriers, apart from moving laterally, have to diffuse a distance of $d$ across the photosensitive film. However, this underestimate of $l_d$ is of the order of $\vartheta\,(d^2/l_d^2)$; in the case under consideration, it does not exceed ~

2–2.5 μm and can be considered small. Another point that should be mentioned here with reference to Fig. 10 is that for low gate voltages $S(x)$-data become less complicated and more prone to be fit, in an extended range of $x$, with a single exponential behavior. Fitting the $S(x)$ data for our photodetectors with a single exponential decay constant yields values of $l_{d\ eff}$ that well coincide (within ~1–1.5 μm) with the values of $l_{d\ eff}$ obtained from the calculations based on formula (3).

### 3.3.2. *The case of large diode photocurrents $j_{ph}$ (normal operational regime of FPA detectors)*

At currents $I_0 \geq 0.5$ nA/pixel in the D-1 detector and ≥1.5 nA/pixel in the D-2 detector, the length $l_{d\ eff}$ becomes comparable with the characteristic lateral size of the diode-array structure $l_{SCR}$, making it necessary to take the nonuniformity of that structure into account in calculating the diode photoresponses $S$. Moreover, at large diode photocurrents, the local effective diffusion length of minority carriers in the region under the back-biased diodes $l_{d\ eff}$ becomes comparable with $d$, and the second applicability condition of the 1D diffusion model, $l_{d\ eff} > d$, also becomes violated. Disregarding, for the time being, the second circumstance, in our analysis of spot-scan data we tried to allow for the discrete structure of the FPA by performing calculations of 2D distributions of the excess carrier concentration $n(x,y)$ in our locally illuminated photodetectors within a 2D diffusion model with $d$=0 presenting an extension of the previously used 1D diffusion model to the case of two lateral dimensions.

#### 3.3.2.1. *2D calculations*

The 2D distribution of the sheet density of photogenerated charge carriers in the photosensitive film of the IR FPA detectors $n_s(x,y)$ with $g(x) \propto I(x)$ was calculated as the solution of the 2D diffusion equation

$$l_d^{\,2} \times \left( \frac{\partial^2 n_s(x,y)}{\partial x^2} + \frac{\partial^2 n_s(x,y)}{\partial y^2} \right) = -g(x) \times \tau + n_s(x,y) + k_{ph} \times P(x,y) \times n_s(x,y). \tag{7}$$

In Eq. (7), the discrete structure of the diode array was taken into account via the function $P(x, y)$, which was assumed to be unity in the region under the FPA diodes and zero in the region outside the FPA diodes. We thereby assumed that, over the area occupied by the FPA, the local diffusion length of charge carriers in the region outside the diodes was equal to $l_d$. In the region under the diodes, a smaller value of the local (effective) diffusion length, $l_{d\ eff}$, defined by a value of $k_{ph}$ to be found via the fit of calculated to experimental spot-scan data, was assumed. [Here, we would like to remind the reader that under the local diffusion length at a point of interest, we everywhere understand the average distance the charge carriers would move in the absorber material if they were spreading from that point in a spatially uniform photosensitive film with recombination properties of the absorber material being everywhere identical to the recombination properties of the material at the point under consideration. Under the effective diffusion length $l_{d\ eff}$, we understand the diffusion length of charge carriers defined, in addition to their bulk recombination in PF (which, taken alone, defines the value of $l_d$), also by the disturbing action due to FPA diodes. The notion of local effective diffusion length of charge carriers, whose meaning can be comprehended by combining the above two definitions, is a notion pertinent to the 2D diffusion model with $d$=0 used in the present analysis].

The equation was solved by using the MathCad function *relax* in a 1024x1024-μm square domain covered by a square grid whose mesh size was 1 μm. The illumination spot with the distribution of illumination intensity $I(x)$ stretched in the middle of the calculation domain along the $y$ axis. At the calculation-domain boundaries parallel to the centerline of the spot, zero boundary conditions were adopted for the sheet density of excess carriers ($n_s$=0). The starting boundary conditions at the domain boundaries normal to the illumination spot were initially taken from the solution of the related 1D diffusion problem, and then, at a second iteration made to refine the solution, from the condition of periodicity of the solution in the spot-parallel direction. Afterward, proper integration (over a 14x14-μm square) was used to calculate the relative values of $S$ for various positions of the illumination spot with respect to the scanning diode. As an example, Fig. 14 shows the 2D plots and contour lines of the solutions $n_s(x,y)$ obtained in calculations with $k_{ph}$=40 for the case in which the scanning diode in the D-2 detector was at the centerline of the illumination spot (Fig. 14 (a)) and for the case in which its center was displaced from the spot axis by 10 μm (Fig. 14(b)). The purpose of our 2D calculations was to find such values of $k_{ph}$ at which the best coincidence between the normalized calculated and experimental diode photoresponse profiles $S(x)$ could be achieved for the experimentally implemented values of $V_g$, including those for the normal operational regime of the detectors.

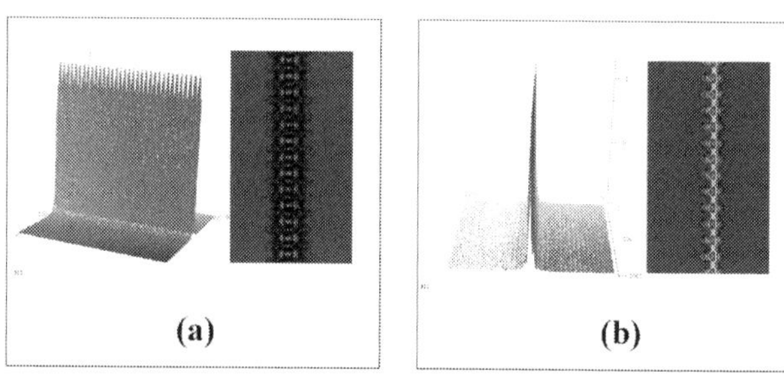


**Figure 14.** 2D plots and contour lines of the sheet density of excess minority carriers $n_s$ for the measuring-diode position at the centerline of the illumination spot (*a*) and for the diode-center position 10 μm aside from the spot axis (*b*). The data were obtained for the D-1 detector in the calculation with $k_{ph}$=40.

### 3.3.2.2. *Calculated data and their comparison with the measured spot-scan profile S(x)*

The found values of $k_{ph}$, $k_{ph\ opt}$, are listed in the last column of Table 2. The values of $k_{ph\ opt}$ found for the D-1 detector in operation with $V_g$≤0.90 V and for the D-2 detector in operation with $V_g$≤0.95 V proved to be 4–5.5 times greater than the optimal values of the coefficient $k$, $\bar{k}$, found in the continuum model of [18] with uniform current suction. With the factor 4–5.5 being close to the ratio between the pixel and diode areas, it can be concluded that both models, the model with continuous photocurrent suction and the model with structured photocurrent suction, yield adequate descriptions to the data measured at small values of $V_g$ providing that one and the same average photocurrent density on the area occupied by the FPA is assumed. The latter result can be regarded as verification of the applicability of the approach of [18] to the analysis of spot-scan data obtained for reduced potentials $V_g$. For values of $V_g$ typical of the normal operational regime of the detectors ($V_g$=1.086 V for D-1 and $V_g$=0.95 V for D-2), the shape of the spatial diode photoresponses $S(x)$ proves to be more complicated in comparison with the profiles $S(x)$ measured at reduced values of $V_g$, with the wings of the profiles $S(x)$ being less prone to fitting with a single-exponential behavior (Fig. 10). A comparison between the spot-scan profile measured at $V_g$=0.95 V for the D-2 detector and the profiles $S(x)$ calculated for the same detector by the 2D model with $k_{ph}$ =20, 40, and 60 is given in Fig. 15. Evidently, with $k_{ph}$ ≈ 20-40, a good agreement between the measured profile and the calculated spot-scan profiles was achieved, which shows that, at least in some practically important cases, the model with $d$=0 is also capable of providing a satisfactory approximation of the measured with calculated profiles $S(x)$ for the values of $V_g$ typical of the normal operational regime of the detectors.

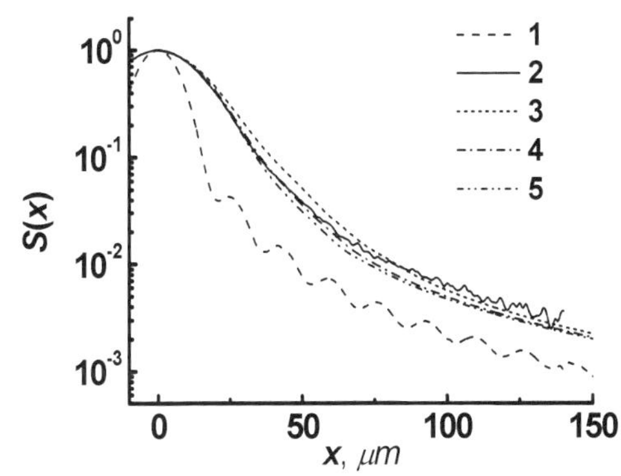


**Figure 15.** Spot-scan profile $S(x)$ in the D-2 detector measured at $V_g$=0.95 V (curve 2) in comparison with the profiles $S(x)$ calculated by the 2D diffusion model with $k_{ph}$=20, 40, and 60 (curves 3, 4, and 5, respectively). Curve 1 is the distribution of the IR radiation intensity.

The above values of $k_{ph}$, $k_{ph\ opt}$ ≈ 20–40, refer to the effective diffusion-length values of charge carriers in the region under the photodiodes $l_{d\ eff} = l_d / \sqrt{1 + k_{ph\ opt}} \approx$ 3.7–5.2 μm. It should be noted here that the above-described fitting procedure yields an upper-type estimate for $l_{d\ eff}$ because, with increasing the magnitude of $k_{ph}$ over $k_{ph\ opt}$ ≈ 20–40, the calculated photoresponses $S$ (and also the shape of the profiles $S(x)$) proved to be weakly dependent on $k_{ph}$, exhibiting saturation in the vicinity of the measured $S$-values. Yet, it can be shown that the obtained values $l_{d\ eff}$ ≈ 3.7–5.2 μm agree well with a theoretical estimate of the length $l_{d\ eff}$ (see below). Such an *a priori* estimate of the length $l_{d\ eff}$ in the region under photodiodes can be obtained due to the fact that the values of the second derivatives of the excess carrier concentration $n$ in the lateral direction normal to the spot axis and in the direction across the absorber layer are interrelated by the 3D diffusion equation, while some plausible assumptions about the shape of the cross-film distributions $n(z)$ can be adopted.

### 3.3.2.3. *A priori estimate of the length* $l_{d\,eff}$ *under the back-biased photodiodes*

The main idea behind the spot-scan method consists in examining the spot-scan profiles $S(x)$ in the film region at the spot edges, where the illumination intensity $I$ rapidly decreases to allow the shape of the formed distributions $n_s(x)$ and $S(x)$ to be largely controlled by the diffusion of charge carriers. A characteristic feature of those areal parts of the photosensitive film is that the excess charge carriers appear in those areal parts predominantly due to their lateral diffusion out of the adjacent film regions (and not due to photogeneration). For those film regions, an *a priori* estimate of $l_{d\,eff}$, which can be used for verification of the found values of $l_{d\,eff}$, can be obtained.

Indeed, let us consider an extended (with a characteristic size $l \gg l_{d\,eff}$) region of an MCT film under a large-area back-biased photodiode. We assume that photogeneration of charge carriers in this region is negligible, and excess carriers enter this region through its lateral boundary parallel to a nearby elongated illumination spot. At a fixed distance from the spot axis ($x$=const), the function that describes the distribution of the excess carrier concentration across the film (over the $z$ coordinate) can be expanded in a series of sines of the type $\sin[\frac{z}{d_{eff}} \times (\frac{\pi}{2} + 2\pi \cdot n)]$, $n$=0,1..., which all satisfy the boundary conditions at the film boundaries ($n \approx 0$ at $z$=0 and $\partial n / \partial z = 0$ at $z \approx d_{eff}$). Here $d_{eff}$ is the effective thickness of the photosensitive film in the region under the FPA diodes (the actual film thickness $d$ minus the sum of the inward extension of the $p$-$n$ junction in depth of the absorber layer ($d_{deep} \approx 2$–3 μm) and a length of order of the optical absorption length of IR radiation in the absorber material $l_{absorp} \approx 2.4$ μm, see above). Solving the 3D analogue of Eq. (7) by the variable separation method, we see that each component in the series decays exponentially in the PF plane along the $x$ coordinate, the characteristic length of the decay being

$$l_{decay}^{(n)} = l_d \Big/ \sqrt{1 + \left[ \frac{l_d}{d_{eff}} \times \left( \frac{\pi}{2} + 2\pi \times n \right) \right]^2} . \qquad (8)$$

The length $l_{decay}$ for the least rapidly decaying component with $n$=0, which can be expected to dominate the whole series far enough from the source boundary, should be identified as the effective lateral diffusion length of charge carriers in the photosensitive film $l_{d\,eff\,lat}$. For several realistic values of $d_{eff}$ in the D-2 detector, $d_{eff}$=6, 7, and 8 μm, formula (8) yields $l_{d\,eff} \approx \frac{2d_{eff}}{\pi} = 3.8$, 4.4, and 5.0 μm, respectively (the approximate equality holds if $(l_d / d_{eff})^2 >> 1$). The above values of $l_{d\,eff\,lat}$ translate into $k_{ph} \approx 22$–39 (here we can use the relation $l_{d\,eff} = l_d / \sqrt{1 + k_{ph}}$ or, alternatively, take into account the fact that $k_{ph}=l_d^2\times\frac{\partial n(z)}{\partial z}\bigg/\int_0^{d_{eff}} n(z)dz\approx(\frac{\pi}{2}\times\frac{l_d}{d_{eff}})^2$ for the harmonic with $n$=0).

In compliance with expectations, the length $l_{d\ eff\ lat}\approx$ 4–5 μm proves to be comparable with the thickness $d_{eff}$. A similar estimate of the length $l_{d\ eff\ lat}$, $l_{d\ eff\ lat}\approx$ 4–5 μm, was also obtained for the D-1 detector. Thus, for the film region under back-biased diodes, we obtain a good agreement between the lateral diffusion-length values $l_{d\ eff\ lat}$ deduced from the analysis of the shape of the photoresponse profile $S(x)$ (namely, from the values of $k_{ph\ opt}$) and the theoretical estimate of this length.

*3.3.2.4. Substantiation of applicability of the 2D diffusion model to the description of spot-scan profiles measured in normal operational regime of the photodetectors*

Below, we present heuristic considerations that explain why the diffusion model with $d$=0 has proved capable of rather adequately reproducing the spot-scan data obtained not only at small, but also at large diode photocurrents $j_{ph}$. As it was noted above, at low diode photocurrents the distribution $n(z)$ flattens across the film, thus making the treatment of the cross-film dimension unnecessary. On the other hand, at high values of $j_{ph}$ (when the photocurrents show saturation with increasing the potential $V_g$), the concentration $n$ in the vicinity of the back-biased $p$-$n$ junctions becomes low, and the distribution $n(z)$ can be no longer considered quasi-uniform.

Substantiation of the 2D model with $d$=0, as used for the description of spot-scan data in the latter case, can be obtained by integrating both sides of the 3D analogue of Eq. (7) across the photosensitive film. For simplicity, we begin our analysis with consideration of a photosensitive film covered by a continuous large-area back-biased photodiode. For the integrated equation to be consistent with Eq. (7), it is required that the term in the integrated equation which results from integration of the Laplacian component $\partial^2 n/\partial z^2$ and takes the boundary condition on the diode side of the film into account would transform to the term of Eq. (7) with the coefficient $k_{ph}$. In turn, for such a transformation, it is required that the gradient of $n$ along the film-normal direction on the diode side of the film would vary everywhere (in the film region under the photodiode) in proportion to $n_s$. Indeed, the local photocurrent is to be calculated in the 3D diffusion model by integrating the film-normal gradient of $n$ over the diode area, whereas the 2D model with $d$=0 assumes that the local photocurrent density varies in proportion to $n_s$. If the normal gradient of $n$ on the diode side of the film were varying everywhere in a strict proportion to $n_s$, then the coefficient $k_{ph}=\frac{j_{ph}\times\tau}{n_s}=\frac{D\times\partial n/\partial z\mid_{z=0}\times\tau}{n_s}=\frac{l_d^2\times\partial n/\partial z\mid_{z=0}}{\int_0^d n(z)dz}$ would retain its constant value throughout the whole film area covered by the photodiode, and the 2D diffusion model would then be capable of reproducing results of spot-scan measurements as adequately as the 3D diffusion model does. However, in reality, perfect proportionality of $\partial n/\partial z\mid_{z=0}$ to $n_s$ everywhere in the film is lacking because different areal parts of the film feature different distributions $n(z)$ formed in compliance with local conditions of the diffusion problem.

With the aim of evaluating the capability of the 2D model in approximating spot-scan data in the film region under a large-area photodiode, let us consider characteristic distributions $n(z)$ typical of different parts of the scanning interval (see Fig. 16). One characteristic case with $n(z) \propto \sin(\frac{\pi}{2} \times \frac{z}{d_{eff}})$ was discussed above; this case will be referred to below as Case 1 (Fig. 16(b)). Another, in a sense, opposite case is represented by a uniformly illuminated film region in which excess carriers appear due to their photogeneration in a thin layer of the absorber material at the film–substrate interface; this case will be referred to below as Case 2. Clearly, in Case 2, the distribution of $n$ across the film is given by a hyperbolic sine, $n(z) \propto \sinh(z / l_d)$ (Fig. 16(a)). Similarly to Case 1, for which we have found that $k_{ph}^{(1)} \approx (\frac{\pi}{2} \times \frac{l_d}{d_{eff}})^2$, we can calculate the value of $k_{ph}$ for Case 2 as well; then, we obtain: $k_{ph}^{(2)} = (2 \times (\sinh(\frac{d_{eff}}{2l_d}))^2)^{-1}$, so that we have $k_{ph}^{(2)} \to 2(\frac{l_d}{d})^2$ in thin films with $d/l_d \ll 1$. The latter value of $k_{ph}$ quite moderately (within a factor $\xi = \pi^2/8$) differs from the value of $k_{ph}$ that was obtained for Case 1. It can be suspected that in intermediate cases, in which photogeneration of charge carriers and their diffusion make comparable contributions to the density $n_s$ in the film region of interest, the coefficient $k_{ph}$ will take values from the interval between $k_{ph}^{(2)}$ and $k_{ph}^{(1)}$. The near-unity magnitude of $\xi$ explains why the 2D diffusion model provides a good approximation to the 3D diffusion model in describing the profiles $n_s(x)$ formed in the film region under a continuous back-biased photodiode far enough from the diode edges.

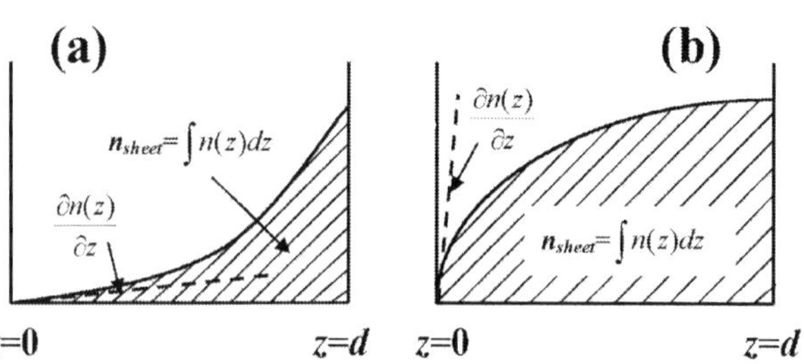


**Figure 16.** Typical distributions of excess carrier concentration across an MCT film with photocurrent suction in a uniformly illuminated film region (*a*) and in a film region where the excess charge carriers appear due to their lateral diffusion over the film by spreading from a nearby illumination spot (*b*).

In a real FPA structure, the film area occupied by the diode array involves regions with suction of charge carriers (under the back-biased diodes) and no-suction regions (in between the diodes). The excess carriers, as they diffusionally spread sideways from the illumination spot, sequentially pass those alternating regions. As a cloud of charge carriers moves, down the gradient of $n_s$, past a no-suction region, the distribution $n(z)$ flattens across the film. This flattening results in that there arises an increased fraction of sine harmonics with high spatial frequencies in the distribution $n(z)$ at the edge of the next diode approached by the cloud of excess carriers; the latter causes an increase in the local photocurrent density at this diode edge in comparison with the value of $j_{ph}$ predicted by the 2D diffusion model as the latter model disregards the evolution of $n(z)$ across the absorber layer. In an improved 2D diffusion model, the increase in the photocurrent density at the diode edge might have been allowed for via a nonuniform distribution of $k_{ph}$ over the diode area, with an increased value of $k_{ph}$ occurring at the diode edge. Apparently, the enhanced rate of withdrawal of charge carriers out of the photosensitive film at the diode edge would result in that, during the further advancement of the cloud in under the photodiode, the extracted local photocurrent will become smaller than the photocurrent predicted by the initial 2D diffusion model with areally uniform value of $k_{ph}$. It seems that the realistic values of the coefficient $k_{ph}$ and length $l_{d\ eff}$ obtained by fitting the measured spot-scan profile $S(x)$ with the profiles $S(x)$ calculated by the 2D diffusion model result from partial compensation of the two above-mentioned phenomena, this compensation occurring as the scanning photodiode "integrates" the photocurrent density $j_{ph}$ over its own area during spot-scan measurements.

This consideration substantiates the use of the 2D diffusion model for approximating spot-scan data obtained for an arbitrary level of diode photocurrents and explains why this model yields rather realistic local diffusion-length values for the analyzed MCT IR FPA detectors.

### 3.3.2.5. *Refined procedure for determining the length $l_d$*

With the found values $k_{ph\ opt}$, using the solutions $n_s(x,y)$ of the 2D diffusion problem obtained for each value of $V_g$ implemented in our experiments, we can calculate the electron photocurrent $I_{o\ calc}$ that flows across the *p-n* junction of the spot-centered measuring diode. In Fig. 17, the calculated values of $I_{o\ calc}$ are compared to the photocurrents $I_{o\ exp}$ that were "measured" by the detector (in Fig. 13, the latter photocurrents, evaluated from the voltage photosignal $\Delta V_0$, were denoted as $I_0$). The values of $I_{o\ calc}$ were calculated as relative values, and those relative values were subsequently normalized to ensure matching between the saturation levels of the photocurrents $I_{o\ calc}$ and $I_{o\ exp}$ at large potentials $V_g$ (see Fig. 17). An appreciable difference between the two photocurrents observed at reduced potentials $V_g$ for the LWIR FPA D-2 detector (with a smaller bandgap energy of the absorber material) can be attributed to noticeable recombination of photoelectrons with photogenerated holes in the *n*-region of the diode, which approached open-circuit conditions as the potential $V_g$ decreased. Indeed, under such conditions the built-in potential for holes in an illuminated diode loaded to a high-resistance external circuit (high FET-channel resistance) becomes reduced due to accumulation of photoelectrons in the *n*-region (which now cannot be withdrawn into the readout circuit in sufficient amounts), and this reduction provides conditions for a nonnegligible flow of excess holes out of the *p*-type MCT film into the *n*-region, where the holes can recombine with photoelectrons. Under the forward-biased junction, the related electron flux removes photoelectrons out of the cloud of excess carriers diffusionally spreading over the *p*-type photosensitive film; yet, it makes no contribution to the photocurrent "measured" by the detector. The latter circumstance (the existence of an ambipolar flux of excess carriers into the *n*-type diode region out of the *p*-type MCT film and subsequent recombination of those carriers in this region) violates the previous tacitly adopted assumption that the photocurrent $I_{o\ exp}$ always presents a good approximation to the electron photocurrent being extracted by the photodiodes out of the photosensitive film. Indeed, due to a nonzero rate of "surface recombination" of excess charge carriers on the film surface covered by the photodiode, the electron flux across the *p-n* junction remains finite even as the net photocurrent through the forward-biased photodiode loaded to the high-resistance load tends to zero. This point can be comprehended by considering the diagram of the photocurrent components in the *p-n* junction shown in Fig. 18.

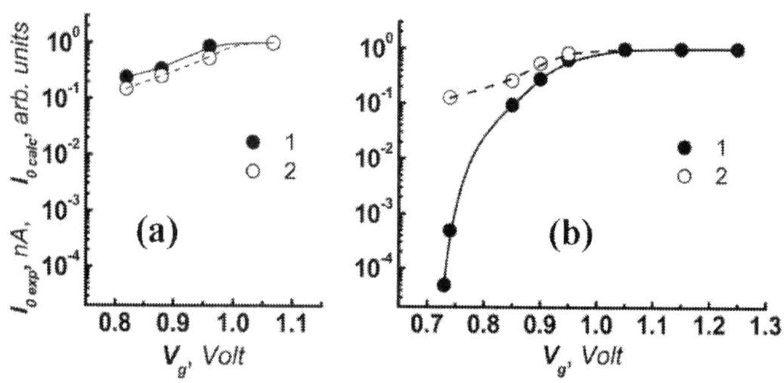


**Figure 17.** Comparison of the photocurrent values $I_{o\ exp}$ evaluated from the measured voltage photosignals $\Delta V$ (*1*) for the spot-centered diode with the photocurrent values $I_{o\ calc}$ calculated by the 2D diffusion model (*2*) for the D-1 (*a*) and D-2 (*b*) detectors. The data for D-2 were normalized to make them refer to the same IR radiation intensity in the illumination spot.

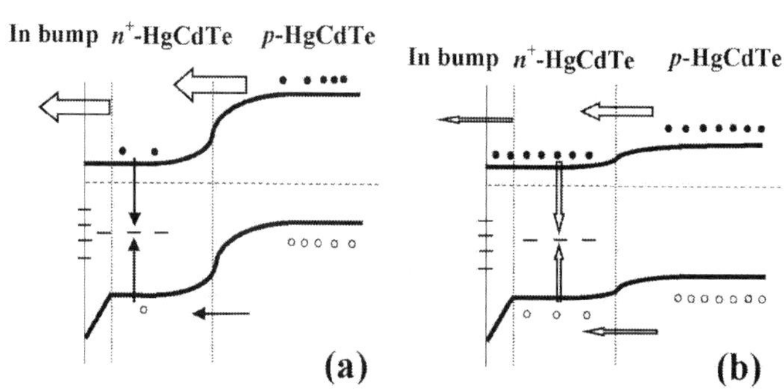


**Figure 18.** Photocurrent components in the vicinity of a diode junction at a high (*a*) and low (*b*) value of $V_g$. The diagram illustrates possible occurrence, at low values of $V_g$, of noticeable recombination of photoelectrons and photoholes in the *n*-type region of the photodiode, the photoelectrons and photoholes having been generated in the *p*-type absorber layer. The arrows show the electron and hole fluxes; the arrow sizes give a rough idea of the magnitude of the shown photocurrent components.

Thus, it is the photocurrent $I_{o\,calc}$ that had to be plotted instead of photocurrent $I_{o\,calc}$ along the abscissa axis in Fig. 13. Within the context of the present study, it is essential that the readily stemming modification of the determination procedure for the length $l_d$ (using $I_{o\,calc}$–values instead of $I_{o\,calc}$–values in plotting the $l_{deff}$-vs-$I_0$ graphs) typically induces quite moderate or even negligible corrections to the initially obtained values of $l_d$. Indeed, for the D-2 detector the modified procedure yields a value $l_d \approx 25.5$ μm, this length being rather close to the previously obtained value of 24 μm. For the D-1 detector, the modified procedure yields a value of $l_d$ being perfectly coincident with the previously obtained estimate $l_d \approx 19.5$ μm.

## 4. Conclusions

Two novel methods for determining the bulk diffusion length of minority charge carriers in photosensitive MCT films have been proposed.

The first method, suitable for determining the length $l_d$ in MCT films with suppressed surface recombination of excess carriers, uses tailored diode structures with photodiodes whose vicinity is shielded from incident radiation by a coaxial metal contact. Photocurrent measurements performed on the diodes with different radii of cap contacts can be used to determine the length $l_d$ via a comparison of measured photocurrents with the photocurrents numerically calculated for examined diode configurations. Experimentally, for an Hg-vacancy-doped $Cd_xHg_{1-x}Te$ film with $x$=0.223 and hole concentration $p$=6.7x10$^{15}$ cm$^{-3}$ $l_d$-values ranging between 19 and 23 μm were obtained.

The second method, which can be used for determining the length $l_d$ in the continuous (thickness-uniform, without mesa-isolation of diodes) absorber layers of MCT 2D IR FPA detectors, is based on an analysis of spot-scan data obtained for such detectors at different levels of diode photocurrents. Experimental data gained for $n$-on-$p$ MCT 2D MWIR and LWIR FPA detectors were analyzed using a 2D diffusion model taking the discrete structure of FPA into account. The properties of the used model were discussed to show that this model could indeed be applied to the analysis of the diffusion process under study. As a result, a general scheme for a comprehensive spot-scan analysis of MCT IR FPA detectors has been proposed. The performed analysis has yielded quite realistic bulk and local diffusion-length values for charge carriers in the film regions under and outside FPA diodes in the examined MWIR and LWIR FPA detectors. Namely, for MCT MWIR and LWIR FPA detectors with long-wave cutoff wavelengths ~5.4 and ~10 μm, for the bulk electron diffusion length the values of $l_d$≈19.5 и 24 μm were obtained. The latter values comply fairly well with the values obtained by the first method, and they are in a good agreement with relevant data which were previously reported in the literature. Simultaneously, the estimated value of the local diffusion length of minority carriers in the film region under the back-biased FPA diodes, $l_{d\,eff\,lat} \approx$ 4-5 μm, proved to be consistent with a theoretical estimate of this length $l_{d\,eff} \approx \frac{2d_{eff}}{\pi}$, where $d_{eff}$ is the effective PF thickness.

We believe that the newly proposed methods for determining the diffusion lengths of charge carriers in MCT films will add to the toolkit of characterization means for MCT-based IR FPA detectors.

## Author details

S.A. Dvoretsky, V.V. Vasil'ev, A.V. Predein, A.V. Vishnyakov, V.A. Stuchinsky*, D.V. Brunev and A.V. Zverev

*Address all correspondence to: stuchin@isp.nsc.ru

Institute of Semiconductor Physics, Russian Academy of Science, Siberian Branch, Novosibirsk, Russia

## References

[1] Rogalski A. *Infrared Detectors*. Boca Raton–London–New York: CRC Press, Taylor & Francis Group; 2011.

[2] Aseev A.L., (ed). *Photodetectors Based on the Cadmium-Mercury-Tellurium Epitaxial System*. Novosibirsk: SB RAS Publishing House; 2012.

[3] Levy D., Schacham S.E. Three-dimensional excess carrier distribution in semiconductor imaging array. *J. Appl. Phys.* 1988; 64(10) 5230-5233.

[4] Holloway H. Theory of lateral-collection photodiodes. *J. Appl. Phys.* 1978; 49(7) 4264-4269.

[5] Musca C.A., Dell J. M., Faraone L., Bajaj J., Pepper T., Spariosu K., Blackwell J., Bruce C. Analysis of crosstalk in HgCdTe p-on-n heterojunction photovoltaic infrared sensing arrays. *J. Electron. Mater.* 1999; 28(6) 617-623.

[6] Boltar K.O., Mansvetov N.G., Stafeev V.I., Yakovleva N.I. Inter-element crosstalk in IR focal plane arrays. *J. Opt. Technol.* 2000; 67(2) 153-156.

[7] Sanders T.J., Caraway E.L., Hess G.T., Modeling and test of pixel crosstalk in HgCdTe focal plane arrays. *Proc. of SPIE* 2001; 4369 458-466.

[8] Karp L., Musca C.A., Dell J.M., Faraone L. Characterization of crosstalk in HgCdTe n-on-p photovoltaic infrared arrays. *Proc. of SPIE* 2004; 5274 183-193.

[9] Yinghui S., Bo Z., Nuifang Yu., Oingjin L., Yan Z., Xin W., Peilu J., Xiaoning H., Ning D. Crosstalk of HgCdTe n-on-p diode arrays. *J. Semicond.* 2009; 30(9) 094007-(1-4).

[10] Pultz G.N., Norton P.W., Krueger E.E., Reine M.B. Growth and characterization of *p*-on-*n* HgCdTe liquid-phase epitaxy heterojunction material for 11-18 μm applications. *J. Vac. Sci. Technol.* B 1991; 9(3) 1724-1730.

[11] Reine M.B., Marschoff K.R., Tobin S.P., Norton P.W., Mroczkowski J.A., Krueger E.E. The impact of characterization techniques on HgCdTe infrared detector technology. *Semicond. Sci. Technol.* 1993; 8 788-804.

[12] Predein A.V., Vasil'ev V.V. Method for measuring the diffusion length of minority carriers in semiconductors and a test structure for its realization. Patent RU 2501116 C1, Application No. 2012124443/28, 13.06.2012 (in Russian).

[13] Predein A.V. Application of a test structure with shielded photodiodes for measuring the electron diffusion length in epitaxial *p*-type MCT layers, Proc. of the 23-th ISEC on Photoelectronics and Night Vision Devices, 28-30 May 2014, Moscow, Russia. Moscow: SPA Orion; 2014 (in Russian).

[14] Osadchy V.M., Suslyakov A.O., Vasil'ev V.V., Dvoretsky S.A. Effective lifetime of charge carriers in HgCdTe variband structures. *Fizika i Tekhnika Poluprovodnikov* 1999; 33(3) 293-296 (in Russian).

[15] Vasil'ev V.V., Predein A.V. Influence of graded p-P heterojunction's potential barrier on characteristics of three-dimensional HgCdTe photodiode. *Proc. SPIE* 2005; 5834 83-91.

[16] Varavin V.S., Dvoretsky S.A., Kostyuchenko V.Ya., Ovsyuk V.N., Protasov D.Yu. Mobility of minority charge carriers in p-HgCdTe films. *Semiconductors*. 2004; 38(5) 514-519.

[17] Schacham S.E., Finkman E. Recombination mechanisms in p-type HgCdTe: freezout and background flux effects. *J. Appl. Phys.* 1985; 57(6) 2001-2009.

[18] Vishnyakov A.V., Stuchinsky V.A., Brunev D.V., Zverev A.V., Dvoretsky S.A. Determination of charge-carrier diffusion length in the photosensing layer of HgCdTe n-on-p photovoltaic infrared focal plane array detectors. *Appl. Phys. Lett.* 2014; 104 092112-(1-4).

[19] Varavin V.S., Dvoretsky S.A., Marchishin I.V., Mikhailov N.N., Predein A.V., Remesnik V.G., Sabinina I.V., Sidorov Yu.G., Suslyakov A.O. 320x256 HgCdTe IR FPA with a built-in shortwave cut-off filter. *Opto-Electron. Rev.* 2010; 18(3) 236-240.

[20] Predein A.V., Sidorov Yu.G., Sabinina I.V., Vasil'ev V.V., Sidorov G.Yu., Marchishin V.I. High-performance 320x256 long-wavelength infrared photodetector arrays based on CdHgTe layers grown by molecular beam epitaxy. *Optoelectron. Instrum. Data Process.* 2013; 49(5) 485-491.

[21] Itsuno A.M. Bandgap-engineered HgCdTe infrared detector structures for reduced cooling requirements. Dissertation for the Degree of Philosophy Doctor (Electrical Engineering). University of Michigan; 2012.

[22] Sivukhin D.V. *The General Course of Physics, Vol. 4, Optics*. Moscow: Nauka, 1985 (in Russian).

[23] Vishnyakov A.V., Varavin V.S., Garifullin M.O., Predein A.V., Remesnik V.G., Sabinina I.V., Sidorov Yu.G. Effect of post-implantation annealing on the current-voltage characteristics of IR photodiodes based on p-HgCdTe. *Opto-Electron. Rev.* 2010; 18(3) 236-240.

[24] Vishnyakov A.V., Varavin V.S., Garifullin M.O., Predein A.V., Remesnik V.G., Sabinina I.V., Sidorov Yu.G. Effect of post-implantation annealing on the current-voltage characteristics of IR photodiodes based on p-HgCdTe. *Optoelectron. Instrum. Data Process*. 2009; 45(4), 308-315.

[25] Jung H., Lee H.C., Kim C.K. Measurement of the steady-state minority carrier diffusion length in a HgCdTe photodiode. *Jpn. J. Appl. Phys.* Part 2, 1996; 35(10B) L1321-L1323.

[26] Ikusov D.G., Sizov F.F., Staryi S.I., Teterkin V.V. Recombination mechanisms of nonequilibrium carriers in the $Cd_xHg_{1-x}Te$ (x=0.20 – 0.32) epitaxial layers. *Fizika Tekhnika Poluprovodnikov* 2007; 41(2) 134-139 (in Russian).

[27] Kinch M.A., Aqariden F., Chandra D., Liao P.-K., Schaake H.F., Shin H.D., Minority carrier lifetime in p-HgCdTe. *J. Electron. Mater.* 2005; 34(6) 880-884.

# Chapter 8

# Polymer Micro/Nanofibre Waveguides for Optical Sensing Applications

Fuxing Gu, Li Zhang and Heping Zeng

Additional information is available at the end of the chapter

http://dx.doi.org/10.5772/60626

**Abstract**

This chapter focuses on polymer micro/nanofibre (PMNFs) waveguides and their applications in sensing applications. The PMNFs are functionalized by doping with dyes or blending with solvated polymers before the drawing process. Based on the evanescent wave-coupling technique, the excitation light is efficiently coupled into the PMNFs using silica-fibre tapers and guided along the long-length PMNF waveguides. Due to the tight confinement, the interaction of light with PMNFs is significantly enhanced. Intriguing advantages such as enhanced excitation efficiency, low excitation power operation and high photostability are obtained. On the basis of the optical response when exposed to specimens, functionalized PMNFs are used for humidity, $NO_2$, and $NH_3$ detection with high sensitivity and fast response. By using a simple and low-cost nanoimprinting technique, PMNF Bragg gratings are also demonstrated for strain sensing with a high sensitivity of −2.5 pm/με.

**Keywords:** Polymer, micro/nanofibre, waveguides, sensors, evanescent wave coupling

## 1. Introduction

In the past few decades, nanotechnology has created a tremendous amount of excitement in various scientific and technological areas, and it is anticipated that it will revolutionize the world in the future. Fundamental to this revolution is the development of novel nanomaterials. Due to size-related effects, nanomaterials exhibit unique property advantages over their bulk forms, including in physical, chemical, optical, electrical, magnetic and other properties, which can be used to develop novel sensors [1–9]. In contrast to electrical schemes, optical sensing offers potentials of high sensitivity, fast response, immunity to electromagnetic interference, and safe operation in explosive or combustive atmospheres, as well as more options for signal retrieval from optical intensity, spectrum, phase, polarization, and fluorescence lifetime. Nanomaterial-based optical sensors can also be expected to exhibit further advantages, such as higher-integration density and higher efficiency of energy utilization, over conventional sensors.

Recently, polymer micro/nanofibres or nanofibres (PMNFs) have been proven to be promising building blocks for integrating nanoscale optoelectronic devices [10–27]. Compared to those of other glass and semiconductor nanowires and nanofibres, polymers [28] offer a number of highly attractive advantages for sensing applications. First, gas molecules to be detected can be either selectively bound to their surface or diffused into the polymer matrix, which is difficult for other materials such as semiconductors and glasses. Second, polymers can be doped with a wide range of functional dopants, from metal oxides and fluorescent dyes to enzymes that can be used to tailor the properties of the PMNFs with greater versatility, which offers plentiful choices for sensing schemes. Thirdly, operations including doping and drawing are easy to carry out at room temperature, and the doping concentrations are higher than those in glass fibres. Other advantages of polymer materials such as mechanical flexibility, biocompatibility, easy processing and surface modification for the attachment of active compounds [29], and low cost, offer further opportunities for PMNFs over semiconductors and glass fibres in optical sensing applications. Due to these advantages, various sensing devices based on PMNFs have been demonstrated. In this chapter we will focus on the PMNF waveguides, in which the light is coupled based on the evanescent wave-coupling technique, and their optical sensing applications.

## 2. Optical guiding in PMNFs

Usually the PMNFs used have lengths of around several hundred micrometres and diameters of around several hundred nanometres. Compared with irradiation excitation, which relies on free-space light (Figure 1a), the optical guiding approach in wavelength-scale waveguides can significantly enhance the interaction of light with materials [30–33]. First, the optical guiding configuration that forces light along the length of the PMNF (~100 μm) offers a much larger absorbing length than the thickness of the PMNF (~100 nm) in the irradiation approach. Second, in the irradiation configuration only a small fraction (less than 1%) of irradiated light can be intercepted by the PMNFs, while the evanescent coupling technique with adiabatic fibre taper condensation ensures high coupling efficiency (up to 50%) for transferring light into the PMNFs much more efficiently. Experimental results show that in a 380-nm-diameter 520-μm-length RhB-PS nanofibre, enhanced emission power as high as ~2000 times greater magnitude over the irradiation approach is observed [31]. Theoretical calculation also shows that a visible light with input power of 1 mW is capable of generating a power density of ~1 GW/m$^2$ inside a 500 nm diameter PMNF. Based on this approach, photodetectors [30,32,33], lasers [34], optical sensors [35–39] and light-emitting nanofibres [31] have been demonstrated in polymer, semiconductor, and glass micro/nano-waveguides. Compared with the irradiation with micrometre-scale light spot, the optical guiding scheme is best adapted for tightly confined excitation with low cross-talk when multiple PMNFs are closely located – particularly desirable for miniaturization and high-density integration of light-emitting devices.

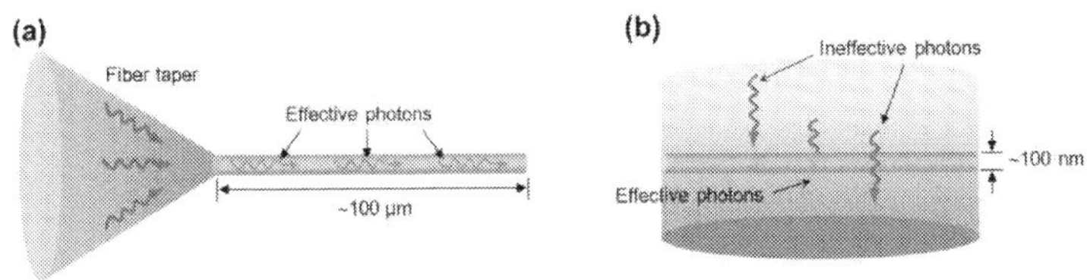


**Figure 1.** Schematic illustration of the optical guiding (a) and irradiation (b) configurations [30].

## 3. Fabrication of PMNFs

A number of different techniques, including electrospinning [23], templating [12], mechanical drawing [21,27,36–39], and thermal evaporation methods [40] have been used to fabricate PMNFs. Here we focus on the direct drawing method due to its advantages of simplicity, low cost, and room-temperature operation. As illustrated in Figure 2a, a tungsten probe with a sharp tip (with a diameter of several micrometres) fabricated using an electrochemical etching method was used to transfer a small droplet out of the polymer solution (packed in a quartz cuvette as shown in Figure 2b) onto a glass slide, and then was quickly moved far from to draw a wire out of the droplet. The solvent evaporates instantaneously, leaving a polymer nanofibre on the glass slide (Figure 2c). For example, by using a tungsten probe, polystyrene (PS, *M*w = 100,000; Alfa Aesar) micro/nanofibres are drawn from a chloroform solution containing 5 wt. % PS. Poly-(methyl methacrylate) (PMMA) nanofibres are drawn from an acetone solution containing 5 wt. % PMMA. Polyacrylamide (PAM) nanofibres are drawn from an aqueous solution containing 2 wt. % PAM (*M*w = 5,000,000-6,000,000; Fluka). If a polymer material can be drawn to nanofibres, a blending method can be used to fabricate composite nanofibres. For example, camphorsulphonic acid-doped polyaniline (PANI) nanofibres cannot be drawn from the chloroform solution, but when PS is added into the solution it is easy to draw PANI/PS nanofibres from it. In addition, doped PMNFs can be drawn from a solution that initially dissolves functional dopants such as pH indicators and fluorescent dyes. For example, 250 mg of PS (*M*w = 100 000, Alfa Aesar) and 0.3 mg of rhodamine B (RhB, Alfa Aesar) are dissolved into 2 g of chloroform, and then the mixture is stirred to form a uniform solution.

Using this drawing method, various functionalized PMNFs are obtained with lengths of up to several millimetres with diameters selectable from 100 to several micrometres (Figure 2d), which shows good wave-guiding performance that is easy to handle and has high coupling efficiency at the range of visible and near-infrared wavelength. Figure 3 shows a typical scanning electron microscope (SEM) image of a 310-nm-diameter PS nanofibre doped with

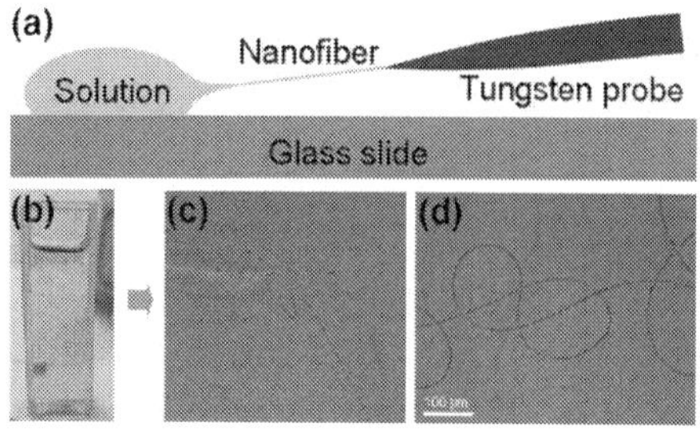


**Figure 2.** (a) Schematic diagram of the fabrication of the PMNFs. (b) The polymer solution packed in a quartz cuvette. (c) The polymer film and nanofibre after the solvent evaporating. (d) The drawn PMNF [31,38].

RhB, in which the excellent uniformity and sidewall smoothness is clearly seen [31]. For individual PMNFs, the variations of diameter $\Delta D$ are very small. For a certain doped PS nanofibre of a diameter of 400 nm, $\Delta D$ is about 30 nm over a 1-mm length. The high-resolution SEM image (inset) shows the surface of a doped PS nanofibre clearly and no obvious defect such as porosity is observed.

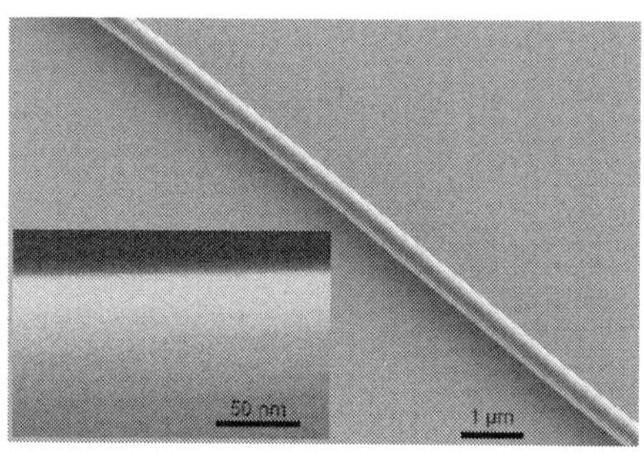


**Figure 3.** SEM image of a 310-nm-diameter RhB-PS nanofibre. Inset, high-resolution SEM image of the nanofibre [31].

## 4. Micromanipulation of PMNFs

As-fabricated PMNFs are tailored and manipulated also using tungsten probes (driven by three-axis precision stages) under an optical microscope equipped with super-long-working-distance objectives [31,38,41]. The tungsten probes, with tip sizes of less than 100 nm, can be used to cut PMNFs (Figure 4a). The fibre tapers with sharp tip diameters of less than 300 nm, fabricated from a standard single-mode silica optical fibre (SMF-28e, Corning), can be used to pick up (Figure 1c), transfer, and deposit the PMNFs onto a low-index $MgF_2$ substrate (refractive index ~1.39), as shown in Figures 4b and 4c.

To obtain a high-stability optical connection and a robust sensing operation, sometimes the PMNFs need to be bonded on a substrate, or their coupling areas with fibre tapers are enclosed to avoid external interference. For example, in humidity sensing when exposed to high/low RH atmosphere, the PAM nanofibre swells/shrinks slightly due to the diffusion of water molecules, which may change the coupling efficiency between the PAM nanofibre and the fibre taper. The fluoropolymer enclosure can isolate the coupling area from the environment, thus eliminating this effect. For this, the PMNFs or the fibre taper to be enclosed is first placed on the surface of the substrate and positioned using tungsten probes driven by micromanipulators. A micro droplet of a UV-curable fluoropolymer is picked up and cast upon the nanofibre and/or the fibre taper, and is then cured for 10 seconds with a mercury lamp under a nitrogen-gas atmosphere. The refractive index of the cured fluoropolymer is around 1.38 within the visible spectral range, making it suitable for optical isolation of polymer PMNFs with considerably higher indices (e.g., index of the PS is about 1.59). For reference, Figure 4d shows a close-up optical microscope image of a typical coupling area of a 400-nm-diameter PAM nanofibre and a silica-fibre taper (about 550 nm in diameter), which is enclosed by a cured fluoropolymer droplet on a $MgF_2$ substrate [38].

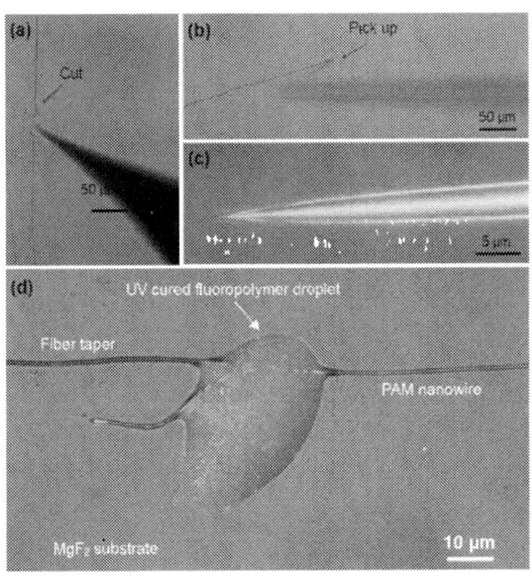


Figure 4. (a) Cutting a nanofibre using a sharp tungsten probe. (b, c) Picking up a nanofibre using a silica-fibre taper with a sharp tip size less than 300 nm. (d) Optical microscope image of a coupling area of a PAM nanofibre and a fibre taper enclosed by a UV-cured fluoropolymer droplet on an $MgF_2$ substrate [31,38].

## 5. Light launching and collection

To couple the excitation light efficiently into the PMNFs, we employed the evanescent wave-coupling technique due to its high efficiency and high compactness [31,38,41–44]. In this approach, light is first lens-coupled into a standard silica fibre and then squeezed into a fibre taper with tip diameter ranging from 300 to about 1 μm. The fibre taper is drawn from a standard optical fibre (SMF-28, Corning) by using a simple flame-heated method. Because the fibre taper is directly connected to the standard optical fibre through the tapering region, the evanescent coupling scheme provides perfect connection between the outer fibre system for processing optical signals and the single micro/nanofibres for optical sensing. As schematically illustrated in Figure 5a, under an optical microscope (Nikon 80i) equipped with super-long-working-distance objectives and using a triple-axis micromanipulator (M-462, Newport), by precisely placing the fibre taper and the PMNF in parallel and close contact within an overlap of a few micrometres, optical near-fields in the fibre taper and the PMNF can strongly overlap, resulting in highly efficient coupling. The close contact between the PMNFs and the fibre tapers can be maintained by van der Waals and electrostatic attraction. The output signals are collected using another fibre taper coupled to the PMNF from the right side as shown.

For active PMNFs, it is convenient to collect outputs such as photoluminescence (PL) using microscope objectives, as illustrated in Figure 5b. The signals are directed to a spectrometer and a CCD camera, respectively. To investigate the polarization behaviours of the emissions, linear polarizers and emission filters are placed between the samples and the detectors.

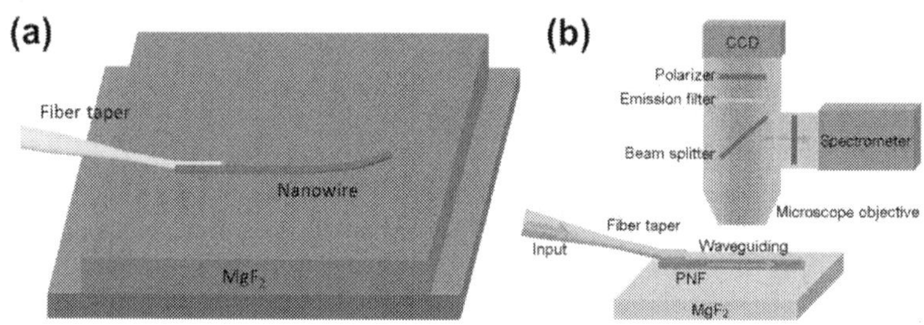


**Figure 5.** (a) Schematic diagram of light coupling into a nanowire using a fibre taper. The excitation light is launched into the nanofibre based on the evanescent wave technique, which is precisely controlled by a micromanipulator. (b) Schematic diagram of an experimental setup for optical guiding of the single nanofibre supported with a low-index $MgF_2$ substrate. The output and PL emissions are picked up using a long-working-distance objective. Polarizers and emission filters are placed between the samples and the detectors (the spectrometer and the CCD) [31,38].

## 6. Optical guiding properties of polymer MNWs

Optical waveguides are the basic elements in optical circuits, so we first investigate the optical guiding properties of PMNFs on the substrate. Figure 6a shows a light from a broadband supercontinuum guided through a 440-nm-diameter PS nanofibre (refractive index ~1.59) supported by an $MgF_2$ crystal (refractive index~1.39) [38]. It is observed that at the coupling area the broad input light is orange, but at the output end of the nanofibre a green light spot is observed, indicating an obvious "filtering effect" [38,45,46]: when light is guided along the nanofibre, the light with longer wavelength will leak a larger fraction of evanescent wave into the substrate, inducing higher optical losses compared with light with shorter wavelength. It is important to point out that although this filter is not helpful for low-loss optical guiding, it can be exploited to design wavelength-dependent sensors. When the PMNF is supported on the substrate, a short-pass filter effect is observed. When a 473-nm light is inputted into the nanofibre (Figure 6b), a big light spot is seen at the nanofibre end and only a small spot at the coupling area. A rough estimation shows that the coupling efficiency is as high as over 90%.

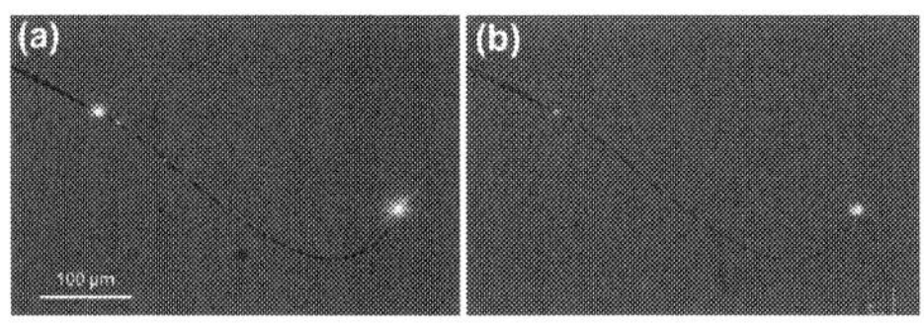


**Figure 6.** (a, b) Optical microscope image of launching a supercontinuum and 473-nm light into an $MgF_2$-supported 440-nm-diameter PS nanofibre using a fibre taper, respectively [38].

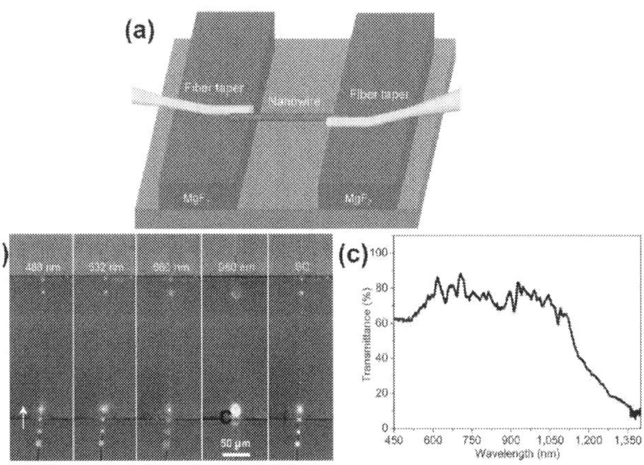


**Figure 7.** (a) Schematic diagram of a microchannel-supported nanofibre with two ends coupled with fibre tapers. (b) Optical micrographs of the nanofibre guiding a broadband supercontinuum (denoted as SC) and monochromatic lasers with wavelengths of 488, 532, 660, and 980 nm, respectively. Scale bar, 50 μm. (c) Broadband transmission spectrum of a 300-nm-diameter PMMA nanofibre [38].

To enable the PMNFs for broadband optical guiding and eliminate the leakage of guided light, we put the PMNFs across a microchannel, as illustrated in Figure 7a [38]. While the two ends of the nanofibre are supported and coupled with fibre tapers on the surface of the substrate, the main part of the nanofibre is placed on the channel. Figure 7b shows the optical images of a 300-nm-diameter PMMA nanofibre (refractive index ~1.49), guiding a broadband supercontinuum and monochromatic lasers (wavelengths of 488, 532, 660, and 980 nm, respectively). It is observed that there are bright light spots at the end of the nanofibre, indicating broadband transmittability of the nanofibre. The measured optical loss of the nanofibre is typically lower than 0.1 dB/mm, which can be neglectable due to the small effective length (e.g., less than 300 μm) of the nanofibre used. In addition, the suspension configuration also makes it possible to guide light using PMNFs with much smaller diameters, and this is very helpful to speed up the diffusion of specimens and subsequently the response of the nanofibre [7,10].

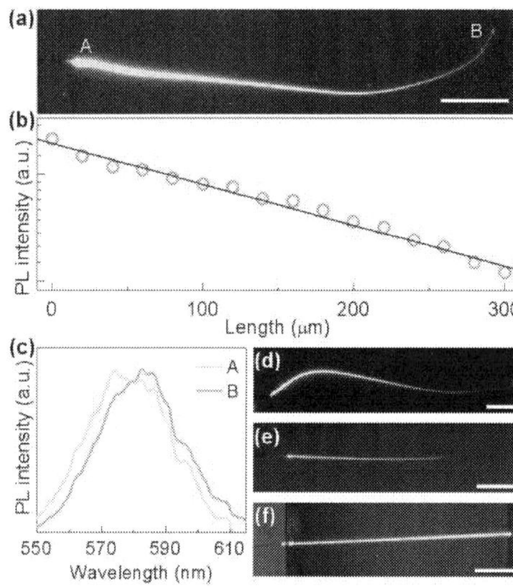


**Figure 8.** (a) PL optical microscope image of the launching of a 473-nm light into a MgF2-supported 470-nm-diameter RhB-PS nanofibre. (b) PL intensity of the RhB-PS nanofibre as a function of nanofibre length. (c) The comparison of the emission peak at the *A* point and the *B* point. (d) PL optical microscope image of the $MgF_2$-supported 600-nm-diameter RhB-PS nanofibre. (e) PL optical micrograph of the microchannel-supported 340-nm-diameter nanofibre. (f) PL optical micrograph of the $MgF_2$-supported 120-nm-diameter nanofibre [31].

Light-emitting sources are also important elements for developing ultra-compactness, so we next investigate the PL of the PMNFs. Figure 8a shows a PL microscope image of a 470-nm-diameter 340-μm-length RhB-PS nanofibre taken with a long-pass emission filter [31]. When a 473-nm laser ($\lambda_{ex}$) is launched from the left side with $P_{ex}$ = 100 nW, bright fluorescent emission is generated and guided along the nanofibre. The PL intensity of the RhB-PS nanofibre decreases exponentially with distance along the NF (Figure 8b), attributed to the exponential absorption of the pump light along the nanofibre by fluorescent dyes obeying the Lambert-Beer law [47]. The measured $\alpha$ of the RhB-PS nanofibre at 473 nm is ~50 cm$^{-1}$. In addition, it is also noticed that the emission peak ($\lambda_{em}$) is around 574 nm at the *A* point and 584 nm at the *B* point, as shown in Figure 8c. The redshift of the peak wavelengths is due to the reabsorption of the PL when guided along the nanofibre, and because of the filtering effect the longer wavelengths also suffer relative larger losses, inducing a weak red spot at the *B* point. When we use nanofibres with relative large diameters, a bright red spot is found at the output end of the nanofibre because the PL can be well guided along the nanofibre, as shown in Figure 8d. In contrast, when we use nanofibres with small diameters, it is found that the PL decays gradually, and no light spot at the output end is observed (Figure 8e) because the excitation light can be well guided along the nanofibre. When using the suspension configuration, the PL can be well excited and guided even using a 120 nm diameter nanofibre, as shown in Figure 8f.

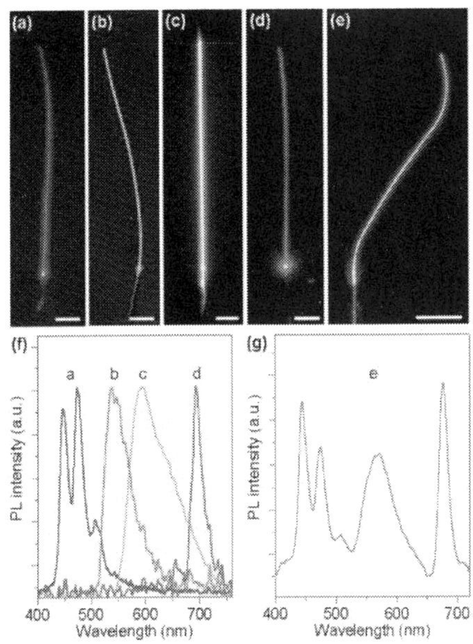


**Figure 9.** PL microscope images of (a) a perylene-PS nanofibre (380 nm in diameter); (b) a FSS-PAM nanofibre (450 nm in diameter); (c) a Ru(bpy)3Cl2-PAM nanofibre (270 nm in diameter); (d) a ZnPc-PEO nanofibre (610 nm in diameter); and (e) a perylene-RhB-ZnPc-codoped PS nanofibre (430 nm in diameter). In panel (a) λex = 355 nm; in panels (b) and (c) λex = 473 nm; in panel (d) λex = 650 nm and in panel (e) λex = 355 nm. The nanofibres in panels (a), (b), and (d) are supported by MgF2 substrate, and in panel (c) the nanofibre is placed over an MgF2 microchannel. Scale bar: (a–e) 50 μm; (f, g) are PL spectra corresponding to the doped PNFs shown in panels (a–e) [31].

We can also incorporate a variety of fluorescent dyes into PMNFs to generate multicoloured emissions. In Figures 9a to d, we show the optical microscope images of nanofibres doped with four fluorescent dyes [31]: (a) a 360-nm-diameter PS nanofibre doped with perylene, (b) a 450-nm-diameter PAM nanofibre doped with fluorescein sodium salt (FSS), (c) a 270-nm-diameter PAM nanofibre doped with tris(2,2′-bipyridine)ruthenium(II) chloride [$Ru(bpy)_3Cl_2$], and (d) a 610-nm-diameter PS nanofibre doped with zinc phthalocyanine (ZnPc). The FSS-PAM nanofibres are drawn from a water solution containing 0.04 wt. % FSS and 5.5 wt. % PAM. The $Ru(bpy)_3Cl_2$-PAM nanofibres are drawn from a water solution containing 0.07 wt. % $Ru(bpy)_3Cl_2$ and 5.5 wt. % PAM. The ZnPc-PS nanofibres are drawn from a tetrahydrofuran solution containing 0.07 wt. % ZnPc and 5.5 wt. % PS. Their corresponding PL spectra are given in Figure 9f, which covers the whole visible spectral range. Multicoloured or white-light emissions can also be obtained by simultaneously doping two or more fluorescent dyes in a single nanofibre. Here we show that white-light emission can be generated in perylene-RhB-ZnPc-codoped PS nanofibres drawn from a chloroform solution containing 0.008 wt. % perylene, 0.05 wt. % RhB, 0.22 wt. % ZnPc and 20 wt. % PS. As shown in Figure 9e, when excited by 355-nm light, white-light emission is observed in a perylene-RhB-ZnPc-codoped (430 nm in diameter) PS nanofibre when excited by 355-nm light. The three dyes are distributed homogenously along the whole NF according to the uniform emission in intensity and colour. Figure 9g shows the measured PL spectrum of the perylene-RhB-ZnPc-codoped PS nanofibre, in which the individual peaks agree well with those of the corresponding components in Figure 9f and also covers the whole visible range from 400 to 710 nm.

## 7. Optical sensing

Next we will introduce two typical PMNF-based sensors: (1) water-soluble polymer-nanofibre humidity sensors, which are based on refractive index changes of the nanofibres [38]; (2) pH-indicator-doped PS-nanofibre ammonia sensors, which are based on absorption changes of the nanofibres [38]. Based on these two underlying mechanisms, many other sensors based on PMNFs have been proposed and demonstrated: by using spectral analysis in the visible/near-infrared region, we introduce selective detection of gas mixtures of $NH_3$ and humidity [37]. Finally, by using a simple nanoimprinting approach, we introduce the PMNF Bragg gratings and their strain-sensing application [39].

### 7.1. Methods for gas sensing

Gas detection is particularly important in both environmental protection and human health. Here the gas-sensing experiments are carried out by placing the PMNFs in a sealed glass chamber (see Figure 10a and b) with a gas-flow system and a hygrothermograph for monitoring the relative humidity (RH) [38]. The analyte gas, diluted with air or nitrogen, is introduced while the transmittance of the nanofibre is measured. The mass-flow rate and concentration of the analyte gas are controlled by mass-flow controllers (MFCs). All experiments are carried out at room temperature and atmospheric pressure.

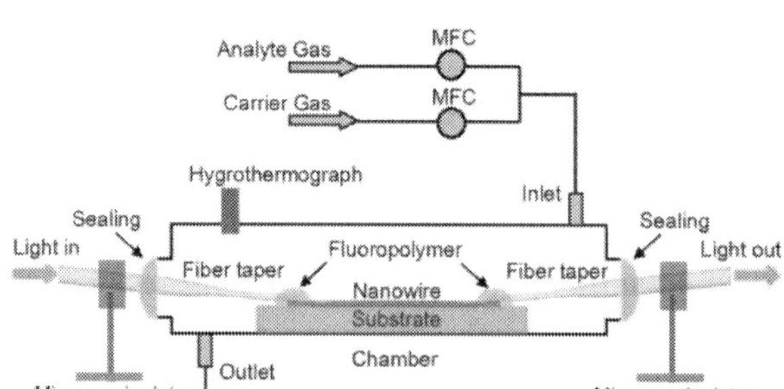


**Figure 10.** Schematic diagram of sensing experiment [38].

### 7.2. PMNF sensors based on refractive index changes

We first employ a PAM nanofibre (drawn from a PAM aqueous solution) for relative humidity (RH) sensing. As shown in Figure 11a, a 410-nm-diameter 250-μm-length PAM nanofibre is supported on an $MgF_2$ substrate, sealed and optically connected to fibre tapers at both ends for evanescent coupling of the probing light. The coupling area is enclosed by a low-index fluoropolymer to isolate it from the disturbing of the water molecules (see inset), because the coupling area is very sensitive to the refractive changes due to the adsorption of water molecules. Figure 11b shows the transmittance of the PAM nanofibre exposed to atmosphere with RH from 35% to 88%, with an average cut-off wavelength of around 545 nm due to the short-pass filter effect. Before the cut-off point, the transmittance decreases monotonously with the increasing RH. The insets show the intensity changes at the wavelength of 532 nm, in which the monotonous dependence can be clearly seen and can be used for RH sensing.

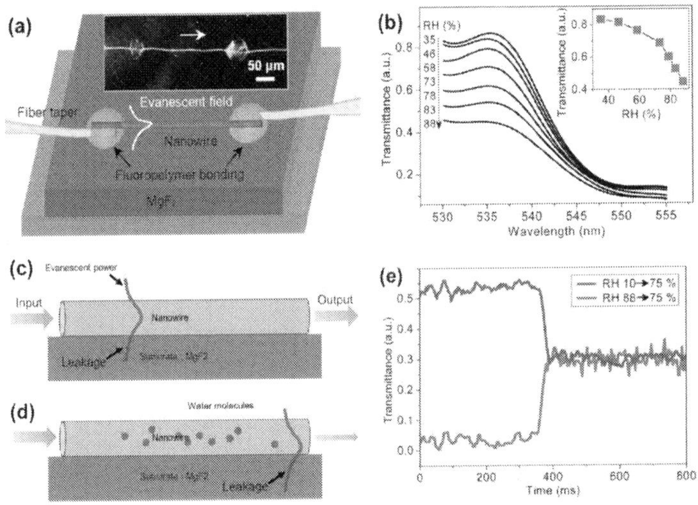


**Figure 11.** PAM single-nanowire humidity sensors. (a) Schematic illustration of the sensor. Inset, optical microscope image of a 410-nm-diameter PAM nanowire supported on $MgF_2$ with a 532-nm-wavelength light injected from the left side. The white arrow shows the direction of light propagation. (b) Transmittances of an $MgF_2$-supported 410-nm-diameter PAM nanowire exposed to an environment of RH from 35% to 88%. Inset, the variation of transmittances at 532 nm wavelength. (c–d) Schematic illustration of the humidity sensor based on evanescent wave leakage. (e) Typical time-dependent transmittance of the sensor reveals the response time of about 24 ms when RH jumps from 10% to 75% and 30 ms when RH falls from 88% to 75% [38].

The underlying mechanism can be explained as follows: when light is guided along the PAM nanofibre, some fraction of the evanescent wave will leak into the substrate [38,43,45,46], as shown in Figure 11c. With the addition of high-RH atmosphere, the refractive index of the PAM nanofibre (about 1.54) decreases due to the diffusion of water molecules, resulting in higher leakage of the guided light nearby and the blue shift of the cut-off wavelength, as shown in Figure 11d. Vice versa, with the addition of low-RH-atmosphere water inside the nanofibre evaporates and the refractive index increases. By alternately cycling 75% and 88%-RH air inside the chamber the nanofibre exhibits an excellent reversibility. By suddenly changing the humidity in the chamber, the response time of the nanofibre-based humid sensor, with typical time-dependent transmittance shown in Figure 11e. The estimated response time (baseline to 90% signal saturation) of the humid sensor is about 24 ms when RH jumps from 10 to 75%, and 30 ms when RH falls from 88 to 75%. The response time is one or two orders of magnitude shorter than that of existing RH sensors. The remarkably fast response of the humid sensor can be attributed to the small diameter of the nanofibres, which enables rapid diffusion or evaporation of the water molecules as well as fast signal retrieval using the optical approach.

### 7.3. PMNF sensors based on absorption changes

Doping chemical indicators is another approach to activate the polymer nanofibres for optical sensing. BTB is a pH indicator that has a pH range of 6.0 to 7.6, and can be used to detect basic materials. Here we demonstrate $NH_3$ gas sensing with a 270-nm-diameter BTB-doped PMMA nanofibre, which is drawn from a chloroform solution containing 0.5 wt. % BTB and 5 wt. % PMMA. The nanofibre is suspended by a 200-μm-width $MgF_2$ microchannel with a probing light of 660-nm wavelength. As shown in Figure 12a, when nitrogen-diluted $NH_3$ gas is introduced and diffuses into the nanofibre, the BTB reacts with the $NH_3$ and changes from acidic form to basic form, resulting in evident absorption of the probing light. Figure 12b shows the optical response of the nanofibre to $NH_3$ gas cycled with concentrations from 3 to 28 ppm at room temperature, showing linear response (see inset) for $NH_3$ sensing below 14 ppm with good reversibility. The response time with ammonia at 14 ppm is about 1.8 s, which is much faster than in conventional ammonia sensors.

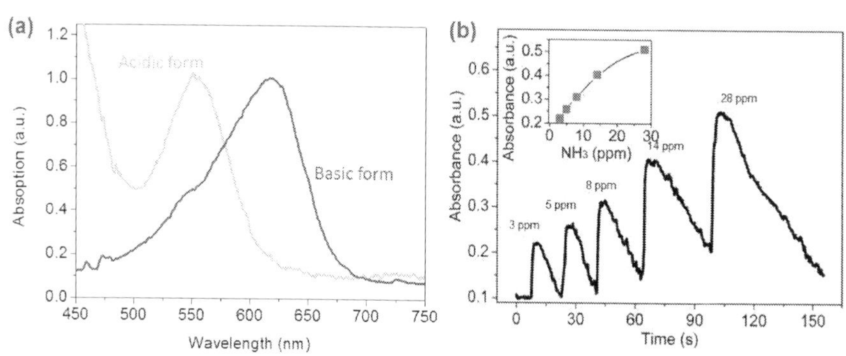


**Figure 12.** (a) Absorption spectrum of the BTB-doped PMMA nanofibre in acidic form and basic form. (b) Time-dependent absorbance of the nanofibre to $NH_3$ gas cycled with concentrations from 3 to 28 ppm. Inset, dependence of the absorbance on $NH_3$ concentration ranging from 3 to 28 ppm [38].

Because of the reversible optical and electrical response to certain gas species, as well as easy processing and environmental stability, doped PANI has been widely used for electrical or optical detection of a variety of gases such as HCl, $NH_3$, $H_2S$ and $NO_2$ [10, 24, 48]. Blending PANI with soluble matrix polymers is essential to fabricate PMNFs, which overcomes its poor mechanical property; the PMNF then shows good mechanical property and low optical loss. Here, PS is used to blend with PANI due to its compatibility with PANI and excellent optical properties, such as good transparency in visible and near-infrared regions and high refractive index (about 1.59). PANI/PS nanofibres are fabricated by direct drawing from polymer-blend solution of 2 wt. % PANI (*Mw* = 50 000; Fluka) doped with 10-camphorsulphonic acid (Alfa Aesar) and 5 wt. % PS (*Mw* = 100 000; Alfa Aesar) in chloroform.

When exposed to $NO_2$, the increase of the oxidation degree of PANI results in changes of the spectral absorption, as shown in Figure 13a. Here, a 250-nm-diameter PANI/PS nanofibre is suspended by a 250-μm-width $MgF_2$ microchannel and optically connected to fibre tapers at both ends. With the addition of 1 ppm $NO_2$ gas, a clear absorbance at the wavelength of 532 nm is observed in a 250-nm-diameter PANI/PS nanofibre. The response time is about 7 s, which is several orders of magnitude shorter than in other $NO_2$ sensors. With $NO_2$ concentration cycled from 0.1 to 4 ppm, the time-dependent absorbance of the nanofibre at room temperature is given in Figure 13b, indicating good reversibility of the nanofibre response.

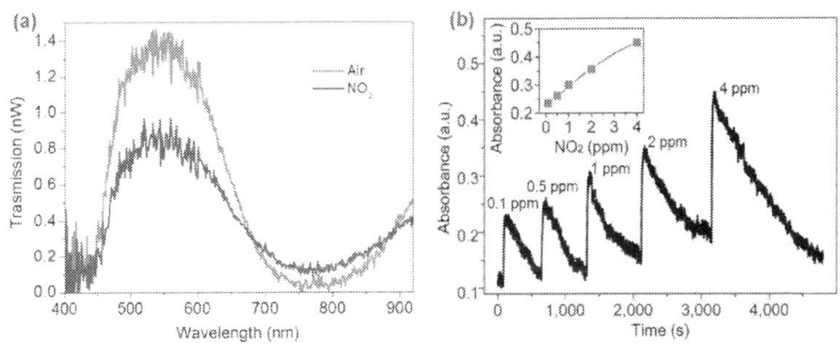


**Figure 13.** (a) Transmission spectrum of the PANI/PS exposed in air and $NO_2$ gas, respectively. (b) Time-dependent absorbance of the nanowire to cyclic $NO_2$/nitrogen exposure with $NO_2$ concentration from 0.1 to 4 ppm. Inset, dependence of the absorbance over the $NO_2$ concentration ranging from 0.1 to 4 ppm [38].

## 7.4. Selective detection of gas mixtures of ammonia and humidity

In electrical sensors relying on the electrical conductance change, it is difficult to determine individual responses in gas mixtures with cross-sensitivity. For example, it is difficult to determine individual responses in gas mixtures of $NH_3$ and humidity with a single PANI-based nanofibre by electrical detection because both of them have similar signal-change trends [49]; however, the spectral selectivity could provide identity information for an analyte at multiple wavelengths, and optical sensing offers the possibility of selective detection of gas mixtures using a single sensing element. Here we demonstrate that by using spectral analysis, highly selective detection of gas mixtures of $NH_3$ and humidity can be realized based on PANI/

PS single-nanofibre devices. When exposed to $NH_3$, camphorsulphonic-acid-doped PANI showing emeraldine salt (ES) form can be transferred to the emeraldine base (EB) form. The ES and EB forms absorb light at different spectral regions, which can be used for optical sensing [50]. For example, Figure 14a shows the output spectra output of a 370-nm-diameter PANI/PS nanofibre in dry air and in 20 ppm-$NH_3$ dry/dry air, respectively. A decrease in absorption ($A_r$) with around 600-nm wavelength and an increase in $A_r$ with around 830-nm wavelength are observed (Figure 14b). Because of the opposite response of the nanofibre to $NH_3$ at these two wavelengths, the differential absorbance can be used to quantify the $NH_3$ concentrations with enhanced sensitivity. The inset of Figure 14b plots the $NH_3$-concentration dependence of $\Delta A_{600\text{-}835}$; the monotonous dependence in relation to the $NH_3$ concentration indicates that the PANI/PS nanofibre could function as an $NH_3$ optical sensor in dry air.

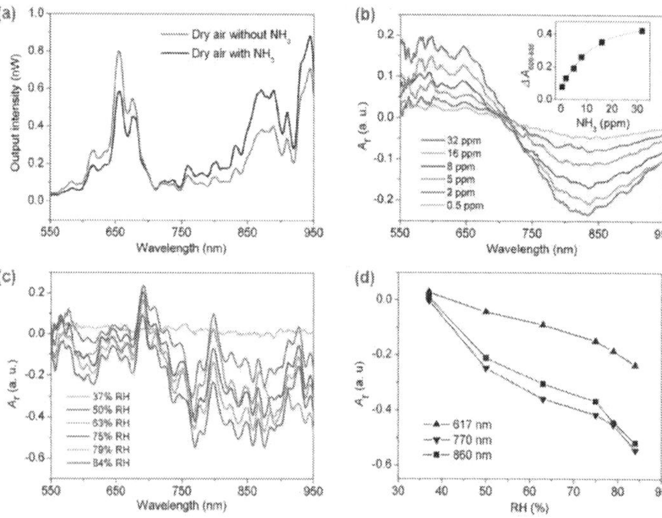


**Figure 14.** (a) Output intensity of a PANI/PS nanofibre (370 nm in diameter) in dry air without $NH_3$ (grey line) and with 20 ppm $NH_3$ (black line). (b) Absorption spectra of the PANI/PS (370 nm in diameter) nanofibre exposed to dry air while the $NH_3$ concentrations vary from 0.5 to 32 ppm. Inset shows the $NH_3$-concentration dependence of $\Delta A_{600\text{-}835}$. (c) Absorption spectra of a 350-nm-diameter PANI/PS nanofibre exposed to air with RH ranging from 37% to 84%. (d) RH-concentration dependence of $A_r$ at wavelengths of 617, 770, and 860 nm, respectively [37].

Figure 14c gives the absorption spectra of a 350-nm-diameter PANI/PS nanofibre with RH increasing from 37 to 84%, in which the valleys in the absorption spectra around the wavelengths of 617, 770 and 860 nm decrease with increasing RH level. The changes in the absorption spectra are due to the diffusion of water molecules, which decreases the refractive index of the PANI/PS nanofibre and subsequently changes the coupling efficiency. Figure 14d shows the RH-dependent $A_r$ of the nanofibre with RH ranging from 37 to 84% at the wavelengths of 617, 770 and 860 nm, respectively. The distinct response spectra and underlying mechanism for humidity and $NH_3$ are apparent and can be used to identify their individual responses using dual-wavelength measurements.

In gas mixtures of $NH_3$ and humidity, the absorbance of the nanowire at a given wavelength is caused by both $NH_3$ and RH. Here, 633-nm- and 808-nm-wavelength lights are used to simultaneously monitor the $A_r$ of the 350-nm-diameter PANI/PS nanofibre. Figure 15a shows the time-dependent response of the nanofibre exposed to (1) 84% RH air, (2) 5 ppm $NH_3$, (3) 5 ppm $NH_3$ with 79% RH air, and (4) 5 ppm $NH_3$ with 84% RH air, respectively. The bar graph in Figure 15b summarizes the response of the nanofibre to the analytic gases at 633-nm and 808-nm wavelengths. The height of each bar stands for the maximum $A_r$ of the responses to analytic gases at corresponding wavelengths, in which each analytic gas can be easily distinguished. By solving the simultaneous equations associated with the nanofibre absorptivity, the concentrations of individual gas components can be calculated, along with the sensing length and the fractional energy confined in the nanofibre at each wavelength.

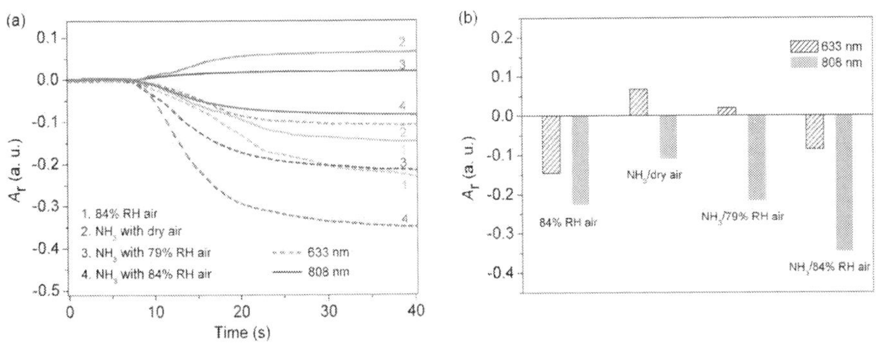


**Figure 15.** (a) Time-dependent response of the nanofibre exposed to (1) 84% RH air, (2) 5 ppm $NH_3$, (3) 5 ppm $NH_3$ with 79% RH air, and (4) 5 ppm $NH_3$ with 84% RH air, respectively, simultaneously monitored with 633 and 808 nm lasers. (b) Bar graph summarizing the optical response of the nanofibre to the analytic gases at wavelengths of 633 and 808 nm [37].

### 7.5. PMNF Bragg gratings and their strain-sensing application

Compared with glass material, polymers offer much higher mechanical pliability and smaller Young's modulus E (e.g., 1 GPa for poly(methyl methacrylate) (PMMA) versus 70 GPa for silica glass) [28, 51, 52]. These properties can be exploited to develop optical sensors with higher flexibility and lower force detection limits. Here, benefiting from the easy processing of polymer materials, we report PMNF Bragg gratings by using a simple and low-cost nanoimprinting approach, which consists of three typical steps, as illustrated in Figure 16 [39]. Firstly, PMNFs fabricated by a direct drawing method are placed on a piece of poly(dimethylsiloxane) (PDMS) film for mechanical supporting. Secondly, the PDMS film is heated to a temperature above the glass transition temperature of the PMNF material. A piece of standard-plane reflection grating is used as a mould and a certain pressure is applied onto it. Finally, after the glass substrate has cooled down to room temperature the mould is mechanically removed, with Bragg gratings imprinted on the surface of the MNFs. To optimize the grating effect for guiding modes of the MNF, the grooves of the grating corrugation are kept perpendicular to the MNF axis. In addition, this nanoimprinting approach can be extended to a variety of thermoplastic polymers.

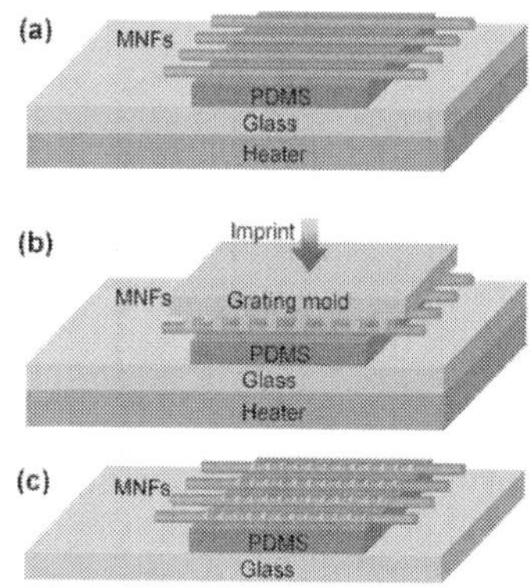


**Figure 16.** Schematic of the nanoimprinting procedure of PMNF Bragg gratings fabrication, including (a) mechanical supporting, (b) imprinting, and (c) mode removal [39].

Figure 17a shows a scanning electron microscopy (SEM) image of the grating area of an imprinted MNFBG from a 1.3 μm diameter PMMA MNF. It explicitly shows that the grating patterns are uniformly spaced without obvious asymmetric deformation, where the grooves of the grating corrugation are perpendicular to the fibre axis. The grating patterns also show quick gradual transition between the imprinted and un-imprinted regions of the MNF (Figures 17b and c) at the starting parts of the Bragg gratings. Figure 17d reveals an atomic-force-microscopy (AFM) image of a grating segment of 1.2-μm-wide Bragg gratings. Compared with the grating mould (Figure 17e), the Bragg gratings well replicate the 555-nm periodicity of the mould (Figure 17f), with a measured period of about 551 nm and a peak-to-peak depth of about 80 nm. In addition, although the groove depth of the PMNF Bragg gratings is smaller than that of the mould (about 270 nm), the index contrast ($\Delta n$ ~0.4) between PMMA and the surrounding air is large enough for the waveguide grating demonstrated here.

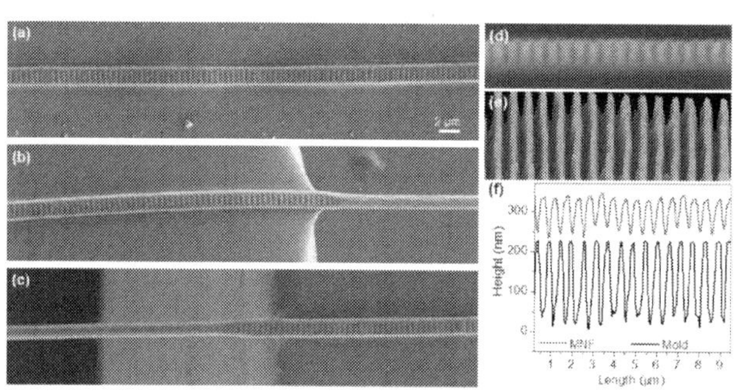


**Figure 17.** Characterization of PMNF Bragg gratings. (a–c) SEM images of the imprinted PMNF Bragg grating from a 1.3-μm-diameter PMMA PMNF. (d) and (e) AFM images of a grating segment of 1.2-μm-wide PMNF Bragg gratings and the grating mould used, respectively. (f) Groove profiles of the PMNF Bragg gratings and the mould [39].

http://dx.doi.org/10.5772/60626

As-fabricated PMMA PMNF Bragg gratings are placed across an $MgF_2$ microchannel by micromanipulation, with two ends coupled to silica fibre tapers for optical input and output coupling. Figure 18 shows the transmission and reflection spectra of a PMNF Bragg grating with thickness, width and grating lengths of 1.4, 2.3 and 200 μm, respectively. The two dips centred at 1517 and 1555 nm in the transmission spectrum agree well with the two peaks centred at 1519 and 1554 nm in the reflection spectrum, respectively. Due to the asymmetry of the rectangle-like cross-section and the large refractive index contrast between the MNFBGs and the surrounding air, the guided light with orthogonal polarizations experience high group birefringence, i.e., different modal effective indices ($n_{eff}$); at the Bragg wavelengths of 1532 and 1510 nm, the calculated quasi-x and quasi-y polarized modes are ~1.39 and ~1.37, respectively, agreeing well with the experimental results at the 1554- and 1519-nm peaks, respectively.

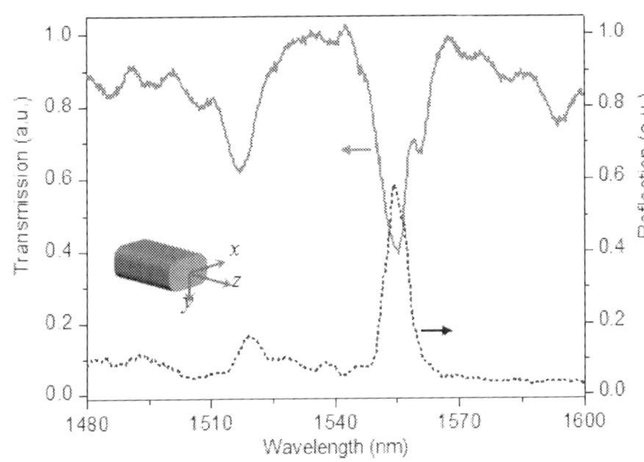


**Figure 18.** Reflection of PMNF Bragg gratings (2.3-μm wide by 1.4-μm thick) [39].

The strain responses of the PMNF Bragg gratings are investigated. As illustrated in Figure 19a, the PMMA PMNF Bragg grating is placed across an $MgF_2$ microchannel, and by moving the stage rightwards a tensile force is applied onto it. For robust operation, the coupling area of the fibre taper and the MNF are bonded together by a low-index UV-cured fluoropolymer. Figure 19b shows the transmission spectra changes of an MNFBG (2.5 μm wide by 1.4 μm thick) while the tensile strain increases from 0% to ~4%. The inset shows the Bragg wavelength's ($\lambda$) monotonous and linear blueshift from 1590 nm to 1485 nm. The $\lambda$ shifts of the polymer MNFBG (> 100 nm) are about 10-fold larger compared with those of silica FBGs (usually less than 10 nm). The strain sensitivity ($\Delta\lambda/\lambda$) at 1590 nm is −2.5 pm/με, which is much higher than that of typical silica FBGs (usually ~1.2 pm/με). The blueshift of the $\lambda$ caused by the tensile strain is different from previous results reported in conventional polymer and silica FBGs, because the grating patterns with a 100-nm depth show a remarkable change in the micro/nanoscale waveguide structures. Upon an axial tensile strain, the MNFBG suffers asymmetric strains and induces large photoelastic effect, which counteracts the elongation-induced redshift effect that is usually dominant in conventional FBGs.

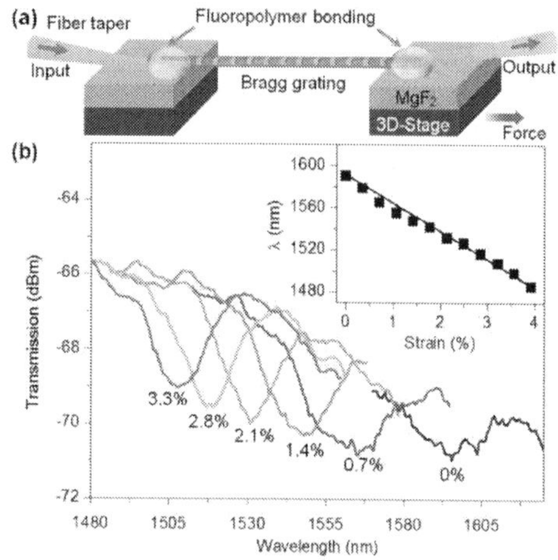


**Figure 19.** (a) Schematic of experimental setup for strain sensing. (b) Transmission spectra of an MNFBG (2.5 $\mu$m wide by 1.4 $\mu$m thick) with tensile strain increasing from 0% to ~4%. Top inset: Bragg wavelength shift of the 1590 nm dip under tensile strain from 0% to ~4% [39].

## 8. Summary

This chapter reviewed PMNF waveguides and their optical sensing applications. The polymer NWs are functionalized by doping with dyes or blending with solvated polymers before the drawing process. Based on the evanescent wave technique using fibre tapers, light is efficiently coupled into PMNFs and guided along the waveguides, with high efficiency and compactness within a broad spectral range. It is shown that in single PMNF waveguides, the tight confinement of light during its propagation along the long length significantly enhances the interaction of light with waveguides by over three orders of magnitude compared with the irradiating approach. Intriguing advantages such as enhanced excitation efficiency, low excitation power operation and high photostability are obtained. On the basis of the optical response when exposed to specimens, functionalized PMNFs are used for humidity, $NO_2$, and $NH_3$ detection with high sensitivity and fast response. Selective detection of gas mixture of ammonia and humidity are also realized by using spectral analysis. By using a simple and low-cost nano-imprinting technique, PMNF Bragg gratings are also demonstrated for strain sensing with a high sensitivity of −2.5 pm/με. In addition, the fibre tapers are connected to a standard optical fibre system and this compatibility may stimulate further exploration of PMNFs and PMNFs-based optical sensors for fast and high-sensitive detection in physical, chemical, and biological applications with remote sensing and multiplexing signals in one optical fibre.

## Acknowledgements

This work was partly supported by the National Natural Science Foundation of China (11304202 and 91221304), Natural Science Foundation of Shanghai (13ZR1458000), National Key Scientific Instrument Project (2012YQ150092), and National Basic Research Programme of China (2011CB808105).

## Author details

Fuxing Gu[1*], Li Zhang[1] and Heping Zeng[1,2,3]

*Address all correspondence to: fuxinggu@gmail.com

1 Shanghai Key Laboratory of Modern Optical System, Engineering Research Centre of Optical Instrument and System (Ministry of Education), School of Optical-Electrical and Computer Engineering, University of Shanghai for Science and Technology, Shanghai, China

2 State Key Laboratory of Precision Spectroscopy, East China Normal University, Shanghai, China

3 Synergetic Innovation Centre of Quantum Information and Quantum Physics, University of Science and Technology of China, Hefei, Anhui, China

## References

[1] Xia Y, Yang P, Sun Y, Wu Y, Mayers B, Gates B, Yin Y, Kim F, and Yan H. One-dimensional nanostructures: synthesis, characterization, and applications. Adv Mater. 2003;15(5):353-89.

[2] Stewart ME, Anderton CR, Thompson LB, Maria J, Gray SK, Rogers JA, et al. Nanostructured plasmonic sensors. Chem Rev. 2008;108(2):494-521.

[3] Soci C, ZA, Bao XY, Kim H, Lo Y, and Wang DL. Nanowire photodetectors. J Nanosci Nanotechnol. 2010;10(3):1430-49.

[4] Yan R, Gargas D, Yang P. Nanowire photonics. Nat Photon. 2009;3(10):569-76.

[5] Brambilla G. Optical fibre nanotaper sensors. Opt Fiber Technol. 2010;16:331-42.

[6] Ramgir NS, Yang Y, Zacharias M. Nanowire-Based Sensors. Small. 2010;6(16): 1705-22.

[7] Yang F, KS-C, Cheng M, Hemminger JC, and Penner RM. Smaller is Faster and More Sensitive: The Effect of Wire Size on the Detection of Hydrogen by Single Palladium Nanowires. ACS Nano. 2010;4(9):5233-44.

[8] Li Y, QF, Xiang J, and Lieber CM. Nanowire electronic and optoelectronic devices. Mater Today. 2006;9(10):18-27.

[9] Senesac L, aTTG. Nanosensors for trace explosive detection. Mater Today. 2008;11(3): 28-36.

[10] Liu HQ, KJ, Czaplewski DA, and Craighead HG. Polymeric nanowire chemical sensor. Nano Lett. 2004;4(4):671-5.

[11] Benedetto FD, CA, Pagliara S, Mele E, Persano L, Stabile R, Cingolani R, and Pisignano D. Patterning of light-emitting conjugated polymer nanofibres. Nat Nanotechnol. 2008;3:614-9.

[12] O'Carroll D, LI, and Redmond G. Microcavity effects and optically pumped lasing in single conjugated polymer nanowires. Nat Nanotechnol. 2007;2(3):180-4.

[13] Di Benedetto F, Camposeo A, Pagliara S, Mele E, Persano L, Stabile R, et al. Patterning of light-emitting conjugated polymer nanofibres. Nat Nanotechnol. 2008;3(10): 614-9.

[14] Camposeo A, Di Benedetto F, Stabile R, Neves AA, Cingolani R, Pisignano D. Laser emission from electrospun polymer nanofibers. Small. 2009;5(5):562-6.

[15] Liu H, EJB, Bellan LM, and Craighead HG. Electrospun polymer nanofibers as subwavelength optical waveguides incorporating quantum dots. Small. 2006;2:495-9.

[16] O'Carroll D, LI, and Redmond G. Melt-processed polyfluorene nanowires as active waveguides. Small. 2007;3:1178-83.

[17] Virji S, FJD, Baker CO, Huang JX, Kaner RB, and Weiller BH. Polyaniline nanofiber composites with metal salts: chemical sensors for hydrogen sulfide. Small. 2005;1(6): 624-7.

[18] Vohra V, CG, Destri S, Pasini M, Porzio W, and Botta C. Toward white light emission through efficient two-step energy transfer in hybrid nanofibers. ACS Nano. 2010;4:1409-16.

[19] Kemp NT, MD, Cochrane JW, and Newbury R. Bridging the gap: polymer nanowire device. Adv Mater. 2007;19(18):2634-8.

[20] O'Brien GA, QAJ, Tanner DA, and Redmond G. A single polymer nanowire photodetector. Adv Mater. 2006;18:2379-83.

[21] Ta VD, Chen R, Sun H. Coupled Polymer Microfiber Lasers for Single Mode Operation and Enhanced Refractive Index Sensing. Adv Optical Mater. 2014;2(3):220-5.

[22] Camposeo A, BFD, Stabile R, Cingolani R, and Pisignano D. Electrospun dye-doped polymer nanofibers emitting in the near infrared. Appl Phys Lett. 2007;90:143115.

[23] Huang ZM, ZYZ, Kotaki M, and Ramakrishna S. A review on polymer nanofibers by electrospinning and their applications in nanocomposites. Compos Sci Technol. 2003;63:2223-53.

[24] Huang JX, VS, Weiller BH, and Kaner RB. Polyaniline nanofibers: facile synthesis and chemical sensors. J Am Chem Soc. 2003;125(2):314-5.

[25] Sun H, ZH, Zhang J, Wei H, Ju J, Li M, and Yang B. White-light emission nanofibers obtained from assembling aqueous single-colored CdTe NCs into a PPV precursor and PVA matrix. J Mater Chem. 2009;19:6740-4.

[26] Moran-Mirabal JM, SJD, DeFranco JA, Verbridge SS, Ilic R, Flores-Torres S, Abruna H, Malliaras GG, and Craighead HG. Electrospun light-emitting nanofibers. Nano Lett. 2007;7:458-63.

[27] Xing X, ZH, Wang Y, and Li B. Ultracompact photonic coupling splitters twisted by PTT nanowires. Nano Lett. 2008;8(9):2839-43.

[28] Ma H, JAK-Y, and Dalton LR. Polymer-based optical waveguides: materials, processing, and devices. Adv Mater. 2002;14:1339-65.

[29] Goddard JM, aHJH. Polymer surface modification for the attachment of bioactive compounds. Prog Polym Sci. 2007;32:698-725.

[30] Gu F, ZL, Yu H, Fang W, Bao J, and Tong L. Large defect-induced sub-bandgap photoresponse in semiconductor nanowires via waveguiding excitation. Nanotechnol. 2011;22(42):425201.

[31] Gu F, Yu H, Wang P, Yang Z, Tong L. Light-emitting polymer single nanofibers via waveguiding excitation. ACS Nano. 2010;4(9):5332-8.

[32] Gu F, WP, Yu H, Guo B, and Tong L. Optical quenching of photoconductivity in CdSe single nanowires via waveguiding excitation. Opt Express. 2011;19(11):10880-5.

[33] Gu F, YH, Fang W, and Tong L. Broad spectral response in composition-graded CdSSe single nanowires via waveguiding excitation. Appl Phys Lett. 2011;99(18): 181111.

[34] Yang Q, JX, Guo X, Chen Y, and Tong L. Hybrid structure laser based on semiconductor nanowires and a silica microfiber knot cavity. Appl Phys Lett. 2009;94:101108.

[35] Sirbuly DJ, TA, Law M, Fan R, and Yang P. Multifunctional nanowire evanescent wave optical sensors. Adv Mater. 2007;19:61-6.

[36] Wang P, GF, Zhang L, and Tong L. Polymer microfiber rings for high-sensitivity optical humidity sensing. Appl Opt. 2011;50:G7-G10.

[37] Gu F, YX, Yu H, Wang P, and Tong L. PANI-PS single-nanowire devices for highly selective optical detection of gas mixtures. Opt Express. 2009;17:11230-5.

[38] Gu F, Zhang L, Yin X, Tong L. Polymer single-nanowire optical sensors. Nano Lett. 2008;8(9):2757-61.

[39] Gu F, Yu H, Fang W, Tong L. Nanoimprinted Polymer Micro/Nanofiber Bragg Gratings for High-Sensitivity Strain Sensing. IEEE Photon Technol Lett. 2013;25(1):22-4.

[40] Zhao YS, PA, Fu H, Ma Y, and Yao J. Nanowire waveguides and ultraviolet lasers based on small organic molecules. Adv Mater. 2008;20:1661-5.

[41] Gu F, Zeng H, Zhu Y, Yang Q, Ang L, Zhuang S. Single-Crystal Pd and its Alloy Nanowires for Plasmon Propagation and Highly Sensitive Hydrogen Detection. Adv Optical Mater. 2014;2(2):189-96.

[42] Gu F, Wu G, Zeng H. Hybrid photon–plasmon Mach–Zehnder interferometers for highly sensitive hydrogen sensing. Nanoscale. 2015;7(3):924-9.

[43] Gu F, Zeng H, Tong L, Zhuang S. Metal single-nanowire plasmonic sensors. Opt Lett. 2013;38(11):1826-8.

[44] Gu F, Zhang L, Wu G, Zhu Y, Zeng H. Sub-bandgap transverse frequency conversion in semiconductor nano-waveguides. Nanoscale. 2014;6(21):12371-5.

[45] Chen Y, MZ, Yang Q, and Tong LM. Compact optical short-pass filters based on microfibers. Opt Lett. 2008;33:2565-7

[46] Law M, Sirbuly DJ, Johnson JC, Goldberger J, Saykally RJ, Yang P. Nanoribbon waveguides for subwavelength photonics integration. Science. 2004;305(5688): 1269-73.

[47] Lakowicz JR. Principles of Fluorescence Spectroscopy; 3rd ed. Berlin, Heidelberg: Springer-Verlag; 2006.

[48] Pud A, ON, Korzhenkob A, and Shapovala G. Some aspects of preparation methods and properties of polyaniline blends and composites with organic polymers. Prog Polym Sci. 2003;28(12):1701-53.

[49] Matsuguchi M, OA, and Sakai Y. Effect of humidity on $NH_3$ gas sensitivity of polyaniline blend films. Sens Actuators B. 2003;94(1):46-52.

[50] Christie S, SE, Persaud K, and Kvasnik F. Remote detection of gaseous ammonia using the near infrared transmission properties of polyaniline. Sens Actuators B Chem. 2003;90:163-9.

[51] Brambilla G, aPDN. The ultimate strength of glass silica nanowires. Nano Lett. 2009;9:831-5.

[52] Peters K. Polymer optical fiber sensors. Smart Mater Struct. 2011;20:013002.

Chapter 9

# High-Speed Single-Photon Detection with Avalanche Photodiodes in the Near Infrared

Yan Liang and Heping Zeng

Additional information is available at the end of the chapter

http://dx.doi.org/10.5772/60481

**Abstract**

As the requisite optical components in quantum information processing, single-photon detectors of high performance at the near-infrared wavelengths are in urgent need. In this paper, we review our recent development in high-speed single-photon detection with avalanche photodiodes, increasing the working repetition frequency up to GHz. Ingenious techniques, such as capacitance-balancing, self-differencing, low-pass filtering, and frequency up-conversion, were employed to achieve high-speed single-photon detection with high detection efficiency and low error counts, offering facility for many important applications, such as laser ranging and imaging, quantum key distribution at GHz clock rate.

**Keywords:** Avalanche photodiode, Single-photon detection, Single-photon frequency up-conversion, Quantum key distribution

## 1. Introduction

Single-photon detectors (SPDs), which are sufficiently sensitive to register single-photon clicks, are widely used in numerous fields of great importance, such as positron emission tomography, optical time domain reflectometry, astronomy and deep-space communication, and biological imaging [1-6]. SPDs are extraordinarily essential not only in fundamental research of quantum physics [7, 8], but also in practical quantum information processing techniques [9-11]. As one of the most commercially successful quantum information applications, quantum key distribution (QKD), which makes it possible for two distant parties to share secret keys via telecommunication, has rapidly progressed since its initial proposal in 1984 [12–16]. In order to achieve efficient QKD of long distances, GHz-clocked QKD systems have been developed. In these schemes, high-speed SPDs of high detection efficiency and low noise at the telecommunication wavelengths are necessary to guarantee the performance of high-speed QKDs [17-18]. Moreover, as for another revolutionary quantum information application, linear optical quantum computing (LOQC), which is a scalable paradigm for quantum information processing and computation, remains difficult to achieve [19-20]. Many efforts have been made worldwide toward this goal. A major factor that limits the advance of LOQC is the performances of the optical components such as SPDs. Significant improvements are still needed in terms of their detection efficiency, error counts, and ability to resolve photon numbers.

Thus far, many concepts and techniques have been proposed to realize high-performance single-photon detection. For instance, single-photon avalanche photodiodes (SPADs), SPDs based on frequency up-conversion, visible-light photon counters, superconducting transition-edge sensors, superconducting nanowire SPDs (SNSPDs), and SPDs based on quantum dots and semiconductor defects, differ in terms of spectral response, quantum efficiency, dark count and afterpulse noise, signal-to-noise ratio, timing jitter, and photon-number-resolving capability, providing optimal choices for various specific applications [21-28]. In consideration of high-speed quantum information processing applications, SPADs, frequency up-conversion technique, and SNSPDs are competitive choices [29-31]. Although SNSPDs possess the characteristic of ultra-low noise and timing jitter, which enables hundreds-of-kilometers QKD system, the requirement of cryogenic cooling system is one of the obvious drawbacks for practical applications. In this chapter, we focus on the recent development of high-speed single-photon detection based on InGaAs/InP SPADs and frequency up-conversion in the near infrared. Since the basic principle of the frequency up-conversion is to translate a near-infrared photon to the visible regime and then detect the photon with silicon avalanche photodiodes, the content presented in the following sections is concluded to be high-speed single-photon detection with avalanche photodiodes in the near infrared.

For their compact structure and low-power consumption, InGaAs/InP SPADs have been used intensively in practical applications at the near-infrared wavelengths, especially in QKD and laser ranging and imaging systems [32-34]. The avalanche photodiode (APD) is reverse-biased above the breakdown voltage (which is called Geiger operation mode), and carriers generated by a single-photon absorption could trigger a detectable macroscopic current after the avalanche gain. To make use of this avalanche propagation progress adequately, the avalanche should be stopped and the APD reset with a peripheral circuit. Generally, InGaAs/InP SPADs are operated in gated Geiger mode that employs gating pulse to determine the switching of the APD's bias voltage between overvoltage and undervoltage. In this operation mode, the dark counts of InGaAs/InP SPADs would be reduced effectively [35-37]. However, since the APD is a capacitive device, the spike noise produced by the gating pulses charging and discharging on the APD's capacitance is an inevitable problem. The weak photon-induced avalanche signals are buried in the spike noise, making the key technique to improve the performance of InGaAs/InP SPADs lie in efficient discrimination of avalanche signals from the spike noise. Furthermore, with the increase of the working repetition frequency, the after-pulsing effect becomes more and more serious, greatly affecting the performance of InGaAs/InP SPADs. The afterpulses are the error counts induced by the release of carriers trapped by defects in the multiplication region during an earlier avalanche event. To solve this issue, the avalanche gain should be decreased correspondingly, unavoidably resulting in the increase of the difficulty in the extraction of the valid avalanche signals.

Recently, some artful techniques, such as sinusoidal gating, self-differencing, and the combination of both, have been demonstrated to suppress the spike noise efficiently while increase the gating repetition rate over 1 GHz [38-42]. In a sinusoidal gating circuit, the spike noise produced by the applied sinusoidal gates was well cancelled by the specific electric band-elimination filters, considering that the APD's capacitive response exhibits a relatively simple frequency spectrum. The InGaAs/InP SPADs that made use of this technique could be operated at 1.5 GHz with the detection efficiency of 10.8% [38]. Comparatively, the spike noise in the self-differencing scheme was eliminated by comparing the output of the APD with that delayed by one gating cycle. This type of SPAD was able to work at 1.25 GHz with the detection efficiency of 10.9% and dark count rate of $2.34\times10^{-6}$ per gate [40]. Besides, the technique of harmonic subtraction was put forward to achieve high detection efficiency and low afterpulse probability for high-speed single-photon detection [43]. With this method, the detection efficiency of InGaAs/InP SPADs could reach ~ 50% with afterpulse probability below $3.5\times10^{-4}$ per gate at 1.25 GHz. Although, all these approaches performed admirably, they are far from mature. There is still room for further enhancement in the respects of dark counts, maximum count rate, and timing jitter.

On the other hand, given the mature silicon SPD with high performance, the frequency up-conversion single-photon detection has shown great potential for many applications. Its basic principle is to translate a near-infrared photon to the visible regime, avoiding the disadvantages of InGaAs/InP APDs [44-46]. This nonlinear optics process requires a large nonlinearity of the nonlinear media and a strong pump field to realize the complete quantum conversion. Generally, the strong pump could be achieved by an external cavity or intracavity enhancement or a waveguide confinement, inevitably bringing about severe background noise because of the parasitic nonlinear interactions. Synchronized single-photon frequency up-conversion was presented to lower the noise. For the improvement of the conversion efficiency, the specific control of the synchronized pulses was required. Recently, efficient single-photon frequency up-conversion detection system operating at tens of MHz has been realized based on the all-optical synchronized fiber lasers, promising its applications in high-speed QKDs.

In this chapter, our recent developments and achievements in high-speed single-photon detection based on InGaAs/InP SPAD and frequency up-conversion single-photon detection were introduced in detail. In Section II, we present the experimental demonstration on some innovative schemes for InGaAs/InP SPAD, such as the optically self-differencing, the low-pass filtering with ultrashort pulses, and the creative combination of the self-differencing and low-pass filtering, to increase the working repetition frequency of the gated SPAD over 1 GHz. Meanwhile, other properties of the SPAD, for instance, the detection efficiency, the timing jitter, and the maximum counts, have been improved as well. Furthermore, a compact synchronized fiber laser system for highly efficient single-photon frequency up-conversion is illustrated in Section III, realizing high conversion efficiency with low background counts. All these high-performance single-photon detectors provide essential facility for high-speed quantum information applications. In Section IV, we discuss the advantage of the high-speed single-photon detection in some applications, such as laser ranging and imaging, quantum key distribution, and so on. Finally, we conclude the chapter in the last section by emphasizing the importance of the high-speed single-photon detection for quantum information applications.

## 2. Single-photon avalanche photodiodes

As mentioned in the introduction, the suppression ratio of the spike noise, which is generated by the capacitive response of the APD to the gating pulses, is quite critical to the performance of InGaAs/InP SPAD operated in gated Geiger mode. In this section, we present several methods to remove spike noise and acquire avalanche signals, achieving high-speed SPAD with excellent properties.

### 2.1. Capacitance balancing technique

Since the spike noise is caused by the capacitance characteristics of the APD, the capacitance-balancing technique employs a capacitor to imitate the APD's response. As shown in Fig. 1 (a), the InGaAs/InP APD was connected in parallel with a complementary capacitor. The output signals of the APD and capacitor were connected to the 0 and $\pi$ inputs of the magic-T network (MTNT), respectively. The MTNT was used as a subtracter here, subtracting the two spike noises from the APD and the capacitor. A tunable capacitor was chosen for the perfect matching between the APD and the complementary capacitor. Then, at the output of the circuit, only the avalanche signal was extracted and acquired by an oscilloscope after amplification by an RF amplifier.

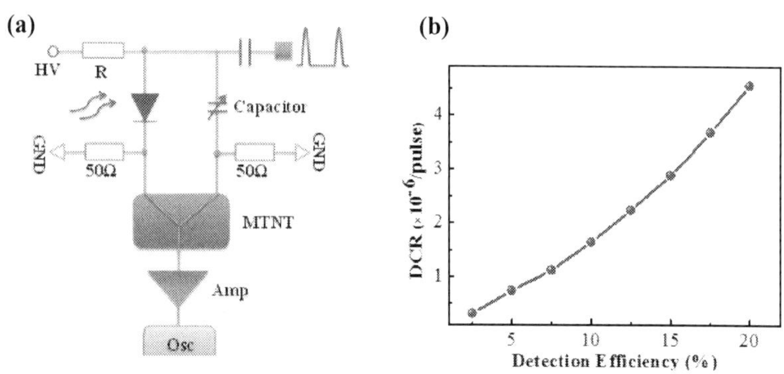


**Figure 1.** (a) Schematic setup of capacitance-balancing InGaAs/InP SPAD. MTNT: a so-called magic-T network consisted with a broadband transformer; Amp: an RF amplifier; Osc: a high bandwidth oscilloscope. (b) Dark count rate of capacitance-balancing InGaAs/InP SPAD as a function of detection efficiency at 100 MHz.

In the capacitance-balancing scheme, the repetition frequency of the gating pulses could be tuned continuously on a large scale. Moreover, the capacitance-balancing InGaAs/InP SPADs are suitable for the applications requiring multi-channel timing acquisition, such as the time-code quantum key distribution. We just need to adjust the tunable capacitor to ensure the suppression ratios of the spike noise. A train of double gating pulses was employed to illustrate the capability of multi-channel detection. By changing the time interval between the double gating pulses, we could obtain that the capacitance balancing technique remained applicable in the single-photon detection with the gating repetition rate no higher than 260 MHz [47].

Here, we examined the performance of the SPADs at 100 MHz. The amplitude of the gating pulses was 4 V with the duration of ~ 1 ns, while the DC bias applied on the APD was varied to obtain different detection efficiencies. The operation temperature of the InGaAs/InP APD was set at -50℃. And a 1550-nm pulsed laser at 10 MHz with full width at half maximum (FWHM) of ~35 ps was attenuated to contain 0.1 photon per pulse before coupling into the APD fiber pigtail as the photon source. The laser pulse was synchronously triggered with the gating pulse, while their delay was adjusted to gain the highest detection efficiency for optimized operation. Figure 1 (b) exhibited the dark count rates as a function of the detection efficiency. The dark count rate increased with the detection efficiency, and we could figure out that the dark count rate was approximately $4.6\times10^{-6}$ with the efficiency of 20%, indicating this SPAD performed well at 100 MHz.

Unlike the double-APD balancing [48], the capacitance-balancing technology using a capacitor instead to imitate the APD was much more economic and practical. The suppression ratio of the spike noise was ~ 19 dB, limiting the working speed. We believe that the capacitance-balancing SPAD would be able to be operated at a higher speed with the advance of semiconductor techniques.

### 2.2. Self-differencing technique

The self-differencing technique first proposed by Z. L. Yuan et al. has shown a great improvement of the detection speed. Recently, the InGaAs/InP SPAD using this method has been shown to perform remarkably with the detection efficiency of 25% and dark count rate of $5.9\times10^{-5}$ per gate at 1 GHz without Peltier cooling [49]. Unlike the traditional self-differencing circuits, we added a tunable phase shifter and attenuator for better suppression of the spike noise, as demonstrated in Fig. 2 (a). The gating pulses were superposed on the reversely biased InGaAs/InP APD. The output signal of the APD was sent to a 50/50 power splitter (DC to 3 GHz), being divided into two identical components. Then one component was delayed by one gating period through the tunable phase shifter, and the tunable attenuator in the other arm was used to guarantee equal amplitudes.

Afterward the two components were combined by a differencer (DC to 3 GHz) before amplification. The output of the circuit was the amplified difference between the two split components, shifted relatively by one gating period. The tunable phase shifter and attenuator precisely controlled the split two components, ensuring the avalanche signal extracted with spike noise suppressed greatly. Furthermore, with the tunable phase shifter, the working repetition frequency of the self-differencing SPAD could be adjusted continuously with ease. And the adjustment range was determined by that of the phase shifter.

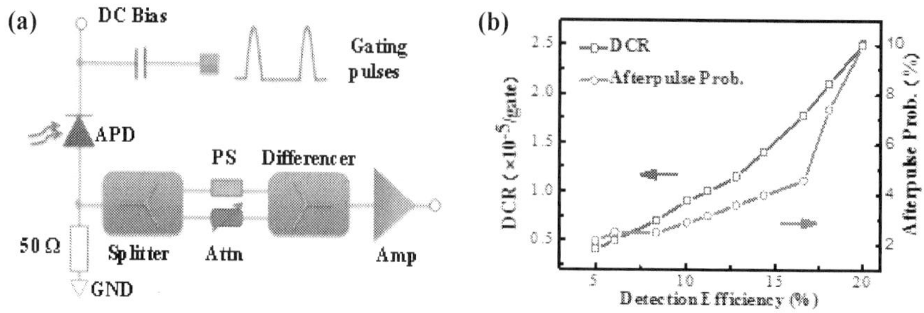


**Figure 2.** (a) Experimental setup of the self-differencing InGaAs/InP SPAD. PS: tunable phase shifter, Attn: tunable attenuator, Amp: RF amplifier. (b) Dark count rate and afterpulse probability as a function of the detection efficiency at –40℃.

The 200-MHz gating pulses with amplitude of 5 V and duration of ~ 1 ns were used to characterize the performance of this InGaAs/InP SPAD. And the InGaAs/InP APD was cooled to be -40℃. Figure 2 (b) displayed the dark count rate and afterpulse probability as a function of detection efficiency. The reverse bias voltage applied on the APD was changed to obtain different efficiencies. The laser source was attenuated to contain 0.1 photons per pulse on average. The detection efficiency was corrected for Poissonian statistics of the photon numbers by the formula

$$1-e^{-\mu\cdot\eta}=R_O\left(1-P_E\right)/R_L\left\langle\mu\right\rangle \tag{1}$$

where η was the detection efficiency, $R_O$ was the overall counting rate, $R_L$ was the repetition rate of the laser pulse, $P_E$ was the error counting probability, and μ was the average photon per pulse. In the experimental measurement, the dark count rate was measured with the laser off. It increased gradually with the detection efficiency.

The afterpulse probability, defined as the ratio of the total afterpulse counts to the photon counts, can be calculated from

$$P_A=\frac{\left(I_{NI}-I_D\right)}{I_{ph}-I_{NI}}R \tag{2}$$

where $I_{Ph}$ and $I_{NI}$ were the count rate per gate at the illuminated and nonilluminated gates, respectively, while $I_D$ was the dark count rate for each gate. R was the ratio of the repetition frequency of the gating pulse to that of the laser pulse. Here we took R=20 for measuring the afterpulse. The afterpulse probability increased with the detection efficiency and began to increase dramatically when the detection efficiency reached 16.7%, greatly impacting the performance of the SPAD. When the detection efficiency was 10.1%, the afterpulse probability was just 2.9% and the corresponding dark count rate was $9.0\times10^{-6}$ per gate.

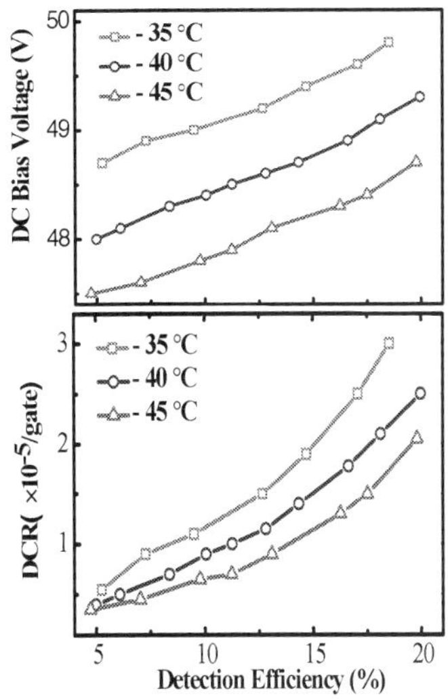


**Figure 3.** DC bias voltage applied on APD and dark count rate as a function of detection efficiency at three different temperatures.

The operation temperature of the InGaAs/InP APD was vital to the SPAD. We Peltier-cooled the APD to work at three different temperatures and tested the parameters of the SPAD. Figure 3 showed the dc voltage and dark count rate as a function of the dc bias voltage. The detection efficiency increased with the voltage. Meanwhile the rising slopes were almost the same at the three temperatures. To achieve the same detection efficiency, we should apply higher voltage at higher temperature. Since the gating pulses were identical, it could be figured out that the breakdown voltage increased with the temperature of the APD, leading to a great influence on the SPAD. Meanwhile, the dark count rate increased with the detection efficiency, while it was higher at the same efficiency at higher temperature. However, cooling the APD to lower temperatures consumes more energy. Therefore, we should choose an appropriate temperature in practical applications.

Furthermore, the InGaAs/InP SPAD using the cascade of self-differencing circuits was demonstrated [50]. By introducing a second self-differencing circuit, the suppression ratio was enhanced up to ~18 dB, making the SPAD more suitable for high-speed applications. Considering the twice splitting of the valid avalanche signal, the signal-to-noise ratio (SNR) was merely improved by ~10 dB.

We also proposed the optically self-balancing technique, as exhibited in Fig. 4 [51-53]. The output of the InGaAs/InP APD was amplified to trigger a laser diode at 1550 nm. Then, an erbium-doped fiber amplifier (EDFA) was used to magnify the transformed optical signal. Afterward, the splitting and the relative delay of the signal were processed through the optical devices. Finally, the avalanche signal was extracted and transformed to the electronic signal

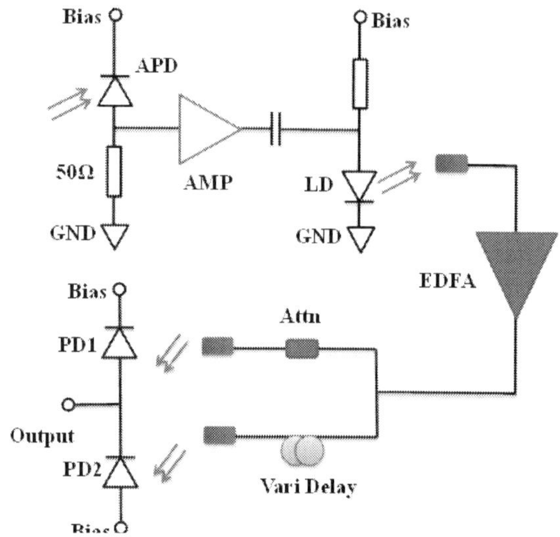


**Figure 4.** Schematic setup of the optically self-balancing SPAD. AMP: RF amplifier; LD: distributed-feedback laser diode at 1550 nm; Attn: tunable optical attenuator; and PD1; 2: pin photodiodes.

by two pin photodiode. Compared to the electronic self-differential technique, the optical method used stable and precisely controllable optical signals, providing immunity to the electromagnetic field of the surrounding circuits. Besides, the impedance matching was not necessary to be considered in this scheme. A 31-dB suppression of the spike noise was obtained, allowing the study on the photon-number resolving dynamics of the InGaAs/InP avalanche photodiode. The detection efficiency reached 22.4% while the afterpulse probability was controlled as low as 2.4% at 25 MHz. However, the transformation between the electronic and optical signal was more and more complicated with the advance of the repetition frequency of the gating pulses superposed on the APD, limiting its applications in high-speed single-photon detection.

The core concepts of the capacitance-balancing and self-differencing techniques are to produce a mimic signal of the spike noise and obtain the valid avalanche signal by making the two signals subtract each other. The performance of the SPAD using those two techniques would be enhanced by improving the similarity of the two signals. In contrast, the technique presented in the next section is to eliminate the spike noise directly by corresponding filters.

### 2.3. Low-pass filtering technique

N. Namekata et al. first put forward the sinusoidal gating technique, employing the sinusoidal gates to control the bias voltage of the InGaAs/InP APD. The frequency distribution of the capacitive response of the APD to the sinusoidal gates was relatively simple, mainly concentrating at the repetition frequency of the gates and its harmonic frequencies. Notch filters were used to eliminate the spike noise and obtain the avalanche signal. The scheme suppressed the spike noise robustly and conveniently. By this means, the detection efficiency of the SPAD reached 10.5% with a dark count rate of $6.1\times10^{-7}$ per gate and afterpulse probability of 3.4% at 2 GHz [39]. However, due to the distortion of the avalanche signal caused by the notch filters, the timing jitter of this SPAD was as large as 180 ps, limiting its applications in high-speed QKD systems or ultra-sensitive long-distance laser ranging. To solve this problem, we proposed the low-pass filtering technology, maintaining the suppression ratio of the spike noise while reducing the timing jitter.

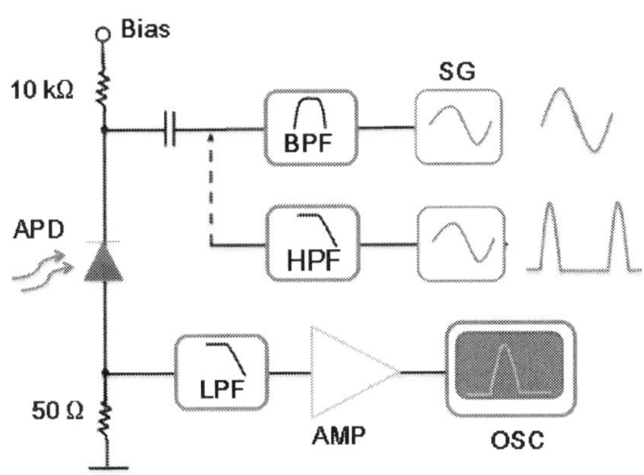


**Figure 5.** Schematic setup of the low-pass filtering SPAD using sinusoidal gates and ultrashort gating pulses. SG: signal generator; BPF: band-pass filter; HPF: high-pass filter; LPF: low-pass filter.

Figure 5 illustrated the low-pass filtering technology. Before being applied on the APD, the sinusoidal gates passed through a band-pass filter to eliminate the sideband noise and harmonic noise. We used 1.5-GHz sinusoidal waves to examine the performance of the low-pass filtering InGaAs/InP SPAD. The output of the APD was filtered by the low-pass filtering cutting off at 1 GHz with the attenuation higher than 40 dB at 1.5 GHz. Since the spectrum of the avalanche signal distributed mostly at low frequency under 1 GHz, while that of the spike noise concentrated at 1.5 GHz and its harmonic frequencies, we could acquire the avalanche signal by the low-pass filter. To obtain higher SNR, we could employ one more low-pass filter.

The operation temperature of the APD was set at -30°C. The laser source was attenuated to contain one photon per pulse on average to shorten the time of data acquisition, and synchronously triggered with the sine wave frequency. The amplitude of the amplified sinusoidal gating waves was fixed at 6 V. Figure 6 (a) illustrated the performance of the SPAD. The dark count rate and afterpulse probability increased with the detection efficiency. While the efficiency exceeded ~27%, the afterpulse probability rose sharply, limiting further increase of the detection efficiency. By this approach, the detection efficiency reached 13.0% with a dark-count rate of $1.5\times10^{-5}$ per gate and afterpulse probability of 1.1%. While the detection efficiency

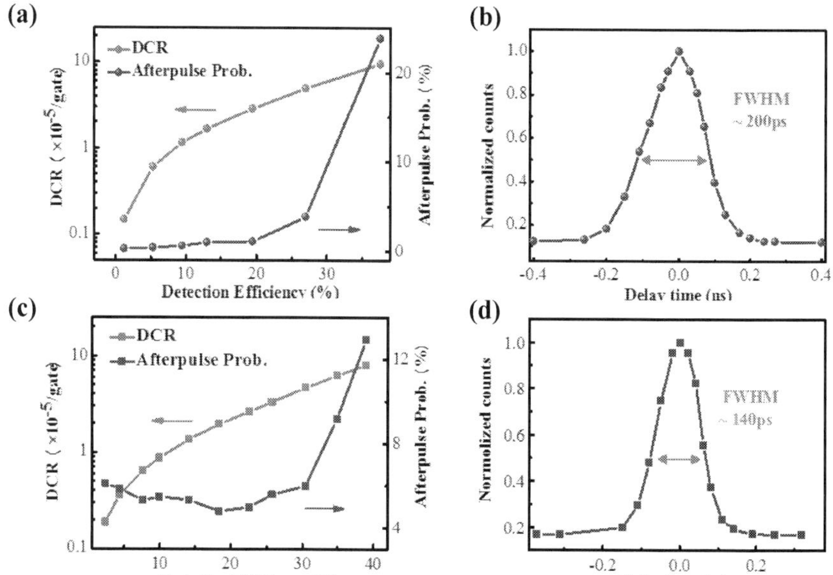


Figure 6. (a) Dark count rate and afterpulse probability of InGaAs/InP SPAD using low-pass filtering technique with sinusoidal gates as a function of the detection efficiency. (b) Count rate dependent on the laser pulse delay in sinusoidally gated SPAD. (c) Dark count rate and afterpulse probability of low-pass filtering SPAD with ultrashort gates as a function of the detection efficiency. (d) Count rate dependent on the laser pulse delay in SPAD with ultrashort gates.

was kept at 10.0%, we adjusted the delay between the laser and sinusoidal gate to get the effective gating width superposed on the APD, as charted in Fig. 6 (b). It was measured to be approximately 200 ps.

Unlike the traditional sinusoidal gating technique, the low-pass filtering was also appropriate for ultrashort gating pulses. We used 1.5-GHz ultrashort gating pulses to characterize the SPAD. The ultrashort pulses were filtered by a high-pass filter (HPF) cutting off at 1 GHz, canceling the noise at low frequencies and ensuring the final SNR of the SPAD. The transmit performance of the HPF remained excellent up to 6 GHz, maintaining the waveform of the ultrashort pulses. The performance of the SPAD was illustrated in Fig. 6 (c). We could find out that the afterpulse probability did not increase obviously until the detection efficiency reached ~ 35%. At the detection efficiency of 35%, the afterpulse probability of the SPAD using ultrashort gating pulses was 9.3% with dark count rate of $6.2\times10^{-5}$ per gate. In comparison, the afterpulse probability of the SPAD using sinusoidal gating pulses was 19.3% with dark count rate of $8.3\times10^{-5}$ per gate. From Fig. 6 (d), it could be noted that the FWHM of the effective gating pulses applied on the APD was 140 ps, less than that in Fig. 6 (b). Since the schematic setup of the two SPADs were exactly the same except the gating signals, we can deduce that ultrashort gating pulses of smaller gating widths improved the SPAD. For a better performance, we could further decrease the gating width.

The timing jitter of the 1.5-GHz SPAD was measured to be 68 ps with a time-correlated single-photon counting (TCSPC) setup (PicoQuant GmbH, PicoHarp 300, Germany). It has been improved a lot by using the low-pass filtering technique, due to the integrity of the avalanche signal preserved well with the low-pass filter. This technology was extraordinarily suitable for high-speed single-photon detection, on account that we could choose low-pass filters cutting off at higher frequencies for better preservation of the avalanche signal. With such a low timing jitter and convenient structure, this type of high-speed SPAD has be widely used in laser ranging systems with high resolution at the near-infrared wavelengths.

### 2.4. Low-pass filtered self-differencing technique

As mentioned in the previous section, the self-differencing and sinusoidal gating techniques offered effective methods for high-speed single-photon detection. To further advance the suppression ratio of the spike noise, there were schemes to combine the two techniques. However, the SNR was not improved as much, due to the splitting of the avalanche signal in the self-differencing circuit. Here, we introduced some combining techniques to take full advantage of both techniques, achieving high-performance GHz InGaAs/InP SPAD.

Figure 7 (a) exhibited the experimental setup of the SPAD using the combining technique. The sinusoidal wave, which came out from the signal generator, was divided into two parts. One part was amplified to serve as the gating signal superposed on the APD. Here, we set the repetition frequency of the sinusoidal signal to be 1 GHz to characterize the SPAD with this scheme. The output of the APD was filtered by a low-pass filter (LPF1) that cut off at 700 MHz with attenuation higher than 40 dB at 1 GHz. Then it was connected to the power combiner, combined with the other part of the 1-GHz sinusoidal signal. The spectrum of the filtered spike noise concentrated at 1 GHz. The phase shifter was used to make the phase difference between the two signal 180°, and the tunable attenuator was employed to ensure the amplitudes equal. Therefore, we could further eliminate the spike noise and get the avalanche signal, improving the suppression ratio of the spike noise by 21 dB. The low-pass filter (LPF2) cutting off at 1.5 GHz was employed to cancel the electronic noise of high frequency of the cascade RF amplifiers. By this means, the SNR of the SPAD could be advanced with the performance of the timing jitter maintained.

We employed a TCSPC with the resolution of 4 ps to test the timing jitter of the SPAD. As shown in Fig. 7 (b), the time histogram of detection events was recorded. The count peak in the illuminated gating pulse was much higher than the other peaks. The residual peaks after the maximum peak, which might be induced by the oscillation, could be neglected by introducing a proper dead time. Here, a 10-ns dead time was applied. The time interval between the peaks was ~1 ns, matching with the 1-GHz repetition frequency of the sinusoidal gating. And the timing jitter of the avalanche signal showed an FWHM of 60 ps, which was extremely low for sinusoidally gated InGaAs/InP SPAD.

With the 10-ns dead time, we could effectively reduce the error counts. However, it would place a limitation for the maximum counts. We measured the linearity and maximum count rate of the SPAD with a continuous wave laser at 1550 nm to illuminate the APD. As charted in the inset of Fig. 8(b), the photon count rate increased linearly as a function of the photon

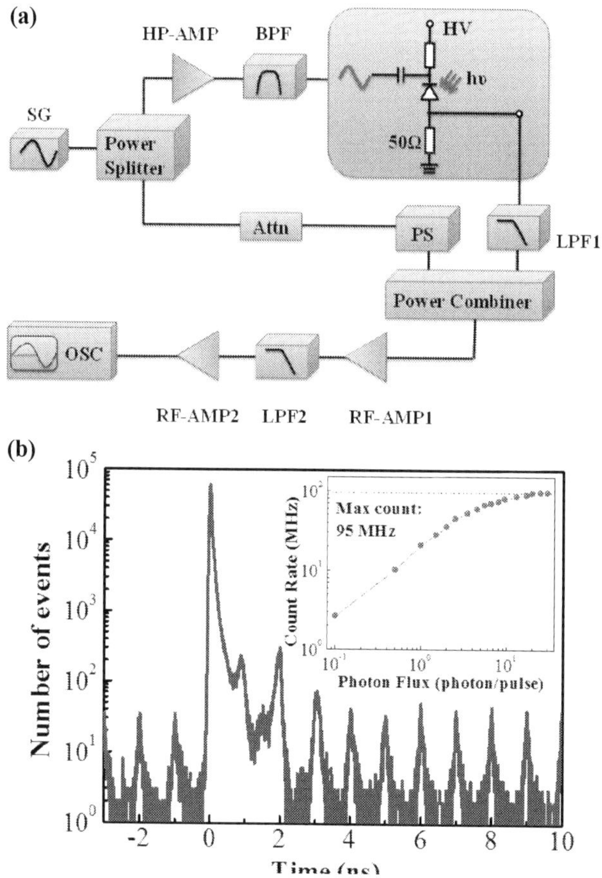


**Figure 7.** (a) Experimental setup of sinusoidally gated SPAD using low-pass filtered self-differencing technique. SG: signal generator; HP-AMP: high-power amplifier; BPF: band-pass filter; Attn: variable attenuator; LPF1, 2: low-pass filter; RF-AMP1, 2: RF amplifier. (b) Time histogram of detection events recorded by the TCSPC. Inset: Photon detection rate as a function of photon flux.

flux while the counting rate was below 80 MHz. Finally the SPAD was saturated at 95 MHz. The SPAD with such a low timing jitter and high maximum counts provided possibilities for the achievement of high-speed QKD.

As mentioned in the previous section, the performance of the SPAD using the ultrashort gating pulses with shorter duration was even more excellent. We proposed a combining method more appropriate for ultrashort gates, as demonstrated in Fig. 8. In consideration of the distribution of the spike noise, the sinusoidal waves at the repetition rates of f and 2f were added by a power combiner to mimic the spike noise. The differential signal of the output of the APD and

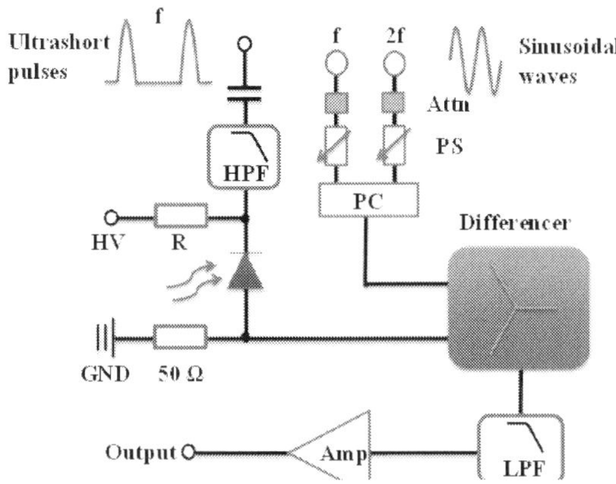


Figure 8. Schematic setup of SPAD using the combining technique with ultrashort gates. HPF: high-pass filter; Attn: tunable attenuator; PS: phase shifter; PC: power combiner; LPF: low-pass filter.

the mimic one was gained and then filtered by a low-pass filter whose cutting-off frequency could be set between 2f and 3f. By this means, the integrity of the avalanche signal could be maintained to the best of the capability of the filtering technique, further reducing the timing jitter of the SPAD. Moreover, the tunable attenuator and phase shifter were used here to guarantee the waveform of the synthetic signal resembles the spike noise as closely as possible. The similarity would advance with the increase of the gating repetition rate, making the scheme quite suitable for high-speed single-photon detection.

The combining technique illustrated earlier took advantages of self-differencing and low-pass filtering technique, achieving high-speed single-photon detection with high suppression ratio and low-timing jitter. With the blossom of the quantum information applications, the SPAD using this technology would be implemented more and more widely.

## 3. Frequency up-conversion

Frequency conversion plays quite an important role in nonlinear optical signal processing. Infrared single-photon up-conversion based on sum frequency generation was put forward to realize single-photon detection at the infrared wavelengths with existing high-performance Si APDs. Recently, the technique has been successfully used in various applications, including infrared imaging, QKD, and infrared ultra-sensitive spectroscopy [54-56]. More and more interest has been focused on proposing novel schemes for achieving single-photon frequency up-conversion with high efficiency and low noise.

The nonlinear optical media of large nonlinearity and a sufficiently strong pump field were necessary to enforce the complete quantum conversion. Generally, periodically poled lithium niobate (PPLN) crystal was used as the nonlinear media for nonlinear interaction, considering its relatively large effective nonlinear coefficient and long interaction length. For the requisite strong pump, schemes using an external cavity or intracavity enhancement or a waveguide confinement have been proposed. With such high-intensity pump, frequency up-conversion has been carried out with almost 100% conversion efficiency. However, a strong pump field would unavoidably bring about severe background noise. To solve this problem, synchronized single-photon frequency up-conversion was presented. In the scheme, each signal photon was synchronized to a pump pulse, achieving high efficiencies of frequency up-conversion with quite low noise.

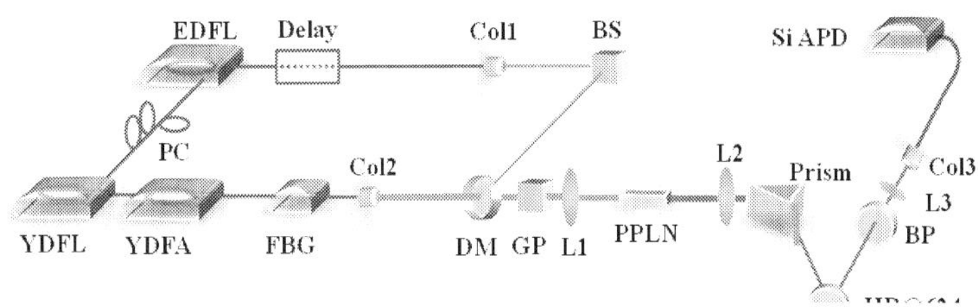


**Figure 9.** Experimental setup of the synchronous single-photon frequency up-conversion detection. EDFL: erbium-doped fiber laser; YDFL: ytterbium-doped fiber laser; YDFA, ytterbium-doped fiber amplifier; FBG: fiber Bragg grating; Col1, 2, 3: collimator; BS: beam splitter; DM: dichroic mirror; GP: Glan prism; L1, 2, 3: lens; PPLN, periodically poled lithium niobate crystal; BP: optical band-pass filter.

Figure 9 exhibited the experimental setup of the synchronous single-photon frequency up-conversion detection. The signal and pump sources were synchronized in master–slave configuration. The repetition frequency of the synchronized system was about 20.3 MHz. We optimized the spectrum and pulse duration to improve the frequency up-conversion system by managing the intracavity dispersion of the two lasers. The narrow spectrum was required to match with the bandwidth of the PPLN crystal. And the pulse duration of the pump should be a little longer than that of the signal to include all the signal photons within the pump envelop, ensuring the final conversion efficiency.

Figure 10 gave the typical spectra and durations of the signal and pump source. The signal source was provided by an Er-doped NPR locking fiber laser (EDFL), whose output centered at 1562.1 nm with a full width at half maximum (FWHM) of 1.0 nm. The pulse duration was measured to be 1.3 ps by two-photon absorption. The pump source was generated by using an Yb-doped NPR locking fiber laser (YDFL), centering at 1041.1 with an FWHM of 0.7 nm. By self-correlation, the pulse duration was measured to be 34.3 ps. Finally, the 1562.1-nm signal source was up-converted through sum-frequency generation and the up-converted photons at 624 nm were detected by using a standard Si-APD based single-photon detector. The quantum conversion efficiency of infrared photons was up to ~80% with the corresponding background noise of ~300 counts per second. Compared with the background counts of CW pumping, the background was about two orders smaller in synchronously pulsed pumping

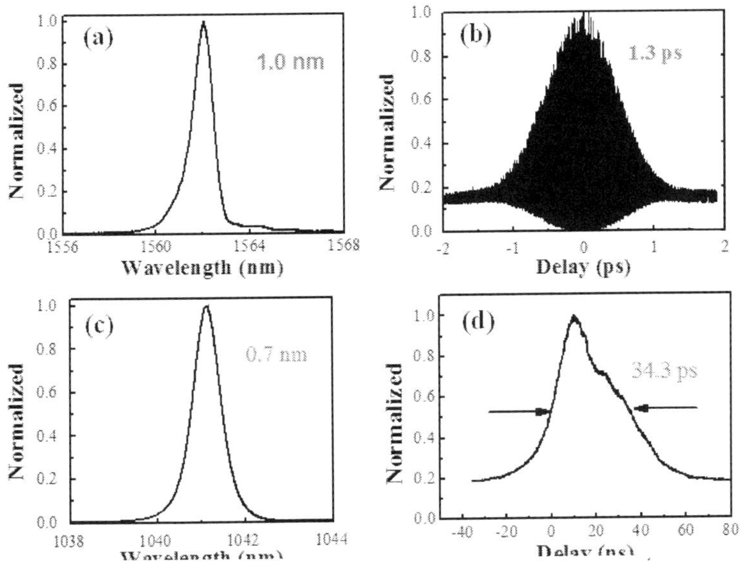


**Figure 10.** (a) (b) Spectrum and duration of the signal source. The pulse duration was measured by two-photon absorption. (c) (d) Spectrum and duration of the pump source. The pulse duration was measured by self-correlation.

scheme. However, for satisfying the applications of GHz QKD, the single-photon frequency up-conversion technique requires enormous development.

## 4. Applications in quantum key distribution

Quantum key distribution (QKD) has nowadays been demonstrated as a cryptographic approach to provide absolute security between the sender (Alice) and the receiver (Bob), according to the fundamental laws of quantum mechanics. Fiber-based systems have been implemented in prototype QKD experiments, with practical stabilities in long-distance telecom fibers. The separation distance between Alice and Bob has achieved tens of kilometers in field trials. However, despite these significant advances over recent years QKD's primary challenge is still to obtain higher bit rates over longer distances. The major factors that limit the performance of the QKD are due to the immaturity of single-photon detectors at telecom-wavelengths. The probability of detection decreasing at long distance because of the high losses, while on the other hand, the noise rate of the detector being constant leads to a too high error rate above a certain distance, making it no longer possible to exchange secret keys. Therefore low-noise detectors are essential for long-distance QKD. It could be figured out that with the high-speed SPADs mentioned in the previous section, the performance of QKD systems would be further improved.

Among the fiber-based QKD systems, the polarization-encoding and phase-encoding methods are most widely implemented [57, 58]. Figure 11 showed the schematic setup of polarization-encoding QKD system based on the BB84 four-state protocol. The single-photon signals generated from attenuated laser pulses. Each laser diode could produce only one state of the BB84 protocol. The single-photon signals transmitted through the quantum channel after attenuation and reached Bob's side. As Bob tried to decode the polarization information, he randomly chose HV base or QR base to measure to polarization. Four single-photon detectors (SPD1~ SPD4) were used to detect the single-photon signals. However, in the polarization-encoding QKD system, the polarization states must be aligned and kept aligned due to the imperfection of the optical fiber and the disturbance of the environment. Thus, the polarization real-time control was needed, which presented the main difficulty for the implementation of polarization-encoding QKD system.

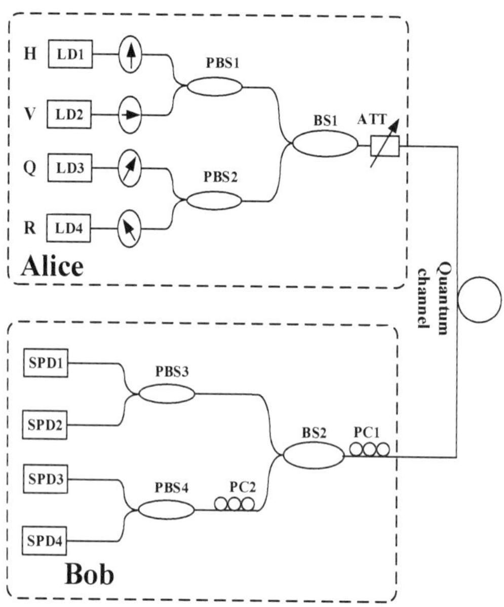


**Figure 11.** Schematic setup of polarization-coding QKD system. ATT: attenuator, BS: beam splitter, PC: polarization controller, PBS: polarization maintaining beam splitter, SPD: single-photon detector, LD: laser diode.

To examine the applications of the high-speed SPDs in QKD systems, we instead used 1.25-GHz sinusoidally gated InGaAs/InP SPADs to investigate the characteristics of a 1.25 GHz light source at 1550 nm. The laser pulse with the duration of ~20 ps was attenuated to contain 0.01 photon per pulse on average. The dark count rate of the 1.25-GHz SPAD using the low-pass filtering technique was $8\times10^{-6}$ per gate at the detection efficiency of 10%. The afterpulse probability was ~3.5% with the 10-ns dead time. Figure 12 illustrated the counting rate recorded by the SPAD by changing the delay between the light source and the detector. The extinction ratio of the light pulse was up to 33 dB with the efficient gate width of ~ 140 ps. It could be noted that this SPAD could efficiently achieve high-speed single-photon detection with low error counts, promising its applications in high-speed QKD systems.

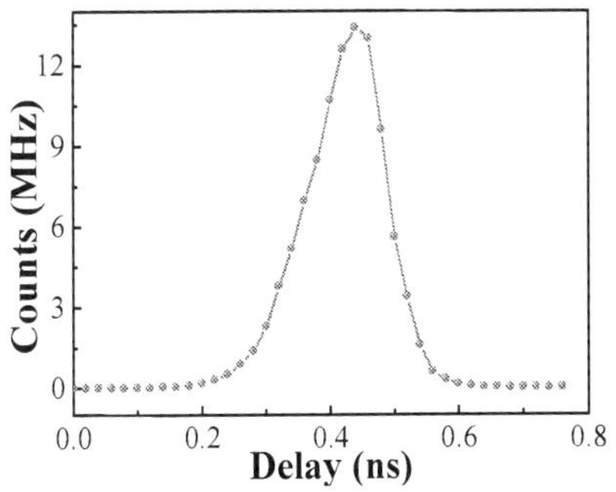


**Figure 12.** Counting rate depends on the pulse laser delay.

## 5. Conclusion

In this chapter, we mainly introduced several techniques, such as capacitance-balancing, self-differencing, low-pass filtering, and the combination techniques, to achieve high-speed single-photon detection based on InGaAs/InP SPAD. The spike noise produced by the capacitive response of the APD was well detached, maintaining high efficiency and reducing the error counts correspondingly at GHz working repetition frequency. Furthermore, the frequency up-conversion technique was used to realize infrared single-photon detection with high conversion efficiency and low background noise at ~ 20 MHz. The advance of single-photon detectors highly supported the development of QKD systems, because both the key generation rate and the key distribution distance were mainly limited by the performance of SPDs thus far.

## Acknowledgements

This work was supported by the National Natural Science Fund of China (Grant No. 61127014) and the National Key Scientific Instrument Project (Grant No. 2012YQ150092), General Financial Grant from the China Postdoctoral Science Foundation (Grant No. 2014M560347), and the Hujiang Foundation of China (B14002/D14002).

## Author details

Yan Liang[1] and Heping Zeng[1,2*]

*Address all correspondence to: yanliangspd@163.com

1 Shanghai Key Laboratory of Modern Optical System, Engineering Research Center of Optical Instrument and System, Ministry of Education, School of Optical-Electrical and Computer Engineering, University of Shanghai for Science and Technology, Shanghai, China

2 State Key Laboratory of Precision Spectroscopy, East China Normal University, Shanghai, China

## References

[1] Moehrs S., Guerra A., Herbert D., and Mandelkern M. A detector head design for small-animal PET with silicon photomultipliers (SiPM). Phys. Med. Biol. 2006; 51(5): 1113–1127.

[2] Legre M., Thew R., Zbinden H., and Gisin N. High resolution optical time domain reflectometer based on 1.55um up-conversion photon-counting module. Opt. Exp. 2007; 15(13): 8237–8242.

[3] Degnan J. J. Photon-counting multikilohertz microlaser altimeters for airborne and spaceborne topographic measurements. J. Geodyn. 2002; 34(3–4): 503–549.

[4] Buller G. S., and Wallace A. M. Recent advances in ranging and three-dimensional imaging using time-correlated single-photon counting. IEEE J. Sel. Top. Quantum Electron. 2007; 13: 1006–1015.

[5] Isoshima T., Isojima Y., Hakomori K., Kikuchi K., Nagai K., and Nakagawa H. Ultrahigh sensitivity single-photon detector using a Si avalanche photodiode for the measurement of ultraweak biochemiluminescence. Rev. Sci. Instrum. 1995; 66(4): 2922–2926.

[6] Berglund A. J., Doherty A. C., and Mabuchi H. Photon statistics and dynamics of fluorescence resonance energy transfer. Phys. Rev. Lett. 2002; 89: 68101.

[7] Wu E, Rabeau J., Roger G., Treussart F., Zeng H., Grangier P., Prawer S., and Roch J. Room temperature triggered single-photon source in the near infrared. New J. Phys. 2007; 9: 434.

[8] Eisaman M. D., Fan J., Migdall A., and Polyakov S. V. Invited review article: single-photon sources and detectors. Rev. Sci. Instrum. 2011; 82: 71101.

[9] Hadfield R. H. Single-photon detectors for optical quantum information applications. Nat. Photonics 2009; 3: 696–705.

[10] Gabriel C., Wittmann C., Sych D., Dong R., Mauerer W., Andersen U. L., Marquardt C., and Leuchs G. A generator for unique quantum random numbers based on vacuum states. Nat. Photonics 2010; 4: 711–715.

[11] Jian Y., Ren M., Wu E, Wu G. and Zeng H. Two-bit quantum random number generator based on photon-number-resolving detection. Rev. Sci. Instrum. 2011; 82: 073109.

[12] Ekert A. K. Quantum cryptography based on Bell's theorem. Phys Rev Lett 1991; 67: 661–663.

[13] Gisin N., Ribordy G., Tittel W., and Zbinden H. Quantum cryptography. Rev. Mod. Phys. 2002; 74: 145–195.

[14] Hwang W. Quantum key distribution with high loss: toward global secure communication. Phys. Rev. Lett. 2003; 91: 57901.

[15] Wang X. Beating the photon-number-splitting attack in practical quantum cryptography. Phys. Rev. Lett. 2005; 94: 230503.

[16] Takesue H., Harada K., Tamaki K., Fukuda H., Tsuchizawa T., Watanabe T., Yamada K., and Itabashi S. Long-distance entanglement-based quantum key distribution experiment using practical detectors. Opt. Express 2010;18: 6777–6787.

[17] Knill E., Laflamme R., and Milburn G. J. A scheme for efficient quantum computation with linear optics. Nature 2001; 409: 46–52.

[18] Ladd T. D., Jelezko F., Laflamme R., Nakamura Y., Monroe C., and Brien J. L. Quantum computers. Nature 2010; 464: 45–53.

[19] Yuan Z. L., Dixon A. R., Dynes J. F., Sharpe A. W., and Shields A. J. Gigahertz quantum key distribution with InGaAs avalanche photodiodes. Appl. Phys. Lett. 2008; 92(20): 201104.

[20] Rosenberg D., Harrington J. W., Rice P. R., Hiskett P. A., Peterson C. G., Hughes R. J., Lita A. E, Nam S. W., and Nordholt J. E. Long-distance decoy-state quantum key distribution in optical fiber. Phys. Rev. Lett. 2007; 98: 010503.

[21] Cova S., Ghioni M., Lotito A., Rech I., and Zappa F. Evolution and prospects for single-photon avalanche diodes and quenching circuits. J. Mod. Optics 2004; 51: 1267–1288.

[22] Wu G., Zhou C. Y., Chen X. L., and Zeng H. P. High performance of gated-mode single-photon detector at 1.55 μm. Opt. Commun. 2006; 265: 126–131.

[23] Gu X., Huang K., Li Y., Pan H., Wu E., and Zeng H. Temporal and spectral control of single-photon frequency upconversion for pulsed radiation. Appl Phys Lett 2010; 96: 131111.

[24] Waks E., InoueK., Oliver W. D., Diamanti E., and Yamamoto Y. High-efficiency photon-number detection for quantum information processing. IEEE J. Sel. Top. Quantum Electron. 2003; 9(6): 1502–1511.

[25] Lita A. E., Miller A. J., and Nam S. W. Counting near-infrared single-photons with 95% efficiency. Opt. Express 2008; 16(5): 3032–3040.

[26] Shibata H., Shimizu K., Takesue H., and Tokura Y. Superconducting nanowire single-photon detector with ultralow dark count rate using cold optical filters. Appl. Phys. Express 2013; 6: 072801.

[27] Chen S., Liu D., Zhang W., You L., He Y., Zhang W., Yang, X. Wu G., Ren M., Zeng H., Wang Z., Xie X., and Jiang M. Time-of-flight laser ranging and imaging at 1550 nm using low-jitter superconducting nanowire single-photon detection system. Appl. Optics 2013; 52(14): 3241–3245.

[28] Li H. W., Kardynal B. E., See P., Shields A. J., Simmonds P, Beere H. E., and Ritchie D. A. Quantum dot resonant tunneling diode for telecommunication wavelength single photon detection. Appl. Phys. Lett. 2007; 91(7): 073516.

[29] Ren M., Wu G., Wu E, and Zeng H. Experimental demonstration of counterfactual quantum key distribution. Laser Phys. 2011; 21(4): 755–760.

[30] Liu Y., Chen T., Wang L, Liang H., Shentu G., Wang J., Cui K., Yin H., Liu N., Li L., Ma X., Pelc J. S., Fejer M. M., Peng C., Zhang Q., and Pan J. Experimental measurement-device-independent quantum key distribution. Phys. Rev. Lett. 2013; 111: 130502.

[31] Takesue H., Nam S. W., Zhang Q., Hadfield R. H., Honjo T., Tamaki K., and Yamamoto Y. Quantum key distribution over a 40-dB channel loss using superconducting single-photon detectors. Nat. Photonics 2007; 1: 343–348.

[32] Ren M., Gu X., Liang Y., Kong W., Wu E, Wu G., and Zeng H. Laser ranging at 1550 nm with 1-GHz sine-wave gated InGaAs/InP APD single-photon detector. Opt. Express 2011; 19(14): 13497–13502.

[33] McCarthy A., Ren X., Frera A. D., Gemmell N. R., Krichel N. J., Scarcella C., Ruggeri A., Tosi A., and Buller G. S. Kilometer-range depth imaging at 1550 nm wavelength using an InGaAs/InP single-photon avalanche diode detector. Opt. Express 2013; 21(19): 22098–22113.

[34] Liang Y., Huang J., Ren M., Feng B., Chen X., Wu E, Wu G., and Zeng H. 1550-nm time-of-flight ranging system employing laser with multiple repetition rates for reducing the range ambiguity. Opt. Express 2014; 22(4): 4662–4670.

[35] Mora A. D., Tosi A., Zappa F., Cova S., Contini D., Pifferi A., Spinelli L., Torricelli A., and Cubeddu R. Fast-gated single-photon avalanche diode for wide dynamic range near infrared spectroscopy. IEEE J. Sel. Top. Quantum Electron. 2010; 16(4): 1023–1030.

[36] Xu L., Wu E, Gu X., Jian Y., Wu G. and Zneg H. High efficiency InGaAs/InP-based single photon detector with high speed. Appl. Phys. Lett. 2009; 94(16): 161106.

[37] Chen X., Wu E, Xu L., Liang Y., Wu G. and Zneg H. Photon-number resolving performance of the InGaAs/InP avalanche photodiode with short gates. Appl. Phys. Lett. 2009; 95(13): 131118.

[38] Namekata N., Adachi S., and Inoue S. 1.5 GHz single-photon detection at telecommunication wavelengths using sinusoidally gated InGaAs/InP avalanche photodiode. Opt. Exp. 2009; 17(8): 6275–6282.

[39] Namekata N., Adachi S., and Inoue S. Ultra-low-noise sinusoidally gated avalanche photodiode for high-speed single-photon detection at telecommunication wavelengths IEEE Photon. Tech. Lett. 2010; 22(8): 529–531.

[40] Yuan Z. L., Kardynal B. E., Sharpe A. W., and Shields A. J. High speed single photon detection in the near infrared. Appl. Phys. Lett. 2007; 91(4): 041114.

[41] Yuan Z. L., Sharpe A. W., Dynes J., Dixon A., and Shields A. J. Multi-gigahertz operation of photon counting InGaAs avalanche photodiodes. Appl. Phys. Lett. 2010; 96: 071102.

[42] Zhang J., Eraerds P., Walenta N., Barreiro C., Thew R., and Zbinden H. 2.23 GHz gating InGaAs/InP single-photon avalanche diode for quantum key distribution. Proc. SPIE, 2010; 7681: 76810Z.

[43] Restellil A., Bienfang J. C., and Migdall A. L. Single-photon detection efficiency up to 50% at 1310 nm with an InGaAs/InP avalanche diode gated at 1.25 GHz. Appl. Phys. Lett. 2013; 102: 141104.

[44] Gu X., Huang K., Pan H., Wu E, and Zeng H. Photon correlation in single-photon frequency upconversion. Opt. Express 2012; 20(3): 2399–2407.

[45] Gu X., Li Y., Pan H., Wu E, and Zeng H. High-speed single-photon frequency upconversion with synchronous pump pulses. IEEE J. Sel. Top. Quantum Electron. 2009; 15: 1748–1752.

[46] Huang K., Gu X., Pan H., Wu E, and Zeng H. Synchronized fiber lasers for efficient coincidence single-photon frequency upconversion. IEEE J. Sel. Top. Quantum Electron. 2012; 18: 562–566.

[47] Liang Y., Wu E, Chen X., Ren M., Jian Y., Wu G., and Zeng H., Room-temperature single-photon detector based on InGaAs/InP avalanche photodiode with multichannel counting ability. IEEE Photon. Tech. Lett. 2011; 23(2): 115–118.

[48] Tomita A., and Nakamura K. Balanced, gated-mode photon detector for quantum-bit discrimination at 1550 nm. Opt. Lett. 2002; 27: 1827–1829.

[49] Comandar L. C., Frohlich, B. Lucamarini M., Patel, K. A. Sharpe A. W., Dynes J. F., Yuan Z. L., Penty R. V., and. Shields A. J. Room temperature single-photon detectors for high bit rate quantum key distribution. Appl. Phys. Lett. 2014; 104: 021101.

[50] Chen X. L., Wu E, Wu G., and Zeng H. P. Low-noise high-speed InGaAs/InP-based single-photon detector. Opt. Exp. 2010; 18(7): 7010–7018.

[51] Wu G., Jian Y., Wu E, Zeng H. Photon-number-resolving detection based on InGaAs/InP avalanche photodiode in the sub-saturated mode. Opt. Exp. 2009; 17(21): 18782–18787.

[52] Jian Y., Wu E, Wu G., and Zeng H. P. Optically self-balanced InGaAs-InP Avalanche photodiode for Infrared single-photon detection. IEEE Photon. Tech. Lett. 2010; 22(3): 173.

[53] Jian Y., Wu E, Chen X., Wu G., and Zeng H. Time-dependent photon number discrimination of InGaAs/InP avalanche photodiode single-photon detector. Appl. Opt. 2011; 50(1): 61–65.

[54] Huang K., Gu X., Ren M., Jian Y., Pan H., Wu G., Wu E and Zeng H. Photon-number-resolving detection at 1.04 μm via coincidence frequency upconversion. Opt. Lett. 2011; 36(9): 1722–1724.

[55] Huang K., Gu X., Pan H., Wu E, and Zeng H. Few-photon-level two-dimensional infrared imaging by coincidence frequency upconversion. Appl. Phys. Lett. 2012; 100(15): 151102.

[56] Zhou Q., Huang K., Pan H., Wu E, and Zeng H. Ultrasensitive mid-infrared up-conversion imaging at few-photon level. Appl. Phys. Lett. 2013; 102(24): 241110.

[57] Chen J., Wu G., Li Y., Wu E and Zeng H. P. Active polarization stabilization of optical fibers suitable for quantum key distribution. Opt. Express 2007; 15(26): 17928–17936.

[58] Chen J., Wu G., Xu L. L., Gu X. R., Wu E, Zeng H. P. Stable quantum key distribution with active polarization control based on time-division multiplexing. New J. Phys. 2009; 11: 065004.

Chapter 10

# Organic Semiconductors for Non-Linear Optical Applications

Anca Stanculescu and Florin Stanculescu

Additional information is available at the end of the chapter

http://dx.doi.org/10.5772/60926

**Abstract**

The organic molecules represent a promising research field because they show special properties which are determined by the conjugated system of $\pi$ electrons and the groups substituted to the aromatic nucleus offering perspectives for a large area of applications, including the non-linear optics.

We have investigated two types of molecular crystalline material prepared from aromatic derivatives, meta-dinitrobenzene (m-DNB) and benzil (Bz), and three types of organic thin films prepared from arylenevinylene (triphenylamine, carbazole) compounds, maleimidic derivatives and anilinic derivatives functionalised copolymers using different methods (matrix-assisted pulsed laser evaporation, vacuum evaporation, spin coating). The effect of the experimental conditions on the morphological and structural particularities of the films has been emphasised. New results are bought about the systems formed from an aromatic derivative (m-DNB and benzil) crystal and inorganic (organic) dopant studied from the point of view of the dopant incorporation considering the stability of the growth interface and its effect on the optical band gap and optical non-linear (ONL) properties of the m-DNB and Bz crystals. The relationship between the morphology and structural order of the maleimidic derivatives monomers in polycarbonate matrix composite layer and the ONL properties is discussed. We analysed the correlation between the molecular structure of organic compound, particularities of the macroscopic order (influenced by the crystal growth and thin-film deposition methods) and optical non-linear (second harmonic generation, two-photon fluorescence) properties.

**Keywords:** Organic materials, optical non-linear properties, thin films, organic crystals

## 1. Introduction

The past decades indicated a large-scale development of the optical telecommunications conditioned by the improvement of the optical fibres performances. The research activity in the field of non-conventional optical effects in a variety of materials has been stimulated by the discovery of the laser emission. The information transmission velocity using light determines an increase in the efficiency of the systems used in a large area of applications from communications to medicine and military domain as well as a direct control at the optical signal level, without the necessity of an optical-electronic-optical conversion. The examples include the liquid crystals for displayers, piezoelectric plastic materials for microphones, active organic materials for luminescent displayers (OLED), as well as the organic materials with special optical properties in solid state for applications in photonics. The optical non-linear materials are assumed as the key elements for further photonic applications, because these allow the control on the luminous beam using another luminous beam.

The exploitation of the remarkable properties of the organic molecules represents a new and promising technological field, because the specific properties of the molecular structure determines the response (selective, fast, intense) at macroscopic level, under some external stimulus action, like luminous radiation. The organic compounds are intensively studied because they show special properties which are associated with the conjugated system of $\pi$ electrons and the groups substituted to the aromatic nucleus generating intra-molecular charge transfer properties associated with a large area of applications, including the non-linear optics and offering an alternative to the inorganic compounds.

In the field of optical non-linear (ONL) applications, the interest is correlated with the preparation of crystals and thin films from molecular, oligomeric and polymeric materials with conjugated systems, extended $\pi$ electrons systems and active optical non-linear chromophoric groups, showing an arrangement of the chromophores adequate for the appearance of some favourable physical properties. The high non-linearity of (molecular and oligomeric/polymeric) organic materials is due to the large delocalisation of $\pi$ electrons cloud, which can be determined by the electronegativity of the functional groups, chromophores, substituted to the aromatic nucleus and operating as electron donors or acceptors. Supplementary, in order to generate optical non-linear effects, it is necessary to pack the organic molecules in a non-symmetrical macroscopic form. These conditions could be satisfied in bulk molecular crystal or in polymers/monomers thin-films materials because they are characterised by a high molecular order and density of chromophoric groups. In the case of molecular crystals special requirements concern the crystallinity, and in the case of monomers and polymeric materials special requirements concern the quality of the thin films. The main characteristics making favourable the utilisation of the organic molecular crystals for the mentioned applications are: large transparency domain, high birefringency, high non-linear coefficients, high damage thresholds in laser beam, versatility of the molecular structure which can be changed by molecular engineering for optimising the properties of interest [1, 2]. Since the discovery of the high optical non-linearities of the organic materials, the interest for these materials considerably grew up, opening new directions for studying and identifying new applications in the photonics field.

Lately, there is significant growth of the interest in analysing the potential of organic materials for new applications in the dynamic processing of image implying: optical conversion of frequency [3]; electro-optical modulation [4]; dynamic holography [5]; real-time interferometry [6]; optical storage of information; optical conjugation of phase [7]; optical amplification of image [8]; spatial modulation of the light [9]; optical interconnection [10]. Organic materials can be also utilised in the field of second harmonic generation (SHG) for medical applications and two-photon fluorescence (TPF) for biological imaging microscopy [11, 12].

In the field of non-linear optics, the two photons absorption (TPA) is a third-order process correlated with the extended $\pi$-conjugated electronic structures and charge transfer properties, which can be indirectly evaluated by two photons absorption emission fluorescence (TPF). Organic compounds showing TPA are interesting for potential applications in frequency up conversion lasing, optical power limiting, 3D fluorescence imaging, 3D optical data storage, 3D lithographic microfabrication and photodynamic therapy [13-15].

There is a tight bond between the material properties and the preparation conditions (synthesis, crystal growth and films deposition) just like between the molecular structure and the properties of interest. The properties of these materials are strongly affected by the presence of structural and/or chemical defects. Therefore, the special promises associated with the potential performances of these materials related to intrinsic properties of material could be valorised only solving the concrete problems involved by the manufacture of a material having the desired form, dimension, quality and properties adequate for a target application. For this reason of great importance is the identification of the optimal methods to prepare these materials both in crystalline form (molecular organic compounds) and as thin films (monomeric/polymeric compounds). In this context, a special attention must be paid to the design, synthesis and characterization of the organic compound, preparation and evaluation of the optical non-linear properties of the organic material for photonic/optoelectronic applications. Till now, a satisfactory compromise between the theoretically predicted non-linearity and the quality of the materials related to preparation and processing challenges has not been obtained for organic materials.

The purpose of this chapter is to enlarge the knowledge about the optical properties and processes involved in the interaction of light with the organic solid state showing new molecular structures and about the correlation between the preparation conditions of the molecular/oligomeric/polymeric materials and their physical characteristics. This will offer a way to restrain the organic compounds by selecting the adequate materials for target photonic applications focusing on non-linear optics. The correlation between the molecular structure of monomer, oligomer and polymer, type of macroscopic order (associated with the growth method for crystal and deposition method for thin film) and optical non-linear (SHG, TPF) properties (involving interaction phenomena of radiation with organic compounds) will be emphasised.

## 2. Bulk organic semiconductors for optical non-linear applications

The aromatic derivatives such as meta-dinitrobenzene (m-DNB) and benzil (Bz) are organic compounds of interest in preparing bulk crystal materials showing large transparency domain, large optical band gap and high non-linear coefficients. These two compounds show important optical non-linear phenomena despite the differences in their chemical structure at the molecular level. In the context of special attention paid to the investigation of bi-component organic systems for optical non-linear (ONL) applications involving the synthesis of the organic compounds, organic crystal growth and characterisation, the interest in studying m-DNB and Bz bulk crystals is justified by the perspective to use these materials as a crystalline host matrix for both organic and inorganic guests.

m-DNB ($C_6H_4N_2O_4$) is a negative bi-axial crystal that belongs to the Pbn21 space group and the point symmetry group mm2 which crystallises in the orthorhombic system at room temperature showing a pyramidal shape. The unit cell contains four molecules and has the following dimensions: $a$ = 13.20 Å, $b$ = 13.97 Å and $c$ = 3.80 Å. The transparency range of m-DNB crystalline materials is 0.4-2.5 μm [16, 17, 19]. Bz ($C_6H_5COCOC_6H_5$) is an uni-axial crystal that belongs to the trigonal space group $P3_12_1$ and the point symmetry group $D_3^4(D_3^6)$. The unit cell contains three molecules which are helically disposed and closely packed around the 31 axis and the hexagonal unit cell has the following dimensions: $a$ = 8.42 Å and $b$ = 13.75 Å. The transparency range of Bz crystalline materials is from UV through near IR [18, 19].

Pure m-DNB and Bz crystals have been grown from the melt by the Bridgman-Stockbarger method using the same experimental set-up and the same experimental conditions defined by a temperature gradient ($\Delta T$) of 25°C cm$^{-1}$ and growth speed ($v$) of 1.7 mm h$^{-1}$ [19]. The crystal of m-DNB doped with iodine has also been grown in the same experimental set-up and the same experimental conditions such as the crystal of Bz doped with iodine or m-DNB: $\Delta T$ = 30°C cm$^{-1}$ and $v$ = 2.0 mm h$^{-1}$ [19].

We have developed some studies about the influence of the crystal growth conditions on the incorporation of dopant and the effect of dopant on the optical properties of pure m-DNB and Bz crystals. We have analysed the two aromatic derivatives crystals from two different perspectives: (1) the effect of iodine ($I_2$) dopant concentration (1 wt% and 2 wt%) on the two different matrix (m-DNB and Bz) and (2) the behaviour of the same aromatic derivative, m-DNB, as matrix and dopant [19].

The X-ray diffraction study of pure and doped m-DNB and Bz crystals has offered data which have been analysed in the Pawley approach with the TOPAS software emphasising the effect of dopant on the lattice parameters [19].

The study has been realised on slices cut perpendicularly to the growth direction of the crystal. The sharp diffraction peaks shown in Figure 1 sustain a good crystallinity of the bulk materials and these peaks have been attributed using the Power Diffraction File for m-DNB and Bz which contains the reference values of the lattice parameters. The crystallographic planes which give the strongest reflections are parallel to the surface of the slice and therefore, the strongest reflection is given in Bz by the (003) plane, the growth direction of the crystal being [001] and the strongest reflections in m-DNB are given by the (111), (311) and (002) planes confirming the polycrystalline nature of the grown crystal showing three regions of three different orientations [19].

The values of lattice parameters obtained for the hexagonal benzil crystal, $a = b = 8.350$ Å and $c = 13.557$ Å, are in agreement with the reference value, $a = b = 8.410$ Å and $c = 13.679$ Å [19]. The lattice parameters evaluated for m-DNB orthorhombic crystal, $a = 13.246$ Å, $b = 14.029$ Å and $c = 3.807$ Å, are also in agreement with the reference values, $a = 13.290$ Å, $b = 14.070$ Å and $c = 3.813$ Å [19]. The slight smaller experimental values of $a$ and $b$ compared to the reference values suggest a compression of the lattice and the presence of some internal stress/strain. Considering the effect of dopant on the lattice parameter of the matrix, we can obtain information on the type of mechanism which could be involved in the incorporation of dopant: in interstitial or in substitutional positions. By doping with $I_2$ the lattice parameters of Bz have increased from $a = b = 8.350$ Å to $a = b = 8.453$ Å and from $c = 13.557$ Å to $c = 13.622$ Å. These values confirm an interstitial incorporation of $I_2$ dopant in Bz matrix [19]. A similar behaviour and interstitial incorporation is also shown by $I_2$ in m-DNB matrix, the lattice parameters of pure m-DNB increasing by doping with $I_2$ from $a = 13.246$ Å to $a = 13.345$ Å, from $b = 14.029$ Å to $b = 14.186$ Å and from $c = 3.807$ Å to $c = 3.821$ Å [19].

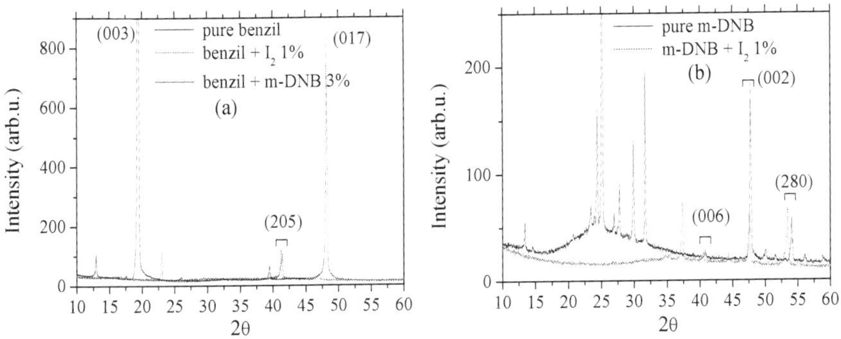


**Figure 1.** XRD patterns for pure and doped Bz (a) and pure and doped m-DNB (b) [19].

On the contrary, by doping with m-DNB the Bz crystal lattice parameters $a$ and $b$ have decreased from $a = b = 8.350$ Å to $a = b = 8.267$ [19] confirming a substitutional incorporation of m-DNB molecule in Bz, the molecule of m-DNB being smaller than the molecule of benzil.

The optical properties of the organic crystal are affected by the quality of the crystal determined by the shape of the solid-liquid interface through the growth conditions. Additionally, the homogeneity is very important for the doped crystals and is directly connected to the dopant incorporation determined also by the solid-liquid interface shape.

In this context, new results have been bought about the systems aromatic derivative/dopant studied from the point of view of the growth interface stability criterion represented by eq. (1) with the purpose to analyse the incorporation of dopant and its effect on the quality of the crystal. The stable growth limits and the experimental conditions related to temperature gradient at the growth interface and moving speed of the ampoule in the furnace have been investigated.

The previous results on growth interface stability for Bz crystal [20, 21] have been completed with new conclusions on the stable growth of m-DNB crystals. We will emphasise the conditions favouring the initiation of instabilities, which are associated with structural defects and compositional non-homogeneities in the crystal of m-DNB affecting the optical properties including the ONL properties.

A slightly convex liquid-solid interface is necessary to obtain good-quality crystals, but during the growth process the shape of the interface changes as a consequence of the variation in the thermal conductivity of the organic compound in the liquid and solid state. Because the ratio between the thermal conductivity coefficient in melt and solid is ≤1 both for m-DNB and Bz, it is predicted a convex growth interface with a lower deviation from the planar growth interface in the case of Bz. This conclusion confirms that it is easier to grow good-quality Bz than m-DNB crystals [19].

The stability of the growth interface in the binary system m-DNB doped with iodine has been considered as an example for analysing the solidification process in doped m-DNB crystals [22-24]. Starting with the stability criterion we have drawn the curves $\Delta C \times \cdot m = f(\Delta T)$ at $v$ = const. and $\Delta C \times m = f(v)$ at ΔT=const, where

$$\Delta C \cdot m \le \frac{\rho_m \cdot \Delta_f H}{k_m + k_s} \cdot v - \frac{2 \cdot k_m}{k_m + k_s} \cdot \Delta T \tag{1}$$

and $\Delta T$ = thermal gradient at the growth interface; $\Delta C$ = gradient of concentration in the liquid at the growth interface; $m$ = liquidus line slope; $v$ = moving speed of the interface during the solidification process; $\rho_m$ = melt density; $\Delta_f H$ = enthalpy of fusion; $k_s$ = thermal conductivity of the solid; $k_m$ = thermal conductivity of the melt. For m-DNB we have used the following values for constants [19]: $\Delta_f H$ = 17.30 kJ mol$^{-1}$ [25], $\rho_m$ = 1.3644 g cm$^{-3}$ [26], $k_m = 0.183\times10^{-2}$ Wcm$^{-1}$K$^{-1}$ [27] and $k_s = 0.4\times10^{-2}$ W cm$^{-1}$K$^{-1}$ [28].

In the Bridgman-Stockbarger configuration, above these curves presented in Figure 2, it is situated the region of stable growth and below, the region of unstable growth. For the curves corresponding to $\Delta C \times \cdot m = f(\Delta T)$ at $v$ = const. shown in Figure 2a and high concentration gradients at the interface, the stable growth can not be assured at low-temperature gradients even for high moving speed. For the curves corresponding to $\Delta C \times m = f(v)$ at $\Delta T$ = const. shown in Figure 2b, the area of the stable growth region increases with the increase in thermal gradient, the increase in the interface moving velocity being not critical [19]. Therefore, to be situated in the stable growth region for high values of the product $\Delta C \times m$, we have to assure high thermal gradient by adequately positioning the ampoule in the thermal profile of the heater.

As a consequence, the experimental conditions used to grow pure m-DNB crystals ($\Delta T$ = 25°C, $v$ = 1.7 mm h$^{-1}$) and iodine doped m-DNB crystals ($\Delta T$ = 30°C, $v$ = 2.0 mm h$^{-1}$) are fixing the limits for the product $\Delta C \times m$ that assure a stable growth and are compatible with the experimental conditions for a stable growth of pure and doped benzil crystals, previously presented [20, 21]. Therefore, it is possible to grow m-DNB and Bz crystals in the same experimental

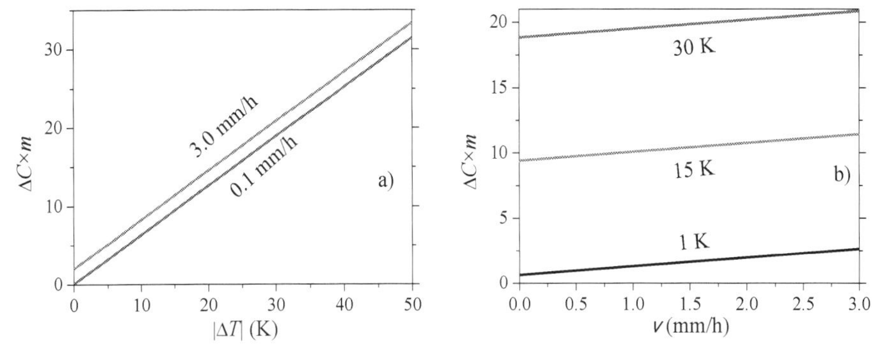


**Figure 2.** Stable and unstable growth region for m-DNB: $\Delta C \times m = f(v)$ at $\Delta T$ = const. (a); $\Delta C \times \cdot m = f(\Delta T)$ at $v$ = const. (b) [19].

conditions, such as the selected conditions which are characterised by high $\Delta T$ and $v$, even for high concentration gradients at the growth interface [19]. The shape of the growth interface is correlated with the generation of defects which induce a certain degree of disorder in crystal. New contributions will concern the effect of the growth interface stability on the particularities of dopant incorporation and, as a consequence, on the optical band gap and crystal quality.

Using the Tauc plot, $(\alpha h\nu)^{1/2}$ against the photon energy ($h\nu$), presented in Figure 3, it has been evidenced the classical, large band gap semiconductor behaviour of m-DNB ($E_g$=2.69 eV) and Bz ($E_g$=2.68 eV) [19]. As a consequence of the disorder induced during the growth of the crystals, the effect of inorganic dopant ($I_2$) on the band gap of m-DNB and Bz is different: an increase in m-DNB band gap, tendency maintained with increasing iodine concentration and a decrease in Bz band gap, tendency maintained also for organic dopants (m-DNB).

The m-DNB crystals have shown a higher disorder (evidence by Urbach law) compared to benzil crystal grown in the same experimental conditions [19]. The Urbach energy has been evaluated from the slope of the plot ($\ln\alpha$) against ($h\nu$) in the region of low energy. The m-DNB crystals doped with $I_2$ have shown a lower disorder than the undoped crystals. While $I_2$ dopant increases the degree of disorder in Bz crystal, the effect of organic dopant (m-DNB) is, on the contrary, smaller [19].

A special configuration presented in Figure 4 has been used for the excitation of ONL phenomena in organic materials samples both bulk crystals (mentioned above) and thin films (will be introduced later). The experimental set-up is based on an ultra short pulsed laser Spectra Physics 'Tsunami' characterised by a maximum wavelength of 800 nm, pulse duration of 60 fs, frequency of 80 MHz and medium power of 780 mW [29].

A high N.A. Mitutoyo microscope objective was used to focus the laser beam on the sample mounted on a motorised Thorlabs XYZ stage. The same microscope objective used for irradiation is also used for collecting the light emitted by the sample in 180 degrees geometry.

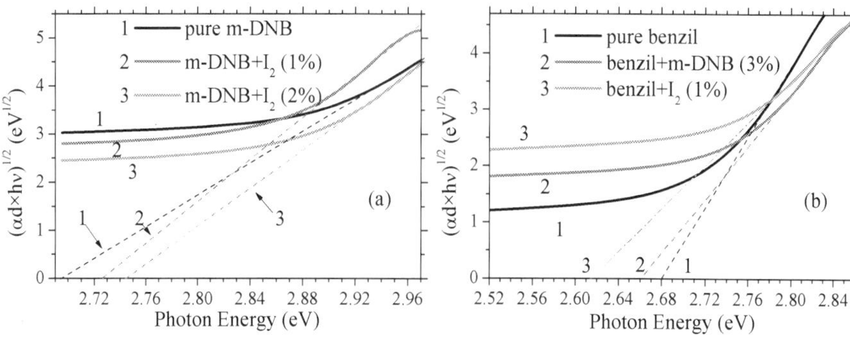


**Figure 3.** Tauc plots for: pure and doped m-DNB (a); pure and doped Bz (b) [19].

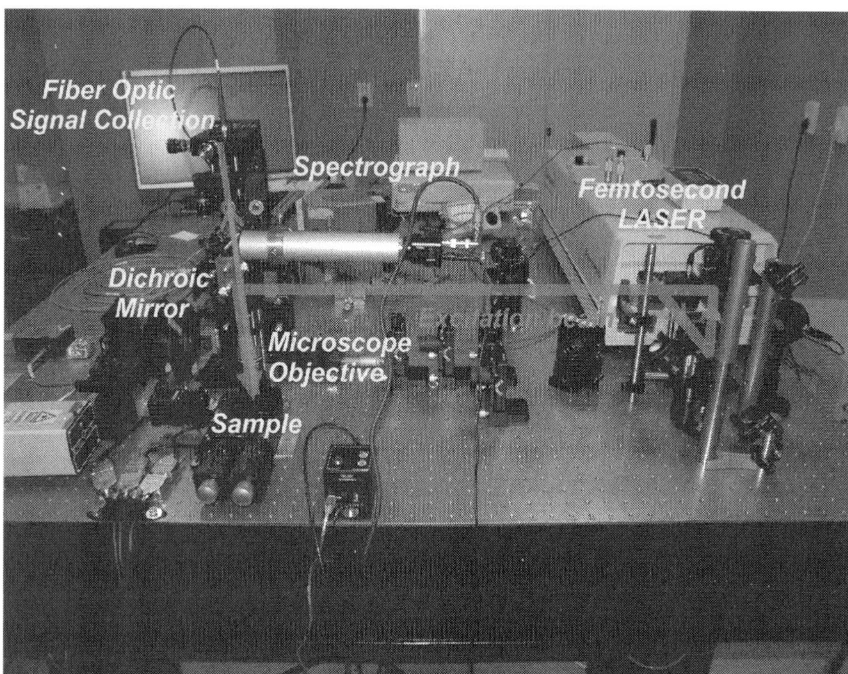


**Figure 4.** Experimental set-up for ONL phenomena measurements.

The detection system is optimised to be sensible in the visible range. The emitted light is passing through a dichroic mirror and, after this, is focused by a lens on the collecting optical fiber connected with a spectrograph ANDOR Shamrock 163i. An ANDOR IDUS CCD camera

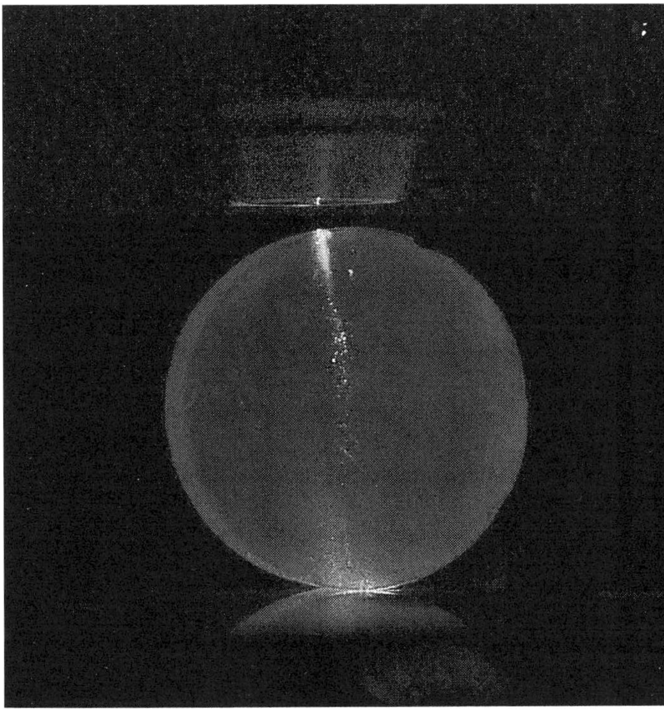

Figure 5. Image of the ONL signals obtained in a crystal of Bz.

cooled at 213 K is used to measure the dispersed light. The excitation light is blocked and its effect is minimised with a NIR cut-off filter placed in front of the focalisation lens. The amplification of the CCD detector can be increased without the necessity to increase the signal at longer wavelengths [29].

The intensity of optical non-linear phenomena, like SHG and TPF, associated to green light in Figure 5, produced in aromatic derivative crystals is also affected by the stability of the growth interface and the presence of defects.

The two types of aromatic derivatives crystals grown in the same experimental conditions show different dominant ONL phenomenon. In m-DNB, SHG dominates while in Bz, TPF dominates as can be seen in Figure 6.

In two-photon absorption process (TPA) two photons combine to bridge a gap of energy larger than the energy corresponding to each individual photon. This process is not involving any intermediate or virtual state created by the interaction of the photons with the organic molecules, the two photon absorption emission fluorescence (TPF) being an indirect measure of this process.

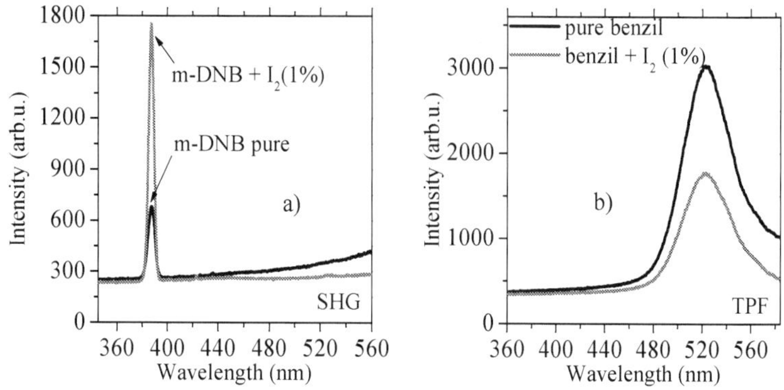


**Figure 6.** ONL phenomena in organic molecular crystalline materials: m-DNB (a); Bz (b) [19].

The SH associated with the peak situated around 400 nm is determined in m-DNB by the inductive and mezomeric effects associated with the strongly polarisable substituent groups of -$NO_2$. The ONL phenomena in Bz are determined by the long conjugated chain causing a displacement of the charge inside the delocalised π-electrons over two aromatic rings and two carbonyl groups. The broad emission band centred around 540 nm is determined by the luminescence obtained by two-photon absorption because Bz has an electronic excited level corresponding to singlet excited state (n, π*) in skew configuration situated at 3.2 eV [30]. This energy is close to that corresponding to a photon with twice the frequency of the laser beam and generates the excitation at the edge of the absorption band. After excitation, the relaxation of the Bz lattice implies three steps: (1) the planar rearrangement of molecules and redistribution of energy, because the carbonyl groups emit only from the planar configuration characterised by an energy level situated at 2.6 eV; (2) radiationless relaxation by a process of intersystem crossing to the first triplet excited state situated at 2.3 eV; (3) radiative de-excitation generating the emission band centred on 540 nm, the position and shape of this band corresponding to the usual fluorescence in Bz at linear excitation [19].

The intensity of SH signal increases in m-DNB doped with $I_2$.compared to pure m-DNB, in concordance with the lower degree of disorder associated with lower Urbach energy in the grown doped crystal. ONL properties are enhanced by the deformation of the m-DNB molecule in the presence of $I_2$ determining an asymmetry in the electronic cloud. The intensity of TPF signal decreases in Bz crystals doped with $I_2$ grown in the same experimental conditions as a consequence of the increased disorder associated with higher Urbach energy favouring absorption/scattering of radiation.

## 3. Organic semiconductor thin films for optical non-linear applications

The major problems for large-scale application of the aromatic derivatives as bulk crystals in photonics are associated with the difficulties to grow such crystals (determined by the low melting point, supercooling, low thermal conductivity) and to process the crystals (determined by the low mechanical properties imposed by the weak bonding forces between molecules). Other difficulty is related to the homogeneous distribution of the dopant within the matrix. Therefore, the thin films are preferred, but the same problems correlated with the properties of the organic compounds still remain and involve adequate method for thin-films preparation, like vacuum evaporation deposition, matrix-assisted pulsed laser evaporation (MAPLE) and spin coating, in correlation with the particularities of the organic molecule. We have selected and investigated three types of organic thin films prepared from arylenevinylene compounds, maleimidic derivatives and anilinic derivatives functionalised copolymers.

### 3.1. Thin films based on arylenevinylene compounds

From arylenevinylene compounds we have selected triphenylamine- and carbazole-based compounds whose chemical structures are presented in Figure 7, two aromatic amines, which could be used as building blocks for the preparation of conjugated oligomers and polymers, potential candidates for ONL applications. By the combination of these aromatic amines showing hole transporting properties with vinyl segments with strong emissive properties, a new class of electroactive conjugated materials can be generated. The tertiary amine units are not interrupting the conjugation of the principal chain in these compounds, the pair of electrons of the nitrogen atom contributing to the increase in effective conjugation and the double vinyl bonds creating 'conjugated bridges' between the aromatic units [31, 32].

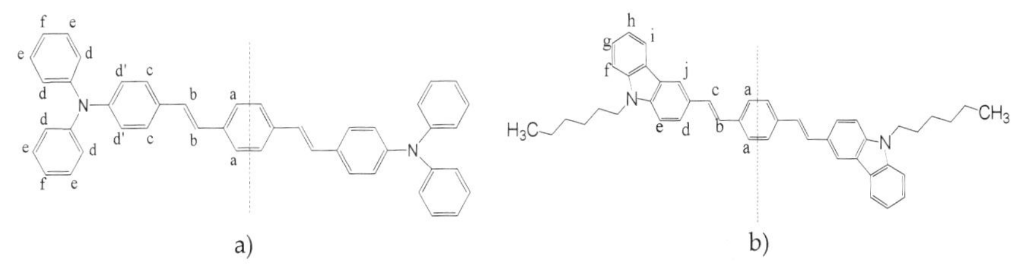

**Figure 7.** Chemical structures of the oligomers L78: 1, 4-bis [4-(N, N'-diphenylamino)phenylvinyl] benzene (a); L13: 3, 3'-bis (N-hexylcarbazole)vinylbenzene (b) [31].

Some recent works have emphasised that symmetrically substituted organics D-π-D with donating groups situated at the ends linked by conjugated bridges, similar to those mentioned above, show important advantages, among which the large TPA cross-section [33, 34] and high transmission in visible range sustaining the generation of significant ONL phenomena.

Two arylenevinylene oligomers, L78 and L13, have been synthesised by Witting condensation using the following starting reagents: 4-formyltriphenylamine and N-hexyl-3-formylcarbazole and phosphonium salt of 1, 4-bis(chloromethyl)benzene [31, 35].

The chemical structure of the synthesised trans-arylenevinylene oligomers L78 and L13 has been evidenced correlating the FTIR measurements data presented in Figure 8 and Raman measurements data presented in Figure 9. The FTIR spectra shown in Figure 8 evidence the characteristic bands for the chromophoric groups in oligomers powders: ring in-plane bending vibration of the mono-substituted phenyl ring ($\nu_C$-$\nu_C$) situated at 1590-1598 $cm^{-1}$; phenyl rings (C=C) vibration situated at 1505-1586 $cm^{-1}$; normal vibration with important contribution from C-N stretching coordinates situated at 1316-1332 $cm^{-1}$ and 1030-1190 $cm^{-1}$; out-of-plane (wagging) vibration of the hydrogen atoms linked to phenyl ring situated at 695-697 $cm^{-1}$; out-of-plane (puckering) vibration of phenyl ring situated at 747-752 $cm^{-1}$; deformation vibration of the C-H bonds of the p-substituted phenyl ring situated at 814-827 $cm^{-1}$. The trans-arylene-levinylene structure of the oligomers is sustained by the absorption peaks situated between 960 and 962 $cm^{-1}$ corresponding to the out-of-plane bending vibration of HC=CH [35].

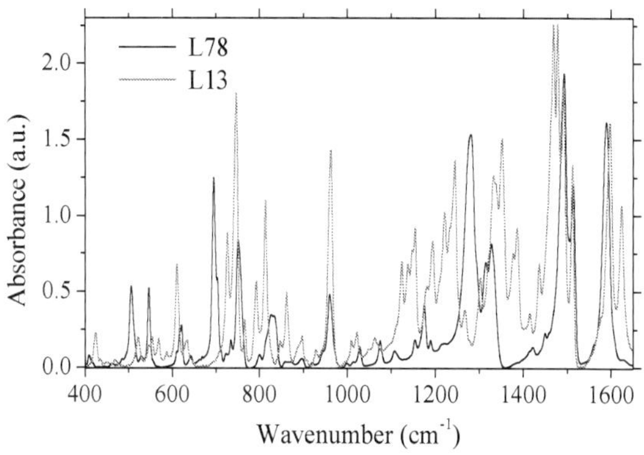


**Figure 8.** FTIR spectra of L78 and L13 oligomers powder in KBr [35].

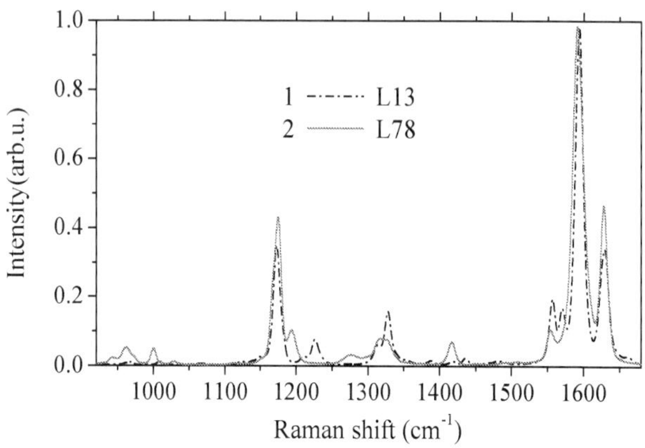


**Figure 9.** Raman spectra of L78 and L13 oligomers powder [31].

In Figure 9 are presented the Raman spectra which have offered complementary details about the characteristic bands for the chromophoric groups in oligomers powders: C-N stretches with associated C-C stretches in both L78 and L13 attributed to 970 $cm^{-1}$; C-N stretches with associated large C-C stretches in both L78 and L13 attributed to 1174 $cm^{-1}$; large C-N stretches with associated C-C stretches in L78 attributed to 1280 $cm^{-1}$; quadrant stretch of phenyl rings with associated C-N stretches in both L78 and L13 attributed to 1590 $cm^{-1}$. In carbazole-based compounds like L13, supplementary peaks attributed to $C_{carbazole}$-N stretches appeared at: 1123 $cm^{-1}$, 1132 $cm^{-1}$ and 1230 $cm^{-1}$ [36]. The Raman spectra show most of the sharp narrow well-defined peaks in spectral range 900-1750 $cm^{-1}$ and are determined by scattering on the internal vibration modes [31].

Thin films of the mentioned arylenevinylene oligomers have been prepared on different substrates (quartz, silicon/Si, titanium/Ti) using two methods: vacuum evaporation and MAPLE. The properties of the layers have been analysed considering both the particularities of the molecular structure and solid state packing.

The X-ray diffraction spectra drawn in Figure 10 show no diffraction peaks confirming that the film deposited by vacuum evaporation is mostly amorphous, the molecule being randomly oriented [31]. We can consider a very reduced degree of order (more significant in L13), associated with the presence of the small, broad peak situated at $2\theta=20^{\circ}$. Because the peak is broad, a small grain structure is attributed to the film [31].

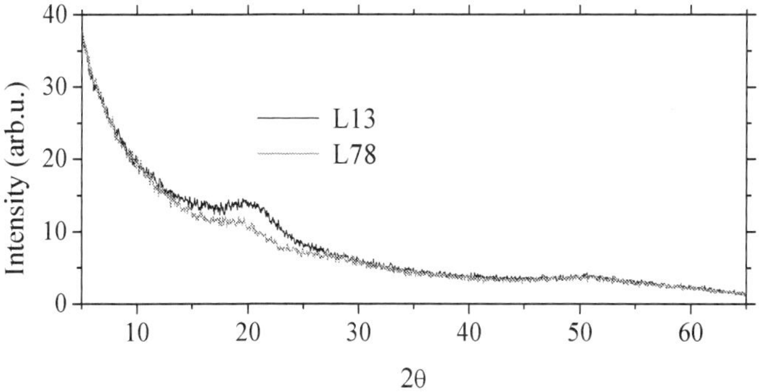


**Figure 10.** X-ray diffractograms for L78 and L13 oligomers thin films deposited by vacuum evaporation [31].

The SEM images of the oligomers films deposited by vacuum evaporation on Si presented in Figure 11 confirm the compact, small grain morphology evidenced by the broad XRD peak [31]. Some particular structures and aggregates, with acicular shape embedded in the L78 films and platelet shape in L13 films have also been evidenced [31]. This low density of randomly distributed structures has been developed during the deposition process.

**Figure 11.** SEM images of L13 (a) and L78 (b) oligomer thin film deposited by vacuum evaporation [31].

**Figure 12.** SEM images of L78 deposited by MAPLE from different solvents on different substrates: chloroform/quartz (a); dimethylsulphoxide (DMSO)/quartz (b); chloroform/Ti (c) [35].

The SEM images of the oligomers films deposited by MAPLE presented in Figures 12 and 13 have revealed different morphologies. The wrinkle morphologies of oligomers is generated by the inclusion of volatile solvent in the oligomer-rich surface favoured by the oligomer-solvent matrix droplets which appeared during the disintegration of the heated target [37].

In Figure 12 is evidenced the morphology of L78 thin films deposited by MAPLE both on quartz and on Ti using chloroform as solvent. L78 shows grains with a typical dimension up to 200 nm and larger-scale structures with different dimensions having spherical morphology randomly distributed in the layer [35]. When DMSO is used as solvent, the morphology of L78 on quartz is characterised by a lower density of spherical structures [35].

On the contrary, the films of L13 deposited on quartz have shown a morphology characterised by grains randomly distributed, as it is presented in Figure 13. The density of these grains has increased when the solvent was DMSO. A lower density of large spheres has been evidenced in the films of L13 deposited on Ti substrate using chloroform [35]. We can conclude that the morphology of the oligomers thin films depends on the deposition method, solvent and substrate.

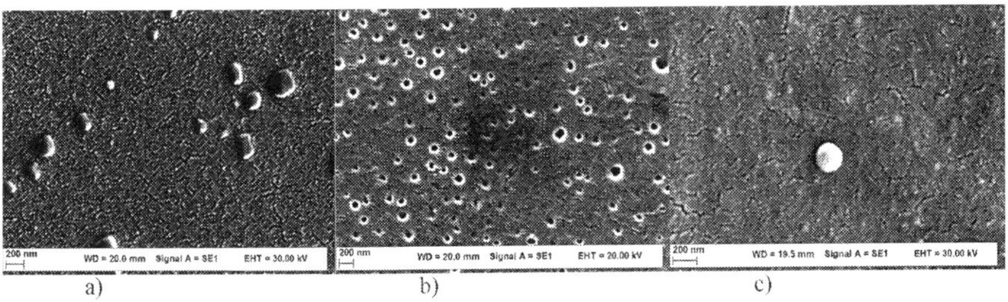

Figure 13. SEM images of L13 deposited by MAPLE from different solvents on different substrates: chloroform/quartz (a); DMSO/quartz (b); chloroform/Ti (c) [35].

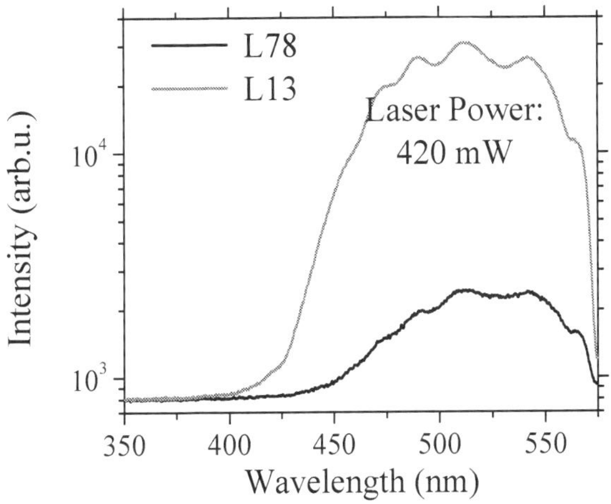


Figure 14. Fluorescence emission induced by two-photon absorption in oligomers L78 and L13 thin films deposited by vacuum evaporation on Si [31].

The shape of the TPF spectra and the position of the emission peaks correspond to those associated with the usual linear fluorescence spectra. This behaviour confirms that linear fluorescence and the TPF are generated by the relaxation of the system from the same excited state [31].

The optical properties of the MAPLE layers have been analysed in correlation with the molecular structure of the compounds, and the effects of the layer morphology determined by the solvent on the deposition process and on the resultant ONL processes have been emphasised.

Figure 14 revealed a strong TPF band centred on 525 nm in both oligomers, L78 and L13, thin films deposited by vacuum evaporation on Si. For high-intensity laser beam, the absorption of two photons with $\lambda$ = 800 nm corresponding to the excitation wavelength of the laser, $\lambda$ =

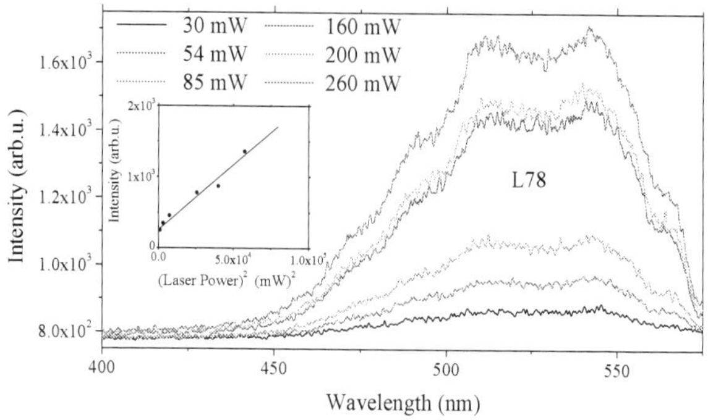


**Figure 15.** Dependence of TPF on laser power for L78 films deposited by vacuum evaporation on Si [31].

800 nm, could be replaced by the absorption of one photon with $\lambda = 400$ nm. After the absorption of one photon of 800 nm, the molecules of L78 and L13 do not have the time to relax because the duration of the laser pulse is very short, in the fs domain. Meanwhile, it absorbed the second photon of 800 nm and the molecular system was excited in a state from which it relaxed radiatively by TPF. For L78 and L13 using the same power of the excitation laser, 420 mW, we have obtained different intensity of TPF band, as a consequence of different focalisation regime of the laser beam [31].

The relation $I(2\omega) \propto (\chi^{(2)})^2 P_{(\omega)}^2$ [39] where P is the average power of the pulsed laser gives a simplified expression of the SH intensity. Therefore, a measure of the ONL behavior of the material can be given by the slope of the plot $I(2\omega)$ versus $P_\omega^2$ presented in Figure 15 and 16.

The quadratic dependence between the fluorescence signal of the L78 thin films and the laser excitation power, as emphasised in Figure 15, with the slope ~2 at laser power <260 mW, confirms the presence of the ONL phenomenon and the two-photon nature of the excitation [31, 38].

The SHG, presented in Figure 16, induced by the ultra-intense focused beam of a femtosecond laser in a reflection configuration, has been used to compare the second-order ONL coefficients of the oligomer L78 and L13 thin films deposited on optically inactive Ti substrates by MAPLE [35].

Both oligomers thin films deposited on an optically inactive substrate have shown nearly equal values of the second-order ONL coefficients and these oligomers are similar from this point of view, when are irradiated with a low-power laser beam <100 mW. At laser powers higher than 100 mW was emphasised a deviation from the linear behaviour and the saturation of the SH attributed to significant changes in polarisability determined by thermally induced conformational changes of the side chain in oligomer molecules. These changes induced under the effect of absorbed laser radiation determine the reorientation of the dipoles and changes in polarisability of the material.

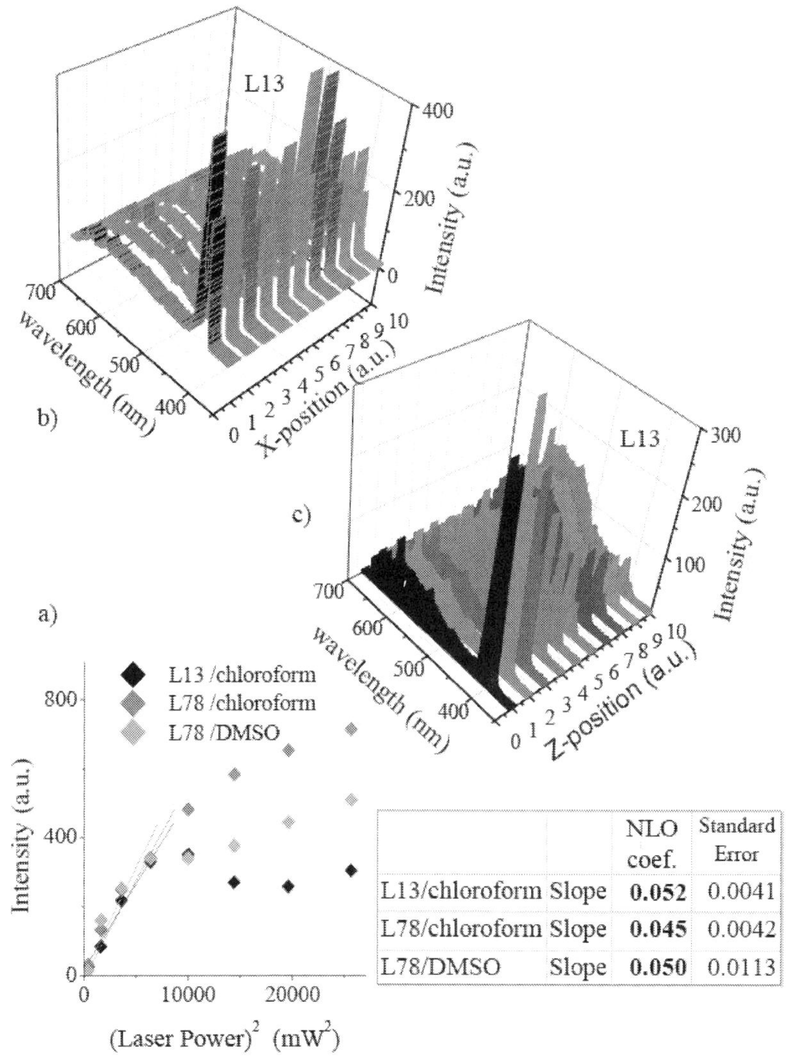


| | | NLO coef. | Standard Error |
|---|---|---|---|
| L13/chloroform | Slope | **0.052** | 0.0041 |
| L78/chloroform | Slope | **0.045** | 0.0042 |
| L78/DMSO | Slope | **0.050** | 0.0113 |

**Figure 16.** SH measurements in oligomers L78 and L13 thin films deposited by MAPLE on Ti substrate [35].

Using the scanning in z and x directions (z-scan and x-scan representation) for oligomer L13 thin films, a broad TPF band with a peak at 650 nm which became more intense when the laser spot is focused inside the sample has been revealed. This confirms that TPF is a process taking place in the volume of the sample. The SH signal is stronger at the surface of the sample and therefore could be altered by the surface morphology determined by the deposition experimental conditions including the solvent selected for the preparation of the MAPLE target [35].

### 3.2. Thin films based on maleimidic derivatives

Theoretical studies of the molecular structures with $\pi$-conjugation, containing functional groups having donor and acceptor properties, have revealed high first-order molecular hyperpolarisabilities and therefore these compounds seem adequate for the generation of ONL phenomena and manufacturing of thin-films-based optical devices for low-cost technology related to emerging photonic data processing, such as tunable wavelength systems filters, remote sensing, small-angle beam steering, high-bandwidth optical switches and modulators [40]. There are different ways to improve the generation of ONL phenomena from the increase in the strength of the electron donor and acceptor groups to increase in the length of the conjugated core and decrease of the ground state aromatic character [41]. The polarity of the monomers depends on the groups that substitute hydrogen in the benzenic nucleus and on the difference in electronegativities between these groups [42]. The selected maleimidic monomers contain dipolar segments determined by intramolecular electron withdrawing (donor) group and electron acceptor group and are characterised by the lack of balance between the effects of the substituent groups which determines the asymmetry in the distribution of electrons.

The amidic derivatives monomers with the chemical structure presented in Figure 17 have been synthesised from maleic anhydride and a nitro aniline derivative with substituent groups showing different electronegativities, R [-NH-; -NH-NH-]; R1 [-COOH; $-NO_2$; -CN]; R2 [$-NO_2$]. The synthesis has implied the opening of the anhydridic cycle in dimethylformamide (DMF) at room temperature (A1, A2, A3, A5 and A6) and at higher temperature, 50°C, (A7) under weak stirring for 2 hours. The monomers have precipitated in ice and have been separated by filtration and recrystallisation from methanol [29].

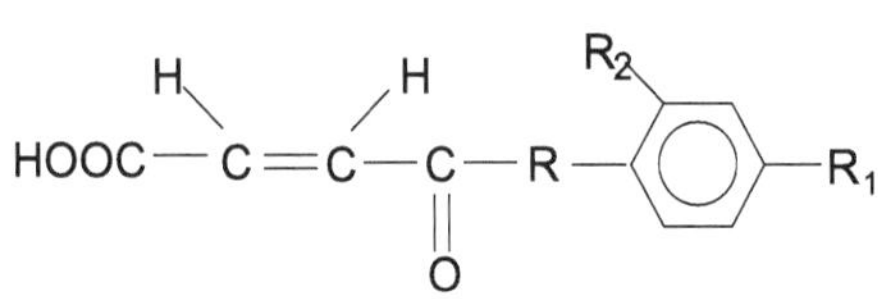

(A1) R: -NH-; $R_1$=-COOH; $R_2$:-H;
(A2) R: -NH-; $R_1$=$-NO_2$; $R_2$:-H;
(A3) R: -NH-; $R_1$=-CN; $R_2$:-H;
(A5) R: [-NH-]; $R_1$ and $R_2$: [$-NO_2$];
(A6) R: [-NH-NH-]; $R_1$ and $R_2$: [$-NO_2$]
(A7) R: [-NH-NH-]; $R_1$ and $R_2$: [$-NO_2$].

**Figure 17.** Chemical structure of amidic derivatives monomers [29, 40].

The strong acceptor character of the two [$-NO_2$-] substituent groups compared to donor character of [-NH-NH-] group determines the high polarity of the molecules in monomers (A6) and (A7).

### 3.2.1. Monomeric maleimidic derivatives thin films

Thin films of the selected monomers, presented in Figure 17, have been prepared by MAPLE, this method assuring a better control of the film thickness by the number of pulses, surface morphology by the fluence, amplified adhesion film/substrate and elimination of the possible source of contamination because it is a non-contact procedure which can be integrated with other non-contaminant processes [29]. For the preservation of molecular structure and deposition of clean films low fluences of 350-430 mJ cm$^{-2}$ and a dimethylsulphoxide (DMSO) matrix have been used. The thermalisation of the organic molecules was assured by a background gas, nitrogen at pressure between 0.05 and 0.3 torr. The quality of the deposited layer was increased by heating the substrate at 150 °C or 250 °C during the deposition [29].

In the FTIR spectra of the powders of the above mentioned monomers have been evidenced strong and moderate characteristic bands as shown in Figure 18 attributed to the following sequences in aromatics, carboxyl acids and amides/imides: 1640 cm$^{-1}$, 1540 cm$^{-1}$ to [-NH-] group; 1250 cm$^{-1}$ to [-C=O] group; 1240 cm$^{-1}$ to [-CN] group; ~840 cm$^{-1}$ to [-CH=CH-] double bond; ~640 cm$^{-1}$ to [-C-H-] single bond. The presence of the bands situated at 1390 cm$^{-1}$ characteristic for amide and of the bands situated at 1100 cm$^{-1}$ and 730 cm$^{-1}$ corresponding to imides ring deformation confirm the simultaneous presence of amidic and imidic forms; bands situated in the range 650-800 cm$^{-1}$ correspond to substituted benzene; bands situated in the range 1250-1300 cm$^{-1}$ to symmetric stretching vibration of [-$NO_2$] group [40]. Most of the absorption peaks present in powder sample of a monomer have disappeared in thin-films sample of the same monomer because the films deposited by MAPLE are very thin (10-140 nm evaluated by spectroscopic ellipsometry [40]). As can be seen in Figure 18, most of the above mentioned absorption peaks have been evidenced in the thicker films of monomer A2 confirming the preservation during the MAPLE deposition of the monomer chemical structure [40].

The FTIR spectra of monomers A5 and A6 powder shown in Figure 19 have also revealed the presence of the characteristic bands situated at 3500-3200 cm$^{-1}$ and 1620 cm$^{-1}$ for [-COOH] group, 1580 cm$^{-1}$ for [-NH-] group, 1250 cm$^{-1}$ for [-C=O] group, ~840 cm$^{-1}$ for [-CH=CH-] double bond and ~640 cm$^{-1}$ for [-C-H-] single bond [29]. The disappearance, in the FTIR spectra of maleic anhydride, of the carbonylic band situated at 1860 cm$^{-1}$ and the decrease, until disappearance, of the band situated at 1774 cm$^{-1}$ sustain the formation of the maleamic acid and the presence of both amidic and imidic form of the monomer [29].

The well-defined peaks of Raman spectra in DMSO presented in Figure 20 situated between 800 cm$^{-1}$ and 1750 cm$^{-1}$ are atributted to scattering on internal vibrational modes and could be associated with the deformation of the molecules. The Raman shift peaks for different monomers corresponding to different substituent groups are: stretching vibration of [-$NO_2$] group in A2 and A7 situated at ~1340 cm$^{-1}$; stretching vibration of [-CN] group in A3 situated at 2220 cm$^{-1}$. The tri-substituted nucleus in A7 shows Raman shift peaks at 834 cm$^{-1}$ and 1530 cm$^{-1}$. The peak situated at ~1600 cm$^{-1}$ corresponds to stretching vibration of double bond -C=C-. Peaks situated at 75 cm$^{-1}$ and 100 cm$^{-1}$, could be correlated with the external vibrations related to relative rotational or translational movement of molecules in solution [40]. The Raman spectra of A5 and A6 powders in DMSO drawn in Figure 21 show no significant differences,

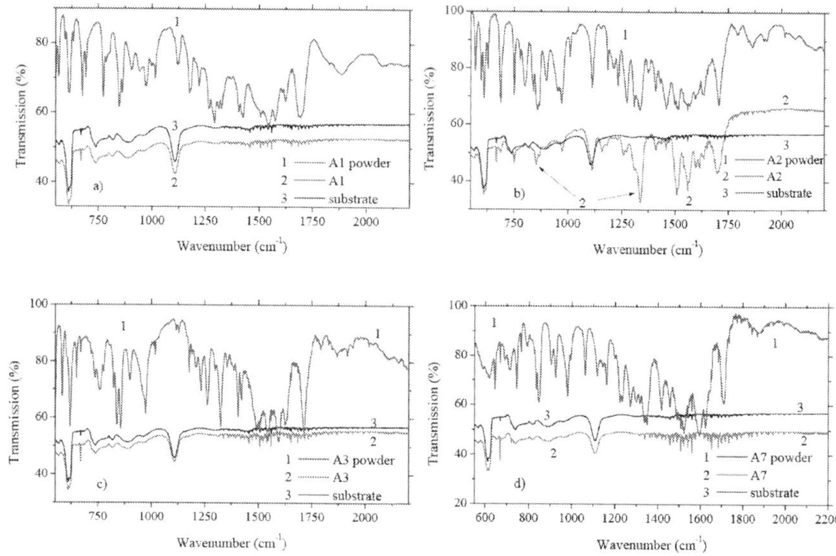


**Figure 18.** FTIR spectra of monomers powders and thin films deposited by MAPLE: monomer A1 (a); monomer A2 (b); monomer A3 (c); monomer A7 (d) [40].

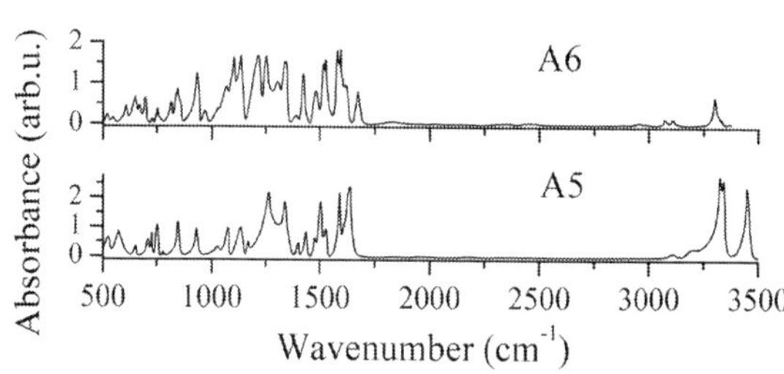


**Figure 19.** FTIR spectra of monomers A5 and A6 powder [29].

most of well-defined peaks being situated in the range 800-1700 cm$^{-1}$ and being determined by the scattering on the internal vibration modes correlated with deformation of the molecules. The spectra are different in the range 1250-1350 cm$^{-1}$ because of the stretching vibration of the C-C skeleton [29].

The peaks of 1, 2, 4-trisubstituted nucleus situated at 220 cm$^{-1}$, 375 cm$^{-1}$, 400 cm$^{-1}$, 480 cm$^{-1}$ in Raman spectra of monomers are masked by the peaks corresponding to the solvent [29].

In the monomer thin films we have not evidenced Raman shift peaks because these films are very thin and can not generate a significant signal [40].

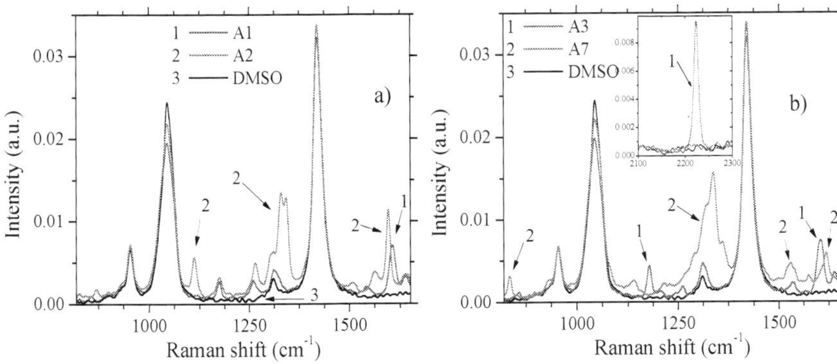


**Figure 20.** Raman spectra of monomers A1 and A2 (a), A3 and A7 (b) powders in DMSO [40].

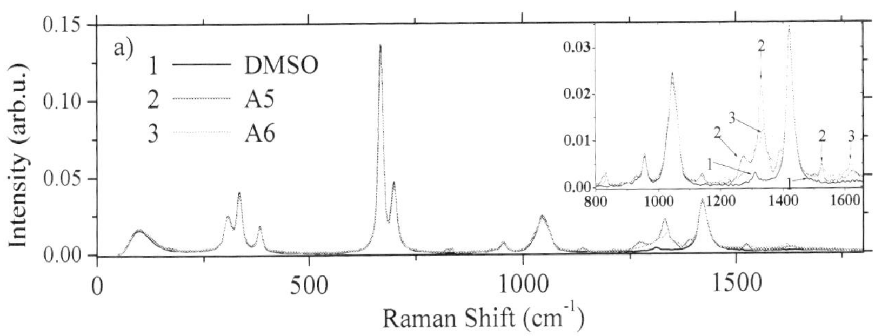


**Figure 21.** Raman spectra of monomers A5 and A6 powders in DMSO [29].

The FTIR spectroscopic data presented in Figure 22 offer information about the influence of the deposition methods on the chemical composition of the monomers thin films deposited by MAPLE [29]. Most of the characteristic bands corresponding to chromophoric groups in powder spectra of synthesised compounds A5 and A6, presented in Figure 19, have also been evidenced in thin films of compounds A5 and A6, while only some of the weakest peaks have disappeared in these thin films [29]. The effect of the deposition conditions (nitrogen pressure) is evidenced in FTIR spectra at $\lambda < 450$ cm$^{-1}$ being correlated to stress generation during the layer deposition [29].

Optical non-linear properties are shown in Figure 23 and Figure 24 for the monomers A6 and A7 thin films deposited by MAPLE and characterised by a maleamic acid structure containing one [-NH-NH-] donor group and two [$-NO_2$] acceptor groups. These compounds combine the high polarisability and high dipolar momentum determined by the molecular structure with the good transparency and weak photoluminescence emission in the visible range preventing the re-absorption of the SH radiation and the decrease/screening of the SH signal [40]. In

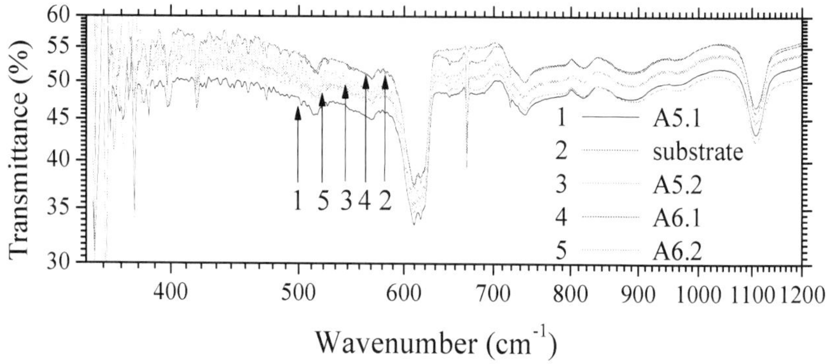


**Figure 22.** FTIR spectra of A5 and A6 thin films deposited at different nitrogen pressure: 0.05 torr A5.1 and A6.1; 0.2 torr A5.2 and A6.2 [29].

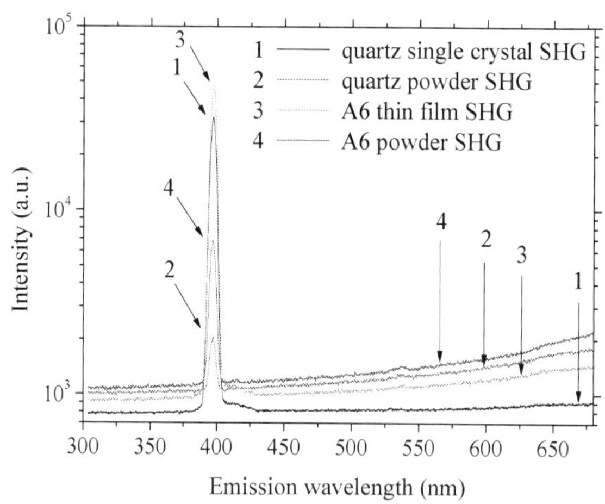


**Figure 23.** ONL effects in monomer A6 [29].

general, the para substitution in aromatic nucleus leads to highly polarisable charge transfer molecule with the tendency to crystallise in a centrosymmetric structure which do not favour the SH emission. In A6 and A7, the high difference in electronegativity of the substituent groups situated in position 1 [-NH-NH-] and position 4 [-$NO_2$] and the supplementary group situated in position 2 [-$NO_2$] acting on the electronic cloud broke the centrosymmetry and favour the SH generation [29]. This combined effect of the asymmetry of the side groups and of delocalized, highly polarisable π electrons determines an acentric alignment of the aromatic nucleus in a special packing [43].

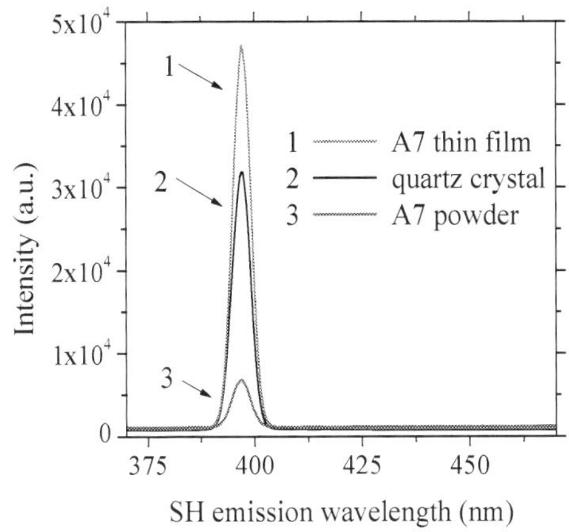


**Figure 24.** SH spectra of monomer A7 [40].

The SH signal of A6 and A7 thin films is compared with the signal given by A6 and A7 powders deposited on a substrate of aluminium which has no contribution to the SH emission. A quartz sample (crystal and powder) is used for comparison.

### *3.2.2. Maleimidic derivatives monomers in polycarbonate matrix as organic/organic composite thin films*

The research in the field of composite materials based on vitreous (inorganic) and molecular/ polymer (organic) matrix and organic/inorganic inclusions have received a special attention in the past years with the purpose to obtain a class of materials combining the advantageous properties of both components [44, 45]. Considering the advantage of organic compounds related to inexpensive thin-films preparation methods such as the deposition from solution by spin coating, the organic/organic composite materials offer an alternative for a large area of applications, including photonics.

The polymeric matrices are easy to process because most of the polymers are soluble in ordinary solvents, show thermal and mechanical stability and are characterised by a large range of good transparency, homogeneous refraction index and efficient fluorescence mechanism. The interest is focused on the identification of $\pi$-conjugated systems with functional groups which improve the properties of the selected matrix considering the extension of the delocalised system of $\pi$-electrons, length of the conjugated chain and planarity of the molecule, factors involved in the generation of ONL phenomena [46]. The properties of the composite materials of the type 'host/guest' are determined by the solubility of the 'guest' molecule into the 'host' matrix.

We have selected a polymeric matrix of polycarbonate of bisphenol A showing a large transparency domain between 300 and 1100 nm and inclusions of monomers with maleamic acid structure such as A3 and A5, indicated in Figure 17. The composite films have been prepared from solution by spin coating on glass substrate successively cleaned with acetone and distillate water, using a solution containing both components: polymer in concentration of 0.09 g $cm^{-3}$ or 0.250 g $cm^{-3}$ and monomer in concentration varying from 0.01 g $cm^{-3}$ to 0.25 g $cm^{-3}$ in dimethylformamide (DMF) [47].

The solvent DMF has been selected considering the surface energy determined by contact angle measurements, the high solubility of both polycarbonate and mentioned monomers in DMF compared to DMSO and the volatility at room temperature. The experimental conditions for the preparation of films have been selected from macroscopic morphological consideration referring to the uniformity of the layer: $t_1 = 6$ s and $v_1 = 1.13$ krpm in the spreading step and $t_2 = 10$ s and $v_2 = 2.7$ krpm in the homogenisation step, for a weight ratio between monomer and polymer of 1/3 [47]. The effect of the matrix on the properties of monomer inclusions are evidenced comparing the properties of these films with those of the reference samples of monomers thin films prepared by vacuum evaporation.

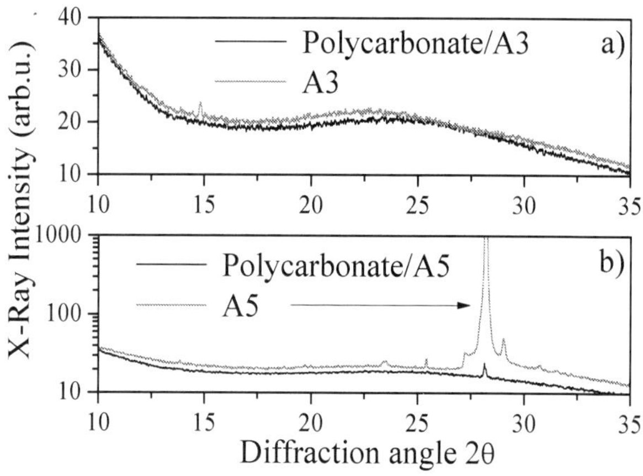


**Figure 25.** X-ray diffraction spectra of monomers A3 and A5 and polycarbonate/monomer (A3, A5) deposited on glass substrate [47].

An increased disorder in polycarbonate/monomer Ax with $x = 3, 5$ layer deposited by spin coating compared to monomer Ax with $x = 3, 5$ layer deposited by vacuum evaporation is confirmed by the X-ray diffractograms in Figure 25. The peak situated around 28° corresponding to dinitroaniline is present both in the spectrum of monomer A5 and in the composite based on the same monomer. The film of polycarbonate/A3 deposited on glass is mostly amorphous compared to the vacuum evaporated A3 film. The A3 film shows a small peak situated at 13° corresponding to cyano aniline and indicating a certain small degree of order (crystallisation) in the vacuum-evaporated film. The X-ray diffractograms have evidenced a preferential orientation of A5 monomer film which is partially preserved in composite polycarbonate/A5 film, compared to A3 and polycarbonate/A3 films showing a substantial disorder [47].

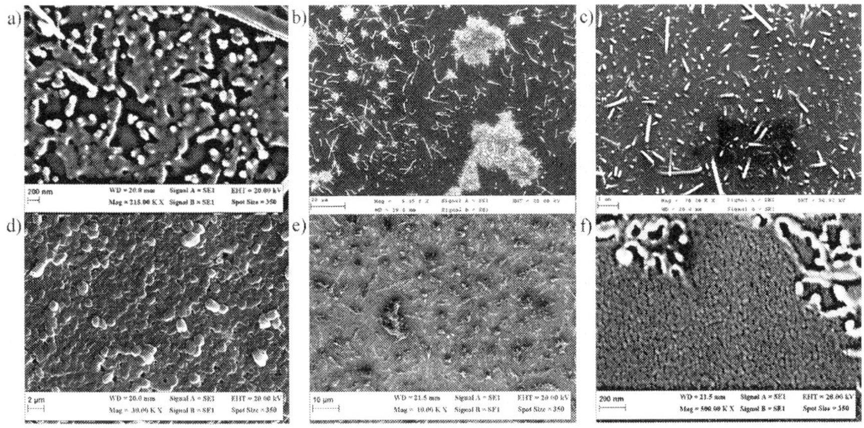


**Figure 26.** SEM images of monomers A3 (a) and A5 (d) films vacuum evaporated on glass and composite polycarbonate/A3 (b, c); A5 (e, f) layers deposited by spin coating on glass [47].

The SEM images presented in Figure 26 have evidence a continuous, compact organic layer deposited on glass, with large grain morphology for A5 and smaller grain morphology with lower packing density for A3 [47]. The composite layers have shown morphology characterised by 'grains' and 'grains clusters' which are homogeneously developed in the polymeric matrix that wets the glass substrate [47].

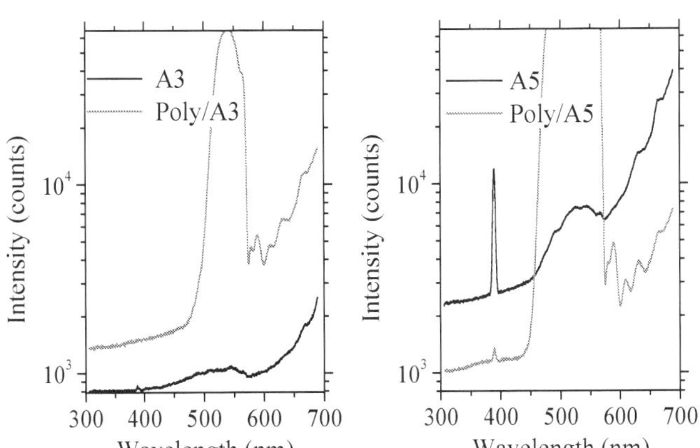


**Figure 27.** ONL effects in monomers (A3, A5) and polycarbonate/monomer (A3, A5) deposited on glass [47].

The emission spectra of the monomers thin films at illumination with a laser beam with $\lambda$ = 800 nm presented in Figure 27 show an emission peak situated at 400 nm corresponding to SH. This signal is strongly affected by the surface properties like morphology. The larger grain morphology of A5, as shown in Figure 26, is associated with a stronger SH signal because of the lower loss by scattering compared to A3 showing smaller grain morphology. The SH emission is explained by an intense mezomeric effect determined by the difference in electronegativity of the substituent groups and a strong inductive effect between the strongly polarisable chromophoric groups -$NO_2$ and -CN coupled with the polarisable aromatic nucleus [47].

Polycarbonate/A3 and polycarbonate/A5 films show a strong signal situated between 475 and 600 nm attributed to TPF emission together with a decrease in SH signal which is affected by surface morphology of the layers deposited by spin coating. TPF is a volume process and it is not affected by the morphology of the surface [47].

### 3.3. Thin films based on functionalised copolymers

Another alternative to the inorganic compounds for ONL applications is represented by the polymers showing high non-resonant ONL properties. In polymeric materials containing dipolar chromophores, the macroscopic second-order optical non-linearity could be induced by intramolecular donors-acceptors groups. In these materials, either molecular fragments showing ONL properties are spread in a polymeric matrix or highly colored chromophores are attached covalently in two possible ways, as side chain or intercalated in the backbone of the polymer's chain [48]. High values of the non-linear coefficients can be obtained in a non-centrosymmetric packing by an acentric alignment of the molecules favoured by the structure of the side groups and/or polymer. This condition can be accomplished by an increased density of chromophores assured by the functionalisation of polymers (copolymers) with molecules containing highly polarisable conjugated electronic clouds [49, 41].

The structural and morphological particularities of the materials depend on the method selected for preparation (MAPLE and vacuum evaporation; solution deposition such as spin coating) and affect the properties of the materials.

We have studied two polymeric structures introduced in Figure 28, P1: MA-CMS-DNA synthesised by the copolymerisation of maleic anhydride (MA) with vinyl benzil chloride (MA-CMS) and subsequent functionalisation with 2, 4-dinitroaniline (DNA) and P2: MA-MMA-DNA synthesised by the copolymerisation of maleic anhydride and methylmethacrilate (MA-MMA) and subsequent functionalisation with 2, 4-dinitroaniline (DNA). Functionalised copolymers have been synthesised in a two-stages process: (1) polymerisation by precipitation initiated by azoizobutylnitrile in toluene at 80°C for 2 h; (2) polymers grafting with ortho-, para-aniline derivatives in DMF by opening and closing of the anhydridic cycle at 120°C for 2 h [50-52, 29].

Thin films of grafted copolymers P1 and P2 have been prepared using different methods on different substrates like glass, cleaned in acetone, and etched/etched single silicon wafers, successively cleaned ultrasonically with acetone, hydrofluoric acid and distillate water [50].

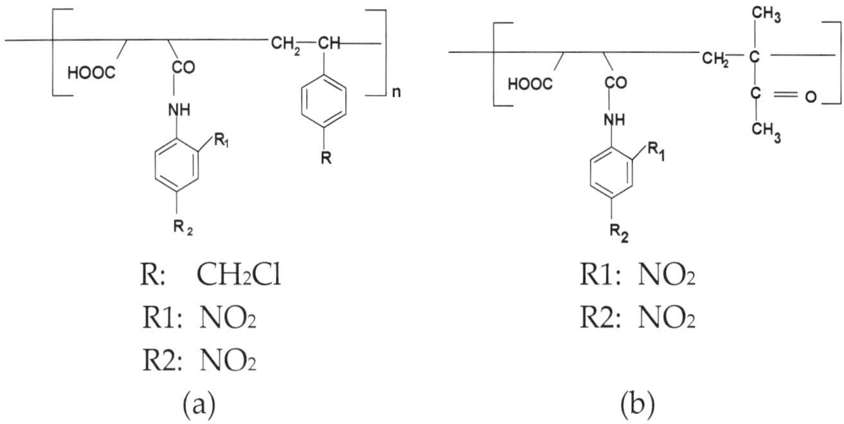


**Figure 28.** Selected copolymers: P1: MA-CMS-DNA (a); P2: MA-MMA-DNA (b) [50].

Polymeric films have been deposited from solution by spin coating in a process involving two steps: (1) preparation of polymer solution in an adequate solvent; (2) deposition of the film by solution spreading (spinning time 9 s for P1 and 3 s for P2; rotational speed 0.75 krpm for P1 and 0.65 krpm for P2 [50]) followed by homogenisation (spinning time 20 s for P1 and 10 s for P2; rotational speed 2.75 krpm for P1 and 1.70 krpm [50]). We have selected DMF as solvent, because it is characterised by a low evaporation rate and strong interactions with the native layer of $SiO_2$ covering the Si surface.

Polymeric thin films have also been prepared by vacuum evaporation at a temperature of 100-110°C for P1 and 80-90°C for P2 and a pressure between $1.2\text{-}1.9\times10^{-5}$ mbar for P1 and $1.0\text{-}1.7\times10^{-5}$ mbar for P2 [50]).

The characteristics of the films prepared by spin coating and vacuum evaporation have been compared with those of the films prepared by a laser-based technique, MAPLE [51, 29]. To avoid the deterioration in laser beam of the polymer characterised by weak chemical bonds, it is used a matrix from a solvent with high molecular weight (100-300 Da) and a good absorption at the laser wavelength, in our experimental configuration $\lambda$=248 nm, namely DMSO [51]. The solution of polymer in DMSO is frozen in liquid nitrogen and this target is placed in the deposition chamber. The energy of the incident laser beam is absorbed mostly by the molecules of solvent, the target heats and the components of the target are simultaneously evaporated. The polymer deposits on a heated substrate while the volatile solvent is pumped away from the chamber. Low values of the fluencies (~0.4 J $cm^{-2}$) are necessary to avoid the fragmentation of the polymeric chain and decomposition of the photosensible organic compounds [53]. Beside fluence, other important experimental parameters are presented in Table 1 concentration of the target, nitrogen pressure, substrate temperature and number of pulses [51, 29].

| Sample | Target | Temperature (°C) | Pressure (Pa) | Thickness (nm) |
|---|---|---|---|---|
| P1.1 | 2% MA-CMS-DNA in DMSO | 250 | 30 | 41 |
| P1.2 | 2% MA-CMS-DNA in DMSO | 150 | 5 | 84 |
| P1.3 | 3% MA-CMS-DNA in DMSO | 250 | 10 | 70 |
| P1.4 | 3% MA-CMS-DNA in DMSO | 150 | 5 | 105 |
| P2 | 3% MA-MMA-DNA in DMSO | 250 | 3 | 87 |

**Table 1.** Conditions for MAPLE deposition of copolymer thin films [51].

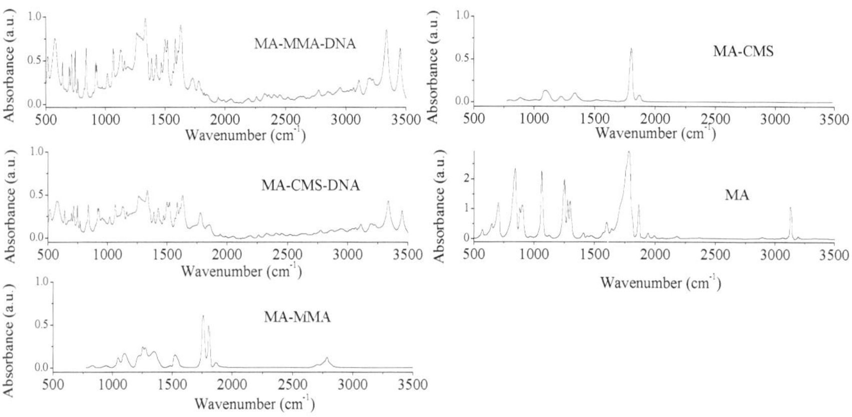


**Figure 29.** FTIR spectra of powder maleic anhydride (MA), copolymeric support (MA-CMS and MA-MMA) and dinitroaniline derivative functionalised copolymers (MA-CMS-DNA:P1 and MA-MMA-DNA:P2) [29].

In the FTIR spectra of the copolymers powders shown in Figure 29, have been identified the following characteristic bands: for MA at 1850 $cm^{-1}$, for MA in MA-CMS at 1780 $cm^{-1}$, for carbonyl in MMA at 1720 $cm^{-1}$ and for $CH_2$-Cl in CMS at 1220 $cm^{-1}$ [29].

The FTIR spectra of copolymers functionalised with dinitroaniline powders are presented in Figure 29. The position and intensity of the characteristic band situated at 1620 $cm^{-1}$ confirms the formation of amide-imide structure and the band situated at 3300 $cm^{-1}$ confirms the presence of [-COOH] group obtained by opening the MA cycle [29].

At an increased degree of transformation of copolymers, the support copolymers can affect the FTIR spectra by significantly shifting or broadening the bands of 2, 4-dinitroaniline used as grafting entity. Despite the high volume of aniline molecule, the transformation degree of MA-MMA copolymer is higher ~16% compared to MA-CMS ~9% and can be explained by the higher accessibility of MA entities to the grafting units determined by the structure of the copolymer [29].

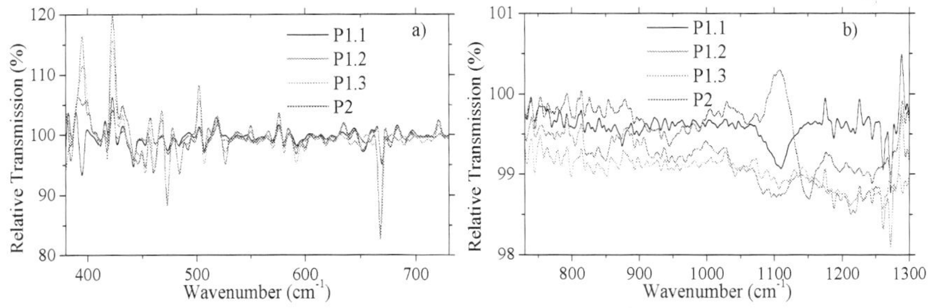


**Figure 30.** FTIR spectra of P1 and P2 films deposited on Si by MAPLE from DMSO in different spectral ranges: 380-730 nm (a); 730-1300 nm (b) [51].

The FTIR spectra presented in Figure 30 have evidenced the characteristic bands in P1 and P2 films deposited by MAPLE, attributed to 2, 4-dinitroaniline which has been used for the functionalisation of copolymers: 1270 cm$^{-1}$ carbonyl (C=O); 900 cm$^{-1}$ double bond (CH=CH); 675 cm$^{-1}$ single bond (C-H); 1390 cm$^{-1}$ amide group; 1100 cm$^{-1}$ imide group (ring deformation). The last two bands confirm the presence of amidic and imidic forms. The strong or moderate absorption band situated between 900 and 600 cm$^{-1}$ is attributed to bending vibration in aromatics, carbonyl acids and amides. The absorption band situated between 600 and 400 cm$^{-1}$ can be attributed to meta-substituted aromatics. Other absorption bands are: 1250-1300 cm$^{-1}$ the symmetric stretching vibration of $NO_2$; ~1240 cm$^{-1}$ aromatic CN bonding; 1175 cm$^{-1}$ CH in plane deformation. The formation of copolymer MA-CMS is confirmed by the presence of the absorption band situated at ~1210 cm$^{-1}$ corresponding to $CH_2$-Cl bond in CMS. The bands situated at ~1150 cm$^{-1}$ and ~1260 cm$^{-1}$ corresponds to maleic anhydride in methyl methacrylate. The band characteristic for 2, 4-dinitroaniline could be broadened or shifted under the effect of the support copolymers depending on the degree of transformation [50, 51]. The support copolymers can affect as well the FTIR spectra of MAPLE deposited films by significantly shifting or broadening the bands of the 2, 4-dinitroaniline at high degree of transformation of copolymers.

There are no significant differences between the functionalised copolimers films deposited by vacuum evaporation and spin coating, excluding the possibility of polymer decomposition during the evaporation process, a high-temperature process compared to spin coationg, a room temperature process. New absorption peaks corresponding to new bonds have not been evidenced in the polymeric films deposited by vacuum evaporation. Even no chemical reaction took place during the evaporation, it could be possible a fragmentation of the copolymeric support, maintaining intact the chromophoric groups. This is sustained by the similarity between the shape of the FTIR spectra of the evaporated polymeric films and of the MA-DNA monomer fragment [29, 50].

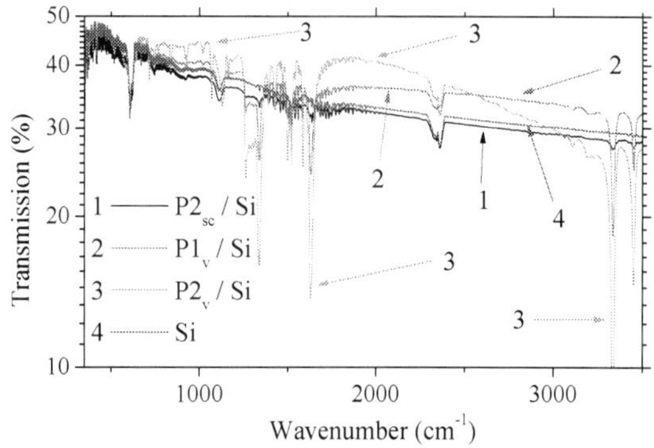


**Figure 31.** FTIR spectra of P1 and P2 deposited on Si by vacuum evaporation (v) and spin coating (sc) [50].

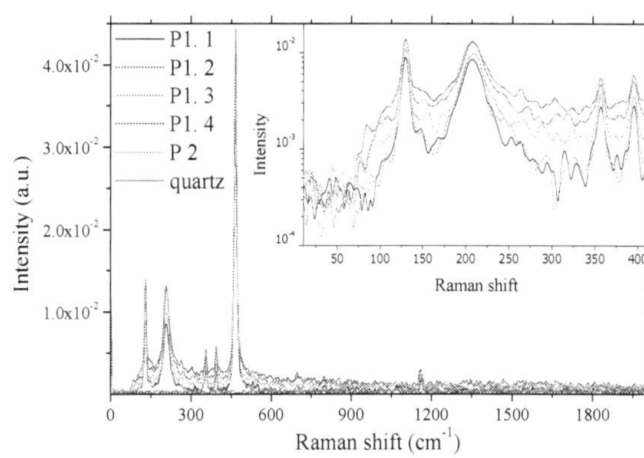


**Figure 32.** Raman spectra of P1 and P2 functionalised copolymers deposited on quartz by MAPLE from DMSO (Table 1) [29, 51].

As well as for monomers A5 and A6, the Raman spectra of functionalised copolymers thin films deposited by MAPLE drawn in Figure 32 show well-defined peaks in the range 200-1200 $cm^{-1}$ being also correlated with the scattering on internal vibration modes which are associated with deformation of the molecules. The peaks situated at 220 $cm^{-1}$, 375 $cm^{-1}$, 400 $cm^{-1}$, 480 $cm^{-1}$ correspond to the unit 1, 2, 4-trisubstituted nucleus unit of the polymers chemical structure [51, 29].

The band situated between 1050 and 1200 $cm^{-1}$ could be associated with the stretching vibration of the C-C skeleton and the bands situated at low wavenumbers between 10 and 200 $cm^{-1}$ could be correlated to external vibrations and to the relative in the MAPLE deposited thin films [51].

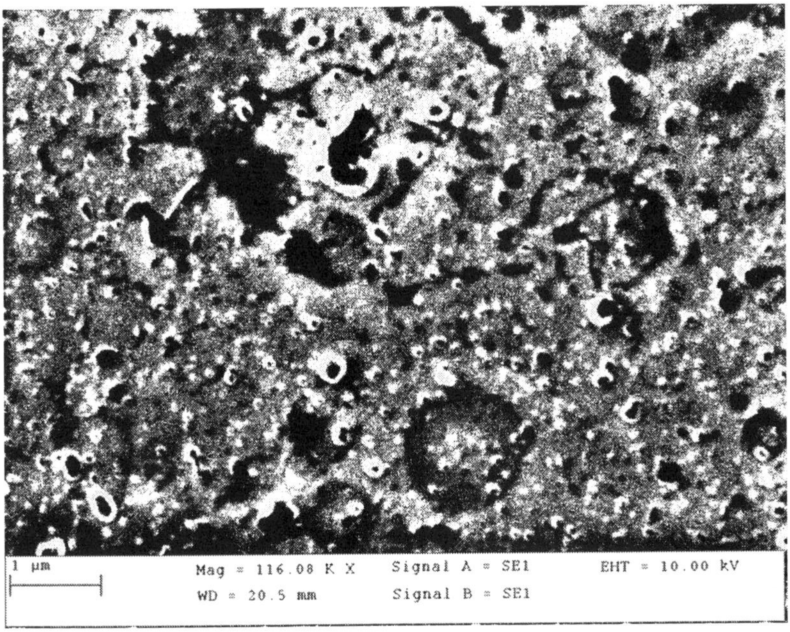


**Figure 33.** SEM image of P1 film deposited by MAPLE on Si from DMSO [51].

The FTIR spectra presented in Figure 29, Figure 30 and Figure 31 and Raman spectra presented in Figure 32 have not evidenced any change or damage in the chemical structure of the polymer and generation of compositional non-homogeneities in the film under the effect of laser beam.

The morphology of the MAPLE deposited P1 films evidenced by SEM in Figure 33 is relatively uniform on large area, with droplets type structures having dimension <0.1 μm [51]. During the thermal evaporation of the target, fractions of solvent matrix could be retained within the ejected matrix-polymer droplets [54].

The morphology and uniformity of the deposited films is determined by the mechanism of droplets evaporation [51]. The solvent could evaporate partially or totally during the deposition on heated substrate and this process led to the mentioned morphology [51, 54].

Different morphologies, compared to MAPLE deposited functionalised copolymers P1 and P2 thin films, have been revealed for samples of the same functionalised copolymers prepared on Si by vacuum evaporation and by spin coating in Figures 34 a and b.

A special morphology showing two types of local organisation as granules and crystals randomly oriented has been emphasised for the films deposited by vacuum evaporation. The morphology of the spin-coated films is also characterised by the presence of local organisation as granules dispersed in a uniform layer.

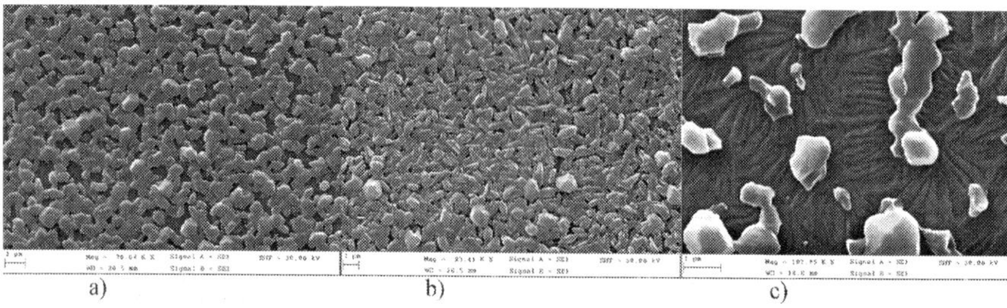


**Figure 34.** SEM image of: P1 deposited on Si by vacuum evaporation (a); P2 deposited on Si by vacuum evaporation (b); P1 deposited on Si by spin coating [50].

As mentioned above, the temperature involved in the vacuum deposition process can determine the fragmentation of the polymeric chain leading to entities which maintain intact the functional groups implied in the ONL phenomena. This explanation is sustained by the presence of the second, granular, phase in the vacuum evaporated films evidenced by SEM.

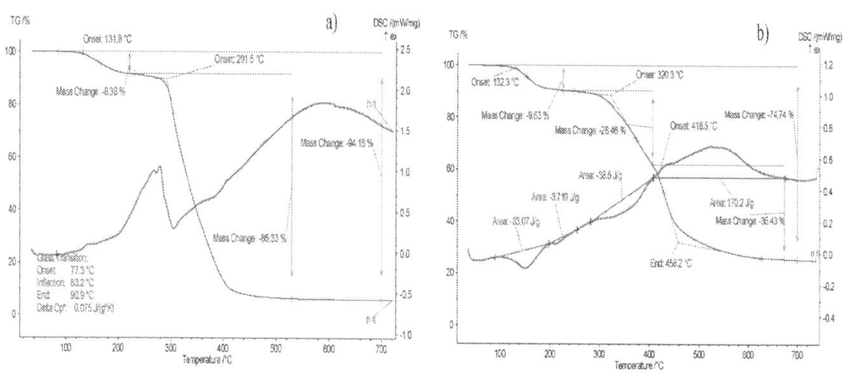


**Figure 35.** Thermogravimetric analysis of functionalised copolymers: P1 (a); P2 (b) [29].

The distinct crystals phase has been associated with the presence of 2, 4-dinitroaniline in excess remained ungrafted and crystallising in the films during the deposition process. The copolymer degradation process is not involved in the vacuum deposition because the temperature of 110°C is lower than the temperature of ~130°C revealed by the thermogravimetric analysis, presented in Figure 35, as the starting point for a chemical degradation of the copolymeric support. Starting with this temperature, the [-CH-CH-] groups and MA units are simultaneously lost in the both components [50].

A certain degree of short-range order in the polymeric films prepared by vacuum evaporation has been also evidenced by X-ray diffraction measurements indicated in Figure 36. This means that the fragments of the polymeric chain with anilinic moieties are ordered and the strong

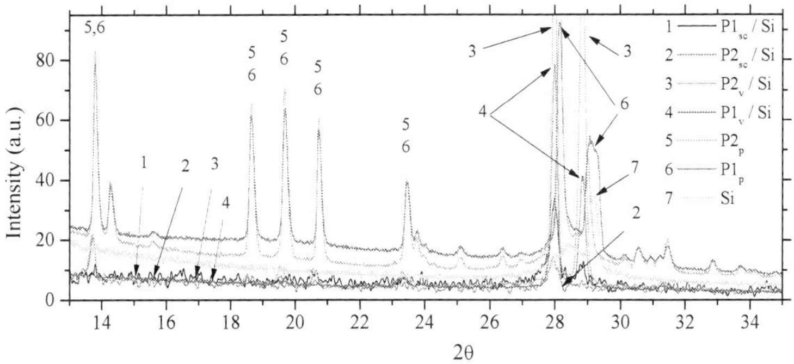


**Figure 36.** X-Ray diffractograms for films of P1 (a) and P2 (b) prepared on Si by vacuum evaporation (v) and spin coating (sc). The spectra of P1 and P2 powder (p) on Si are given as references [50].

peaks are assigned to the $d_{Bragg}$ parameter for 2, 4-dinitroaniline [50], using the Powder Diffraction Files. The films of P1 and P2 prepared both by spin coating and vacuum evaporation have shown the same shape of the XRD spectra. The spin coating is a room temperature method and the fragmentation of the polymer chain could not be considered. The similarity between the films prepared by these two methods confirms a certain degree of order in the spin-coated layer induced by the grafted entities, the d spacing corresponding to 2, 4-dinitroaniline.

ONL phenomena evidenced in polymeric films deposited by vacuum evaporation and spin coating are presented in Figure 37a and by MAPLE in Figure 37b: the SH emission situated at 400 nm and a strong TPF emission band with a peak situated between 500 and 600 nm [50].

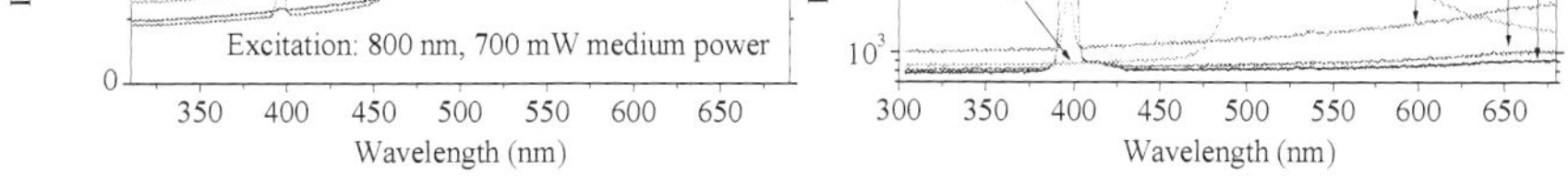


**Figure 37.** ONL effects (SHG and TPF) in functionalised copolymer P1 and P2 films prepared by vacuum evaporation (v) and spin coating (sc) (a); ONL effects (SHG and TPF) in functionalised copolymer P1 powders and thin films deposited by MAPLE (b) [50].

There is a correlation between the preparation method (vacuum evaporation, spin coating, MAPLE) of the thin films and the properties (SH and TPF emissions) of these types of maleic anhydride based copolymers functionalised with the same polar chromophoric groups bearing anilinic moieties. The SHG is obtained in these thin films showing asymmetric $\pi$ electron clouds generated by the high difference in the electronegativities of the substituent groups. The ONL properties are preserved despite the potential fragmentation of the polymeric chain during the evaporation process because the potential resulting mononers/oligomers maintain intact the functional groups implied in the generation of ONL phenomena [50].

The SH measurements have been realised on thin films of P1 deposited by MAPLE on quartz and compared with that realised on powder placed on an aluminium substrate with no contribution to ONL phenomena. A broad-structured emission band situated between 500 and 600 nm was evidenced in polymer thin film and was determined by a strong luminescence emission which has been generated by a two-photon absorption process as previously mentioned [29]. As already emphasised, the functionalised copolymers prepared by MAPLE have shown SHG because of the acentric alignment of the organic molecules [43].

## 4. Conclusions

We summarise in this chapter some of the most significant results of our research in the field of new organic materials for optical non-linear applications. We have selected and investigated three types of organic thin films prepared from arylenevinylene compounds, maleimidic derivatives and anilinic derivatives functionalised copolymers and two types of molecular crystalline materials prepared from aromatic derivatives m-DNB and Bz.

In the context of special attention paid to the investigation of bi-component organic systems for ONL applications involving the preparation and characterisation of organic crystals and organic/organic composites, our interest was focused on studying pure and doped m-DNB and Bz bulk crystals and maleimidic derivatives monomers in polycarbonate matrix composite layers. The previous results on growth interface stability for Bz crystals have been completed with new conclusions on the stable growth of m-DNB crystals. We have emphasized the conditions favouring the initiation of instabilities, which are associated with structural defects and compositional non-homogeneities in the crystal of m-DNB affecting the optical properties including the ONL properties. New results have been bought about the systems aromatic derivative crystal/dopant studied from the point of view of the inorganic/organic dopant incorporation and its effect on the optical band gap and ONL properties of the m-DNB and Bz crystals.

The correlation between the morphology and structural order of organic/organic composite layer and the ONL properties has been discussed.

A special attention was paid to chose the adequate method for thin-films preparation, like vacuum evaporation deposition, matrix-assisted pulsed laser evaporation (MAPLE) and spin coating, in correlation with the particularities of the organic molecule: arylenevinylene (triphenylamine, carbazole) based oligomers, maleimidic derivatives monomers, aniline derivative functionalised copolymers (maleic anhydride-vinyl benzyl chloride and maleic anhydride-methylmethacrilate supports functionalised with 2, 4-dinitroaniline).

We have analysed the correlation molecular structure-preparation conditions of the organic films determining their morphology and structure-optical non-linear (SHG and TPF) properties.

Important results have been bought in the field of maleic anhydride-based copolymers functionalised with polar chromophoric groups bearing anilinic moieties. Despite the potential fragmentation of the polymeric chain during the deposition process, the ONL properties are preserved because the potential resulting entities maintain intact the functional groups implied in the generation of these phenomena.

Further research will focus on the improvement of the organic material quality by a more rigorous control of the deposition conditions with the purpose to increase the efficiency of the ONL phenomena.

## Author details

Anca Stanculescu[1*] and Florin Stanculescu[2]

*Address all correspondence to: sanca@infim.ro

1 National Institute of Materials Physics, Bucharest-Magurele, Romania

2 University of Bucharest, Bucharest-Magurele, Romania

## References

[1] Freyhardt HC. ed. Crystals: Growth, Properties and Applications, Vol.4. Berlin: Springer; 1980.

[2] Babu GA, Ramasamy RP, Ramasamy P, Kumar VK. Synthesis, crystal growth, and characterization of an organic nonlinear optical donor-π-acceptor single crystal: 2-amino-5-nitropyridinium-toluenesulfonate. Crystal Growth Design 2009;9(7):3333-7.

[3] Fejer MM. Nonlinear optical conversion. Phys Today 1994;47(5):25-32.

[4] Munn RW, Miniewicz A, Kuchta B, eds. Electrical and Related Properties of Organic Solids: Dordrecht Netherlands: Kluwer Academic Publishers, 1997.

[5] Chen CP, Su Y, Jhun CG. Recent advances in holographic recording media for dynamic holographic display. J Optics Photonics 2014;1:article number 1, doi: http://dx.doi.org/10.7243/2057-1569-1-1.

[6] Agulló-López F, Cabrera JM, Agulló-Rueda F. Electrooptics. Phenomena, materials and Applications, London: Academic Press; 1994.

[7] Hillman TR, Yamauchi T, Choi W, Dasari RR, Feld MS, Park YK, Yaqoob Z. Digital optical phase conjugation for delivering two-dimensional images through turbid media. Sci Rep 2013;3:article number:1909, doi:10.1038/srep01909.

[8] Choi S-K, Vasilyev M, Kumar P. Noiseless optical amplification of images. Phys Rev Lett 1999;83(10):1938-41.

[9] Krenn M, Robert Fickler R, Fink M, Handsteiner J, Malik M, Scheidl T, Ursin R, Zeilinger A. Communication with spatially modulated light through turbulent air across Vienna. New J Phys 2014;16:113028, http://iopscience.iop.org/1367-2630/16/11/113028.

[10] Biberman A, Bergman K. Optical interconnection networks for high-performance computing systems. Rep Prog Phys 2012;75:046402, http://iopscience.iop.org/0034-4885/75/4/046402.

[11] Winter PW, Shroff H. Faster fluorescence microscopy: advances in high speed biological imaging. Curr Opin Chem Biol 2014;20:46-53.

[12] Winter PW, York AG, Nogare DD, Ingaramo M, Christensen R, Chitnis A, Patterson GH, Shroff H. Two-photon instant structured illumination microscopy improves the depth penetration of super-resolution imaging in thick scattering samples. Optica 2014;1(3):181-91.

[13] Sun H-B, Kawata S. Two-photon photopolimerization and 3D lithographic microfabrication. APS 2004;170:169-273.

[14] Goradia I, Doshi J, Deulkar KA. Review paper on 3D optical storage data. Int J Curr Eng Technol 2014;493:2933-6.

[15] Tang X-J, Wu L-Z, Zhang L-P, Tung C-H. Two-photon-pumped frequency-upconverted lasing and optical power limiting properties of vinylbenzothiazole-containing compounds in solution. Phys Chem Chem Phys 2002;4(23):5744-7.

[16] Stanculescu A, Stanculescu F. Some optical investigations on crystalline doped meta-dinitrobenzene. J Optoelectronics Adv Mater 2000;2(5):536-41.

[17] Stanculescu A, Stanculescu F, Alexandru H. Melt growth and characterization of pure and doped meta-dinitrobenzene crystals 1999;198/199(1-4):572-7.

[18] Stanculescu A, Tugulea L, Alexandru HV, Stanculescu F, Socol M. Molecular organic crystalline matrix for hybrid organic-inorganic (nano) composite materials. J Crystal Growth 2005;275(1-2):e1779-86.

[19] Stanculescu F, Ionita I, Stanculescu A. Organic/inorganic-doped aromatic derivative crystals: growth and properties. J Cryst Growth 2014;401:215-20.

[20] Stanculescu A. Investigation of the growth process of organic/inorganic doped aromatic derivatives crystals. J Optoelectronics Adv Mater 2007;9(5):1329-36.

[21] Barvinschi F, Stanculescu F, Stanculescu A. Heat transfer process during the crystallization of benzil grown by the Bridgman-Stockbarger method. J Crystal Growth 2011;317(1):23-7.

[22] Sekerka RF. Morphological stability. J Cryst Growth 1968;3(4):71-81.

[23] Mullins WW, Sekerka RF. Morphological stability of a particle growing by diffusion or heat flow. J Appl Phys 1963;34(2):323-9.

[24] Mullins WW, Sekerka RF. Stability of a planar interface during solidification of a dilute binary alloy. J Appl Phys 1964;35(2):444-51.

[25] Acree WE Jr. Thermodynamic properties of organic compounds: enthalpy of fusion and melting point temperature compilation. Thermochim Acta 1991;189(1):37-56.

[26] Urbanski T. Chemistry and technology of explosives.Vol. 3. Elsevier Sci Technol 1963:234-235.

[27] Yaws CL. Organic Compounds C5 to C7 in Handbook of Thermal Conductivity, Vol. 2, Houston: Gulf Publishing Company; 1995.

[28] Lin Y, Shi Z, Wildfong PL. Thermal conductivity measurements for small molecule organic solid materials using modulated differential scanning calorimetry (MDSC) and data corrections for sample porosity. J Pharm Biomed Anal 2010;51(4):979-84.

[29] Stanculescu A, Rasoga O, Preda N, Socol M, Stanculescu F, Ionita I, Albu A-M, Socol G. Preparation and characterization of polar aniline functionalized copolymers thin films for optical non-linear applications. Ferroelectrics 2009;389(1):159-73.

[30] Stanculescu A, Antohe S, Alexandru HV, Tugulea L, Stanculescu F, Socol M. Effect of dopant on the intrinsic properties of some multifunctional aromatic compounds films for target applications. Synthetic Metals 2004;147(1-3):215-20.

[31] Rasoga O, Vacareanu L, Grigoras M, Enculescu M, Socol M, Stanculescu F, Ionita I, Stanculescu A. Optical and electrical properties of arylenevinylene compounds thin films prepared by vacuum evaporation. Synthetic Metals 2012;161(23-24):2612-7.

[32] Grigoras M, Antonoaia NC. Synthesis of isoelectronic polyazomethine and poly(arylene vinylene) by a palladium-catalyzed Suzuki coupling method. Polym Int 2005;54(12):1641-6.

[33] Wang X, Yang P, Xu G, Jiang W, Yang T. Two-photon absorption and two-photon excited fluorescence of triphenylamine-based multibranched chromophores. Synthetic Metals 2005;155(3):464-73.

[34] Yoon KR, Ko S-O, Lee SM, Lee H. Synthesis and characterization of carbazole derived nonlinear optical dyes.Dyes Pigments 2007;75(3):567-73.

[35] Stanculescu A, Vacareanu L, Grigoras M, Socol M, Socol G, Stanculsecu F, Preda N, Matei E, Ionita I, Girtan M, Mihailescu IN. Thin films of arylenevinylene oligomers prepared by MAPLE for applications in non-linear optics. Appl Surface Sci 2011;257(12):5298-302.

[36] Mailhot-Jensen B, Robu S, Rivaton A, Pilichowski J-F, Chirita A, Chilat E, Dragalina G. Carbazole containing copolymers: synthesis, characterization, and applications in reversible holographic recording. Int J Photoenergy 2010;Article ID 945242, 11 pages. doi:10.1155/2010/945242.

[37] Leveugle E, Zhigilei LV, Sellinger A, Fitz-Gerald JM. Computational and experimental study of the cluster size distribution in MAPLE. Appl Surf Sci 2007;253(15): 6456-60.

[38] Hell SW, Booth M, Wilms S, Schnetter CM, Kirsch AK, Arndt-Jovin DJ, Jovin TM. Two-photon near- and far-field fluorescence microscopy with continuous-wave excitation. Optics Lett 1998;23(15):1238-40.

[39] Rulliere C, ed. Femtosecond Laser Pulses: Principles and experiments, N.Y.: Springer Science; 2005.

[40] Stanculescu A, Albu A-M, Socol G, Stanculescu F, Socol M, Preda N, Rasoga O, Girtan M, Iulian I. MAPLE deposited thin monomer films of maleimidic derivatives for photonics. J Optoelectronics Adv Mater 2010;12(3):731-9.

[41] Wong MS, Meier U, Pan F, Gramlich V, Bosshard C, Günter P. Five-membered hydrazone derivatives for second-order nonlinear optics. Adv Mater 1996;8(5):416-20.

[42] Chemla DS, Zyss J. Nonlinear Optical Properties of Organic Molecules and Crystals. Orlando USA: Academic Press; 1987.

[43] Kim WH, Jiang XL, Kumar J, Tripathy SK. Design of new polydiacetylenes as self assembling second order nonlinear optical polymers. Pure Appl Chem 1995;67(12): 2023-30.

[44] Lee KJ, Oh JH, Kim Y, Jang J. Fabrication of photoluminescent-dye embedded poly(methyl methacrylate) nanofibers and their fluorescence resonance energy transfer properties. Adv Mater 2006;18(17):2216-9.

[45] Rosemary MJ, Suryananayanan V, MacLaren I, Pradeep T. Aniline incorporated silica nanobubbles. J Chem Sci 2006;118(5):375-84.

[46] Bosshard Ch, Sutter K, Prêtre Ph, Hullinger J, Flörsheimer M, Kaatz P, Günter P. Organic Nonlinear Optical Materials, vol.1Switzerland: Gordon Breach Publishers; 1996.

[47] Stanculescu F, Stanculescu A, Girtan M, Socol M, Rasoga O. Effect of the morphology on the optical and electrical properties of polycarbonate film doped with aniline derivatives monomers. Synthetic Metals 2012;161(23-24):2589-97.

[48] Nalwa HS. Nonlinear Optics of Organic Molecules and Polymers. New York: CRC; 1997.

[49] Abbotto A, Beverina L, Chirico G, Facchetti A, Ferruti P, Pagani GA. Design and synthesis of new functional polymers for nonlinear optical applications. Synthetic Metals 2003;139(3):629-32.

[50] Stanculescu A, Socol M, Albu A-M, Rasoga O, Stanculescu F, Ionita I, Enculescu I. Investigation of the correlation between the preparation method and the properties of anilinic derivatives functionalised polymer thin films for non-linear optical applications. Mater Sci Forum 2010;636-7, 798-804.

[51] Socol G, Mihailescu IN, Albu A-M, Antohe S, Stanculescu F, Stanculescu A, Mihut L, Preda N, Socol M. MAPLE prepared polymeric thin films for non-linear optic applications. Appl Surface Sci 2009;255(10):5611-4.

[52] Patel CB, Malek NI, Oswals SI. Synthesis and Radical Polymerization of N-[4-N'-(Phenylamino-carbonyl)phenyl]maleimide and its Copolymerization with Methyl Methacrylate. J Macromolecular Sci Part A: Pure Appl Chem 2006; 43:289-303.

[53] Ina E, Matsumoto N, Shikada E, Kannari F. Laser ablation deposition of crystalline copper-phthalocyanine thin films. Appl Surface Sci 1998;127-129(1-2):574-8.

[54] Sellinger AT, Leveugle E, Gogick K, Peman G, Zhigilei LV, Fitz-Gerald JM. Ejection of matrix-polymer clusters in matrix-assisted laser evaporation: Experimental observation.J Phys Conf Ser 2007;59:314-7.

Chapter 11

# Physical Characteristics, Sensors and Applications of 2D/3D-Integrated CMOS Photodiodes

Oscal T.-C. Chen, Yi-Yang Lee and Robin R.-B. Sheen

Additional information is available at the end of the chapter

http://dx.doi.org/10.5772/60737

**Abstract**

Two-dimensional photodiodes are reversely biased at a reasonable voltage whereas 3D photodiodes are likely operated at the Geiger mode. How to design integrated 2D and 3D photodiodes is investigated in terms of quantum efficiency, dark current, crosstalk, response time and so on. Beyond photodiodes, a charge supply mechanism provides a proper charge for a high dynamic range of 2D sensing, and a feedback pull-down mechanism expedites the response time of 3D sensing for time-of-flight applications. Particularly, rapid parallel reading at a 3D mode is developed by a bus-sharing mechanism. Using the TSMC 0.35μm 2P4M technology, a 2D/3D-integrated image sensor including P-diffusion_N-well_P-substrate photodiodes, pixel circuits, correlated double sampling circuits, sense amplifiers, a multi-channel time-to-digital converter, column/row decoders, bus-sharing connections/decoders, readout circuits and so on was implemented with a die size of 12mm×12mm. The proposed 2D/3D-integrated image sensor can perceive a 352×288-pixel 2D image and an 88×72-pixel 3D image with a dynamic range up to 100dB and a depth resolution of around 4cm, respectively. Therefore, our image sensor can effectively capture gray-level and depth information of a scene at the same location without additional alignment and post-processing. Finally, the currently available 2D and 3D image sensors are discussed and presented.

**Keywords:** CMOS photodiode, active pixel, Geiger mode, time of flight, image sensor

## 1. Introduction

Nowadays, 3-dimensional (3D) images and videos are widely used in various applications, like games, robotics, cinema and so on. How to effectively capture 3D information becomes a critical issue. In general, stereo images from two slightly different viewpoints are used to establish 3D perception of which mechanism is based on binocular parallax. For example, a 3D camera of Panasonic LUMIX® DMC-3D1K takes two pictures simultaneously and then displays these two pictures to the left and right eyes of a human for 3D perception [1]. Since the parallax of two viewpoints for capturing is fixed, 3D perception may not be easily adapted to different viewpoints.

Another approach also adopts two cameras in which one captures a gray-level image, and the other seizes object depths where an active light source, like infrared, is utilized. Based on the characteristics of light reflected from objects, a 3D image can be built via post-processing computation. For instance, Kinects I and II from Microsoft Xbox use the structured light technique and Time Of Flight (TOF) technique to acquire object depths, respectively [2], [3]. Particularly, the TOF technique is used to estimate positions of objects in space according to the travel time of light emitted from a light source, reaching an object, reflected from an object and arriving at a sensor [4], as depicted in Fig. 1. Based on the different travel time captured by photodiodes, a depth map of a scene can be fairly attained. The information from the luminance (2D) and depth (3D) cameras is used to reconstruct complete 3D pictures which are observed from multiple viewpoints. Since these two cameras may be located at different positions, the capture-point difference need be compensated to yield a correct 3D picture.

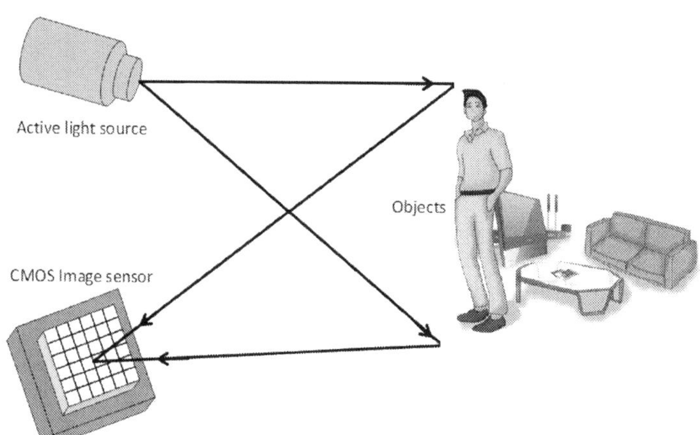


**Figure 1.** Concept of TOF.

In this chapter, physical characteristics of CMOS photodiodes are illustrated first. How to design high-efficiency photodiodes for 2D and 3D sensing is addressed. Second, a 2D/3D-

http://dx.doi.org/10.5772/60737

integrated active pixel is presented where the same photodiodes are shared by 2D and 3D capturing. Additionally, the readout circuits used for 2D and 3D image sensors are integrated. Particularly, two feedback mechanisms are adopted to delay saturation of 2D luminance sensing, and accelerate response time of 3D depth sensing. Third, in order to reach a compact pixel array, high readout speed and the minimized coupling effect between transmission lines and photodiodes, the mechanism associated with bus sharing is developed to effectively accomplish parallel reading. Fourth, based on the TSMC 0.35μm 2P4M CMOS technology, the proposed 2D/3D-integrated image sensor was implemented with a die size of 12mm×12mm where 2D and 3D modes have the resolutions of 352×288 and 88×72 pixels, respectively, under a fill factor of 44%. Finally, the proposed image sensor, and the currently available 2D and 3D image sensors are illustrated, compared and concluded.

## 2. Physical characteristics of CMOS photodiodes for 3D sensing

The *p-n* junction of a diode under a reversely biased voltage is employed to perceive photos. Such a diode is named as a photodiode. Due to the electrical field of the reversely biased voltage, the depletion region between *p* and *n* layers is created. The size of the depletion region is dependent on the amount of the reversely biased voltage, and the doping concentrations of *p* and *n* layers. This depletion region incited by a photo can yield a pair of electron and hole which are drifted and guided by the electrical field to become a photocurrent. When the reversely biased voltage approaches the breakdown voltage, the depletion region becomes large to increase the incitation rate of electrons and holes, and the high electrical field expedites the drifting of electrons and holes and reduces the recombination chance of electrons and holes. This operation manner is called as the Geiger mode [5], which has a rapid response to incident light. Restated, slight light can induce a sufficient photocurrent for detection.

A 3D image can be captured by the time of flight technique where object depths are interpreted by the time of round trips associated with light that originates from a light source, like light emitting diode (LED), shining objects, and is reflected from objects. To effectively seize reflected light, photodiodes must quickly respond to photos owing to high speed of light. Hence, an avalanche photodiode which operates at the Geiger mode can be a good choice to accomplish TOF sensing. An avalanche photodiode has high sensitivity to precisely apprehend slight photos which are early birds. Under such physical characteristics, a 3D depth map can be successfully attained from an array of avalanche photodiodes.

Many avalanche complementary metal-oxide semiconductor (CMOS) photodiodes have been explored and proposed in the literature. For instance, Niclass *et al.* proposed a 3-D imager based on a 2-D array of single-photon avalanche diodes fabricated by the standard CMOS technology where sub-millimetric precision could be achieved under a depth-of-field of several meters [6]. Marshall *et al.* realized a CMOS 64×64 pixel array in which avalanche photodiodes and active pixel sensors were integrated [7]. Zappa *et al.* adopted a standard 0.8-μm CMOS technology to realize an integrated sensor consisting of photodiodes, input sensing circuit, photon-counting and control circuits where an active quenching and active reset circuit were implemented to drive a single-photon avalanche diode [8]. Faramazpour *et al.* developed avalanche photodiodes and their corresponding driving circuits under a standard 0.18-μm CMOS technology where the breakdown voltage was 10.2V and the dead time was 30ns [9]. Atef *et al.* implemented two photodiode structures using the 40-nm standard CMOS technology where one could perform like an avalanche photodiode and the other functioned as a regular photodiode [10]. Pancheri *et al.* presented a low-noise avalanche photodiode based on the graded junction in 0.15-μm CMOS technology [11]. The other studies focused on ameliorating the process, materials and doping concentrations in order to minimize noise, to lower dark currents and to enhance sensitivity [12]-[15].

By using the TSMC 0.35μm CMOS technology, P-diffusion_N-well_P-substrate, N-well_P-substrate and N-diffusion_P-substrate photodiodes are explored to understand the physical characteristics where their cross sections are shown in Fig. 2 [16]. Figures 3 and 4 depict breakdown voltages of P-diffusion_N-well_P-substrate and N-well_P-substrate photodiodes under the same area sizes with different perimeters, respectively. The measurement results reveal that the breakdown voltage becomes larger as the perimeter is decreased. Additionally, the breakdown voltage of a P-diffusion_N-well_P-substrate photodiode is smaller than that of an N-well_P-substrate photodiode. The photodiode with a deeper p-n junction, larger area and smaller perimeter has a higher breakdown voltage. Particularly, the lateral region of a photodiode may form a dead space which brings about the decrease of a fill factor, quantum efficiency and breakdown voltage [16]. From the abovementioned phenomena, a circular photodiode with the maximum ratio of area to perimeter is likely preferred to attain a high breakdown voltage.

As well as the boundary effect, the efficiency of induced photocurrents is examined where a light source adopts a laser at 850nm. Referring to [17], the current gain can be expressed as

$$G = (I_{photo\_Geiger} - I_{dark\_Geiger}) / (I_{photo\_typical} - I_{dark\_typical}) \tag{1}$$

$I_{photo\_typical}$ and $I_{dark\_typical}$ are the induced and dark currents of a photodiode at a typical reversely biased voltage, respectively. This typical voltage can be $V_{dd}$, which is a supply voltage of circuits at the currently used process technology. Restated, it is the maximum reversely biased voltage for 2D sensing of a photodiode. Additionally, $I_{photo\_Geiger}$ and $I_{dark\_Geiger}$ are the induced and dark currents of a photodiode at the Geiger mode, respectively. The measurement results, depicted in Fig. 5, reveal that the current gain of N-diffusion_P-substrate is the best, and the second is P-diffusion_N-well_P-substrate where each of three photodiodes has the area of 30μm×30μm, and $V_{dd}$ is 3.3V. Due to high variations of currents, the logarithmic scale of 'log(fA)' at the left Y axis is adopted to clearly present the measured currents.

In addition to the breakdown voltage of a photodiode, the crosstalk effect, as shown in Fig. 6, need to be well investigated to minimize the interference of the neighboring pixels [18], [19]. Figure 7 displays a top view of 3×3 photodiodes, with and without an N-well guard ring, which were implemented by the TSMC 0.35μm CMOS technology. The photodiode in the center of a 3×3 photodiode array is measured when the neighboring eight photodiodes are biased at different voltages. The measured breakdown voltages of 3×3 P-diffusion_N-well_P-substrate and N-well_P-substrate photodiodes with and without an N-well guard ring are depicted in Fig. 8. The breakdown voltage of a photodiode with a guard ring is greater than that of a photodiode without a guard ring. Notably, the central pixel likely has a higher breakdown voltage as its neighboring pixels are reversely biased to a higher voltage. Figure 9 shows breakdown voltages of 3×3 P-diffusion_N-well_P-substrate, N-well_P-substrate and N-diffusion_P-substrate photodiodes with N-well guard rings while neighboring pixels are reversely biased at voltages from 0V to 9V. The N-well_P-substrate photodiode yields the largest breakdown voltage, the second is the P-diffusion_N-well_P-substrate photodiode and the last one is the N-diffusion_P-substrate photodiode.

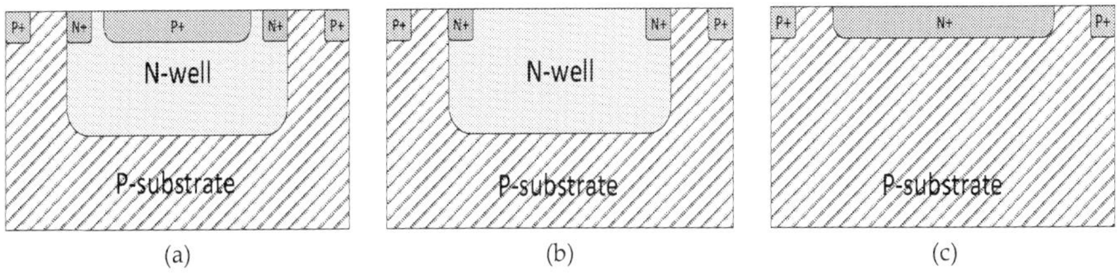


**Figure 2.** Cross sections of photodiodes. (a) P-diffusion_N-well_P-substrate. (b) N-well_P-substrate. (c) N-diffusion_P-substrate photodiodes.

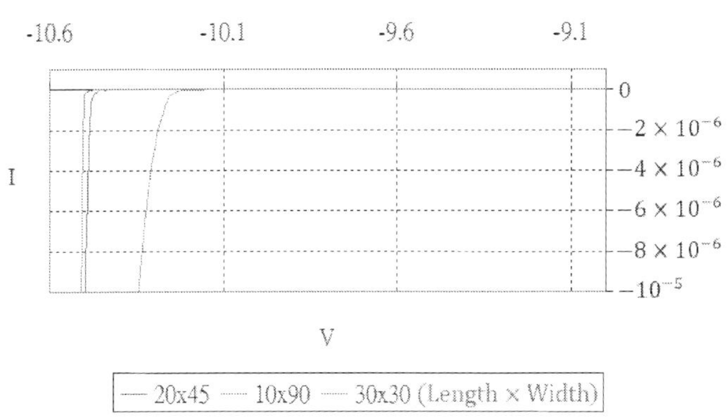


**Figure 3.** Breakdown voltages of P-diffusion_N-well_P-substrate photodiodes.

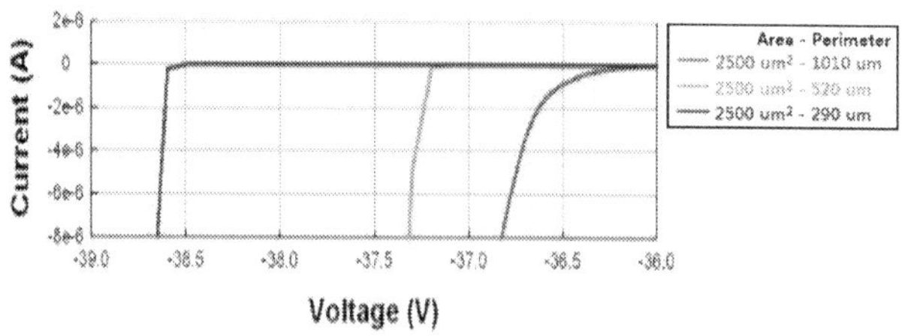


**Figure 4.** Breakdown voltages of N-well_P-substrate photodiodes.

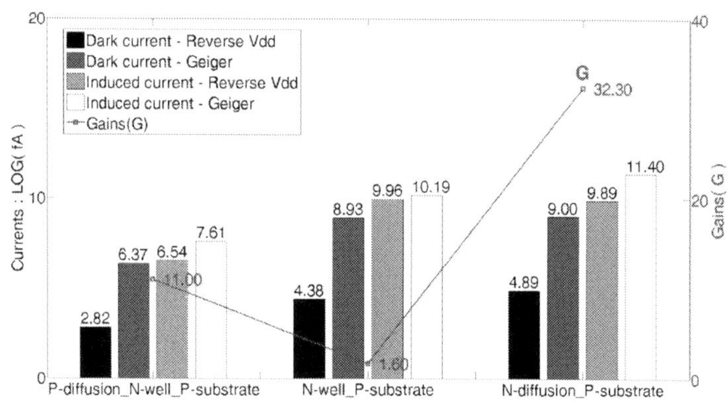


**Figure 5.** Induced currents, dark currents and gains of photodiodes at an 850nm laser.

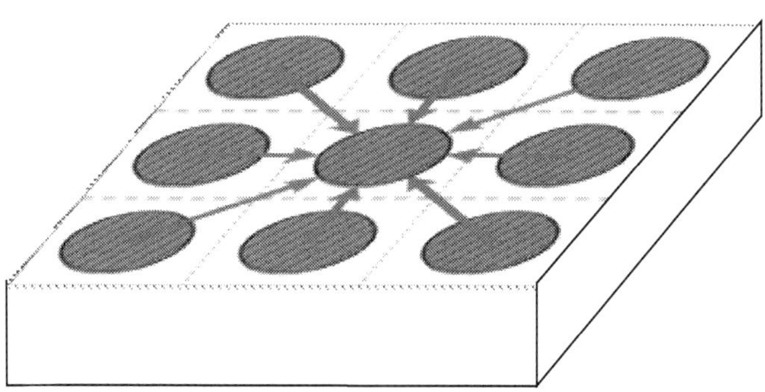

**Figure 6.** Topology of a crosstalk effect.

http://dx.doi.org/10.5772/60737

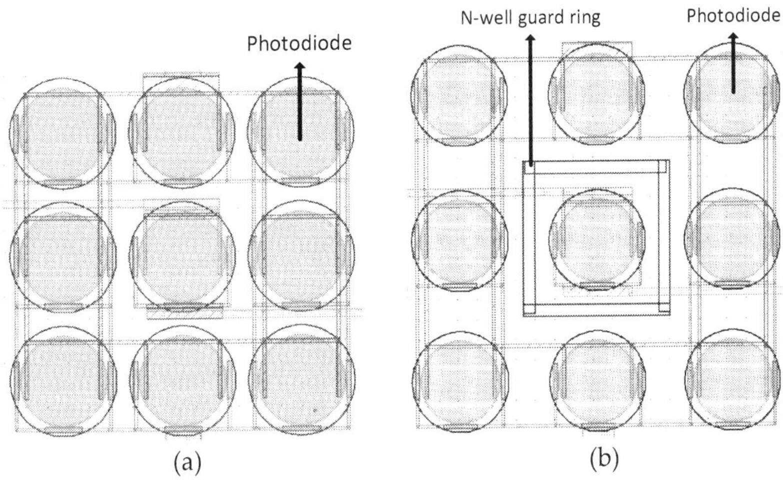


**Figure 7.** Top view of 3×3 photodiode arrays. (a) Without a guard ring. (b) With an N-well guard ring.

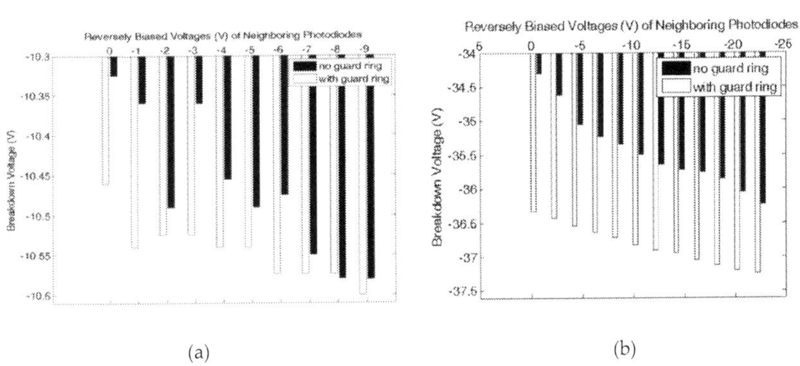


**Figure 8.** Measured breakdown voltages. (a) 3×3 P-diffusion_N-well_P-substrate photodiodes. (b) 3×3 N-well_P-substrate photodiodes.

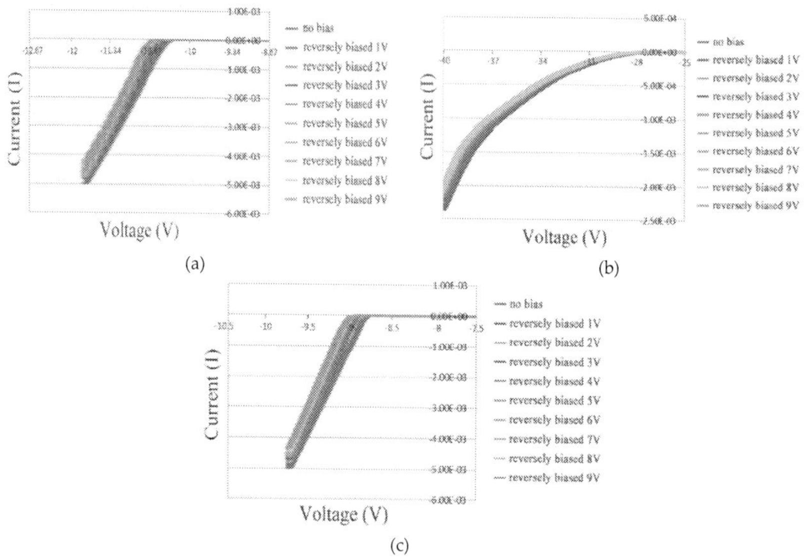


**Figure 9.** Measured breakdown voltages. (a) P-diffusion_N-well_P-substrate. (b) N-well_P-substrate. (c) N-diffusion_P-substrate.

## 3. 2D/3D-integrated pixel

A 3D depth sensor can yield a depth map of a scene but fails to provide delicate gray-scale information. However, a 2D image sensor can effectively interpret a pixel at a fairly fine gray scale rather than depth resolution. Accordingly, there is a demand for using multiple 2D or/and 3D sensors to capture a real-world scene and then to display a 3D picture to a watcher. For example, if two 2D image sensors are employed to mimic binocular vision, 3D perception adheres to a specific viewpoint. Nevertheless, if a 2D image sensor and a 3D depth sensor are used together to capture objects, the difference of viewpoints from these two sensors at different positions has to be amended. Additionally, the camera including two individual sensors takes relatively high hardware cost accompanied with great power consumption. Therefore, we propose to realize 2D luminance and 3D depth perceptions at one sensor of which 2D and 3D operations are alternately executed. The concept of how to design the proposed 2D/3D-integrated image sensor is illustrated in Fig. 10 [20].

The first step of integrating a 2D image sensor and a 3D depth sensor is to design a photodiode that can be shared by these two sensors. Figure 11 depicts the measured spectrum responses of P-diffusion_N-well_P-substrate, N-well_P-substrate and N-diffusion_P-substrate photodiodes using the TSMC 0.35μm CMOS technology where the reversely biased voltage is zero. From the measured results, N-well_P-substrate and P-diffusion_N-well_P-substrate photodi-

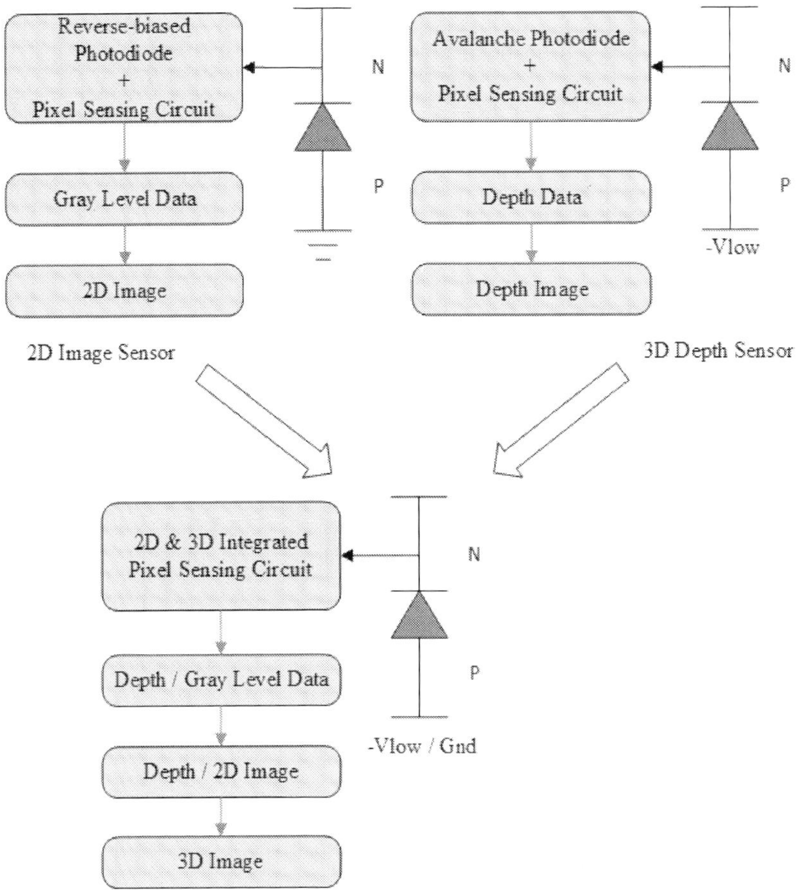


**Figure 10.** Design methodology of the 2D/3D-integrated sensor.

odes are the best and second best, respectively. In the following, the operations of photodiodes associated with 2D image and 3D depth sensors are studied to determine which photodiode structure is preferred.

### 3.1. Photodiodes of 2D and 3D sensors

When a PN photodiode is biased at a reverse voltage, incident light reaches the depletion region of this photodiode, and then incites numerous pairs of holes and electrons which create a photocurrent. The light intensity increases with the induced photocurrent. Based on an exposure period, the sensing circuit of a pixel converts an integrated photocurrent to an analogue voltage which represents gray-level luminance. Such a mechanism accomplishes a

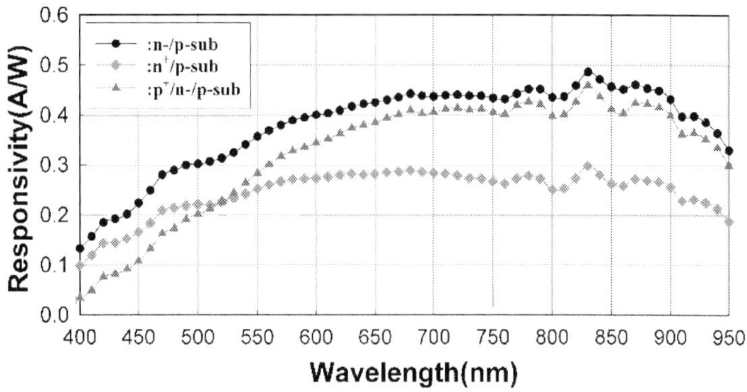


**Figure 11.** Measured spectral responses of three CMOS photodiodes.

2D image sensing. While the reversely biased voltage approaches the breakdown voltage of a photodiode, this photodiode has relatively large electrical field and depletion region in which pairs of holes and electrons are easily generated by few photons. This kind photodiode operating at the Geiger mode is to acquire a 3D depth map where the time of the flight trip associated with light emitted from a source, reaching objects, reflected from objects and arriving at photodiodes is computed.

Physical characteristics of photodiodes with different geometric and junction structures must be well understood in order to effectively integrate 2D and 3D photodiodes. From our previous study [16], the corner in the geometric shape of a photodiode easily leads to breakdown because of charged elements likely gathering at this point. Accordingly, the number of corners in a photodiode increases, resulting in the decrease of a breakdown voltage. Furthermore, the dark current is correspondingly enlarged as the number of corners increases. In the previous section, we conclude that the area and peripheral of a photodiode are somehow proportional and inversely proportional to the breakdown voltage, respectively. The increased number of pixels in an array lowers the breakdown voltage as well. Therefore, a photodiode with a circle shape and a guard ring is chosen to achieve high-efficiency sensing capability.

According to quantum efficiency of photodiodes in Fig. 11, N-well_P-substrate and P-diffusion_N-well_P-substrate photodiodes are good candidates. The reversely biased voltages of a photodiode used for 2D and 3D sensing are quite different. When the N-well_P-substrate photodiode is adopted for 2D and 3D sensors, P-substrate must be driven by two different voltages. However, P-substrate is a common base which is usually shared by transistors of all circuits. Under the standard CMOS technology, P-substrate is always connected to the ground. Hence, the N-well_P-substrate photodiode does not satisfy our demand. In the P-diffusion_N-well_P-substrate photodiode, P-diffusion and P-substrate are addressed by the ground to form two PN junctions for a 2D sensing where P-diffusion_N-well and N-well_P-substrate work for short-wavelength and long-wavelength light receiving, respectively. When a 3D sensing is conducted, P-diffusion and P-substrate are biased by a large negative voltage and 0V, respectively. Such a biasing manner makes P-diffusion_N-well an avalanche photodiode and N-well_P-substrate a photodiode under a normal reversely biased voltage. Once few photos reach the P-diffusion_N-well_P-substrate photodiode, P-diffusion_N-well reacts rapidly, and N-well_P-substrate may not respond too much.

### 3.2. Pixel circuits

To capture 2D gray-level information, the dynamic range is one of the key issues associated with a pixel design. A pixel circuit with a larger dynamic range can interpret a greater range of light luminance. Figure 12 shows the pixel circuit at a 2D mode, which employs an extra path to provide charges [21]. Such a path includes two transistors of $M_2$ and $M_3$ where $M_2$ is manipulated by an external control signal, and $M_3$ functions like an active resistor with gate-node sensing. $I_{charge\text{-}supply}$ goes through this path to compensate the current sink from $I_{photo,}$ and thus to delay the gate node of $M_6$ to become 0V. The simulations in Fig. 13 reveal that the pixel circuit with a charge supply mechanism can reach up to 110dB and greatly extend the dynamic range than without a charge supply mechanism.

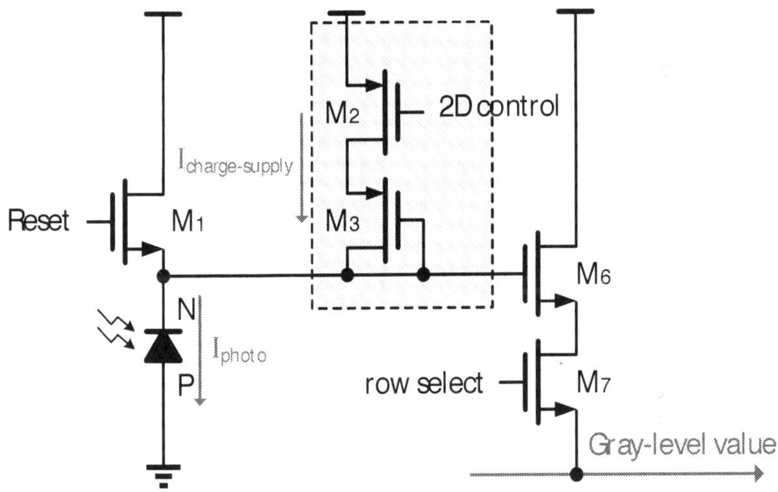


**Figure 12.** 2D pixel circuit.

The pixel circuit at a 3D mode is to discover objects associated with depth information. The photodiode in a pixel is reversely biased close to a breakdown voltage. Once the photodiode is triggered by few photons, a significant current is generated. Such an instant and great current accelerates the response of the pixel circuit. Referring to a simple passive quenching structure [22], we proposed a modified 3D pixel circuit, as depicted in Fig. 14. When a slight light reaches this 3D pixel, a great current is induced to make a voltage drop at the *N* node of a photodiode. This voltage drop is expedited by $M_4$, $M_5$ and an inverter as well. The timing diagram of the

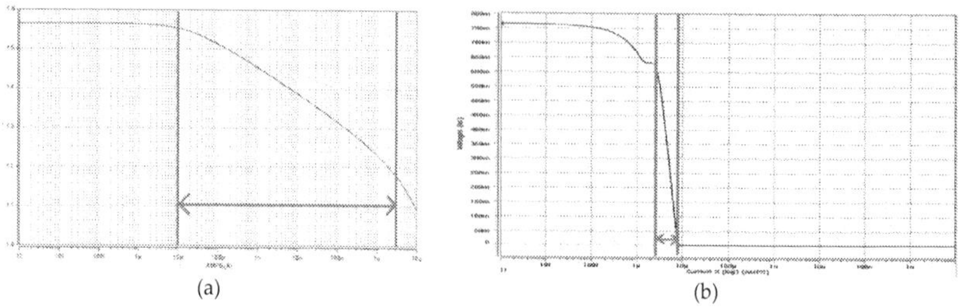

Figure 13. Simulated dynamic range of a 2D pixel circuit. (a) With a charge supply mechanism. (b) Without a charge supply mechanism.

proposed 3D pixel circuit is depicted in Fig. 15. During the reset, the *N* node of a photodiode is charged to a voltage, above $\frac{1}{2}V_{dd}$, which is within the input noise margin of an inverter outputting 0V. The source and gate nodes of $M_4$ are connected to the input and output of an inverter, respectively, to create a feedback loop. Additionally, $M_5$ is linked to the drain node of $M_4$, and is manipulated by a signal of '3D control' to build a pull-down path. After the reset, the photodiode begins to perceive photos. Once few photos are detected, a discharge action is taken. At the same time, $M_4$ and $M_5$ are activated to accelerate the inverter dropping to $1/2V_{dd}$. Such a pull-down path lessens the time interval associated with light detection, and thus raises the depth resolution. Figure 16 shows the circuit diagram and layout of the proposed 2D/3D-integrated pixel that is easily addressed by using the 2D and 3D control signals to switch its operation modes.

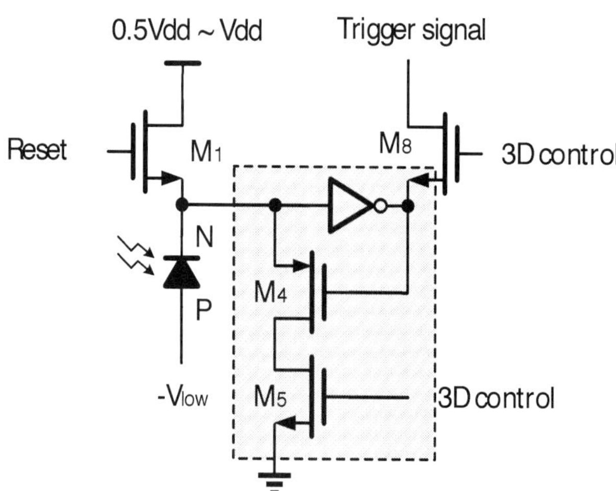


Figure 14. 3D pixel circuit.

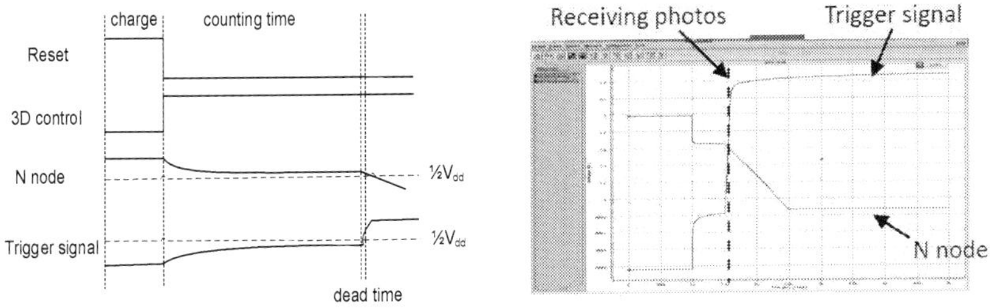


(a) (b)

**Figure 15.** Timing diagram of a 3D pixel circuit. (a) Timing chart. (b) Simulation chart.

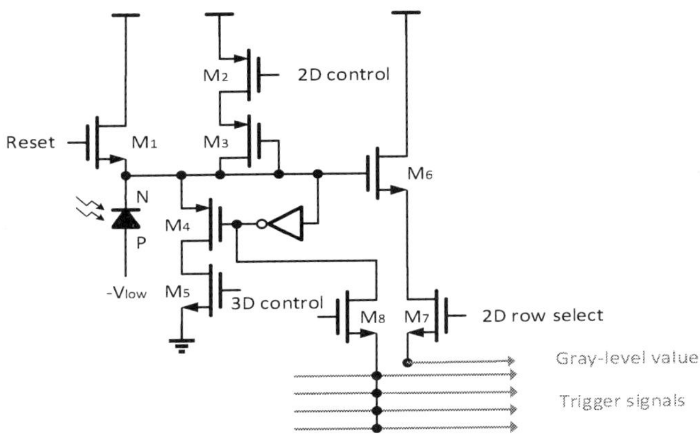


(a)

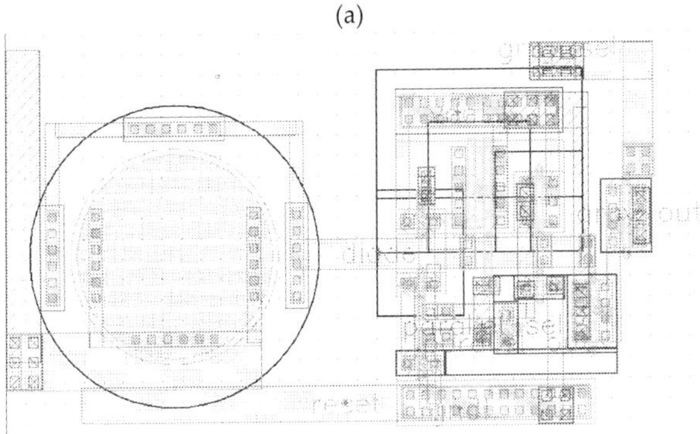

(b)

**Figure 16.** 2D/3D-integrated pixel circuit. (a) Circuit diagram. (b) Layout diagram.

## 4. Parallel reading using the bus-sharing mechanism

When the proposed 2D/3D-integrated pixel circuit is operated at a 3D mode, a pulse signal is triggered by the perceived photos. Afterwards, the time difference between light emitted and detected is calculated, and then the corresponding depth information is derived based on a velocity of light. An independent bus line from each pixel of a large pixel array may not be addressed at low hardware cost. Hence, there is a need for low-cost parallel reading and computation in time. Nevertheless, as the size of a pixel array goes up, the hardware complexity of parallel reading is extremely expanded. Not only is the area increased, but the coupling effect between photodiodes and transmission lines is likely induced as well. According to the conventional work [23], [24], the reading of trigger pulses is fulfilled at a time-multiplexed manner.

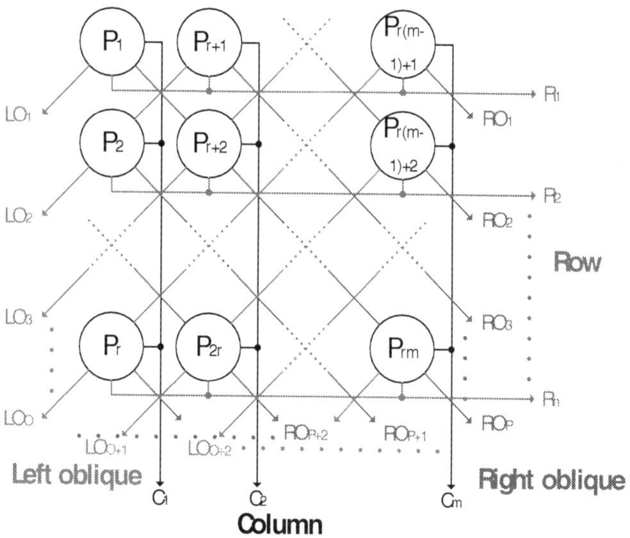


Figure 17. Parallel reading of the bus-sharing mechanism.

To overcome the abovementioned drawbacks, a bus-sharing mechanism is proposed to realize parallel reading at low hardware cost. This bus-sharing mechanism employs a connection topology in which each pixel connects to multiple shared buses, as displayed in Fig. 17. Since each bus is connected to multiple photodiodes, a decoder associated with the shared buses is demanded to determine which photodiode is activated. Restated, when buses are enabled by one or multiple photodiodes simultaneously, these buses become $V_{dd}$. Based on the pattern of the enabled buses, the locations of photodiodes are discovered. For instance, while light is sensed by a pixel, $P_1$, the buses of $LO_1$, $C_1$, $R_1$ and $RO_p$ are pulled up to $V_{dd}$. As $P_2$ also observes light, it enables the buses of $LO_2$, $C_1$, $R_2$ and $RO_{p1}$. Although there is a bus of $C_1$ shared by $P_1$

and $P_2$, the other three buses are quite different, and can be used to distinguish $P_1$, $P_2$ and others. Therefore, according to the enabled buses and triggered time, when and which photodiode(s) are activated can be decoded and ascertained.

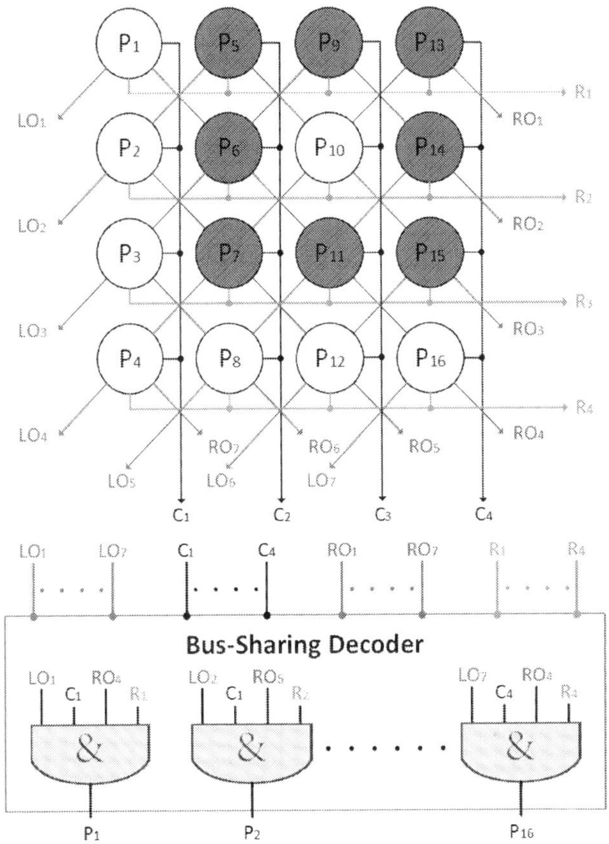


**Figure 18.** Special case of a photodiode mistaken.

Photodiodes triggered at different time can be discovered by the proposed bus-sharing mechanism at an effectual and low-cost manner. However, when many photodiodes receive photos simultaneously, a special condition must be considered. An un-triggered pixel completely encircled by triggered pixels that discover light at once is mistaken for a triggered one. A special case, depicted in Fig. 18, mistakes $P_{10}$ where $P_{10}$ does not capture any light, and photodiodes of $P_5$ ~ $P_7$, $P_9$, $P_{11}$ and $P_{13}$ ~ $P_{15}$ are activated by light. The buses of $LO_4$, $C_3$, $RO_3$ and $R_2$ addressed by $P_{10}$ are enabled by ($P_7$, $P_{13}$), ($P_9$, $P_{11}$), ($P_5$, $P_{15}$) and ($P_6$, $P_{14}$), respectively. During decoding $P_{10}$, an AND gate with inputs of $LO_4$, $C_3$, $RO_3$ and $R_2$ yields logic-1, which is an error solution. Such a situation can be figured out while $P_{10}$ is only enabled at an earlier or later time. Restated, to analyze the activated patterns of $P_{10}$ and its neighboring pixels at different time intervals, the accurate triggering time associated with $P_{10}$ can be attained.

### 4.1. Multi-channel time to digital converter

When an LED generates light projecting upon an object, the time to digital converter (TDC) begins to calculate the time of light travel from an LED to a photodiode. The derived travel time which is multiplied by a velocity of light can present a double distance from a sensor to an object. According to this distance information, the depth of an object apart from a sensor can be attained. Figure 19 shows the block diagram of multi-channel TDC which is an event-driven approach [25]. The counter in TDC is applied to obtain a timescale number which is correlated with time of light flight. Additionally, the delay line circuit is used to interpolate a fractional scale. To attenuate the influence of temperature and process, differential pairs are adopted to realize flip flops of a delay line.

A 3D depth map comes from a pixel array of which pixels demand timing information. If each pixel has the corresponding timing circuit to compute its depth, there are too many timing circuits which take great hardware cost and high power dissipation. To reach the TDC shared by multiple pixels, a multi-channel TDC composed of a ring TDC, a thermal encoder and a 4-bit counter, as displayed in Fig. 19, is adopted. Referring to [25], Fig. 20 shows a 15-stage ring TDC, which is a core of a multi-channel timing circuit. When the signal of 'Start' is active, a NAND gate and 14 inverters build a ring oscillation. This ring TDC produces 15 outputs of $C_1$, $C_2$,... and $C_{15}$, which are compacted by a thermal encoder to give a 4-bit fine result. In the meantime, the counter yields a 4-bit coarse result as well. 4-bit fine and course results form 8-bit timing information which is stored in the latch array, and used to interpret the depth information.

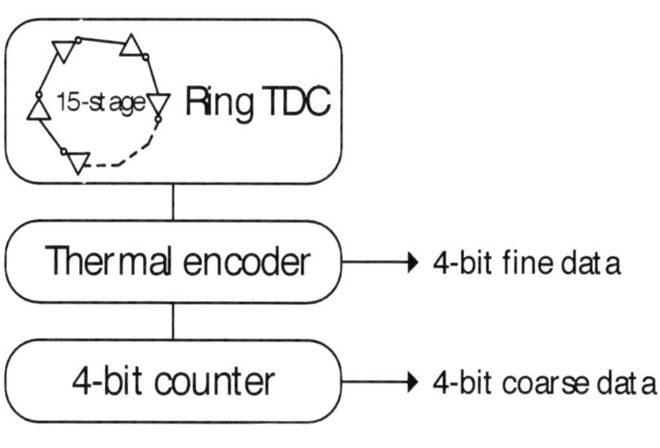


**Figure 19.** Block diagram of multi-channel TDC.

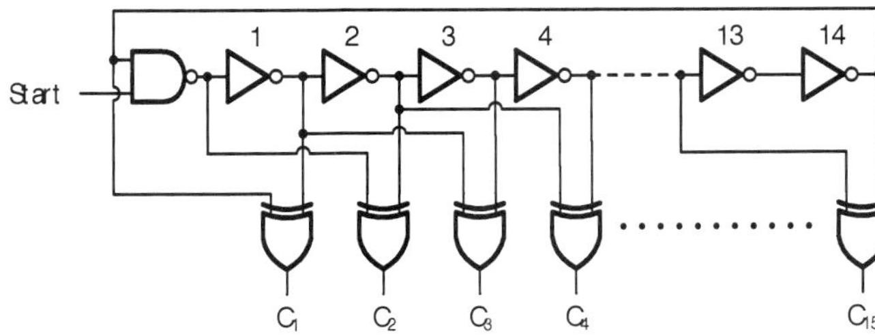


Figure 20. Ring TDC.

## 5. 2D/3D-integrated image sensor

The proposed 2D/3D-integrated image sensor employs P-diffusion_N-well_ P-substrate photodiode, which can be switched to different reversely biased voltages, and then operates at a 2D or 3D photo sensing mode. The Correlated Double Sampling (CDS) circuit and readout circuit used for a 2D mode, and Sense Amplifiers (SA), a multi-channel TDC and a readout circuit used for a 3D mode are implemented. Additionally, sequential and parallel reading is realized by using row and column decoders, and bus-sharing connections and decoders at 2D and 3D modes, respectively. The CDS circuit reduces the fixed pattern noise, and the SA boosts a trigger pulse generated from a pixel to lower the dead time. The block diagram of the proposed 2D/3D-integrated image sensor is depicted in Fig. 21. Since human visual perception has a good resolution in luminance rather than depth, the proposed 2D/3D-integrated image sensor adopts pixel dimensions of 352×288 and 88×72 pixels associated with 2D and 3D sensing, respectively. Restated, the pixel dimension at a 3D mode is one sixteenth of that at a 2D mode to lessen hardware cost. Particularly, to effectively decrease the overhead of parallel reading at a 3D mode, the bus-sharing mechanism can address 88×72 pixels by using 478 lines rather than 6,336 lines. After decoding 478 signal bits, 88×72 bits are stored in latches at every counting-time interval where each bit indicates if a pixel is triggered or not.

The TSMC 0.35μm 2P4M CMOS technology was employed to implement the proposed 2D/3D-integrated image sensor with a die size of 12mm×12mm [26]. The field factor is about 44% where a photodiode has a diameter of 10μm. The peak and average powers are 2.56W and 0.58W, at a supply voltage of 3.3V, respectively. The dynamic range of luminance detection can reach to 100dB. Each stage of the ring TDC was measured to have a delay of 234ps which can interpret a depth resolution of 4cm. When capturing an object, the external start signal resets the pixel circuit, enables the TDC and triggers a light source. The proposed sensor begins to receive the reflected light from an object and calculate the travel time using the TDC. The counted travel time is used to derive object depths. In our measurement, the FPGA board is programmed to manipulate an LED array and provide the timing control signals to read data from the proposed chip, as displayed in Fig. 22, where a cylindrical box is located in front of the sensor. At a light source from an 850nm LED array, Fig. 23 shows the depth map of a

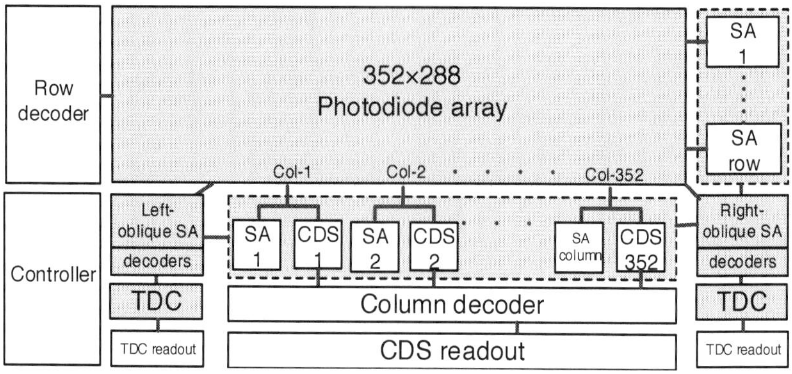


Figure 21. Block diagram of the proposed 2D/3D-integrated image sensor.

cylindrical box. This depth map exhibits a similar cylindrical shape but somehow includes little noise.

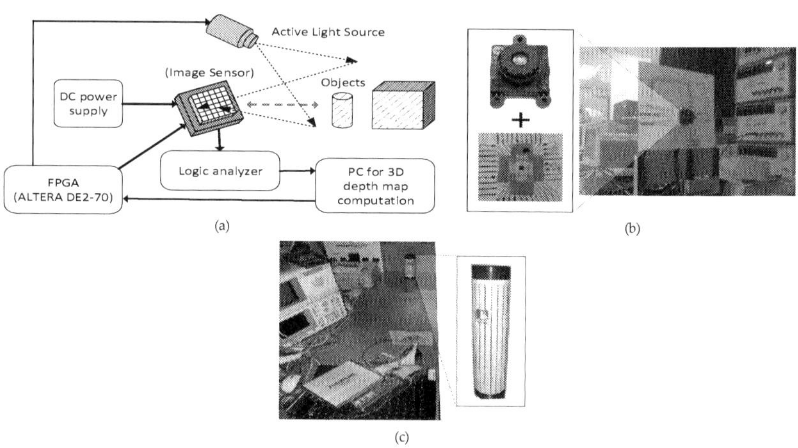


Figure 22. Measurement setup using the FPGA board. (a) Block diagram of measurements. (b) Proposed chip with on-board package and a convex lens. (c) Discovering a cylindrical box.

## 6. Currently available techniques of 2D and 3D image sensors

Nowadays, CMOS image sensors generally have extensive resolutions, high frame rates, large dynamic ranges and low power dissipation. To meet these demands, pixel circuits, read-out structures, analogue-to-digital converter (ADC) architectures and 3D integrated circuits (IC)

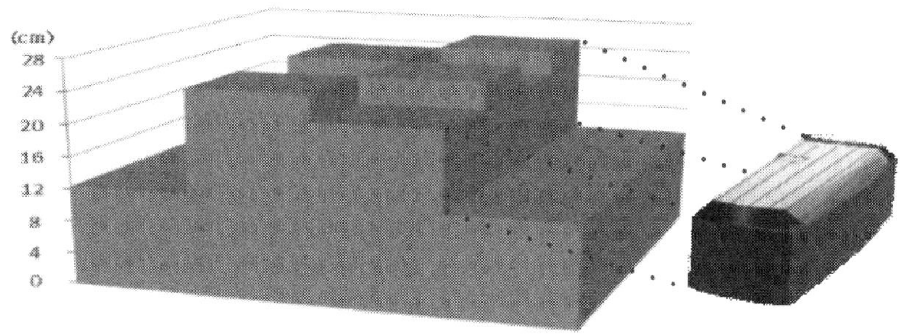


**Figure 23.** Measured depth map of a cylindrical box.

have been well explored as well as high-efficiency photodiodes. For example, Lim *et al.* adopted two column-shared cyclic ADCs arranged at a two-stage stack, to reduce the power consumption of the read-out circuit [27]. Seo *et al.* employed a column-parallel folding integration and cyclic ADC architecture to minimize the amount of noise, and reached to 13-bit or more resolution for each pixel [28]. To enhance the sensitivity of a pixel, Xu *et al.* designed a capacitive trans-impedance amplifier pixel with a tiny metal–oxide–metal (MOM) capacitor. To compensate the mismatch of small integration capacitors across the pixel array, an on-chip calibration scheme with in-pixel circuits was developed [29]. Afterwards, Xu *et al.* further implemented a small multi-layer MOM integration capacitor to achieve high sensitivity and low mismatches across a pixel array. Under such a design, the size of the previous capacitive trans-impedance amplifier pixel was effectively minimized [30]. Sakakibara *et al.* added a floating diffusion node and a storage node in a pixel circuit which supports two-channel read-outs for low and high intensities. Via a single-slope 12-bit ADCs, their pixel with dual storage structure made it possible to obtain up to 83dB dynamic range [31]. Chung *et al.* proposed a 0.5V operated pulse-width modulation CMOS image sensor with threshold-variation canceling and programmable current-controlled threshold schemes in which the fixed-pattern noise could be effectively minimized [32]. Yeh *et al.* used the 3D IC technology to stack a pixel array, an ADC array and an image processing array, which could operate in parallel, to achieve a high frame rate and high spatial resolution [33]. Sukegawa *et al.* combined the back-illuminated photosensitive layer and logic layer to become an image sensor. Such an approach could make the size of 8M pixels down to 1/4 inch. Additionally, the RGBW (Red, Green, Blue and White) coding was added in the color filter of the pixel to improve sensitivity, and binning-SVE was adopted to reach a high dynamic range [34]. The specifications of the abovementioned approaches are summarized in Table 1.

As well as 2D image sensors, there are some special approaches to implement 3D image sensors. For instance, Koyama *et al.* used a camera lens and a single sensor to realize the binocular images for left and right eyes [35]. Particularly, a lenticular lens separates incident light to become two beams for left-eye and right-eye viewing points, which were captured by

| Authors | Xu *et al.* [30] | Sukegawa1 *et al.* [34] | Yeh *et al.* [33] | Chung *et al.* [32] | Sakakibara *et al.* [31] | Seo *et al.* [28] | Lim *et al.* [27] |
|---|---|---|---|---|---|---|---|
| Pixel architecture | CTIA | N/A | 4T-APS PPD | N/A | N/A | 4-TR PPD | 2-T PPD |
| Shutter | Global | N/A | Rolling | N/A | Global | N/A | N/A |
| Process (CMOS) | 0.18 μm | 65nm (1P4M) | 0.18 μm (1P6M) | 0.18 μm (1P6M) | 90nm (1P4M) | 0.18 μm (1P4M) | 0.13 μm (1P4M) |
| Array size | 640×480 | 3280×2464 | 2048×1536 | 64×40 | 5M | 1032×1024 | 1696×1212 |
| Pixel size ( $\mu m^2$ ) | 8.7×8.22 | 1.12×1.12 | 2.8×2.8 | 10×10 | 5.86×5.86 | 7.5×7.5 | 2.25×2.25 |
| Fill factor (%) | 41 | N/A | 28 | 25.4 | N/A | 52 | N/A |
| Frame rate (fps) | 400 | 30 | 100 | 78.5 | N/A | 2.2 | 250 |
| Dynamic range | 50.1dB | 60dB | N/A | 82dB | 83dB | 78dB | 59dB |
| FPN | 0.55% | N/A | 0.43 $mV_{rms}$ | 0.055% | N/A | N/A | 0.1% |
| Sensitivity | 1.89 V/lux · s | 6.7k $e^-$/ lux · s (white pixel) | 3.69 V/lux · s | N/A | 78k $e^-$/ lux · s | 10 V/lux · s | 8.6k $e^-$/ lux · s |
| Temporal noise | 15.6 $e^-_{rms}$ | 2.2 $e^-_{rms}$ | 16 $e^-_{rms}$ | N/A | $4.8e^-_{rms}$ | $\leq 1e^-_{rms}$ | 12.5 $e^-_{rms}$ |
| ADC resolution (bit) | 12 | 10 | 12 | 10 | 12 | 13-19 | 10 |
| ADC type | SAR | N/A | Off-chip | Ramp | Column single-slope | Folding-integration/ Cyclic | Shared Cyclic |
| Power | N/A | 185mW | 19mW | 29.6 μW | 2W | 450mW | 300mW |

Note: CTIA: Capacitive Trans-Impedance Amplifier; PPD: Pinned Photo-Diode

**Table 1.** Specifications of Currently Available 2D Image Sensors.

multiple pairs of neighboring pixels. However, to solve the energy losses due to the beam splitter and the crosstalk and so on, the polycyclic $Si_3N_4$ digital micro lenses were employed to make the beam focused at photodiodes. Wang *et al.* used a light-field image sensor to capture the intensity and angle of the incident light, and then derived depth information where angle sensitive pixels were adopted [36]. Based on the traditional pixel architecture and diffraction gratings, the Talbot effect could be addressed to capture the information of the incident light angle. By calculating the convergence and divergence angles, the depth information is attained. Kim *et al.* use single lens to capture 2D and 3D images via a time-multiplexed manner [37]. The pixel architecture consisted of a 4-transistor active pixel and two floating diffusion nodes where each pixel unit was adjacent to two transfer gates. Through a time-multiplexed manner, the transfer gates are used to switch the operational modes. At a normal mode, the capture manner is identical to that of a traditional CMOS image sensor, and at a 3D mode, all of the transfer gates are switched on to yield many 4×4- pixel units for capturing the depth information. Restated, they enlarged the sensing area of a photodiode to enhance the sensitivity at a 3D mode. However, the increase of the reversely biased voltage usually brings higher sensitivity improvement than the increase of the sensing area. The specifications of the abovementioned 3D image sensors and the proposed 2D/3D-integrated image sensors are listed in Table 2.

| | Authors | Koyama *et al.* [35] | Wang *et al.* [36] | Kim *et al.* [37] | Proposed |
|---|---|---|---|---|---|
| Luminance | Pixel architecture | 3-T APS | 3-T APS | 4-T APS (PPD) | 5-T APS |
| | Process (CMOS) | 0.11μm (1P3M) | 0.18μm (1P6M) | 0.11μm (1P4M) | 0.35μm(2P4M) |
| | Array size | 2.1M | 400×384 | 1920×1080 (2D)<br>480×270 (3D) | 352×288 (2D)<br>88×72 (3D) |
| | Pixel size ( $\mu m^2$ ) | 7.6 | 56.3 | 13.3 (2D)<br>213.2 (3D) | 78.5 |
| | Fill factor (%) | N/A | 58 | 38.5 | 44 |
| | Frame rate(fps) | 60 | 200 | 30 | 30 |
| Depth | Technique | Lenticular lens | Light-field | Time of flight | Time of flight |
| | Measured range (m) | 0.2~1 | 1 | 0.75~4.5 | 1~3 |
| | Non-linearity (%) | ≤5 | 0.25 | 0.93 | N/A |
| | Frame rate (fps) | N/A | N/A | 11 | N/A |
| | Resolution | N/A | 2.5mm | 38mm | 40mm |
| | Calculation | Yes | Yes | No | No |

Note: PPD: Pinned Photo-Diode

**Table 2.** Specifications of currently available 3D image sensors and proposed 2D/3D-integrated image sensors.

## 7. Conclusion

This chapter presents a 2D/3D-integrated image sensor which includes photodiodes, pixel circuits, CDS circuits, sense amplifiers, a multi-channel TDC, readout circuits, row/column decoders, bus-sharing connections/decoders and so on. The luminance and depth information of a scene can be captured by the same pixel at a time-multiplexed manner. Based on the standard CMOS technology, the circular photodiode of P-diffusion_ N-well_ P-substrate is utilized thanks to good quantum efficiency, fair breakdown voltage and easy integration of 2D and 3D pixels. Particularly, the proposed integrated pixel can yield a high dynamic range at a 2D mode using a charge supply mechanism and a high response speed at a 3D mode using a feedback pull-down mechanism. Additionally, the bus-sharing mechanism is employed to diminish hardware cost of parallel reading. By using the TSMC 0.35μm 2P4M CMOS technology, the 352×288-pixel 2D and 88×72-pixel 3D integrated image sensor was designed to have a dynamic range up to 100dB and a depth resolution of around 4cm. The measured results reveal very promising performance. Therefore, the 2D/3D-integrated image sensor proposed herein can be widely applied to various multimedia capturing applications under low hardware cost and low power dissipation.

## Acknowledgements

This work was partially supported by National Science Council (NSC) and Ministry of Science and Technology (MOST), Taiwan, under the project numbers of NSC 97-2221-E-194-060-MY3, and MOST 102-2221-E-194 -047. The authors would like to thank Wei-Jean Liu, Meng-Lin Hsia, Zhe Ming Liu, Ming-Chih Lin, Chieh Ning Chan, Kuan-Hsien Lin, Shu Chun Wang and Hsiu-Fen Yeh who helped in designing, simulating, implementing and measuring the 2D and 3D image sensors. Additionally, Chip Implementation Center (CIC), Hsinchu, Taiwan, providing the services of VLSI CAD tools and chip fabrication is highly appreciated.

## Author details

Oscal T.-C. Chen*, Yi-Yang Lee and Robin R.-B. Sheen

*Address all correspondence to: oscal@ee.ccu.edu.tw

National Chung Cheng University Chiayi, Taiwan

## References

[1] Panasonic. Panasonic Product Support - DMC-3D1K. Available from: http://shop.panasonic.com/support-only/DMC-3D1K.html.

[2] Andersen MR, Jensen T, Lisouski P, Mortensen AK, Hansen MK, Gregersen T. Kinect depth sensor evaluation for computer vision applications. Aarhus University, Department of Engineering, 2012.

[3] Microsoft. Kinect for Windows features. Available from: http://www.microsoft.com/en-us/kinectforwindows/meetkinect/features.aspx.

[4] Charbon E. Introduction to time-of-flight imaging. In: Proc. of IEEE Sensors Conference; 2-5 Nov. 2014; Valencia. p. 610-613.

[5] Aull BF, Loomis AH, Young DJ, Heinrichs RM, Felton BJ, Daniels PJ. Geiger-mode avalanche photodiodes for three-dimensional imaging. Lincoln Laboratory Journal. 2002; 13(2):335-349.

[6] Niclass C, Rochas A, Besse P-A, Charbon E. Toward a 3-D camera based on single photon avalanche diodes. IEEE Journal of Selected Topics in Quantum Electronics. 2004; 10(4):796-802. DOI: 10.1109/JSTQE.2004.833886.

[7] Marshall GF, Jackson JC, Denton J, Hurley PK, Braddell O, Mathewson A. Avalanche photodiode-based active pixel imager. IEEE Trans. on Electron Devices. 2004; 51(3): 509-511. DOI: 10.1109/TED.2003.823051.

[8] Zappa F, Lotito A, Tisa S. Photon-counting chip for avalanche detectors. IEEE Photon Technology Letters. 2005;17(1):184-186. DOI: 10.1109/LPT.2004.838136.

[9] Faramarzpour N, Deen MJ, Shirani S, Qiyin F. Fully integrated single photon avalanche diode detector in standard CMOS 0.18-μm technology. IEEE Trans. on Electron Devices. 2008; 55(3):760-767. DOI: 10.1109/TED.2007.914839.

[10] Atef M, Polzer A, Zimmermann H. Avalanche double photodiode in 40-nm standard CMOS technology. IEEE Journal of Quantum Electronics. 2013; 49(3):350-356. DOI: 10.1109/JQE.2013.2246546.

[11] Pancheri L, Dalla Betta GF, Stoppa D. Low-noise avalanche photodiode with graded junction in 0.15-μm CMOS technology. IEEE Electron Device Letters. 2014; 35(5): 566-568. DOI: 10.1109/LED.2014.2312751.

[12] Kang Y, Mages P, Clawson AR, Yu PKL, Bitter M, Pan Z. Fused InGaAs-Si avalanche photodiodes with low-noise performances. IEEE Photonics Technology Letters. 2002; 14(11):1593-1595. DOI: 10.1109/LPT.2002.803370.

[13] Ong DSG, Jo Shien N, Yu Ling G, Chee Hing T, Shiyong Z, David JPR. InAlAs avalanche photodiode with type-II superlattice absorber for detection beyond 2-μm. IEEE Trans on Electron Devices. 2011; 58(2):486-489. DOI: 10.1109/TED.2010.2090352.

[14] Jun H, Banerjee K, Ghosh S, Hayat MM. Dual-carrier high-gain low-noise superlattice avalanche photodiodes. IEEE Transactions on Electron Devices. 2013; 60(7): 2296-2301. DOI: 10.1109/TED.2013.2264315.

[15] Zhen Guang S, Dun Jun C, Hai L, Rong Z, Da Peng C, Wen Jun L. High-gain AlGaN solar-blind avalanche photodiodes. IEEE Electron Device Letters. 2014; 35(3):372-374. DOI: 10.1109/LED.2013.2296658.

[16] Chan CN, Chen OTC. Physical effects of avalanche CMOS photodiodes. In: Proc. of OptoElectronics and Communications Conference; 5-9 July 2010. p. 824-825.

[17] He J, Xi X, Chan M, Hu C, Li Y, Zhang X. Equivalent junction method to predict 3-D effect of curved-abrupt p-n junctions. IEEE Transactions on Electron Devices. 2002; 4 9(7):1322-1325.

[18] Hsia M-L, Liu ZM, Chan CN, Chen OTC. Crosstalk effects of avalanche CMOS photodiodes. In: Proc. of IEEE Conference on Sensors; 28-31 Oct. 2011; Limerick, Ireland: IEEE; 2011. p. 1689-1692.

[19] Chen OTC, Lin K-H, Liu ZM. High-efficiency 3D CMOS image sensor. In: Proc. of 18th OptoElectronics and Communications Conference; June 30 - July 4, 2013; Kyoto, Japan: Optical Society of America; 2013.

[20] Liu ZM, Lin M-C, Chan CN, Chen OTC. 2D and 3D-integrated image sensor. In: Proc. of IEEE 53rd Midwest Symposium on Circuits and Systems; 1-4 Aug. 2010; Seattle, USA: IEEE; 2010. p. 292-295.

[21] Liu W-J, Yeh H-F, Chen OTC. A high dynamic-range CMOS image sensor with locally adjusting charge supply mechanism. In: Proc. of IEEE 48th Midwest Symposium on Circuits and Systems; 7-10 Aug. 2005: IEEE; 2005. p. 384-387.

[22] Cova S, Ghioni M, Lacaita A, Samori C, Zappa F. Avalanche photodiodes and quenching circuits for single-photon detection. Applied Optics. 1996; 35(12): 1956-1976.

[23] Niclass C, Favi C, Kluter T, Gersbach M, Charbon E. A 128 128 single-photon image sensor with column-level 10-bit time-to-digital converter array. IEEE Journal of Solid-State Circuits. 2008; 43(12):2977-2989.

[24] Gersbach M, Maruyama Y, Labonne E, Richardson J, Walker R, Grant L. A parallel 32× 32 time-to-digital converter array fabricated in a 130 nm imaging CMOS technology. In: Proc. of IEEE European Conference on Solid-State Circuits; 14-18 Sept. 2009; Athens: IEEE; 2009. p. 196-199.

[25] Jianjun Y, Fosterdai F, Jaeger R. 12-Bit vernier ring time-to-digital converter in 0.13 μm CMOS technology. IEEE Journal of Solid-State Circuits. 2010; 45(4):830-842.

[26] Chen OTC, Lin K-H, Liu ZM, Wang SC, Hsia M-L. 2D and 3D integrated image sensor with a bus-sharing mechanism. In: Proc. of IEEE 55th Midwest Symposium on Circuits and Systems; 5-8 Aug. 2012; Boise, Idaho, USA: IEEE; 2012. p. 138-141.

[27] Lim S, Cheon J, Chae Y, Jung W, Lee D-H, Kwon M. A 240-frames/s 2.1-Mpixel CMOS image sensor with column-shared cyclic ADCs. IEEE Journal of Solid-State Circuits. 2011; 46(9):2073-2083.

[28] Seo M-W, Suh S-H, Iida T, Takasawa T, Isobe K, Watanabe T. A low-noise high intrascene dynamic range CMOS image sensor with a 13 to 19b variable-resolution column-parallel folding-integration/cyclic ADC. IEEE Journal of Solid-State Circuits. 2012; 47(1):272-283.

[29] Xu R, Liu B, Yuan J. A 1500 fps highly sensitive 256 256 CMOS imaging sensor with in-pixel calibration. IEEE Journal of Solid-State Circuits. 2012; 47(6):1408-1418.

[30] Xu R, Ng WC, Yuan J, Yin S, Wei S. A 1/2.5 inch VGA 400 fps CMOS image sensor with high sensitivity for machine vision. IEEE Journal of Solid-State Circuits. 2014; 49(10):2342-2351.

[31] Sakakibara M, Oike Y, Takatsuka T, Kato A, Honda K, Taura T. An 83dB-dynamic-range single-exposure global-shutter CMOS image sensor with in-pixel dual storage. In: Technical Digest of IEEE International Solid-State Circuits Conference; 19-23 Feb. 2012; San Francisco, CA: IEEE; 2012. p. 380-382.

[32] Chung M-T, Lee C-L, Yin C, Hsieh C-C. A 0.5 V PWM CMOS imager with 82 dB dynamic range and 0.055% fixed-pattern-noise. IEEE Journal of Solid-State Circuits. 2013; 48(10):2522-2530.

[33] Yeh S-F, Hsieh C-C, Yeh K-Y. A 3 Megapixel 100 Fps 2.8 m pixel pitch CMOS image sensor layer with built-in self-test for 3D integrated magers. IEEE Journal of Solid-State Circuits. 2013; 48(3):839-849.

[34] Sukegawa S, Umebayashi T, Nakajima T, Kawanobe H, Koseki K, Hirota I. A 1/4-inch 8Mpixel back-illuminated stacked CMOS image sensor. In: Technical Digest of IEEE International Solid-State Circuits Conference; 17-21 Feb. 2013; San Francisco, CA: IEEE; 2013. p. 484-485.

[35] Koyama S, Onozawa K, Tanaka K, Kato Y. A 3D vision 2.1 Mpixel image sensor for single-lens camera systems. In: Technical Digest of IEEE International Solid-State Circuits Conference; 17-21 Feb. 2013; San Francisco, CA: IEEE; 2013. p. 492-493.

[36] Wang A, Molnar A. A light-field image sensor in 180 nm CMOS. IEEE Journal of Solid-State Circuits. 2012; 47(1):257-271.

[37] Kim S-J, Kang B, Kim JD, Lee K, Kim C-Y, Kim K. A 1920× 1080 3.65μm-pixel 2D/3D image sensor with split and binning pixel structure in 0.11 pm standard CMOS. In: Technical Digest of IEEE International Solid-State Circuits Conference; 19-23 Feb. 2012; San Francisco, CA: IEEE; 2012. p. 396-398.

Chapter 12

# The Theory of Giant Splash of Photoresponse in Semiconductors at Low-Level Illumination with Increasing Concentration of Deep Recombination Impurity

Viacheslav A. Kholodnov and Mikhail S. Nikitin

Additional information is available at the end of the chapter

http://dx.doi.org/10.5772/61028

**Abstract**

Recombination of excess (nonequilibrium) electrons and holes in semiconductors through impurity recombination centers (traps) known as trap-assisted (Shockley-Read-Hall) recombination is in many cases the dominant process. In this chapter, we develop the general theory of trap-assisted recombination and study in detail two key characteristics: (1) dependences of excess charge carriers' lifetime and photoelectric gain on concentration N of recombination centers and (2) effectiveness of band-to-band photoexcitation of charge carriers and photo-emf in semiconductors at low-level illumination considered outside quasi-neutrality approximation.

We have done systematic mathematical and detailed physical analysis of considered characteristics. Giant splash of photoresponse in semiconductors with increasing recombination center concentration N is caused mainly by the growth of charge carriers' lifetime in orders of magnitude. Also, this factor is sufficient to provide an increase, in order of magnitude and more, in efficiency of charge carriers' photoexcitation with increasing N. Results of strict analytical calculations (i.e., outside commonly used local approximation of quasi-neutrality) show, that, photoinduced local space charge affects substantially on giant splash of semiconductor photoelectric response with increasing concentration of recombination centers.

The theory of giant splash of photoresponse in semiconductors at low-level illumination with increasing concentration of recombination centers could develop further through generalization of boundary conditions on semiconductor surfaces and current contact electrodes, accounting for nonuniformity of charge carriers' photoexcitation along the line of current flow and fluctuation processes. The study of nonstationary (frequency domain and transient) characteristics is of particular interest.

It follows from physical essence of considered effects that similar effects can occur in other mediums with recombination of dissociative or ion-ion type, for example, in multicomponent plasma as well.

**Keywords:** Impurity recombination center (trap), Trap-assisted (Shockley-Read-Hall) recombination, Excess (nonequilibrium) charge carriers, Band-to-band photogeneration of excess charge carriers, Electron lifetime, Hole lifetime, Photoresponse, Photoelectric gain, Photo-emf, Small deviation from equilibrium state in semiconductor, Photoinduced space charge, Approximation of quasi-neutrality, Giant splash of photoresponse

## 1. Introduction

Recombination of excess (nonequilibrium) electrons and holes in semiconductors through impurity recombination centers (traps) referred to in this chapter as trap-assisted (Shockley-Read-Hall) recombination is in many cases the dominant process [1-9]. Hall [10] and Shockley and Read [11] have proposed the theory of trap-assisted recombination as early as 1952. Further, the theory of trap-assisted recombination has been developed in many aspects and details [1-5, 12-14]. At the same time, due attention was not paid to study dependences of lifetimes of excess electrons $\tau_n(N)$ and holes $\tau_p(N)$ on concentration of traps $N$. In some cases, traps are produced intentionally by doping semiconductor (e.g., by bombarding with high-energy ions [15, 16]) to reduce time of transient processes. It seems that lack of attention is caused by traditional understanding that the larger the concentration of traps $N$, the greater the capture rate of excess charge carriers on impurity level traps and, therefore, the shorter the lifetimes of excess charge carriers.

That reasonable understanding is incompletely adequate to reality. As shown below, lifetimes of excess electrons and holes (see our definition of $\tau_n$ and $\tau_p$ in Section 2) may grow strongly (in order of magnitude and more) with increase of concentration $N$.

In this chapter, we generalize the theory and give systematic mathematical and detailed physical analysis of dependences $\tau_n(N)$ and $\tau_p(N)$ on concentration of recombination centers.

## 2. Preliminaries

Let's consider the case of small deviation of free charge carriers' concentrations from equilibrium values. This situation occurs often in semiconductors used for registration of low-level signals, for example, optical signals. It will be shown that both $\tau_n(N)$ and $\tau_p(N)$, under certain conditions, will grow very strongly with increasing $N$ in a particular interval of $N$ values. Completely different physical mechanism causes this increase in lifetimes. It differs from mechanisms available for many years [4], as well as proposed in [17]. We analyze extreme points (corresponding formulas are derived) of dependences $\tau_n(N)$ and $\tau_p(N)$ as functions of semiconductor parameters and temperature. We give detailed physical interpretation of obtained results. In particular, it is shown that physical mechanisms responsible for strong non-monotonic dependences $\tau_n(N)$ and $\tau_p(N)$ differ from each other.

It is reasonable to expect that the growth of lifetimes $\tau_n(N)$ and $\tau_p(N)$ with increasing $N$ will lead to the growth of photoresponse of semiconductor sample (including photoelectric gain $G$). However, specificity of dependences $\tau_n(N)$ and $\tau_p(N)$ does not determine the type of dependence $G(N)$ in total. As it follows from [18, 19], $G$ increases with increasing charge carriers' lifetime, if ambipolar mobility [2, 13] $\mu_a=0$ or there is no recombination at current contact electrodes ($x=0$ and $x=W$ ; see insert in Figure 1a). In reality, recombination at current contact electrodes is always happening to some extent [5-9]. Therefore, in normal conditions ($\mu_a\neq 0$), increase in lifetimes, after reaching some values, does not lead to an increase in photocurrent $I_{ph}$ [5, 9, 18, 19]. Saturation of $I_{ph}$ becomes apparent in the case of high-rate recombination at the contact electrodes (sweep-out effect [5, 9, 18, 19]) when

$$\Delta n(0) = \Delta p(0) = \Delta n(W) = \Delta p(W) = 0, \tag{1}$$

where $\Delta n(x)=n(x)-n_e$ and $\Delta p(x)=p(x)-p_e$ are deviations of electron $n$ and hole $p$ concentrations from their equilibrium values $n_e$ and $p_e$, respectively. In trap-assisted recombination, function $\mu(N)$ can vanish at the same (up to small corrections) concentration $N$, at which dependences $\tau_n(N)$ and $\tau_p(N)$ reach points of maximal extremum $\hat{\tau}_n$ and $\hat{\tau}_p$ (Figure 1b and 1c). Therefore $G$ and, hence, $I_{ph}$ grow to the extent of increase in $\tau_n(N)$ and $\tau_p(N)$. These are physical grounds of giant splash of photoelectric gain with increasing $N$ (Figure 1a).

It was first reported in [20] that vanishing $\mu(N)$ in points of maximal extremum of dependences $\tau_n(N)$ and $\tau_p(N)$ allows avoiding highly undesirable effect – saturation of $G$ in intrinsic photoconductors, when applied bias voltage $V$ increases [5, 9, 18, 19, 21, 22]. As is known [19], this disadvantage is the most evident in photoconductors with sweep-out effect on contact electrodes, i.e., when relations (1) are fulfilled. Result presented in Ref. [20] was obtained in approximation of quasi-neutrality [1-9, 13, 18, 19, 21, 22], which was usually used at moderate electric fields, i.e., when we neglect in Poisson equation by term $\Delta\rho \equiv (\varepsilon \times \varepsilon_0) \times div \Delta E$, which determines the density of photoinduced space charge $\Delta\rho$.

In our case, $\Delta E \equiv E(x) - E_0$ is the variation of electric field caused by deviation of concentrations of free charge carriers and their traps from equilibrium values by reason of band-to-band absorption of radiation, $E(x)$ and $E_0$ are the electric field intensity in the presence and absence of illumination, $\varepsilon$ is the relative dielectric permittivity of semiconductor, and $\varepsilon_0$ is the vacuum permittivity. However, even at moderate electric fields (≈ 1÷10 V/cm), approximation of quasi-neutrality is not always acceptable [23].

Below, in case of single recombination level, we consider in detail the impact of photoinduced space charge $\Delta\rho$ on value $\hat{G}$ of photoelectric gain $G$ in semiconductors with sweep-out effect on contact electrodes at the point of maximal extremum of function $G(N)$ (Figure 1a). Considering semiconductor as base material for making intrinsic photoconductors with threshold electro-optical performance, we assume that photocarriers are excited by photons of low-intensity optical radiation with wavelength range responding to fundamental absorption band of semiconductor. Figure 2a shows that we cannot use approximation of quasi-neutrality, when voltage $V$ across the sample becomes larger than some particular value.

Also, ignoring approximation of quasi-neutrality, we study, at low-level illumination, the effectiveness of band-to-band photoexcitation of charge carriers and photo-emf $V_{ph}$ in semiconductors with dominant trap-assisted recombination. Analytical expressions for photo-emf $V_{ph}$ and mean, with respect to light propagation length, concentrations of photoelectrons $<\Delta n>$ and photoholes $<\Delta p>$ are given. It is shown that target values of $V_{ph}$, $<\Delta n>$, and $<\Delta p>$ can be improved radically by increasing concentration of recombination centers; moreover, approximation of quasi-neutrality can lead to errors of several orders of magnitude.

Analyzing above-mentioned problems, we do not use conventional (Shockley-Read) expression-based form [1-5, 9, 11-14] of generation-recombination rate. This form does not allow to express explicitly dependences $\tau_n$ and $\tau_p$ and, even more, $I_{ph}$ and $V_{ph}$ on $N$. And therefore, because of the need for solving complex transcendental equations, conventional (Shockley-Read) expression-based form leads to serious mathematical difficulties in study (especially analytical) dependences $\tau_{n,p}(N)$ and, even more, $I_{ph}(N)$ and $V_{ph}(N)$.

These difficulties are dramatized by the fact that under certain conditions, $\tau_{n,p}$, $I_{ph}$ and $V_{ph}$ are dependent very strongly on concentration $N$ in a particular interval of $N$ (Figure 1a, 1b, 7). Perhaps, it was the main reason for longtime absence of complete theoretical analysis of lifetime dependences $\tau_{n,p}(N)$, while detailed analysis of lifetime dependences on concentrations of free charge carriers was made in pioneering paper by Shockley and Read [11] concerning the theory of recombination through impurity level trap. In present chapter, we use the method of calculation assuming that $N$ is expressed in terms of the ratio of the number of filled recombination level states to the number of empty. This allows to avoid transcendental equations, in other words, to avoid the need to solve inverse problem. As a result, the described above approach simplifies greatly the mathematical analysis and physical interpretation of calculations of desired parameters.

## 3. Model and basic relations

Consider nondegenerated semiconductor doped by shallow fully ionized single type impurity (for definition donors) with concentration $N_D$. Recombination of excess charge carriers occurs in said semiconductor through the energy level of acceptor impurity atoms with concentration $N$, which can be in two charge states (assume in neutral and singly negatively charged). Concentration of recombination impurity atoms in neutral state corresponds to concentration of acceptor atoms $N_0$, which are simultaneously centers of electron capture and centers of thermal emission of holes. Concentration of recombination impurity atoms in negatively charged state corresponds to concentration of atoms $N_- = N - N_0$, which are capture centers of holes and, at the same time, centers of thermal emission of electrons. Described above is recombination through single-level trap [10-13] (Figure 5b), which is often dominant [1-5, 9, 14] and called Shockley-Read-Hall recombination.

http://dx.doi.org/10.5772/61028

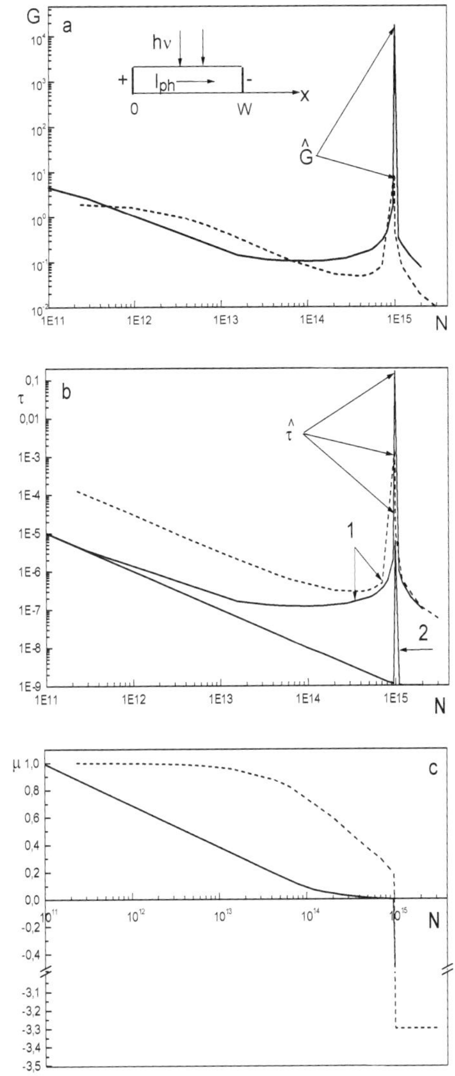


**Figure 1.** Dependences on concentration of single-level recombination centers $N$ (cm$^{-3}$): **(a)** $G$– photoelectric gain; **(b)** $\tau$– lifetime of electrons (curve 1) and holes (curve 2) (s); **(c)** $\mu$– ambipolar mobility of charge carriers (in units of hole mobility). Adopted: room temperature, $W=10^{-1}$cm, $\theta\equiv w_p/w_n=10^2$, $w_n=10^{-8}$cm³/s, $n_i/n_t=10^4$, $N_D=10^{15}$cm$^{-3}$, $E_0$ =10V/cm. Solid curves, $GaAs$, and dashed curves, $Si$. Physical parameters of semiconductors are obtained from monograph [3]. Schematic view of photoconductor on insert in Figure 1a

Let either band-to-band excitation (Figure 5b) or injection on the contacts produce excess electrons and holes. Then, in stationary case, equation

$$R_n = R_p \tag{2}$$

determines the charge state of recombination impurity atoms.

Recombination-generation rates of electrons $R_n$ and holes $R_p$ due to capture of charge carriers by acceptor impurity traps and their thermal emission from recombination level states into permitted conduction or valence bands are equal to

$$R_n = (n \times N_0 - \delta^{-1} \times n_e \times N_-) \times w_n, R_p = (p \times N_- - \delta \times p_e \times N_0) \times w_p. \tag{3}$$

Here, $w_n$ and $w_p$. are electron and hole capture probabilities, respectively, at appropriate recombination level state, $\delta = N_-^e / N_0^e$ (superscript indicates equilibrium values of concentration of recombination impurity atoms $N$ in relevant charge states).

For small deviation of charge carriers' and their capture centers' concentrations $\Delta N_0 = N_0 - N_0^e = -\Delta N_- = N_-^e - N_-$ from equilibrium values, we can linearize relations (2) and (3) with respect to proper deviations. Then taking into account Poisson equation

$$\Delta\rho \equiv \frac{\varepsilon}{4\pi} \times \frac{\partial \Delta E}{\partial x} = q \times \left[\Delta p - \Delta n - \Delta N_-\right], \tag{4}$$

we obtain

$$R_n = \frac{\Delta n}{\tau_n} + a_n \times div\Delta E, \tag{5}$$

$$R_p = \frac{\Delta p}{\tau_p} - a_p \times div\Delta E, \tag{6}$$

$$\Delta p = \frac{\tau_p}{\tau_n} \times \Delta n + (a_n + a_p) \times \tau_p \times \frac{d\Delta E}{dx}, \tag{7}$$

where

$$\frac{1}{\tau_n} = w_n \times N \times \frac{\delta \times \theta}{1+\delta} \times \frac{N + (1+\delta) \times (1+\delta^{-1}) \times (n_e + p_e)}{\delta \times \theta \times N + (1+\delta) \times (1+\delta^{-1}) \times (n_e + \delta \times \theta \times p_e)}, \tag{8}$$

$$\frac{1}{\tau_p} = w_p \times N \times \frac{\delta}{1+\delta} \times \frac{\delta \times N + (1+\delta)^2 \times (n_e + p_e)}{\delta \times N + (1+\delta^2) \times (n_e + \delta \times \theta \times p_e)}, \tag{9}$$

$$a_n = \frac{\varepsilon}{4\pi \times q} \times w_n \times N \times n_e \times \frac{\theta \times (1+\delta)}{\delta \times \theta \times N + (1+\delta) \times (1+\delta^{-1}) \times (n_e + \delta \times \theta \times p_e)}, \tag{10}$$

$$a_p = \frac{\varepsilon}{4\pi \times q} \times w_p \times N \times p_e \times \frac{1+\delta}{N + (1+\delta) \times (1+\delta^{-1})(n_e + \delta \times \theta \times p_e)}, \tag{11}$$

$\Delta E$ is change in electric field caused by deviation of charge carriers' and capture centers' concentrations from equilibrium values, and $q$ is absolute electron charge value and $\theta = w_p / w_n$. First terms in (5) and (6) mean recombination rates of excess electrons and holes (and therefore, symbols $\tau_n$ and $\tau_p$ mean their lifetimes) in quasi-neutrality with respect to electric field $\Delta E$, i.e., at sufficiently small values $div\Delta E$ [1-3, 5, 9-14, 18, 24, 25]. We will use the same terminology for $\tau_n$ and $\tau_p$ in the case of failure to comply quasi-neutrality (see below); therefore values $\tau_n$ and $\tau_p$ will not depend on value $div\Delta E$ in the present study.

High-performance photoconductors operate with extremely low-level illumination. Therefore, linear for $g$ approximation, usually used in the theory of high-performance photodetectors [5-7, 9, 21, 22, 26], is correct in calculation of photoelectric gain $G$, where $g$ is density of charge carriers' photoexcitation rate.

In view of the above provision, we can write expressions for the density of photocurrent components as follows:

$$\Delta I_n = q \times \mu_n \times (E_0 \times \Delta n + n_e \times \Delta E) + q \times D_n \times \frac{\partial \Delta n}{\partial x}, \tag{12}$$

$$\Delta I_p = q \times \mu_p \times (E_0 \times \Delta p + p_e \times \Delta E) - q \times D_p \times \frac{\partial \Delta p}{\partial x}. \tag{13}$$

where $\mu_n$ and $\mu_p$ are electron and hole mobility and $D_n$ and $D_p$ are electron and hole diffusion constants. The density of electron $\Delta I_n$ and hole $\Delta I_p$ components of photocurrent

$$I_{ph} = \Delta I_n + \Delta I_p \tag{14}$$

must satisfy continuity equations:

$$\frac{\partial \Delta I_n}{\partial x} = q \times (R_n - g), \tag{15}$$

$$\frac{\partial \Delta I_p}{\partial x} = q \times (g - R_p), \tag{16}$$

and also

$$\frac{\partial I_{ph}}{\partial x} = 0. \tag{17}$$

Let limit voltage be applied to sample by value

$$V = E_0 \times W \tag{18}$$

that allows to neglect by the dependence of $\mu_n$ and $\mu_p$ on electric field, where $W$ is distance between current contact electrodes (see insert in Figure 1a).

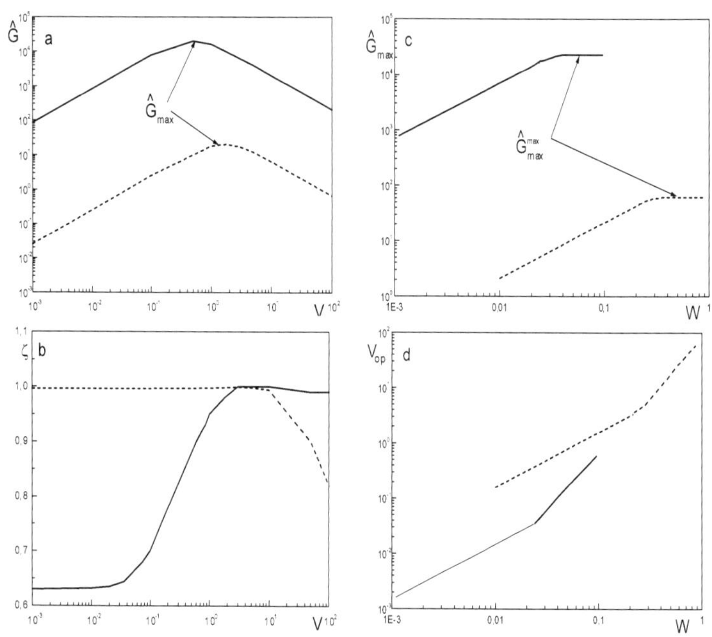

**Figure 2.** Dependences: **(a)**, photoelectric gain $G = \hat{G}$ in point of maximal extremum of function $G(\hat{N})$ (see Figure 1a) on bias voltage across the sample $V$ (distance between current contacts $W = 10^{-1}$ cm); **(b)**, ratio $\zeta \equiv \hat{G}_{appx} / \hat{G}_{exact}$ on $V$ at $W = 10^{-1}$ cm, where $\hat{G}_{appx}$ and $\hat{G}_{exact}$ are approximate and exact values $\hat{G}$, respectively; **(c)**, value $\hat{G}_{max}$ on $W$, where $\hat{G}_{max}$ is maximal value $\hat{G}$ for given $W$ (see Figure 2a); **(d)**, value $V_{op}$ on $W$, where $V_{op}$ is optimal voltage, at which $\hat{G} = \hat{G}_{max}$ (see Figure 2c). Voltage $V$ in V; length $W$ in cm. Physical parameters of semiconductors and temperature are the same as in Figure 1. Solid curves $GaAs$, dashed curves $Si$

## 4. Lifetime of excess charge carriers

Using distribution function of electrons over acceptor level states [12, 27], we can write neutrality equation

$$n_e + N_-^e = p_e + N_D \tag{19}$$

for nondegenerated semiconductor at thermodynamic equilibrium as follows:

$$N = n_t \times \frac{1+\delta}{2\times\delta^2} \times f(\delta), \tag{20}$$

where

$$\delta = \frac{N_-^e}{N_0^e}, f(\delta) = B + A\times\delta - \delta^2, \tag{21}$$

$$A = 2\times\frac{N_D}{n_t}, B = 4\times\frac{p_t}{n_t} = \left(\frac{2\times n_i}{n_t}\right)^2, \tag{22}$$

$n_t$ and $p_t$ are equilibrium concentrations of electrons and holes when Fermi level energy coincides with recombination level energy $E_t$, and $n_i$ is intrinsic charge carriers' concentration. When derived (20), we have adopted that spin degeneracy factor of acceptor state is equal to 1/2 [2, 12, 14, 27]. In considered conditions

$$n_e = \frac{\delta}{2}\times n_t, p_e = \frac{2}{\delta}\times p_t. \tag{23}$$

From expressions (8), (9), (20), and (23), it follows that

$$\tau_n = \frac{2\delta^2}{f(\delta)} \times \frac{\theta\times f(\delta) + (1+\delta)\times(\theta\times B + \delta)}{[\delta\times A + (2+\delta)\times B + \delta^3]w_p \times n_t}, \tag{24}$$

$$\tau_p = \frac{2\delta}{f(\delta)} \times \frac{B + (A+\theta\times B)\times\delta + (\theta\times B + \delta)\times\delta^2}{[\delta\times A + (2+\delta)\times B + \delta^3]w_p \times n_t}. \tag{25}$$

Expressions (20), (24), and (25) determine dependences $\tau_n(N)$ and $\tau_p(N)$ in parametric form. Figure 1b shows that, as usual, dependences $\tau_n(N)$ and $\tau_p(N)$ fall with increased $N$, but in some interval of concentration $N$, dependences can rise up sharply. Further, we give analytical solution of extremum problem for dependences $\tau_n(N)$ and $\tau_p(N)$ at $\theta \geq 1$, because hole is captured on attractive center and electron – on neutral.

### 4.1. Mathematical analysis of hole lifetime

The analysis of equation

$$\frac{\partial}{\partial \delta} \times \frac{1}{\tau_p} = 0, \tag{26}$$

which determines extremum points of dependence $\tau_p(N)$, shows that well-defined non-monotonic behavior of this function occurs at

$$\xi_1 \equiv \frac{3}{\theta \times B} << 1 << \frac{1}{\xi_2} \equiv \frac{A^2}{4 \times B}, \xi_3 \equiv \frac{1}{\sqrt{A}} << 1. \tag{27}$$

To determine minimum point, let us set out equation (26) into the form

$$\delta^2 + 2 \times \theta \times B \times \delta - \theta \times A \times B \times [1 + \Lambda_{1p}(\delta)] = 0, \tag{28}$$

where absolute value of function

$$\begin{aligned} \Lambda_{1p}(\delta) = & \frac{B + (A + \theta \times B) \times \delta}{\theta \times A \times B \times \delta} - \frac{B}{A \times \delta} + \frac{B \times \delta^3 + (\delta^2 + B)[2 \times B + (A + B) \times \delta]}{\theta \times A \times B \times \delta^6} \times \\ & \times [B + (A + \theta \times B) \times \delta + \theta \times B \times \delta^2 + \delta^3] + \frac{B + A \times \delta - \delta^2}{\theta \times A \times \delta^5} \times \\ & \times \{A + (2 \times \theta - 1) \times B + 4 \times \theta \times B \times \delta + [3 + (A + B) \times \theta] \times \delta^2 - 2(\theta - 1) \times \delta^3\} \end{aligned} \tag{29}$$

is much less than unity at

$$\delta = \delta_{1p}^{(0)} \equiv -\theta \times B + \sqrt{\theta \times B \times (A + \theta \times B)}. \tag{30}$$

It means that the first root of equation (26) $\delta_{1p}$ can be found by the method of successive iterations using $\Lambda_{1p}(\delta_{1p})$ as small parameter. Zeroth-order approximation (30) leads to the formula for concentration of recombination centers $N = N_{1p}$, where dependence $\tau_p(N)$ reaches its minimum $\breve{\tau}_p$ (Figure 1b):

$$N_{1p} = N_D + 2\times\theta\times p_t - \sqrt{2\times\theta\times p_t\times(N_D + 2\times\theta\times p_t)}. \tag{31}$$

It follows from this formula that ratio $N_{1p}/N_D$ increases from ½ when $N_D << 2\times\theta\times p_t$ to $(1-\sqrt{2\times\theta\times p_t})\cong 1$ when $N_D >> 2\times\theta\times p_t$ with increased $N_D$ (Figure 3a). Further, it will be demonstrated that extremum, like the maximum of dependence $\tau_p(N)$ as well as $\tau_n(N)$, can occur only at values $N$ closer to $N_D$. Therefore, in expression for upper limit of value $N_{1p}$, small correction has been remained, which is primal.

To determine maximum point of dependence $\tau_p(N)$, let us set out equation (26) into the form

$$[1 - \Lambda_{2p}(\delta)]\times\delta^2 = A + B, \tag{32}$$

where absolute value of function

$$\Lambda_{2p}(\delta) = \frac{1}{\theta\times\delta^4}\times[A + (2\times\theta - 1)\times B + 3\delta^2 - 2\times(\theta - 1)\times\delta^3] + \\ + \frac{B+\delta^2}{\theta\times B\times\delta^5}\times[2\times B + (A+B)\times\delta]\times\left[1 + \frac{\delta\times(\delta+1)\times(\theta\times B+\delta)}{B + A\times\delta - \delta^2}\right] \tag{33}$$

is much less than unity at

$$\delta = \delta_{2p}^{(0)} = \sqrt{A+B}. \tag{34}$$

It means that the second root of equation (26) $\delta_{2p}$ can be found by the method of successive iterations using $\Lambda_{2p}(\delta_{2p})$ as small parameter. It follows from relations (20) and (34) that concentration of recombination centers $N = N_{2p}$, where dependence $\tau_p(N)$ reaches its maximum $\hat{\tau}_p$ (Figure 1b), is determined in first-order approximation for small parameter $\Lambda_{2p}(\delta_{2p})$ by expression

$$\frac{N_{2p}}{N_D} = 1 - \frac{1}{A+B} - \frac{\sqrt{A+B}}{A}\times\Lambda_{2p}\times\left(\sqrt{A+B}\right)\cong 1. \tag{35}$$

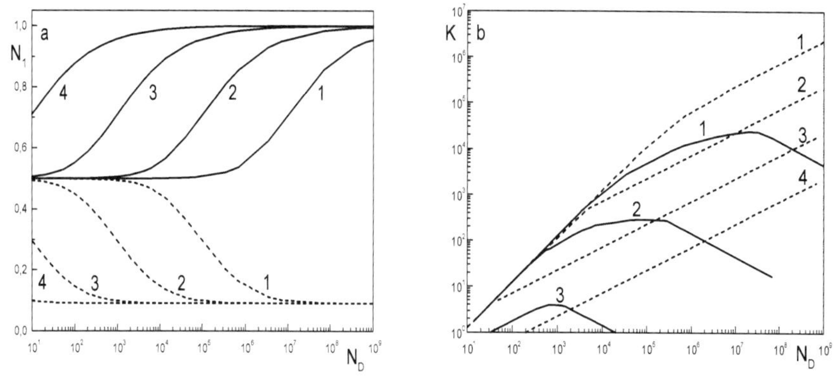


**Figure 3.** Dependences: point of minimum $N_1$ of functions $\tau_p(N)$ и $\tau_n(N)$ **(a)** and ratio $K$ between maximal to minimal charge carriers' lifetimes **(b)** on concentration of shallow donors $N_D$ at different locations of recombination level. Solid curves, holes; dashed curves, electrons. Ratio $n_t / n_i$ values: curve 1, $10^{-4}$; curve 2, $10^{-2}$; curve 3, 1; curve 4, $10^2$. Concentration $N_1$ is measured in units $N_D$, concentration $N_D$ in units $n_i$. Adopted: $\theta = 10^2$

In particular

$$\frac{N_{2p}}{N_D} = \begin{cases} 1 - \dfrac{n_t}{2\times N_D} \times \left(1 + \dfrac{2}{\theta} + \dfrac{\sqrt{2\times N_D \times n_t}}{\theta \times p_t}\right), & for \quad N_D >> 2\times p_t \\ 1 - \left(\dfrac{2\times n_i}{N_D}\right)^2, & for \quad 2n_i << N_D << 2p_t \end{cases} . \tag{36}$$

It follows from (22), (25), (27), (30), and (34) that

$$K_p(N_D) \equiv \frac{\hat{\tau}_p}{\breve{\tau}_p} = \begin{cases} \dfrac{N_D}{8\times n_i}, & for \quad 8\times n_i << N_D << 2\times p_t \\ \sqrt{\dfrac{N_D}{32\times n_i}}, & for \quad N_D << 2\theta \times p_t << \theta \times N_D \\ \dfrac{\sqrt{2}\times \theta \times p_t}{\sqrt{N_D \times n_t}}, & for \quad \sqrt{2\times N_D \times n_t} << 2\times \theta \times p_t << N_D \\ \approx 1, & for \quad 2\times \theta \times p_t << \sqrt{2\times N_D \times n_t} \end{cases} . \tag{37}$$

Relation (37) shows that function $K_p(N_D)$ is non-monotonic and can vary by several orders of magnitude (Figure 3b). Value

$$\hat{\tau}_p = \left(1 + \frac{\sqrt{2} \times \theta \times p_t}{\sqrt{N_D \times n_t + 2 \times n_i^2}}\right) \times \left(N_D \times w_p\right)^{-1} \tag{38}$$

increases with decreasing recombination level energy $E_t$ (Figure 4a).

The dependence of $\hat{\tau}_p$ on temperature $T$ is determined by the location of recombination level in forbidden gap of semiconductor (Figure 4b). Value $\hat{\tau}_p$ decreases with lowering temperature if $E_t \geq E_g/2$ and increases if $E_t \leq E_g/3$ (value $E_t$ is measured from top of valence band and $E_g$ is the energy gap of semiconductor). If $E_g/3 < E_t < E_g/2$, then dependence $\hat{\tau}_p$ has maximum value at

$$T = \tilde{T} \equiv \frac{E_t}{k} \times \ln^{-1}\left(2 \times \frac{N_v}{N_D} \times \frac{E_g - 2 \times E_t}{3 \times E_t - E_g}\right), \tag{39}$$

where $k$ is Boltzmann constant and $N_v$ is effective density of states in valance band.

### 4.2. Mathematical analysis of electron lifetime

The analysis of equation

$$\frac{\partial}{\partial \delta} \times \frac{1}{\tau_n} = 0, \tag{40}$$

which determines extremum points of dependence $\tau_n(N)$, shows that well-defined non-monotonic behavior of this function occurs at

$$2 \times \xi_3 << 1, \xi_4 = 2 \times \frac{\sqrt{B}}{A} << 1. \tag{41}$$

To determine minimum point let us set out equation (40) into the form

$$(\theta - 1) \times \delta^2 - 2 \times \theta \times (A + B) \times \delta + \theta \times A \times (A + B) \times [1 + \Lambda_{1n}(\delta)] = 0, \tag{42}$$

where absolute value of function

$$\begin{aligned} \Lambda_{1n}(\delta) = & \frac{(\theta - 1) \times B}{\theta \times A \times (A + B)} - \frac{2 \times B + A \times \delta}{\theta \times A \times (A + B) \times \delta} \times \\ & \times \left[\left(1 + \frac{2 \times \theta \times B}{\delta}\right) \times \left(1 + \frac{A + B}{\delta^2} + \frac{2 \times B}{\delta^3}\right) + \frac{1}{\delta} \times \left(A + B + \frac{2 \times B}{\delta}\right) \times \left(\theta \times \frac{A + B}{\delta} - \theta + 1\right)\right] + \\ & + \frac{B + A \times \delta - \delta^2}{\theta \times A \times (A + B) \times \delta}\left[2 + (\theta - 1) \times \frac{A + B}{\delta} + \frac{6 \times \theta \times B}{\delta} + \frac{2 \times B}{\delta^2} \times \left(2 \times \theta - 2 - \frac{1}{\delta}\right)\right] \end{aligned} \tag{43}$$

is much less than unity at

$$\delta=\delta_{1n}^{(0)}\equiv\frac{\sqrt{\theta\times(A+B)}}{\theta-1}\times\left(\sqrt{\theta\times(A+B)}-\sqrt{A+\theta\times B}\right). \tag{44}$$

It means that the first root of equation (40) $\delta_{1n}$ can be found by the method of successive iterations using $\Lambda_{1n}(\delta_{1n})$ as small parameter. Zeroth-order approximation (44) leads to the formula for concentration of recombination centers $N=N_{1n}$ where dependence $\tau_n(N)$ reaches its minimum $\breve{\tau}_n$ :

$$N_{1n}=\frac{\sqrt{N_D+2\times\theta\times p_t}}{\theta-1}\times\left(\sqrt{\theta\times(N_D+2\times p_t)}-\sqrt{N_D+2\times\theta\times p_t}\right). \tag{45}$$

It follows, from this formula, that ratio $N_{1n}/N_D$ decreases from ½ when $N_D<<2\times\theta\times p_t$ to $1/(\sqrt{\theta}+1)$ when $N_D>>2\times\theta\times p_t$ with increased $N_D$ (Figure 3a).

We can transform equation (40) to form (32) where $\Lambda_{2p}(\delta)$ will be replaced by function:

$$\begin{aligned}\Lambda_{2n}(\delta)=&\frac{2\times\delta}{A}+\frac{2\times B}{A}\times\frac{\theta\times(A+B)\times\delta+2\times\theta\times B+\delta}{\theta\times(A+B)\times\delta^4}\times\left(2\times A+B+\frac{2\times B}{\delta}\right)+(2\times\theta\times B+\delta)\times\\&\times\frac{A\times(A+B+\delta^2)+2\times B\times\delta}{\theta\times A\times(A+B)\times\delta^3}-2\times\frac{B+A\times\delta-\delta^2}{\theta\times A\times(A+B)\times\delta^4}\times(\delta^3+3\times\theta\times B\times\delta^2-2\times B)-(\theta-1)\times\\&\times\left[\frac{2\times B\times[4\times B+(4\times A+B)\times\delta+(\delta-2)\times\delta^2]+2\times A\times(A+B)\times\delta^2+\delta\times(B-\delta^2)\times(A+B-\delta^2)}{\theta\times A\times(A+B)\times\delta^4}\right].\end{aligned} \tag{46}$$

At value $\delta=\delta_{2n}$, where dependence $\tau_n(N)$ has its maximum $\hat{\tau}_n$, absolute value $|\Lambda_{2n}(\delta)|<<1$. Therefore, in zeroth-order approximation for small parameter $\Lambda_{2n}(\delta_{2n})$, value $\delta_{2n}=\delta_{2p}^{(0)}$ and concentration $N=N_{1n}$, where $\tau_n(N)=\hat{\tau}_n$ equals to $N_D$ (as for holes). And

$$K_n(N_D)\equiv\frac{\hat{\tau}_n}{\breve{\tau}_n}=\begin{cases}\dfrac{N_D}{8\times n_i}, & for\quad 2\times n_i<<N_D<<2\times p_t\\ \dfrac{1}{\kappa}\times\sqrt{\dfrac{N_D}{2\times n_t}}, & for\quad N_D>>2\times p_t\end{cases} \tag{47}$$

where $\kappa=4$ at $\theta=1$ and $\kappa=1$ at $\theta>>1$. It follows from relation (47) that function $K_n(N_D)$, in contrast to $K_p(N_D)$, grows monotonically with increased $N_D$ and this growth can be many orders of magnitude (Figure 4b).

Value

$$\hat{\tau}_n = \sqrt{\frac{N_D + 2 \times p_t}{2 \times n_i}} \times \left(N_D \times w_n\right)^{-1} \tag{48}$$

increases, as for $\hat{\tau}_p$, with the decrease of recombination level energy (Figure 4a) and, in contrast to $\hat{\tau}_p$, always falls with temperature rise (Figure 4b).

### 4.3. Physical interpretation

Let's explain physical mechanisms of the above regularities.

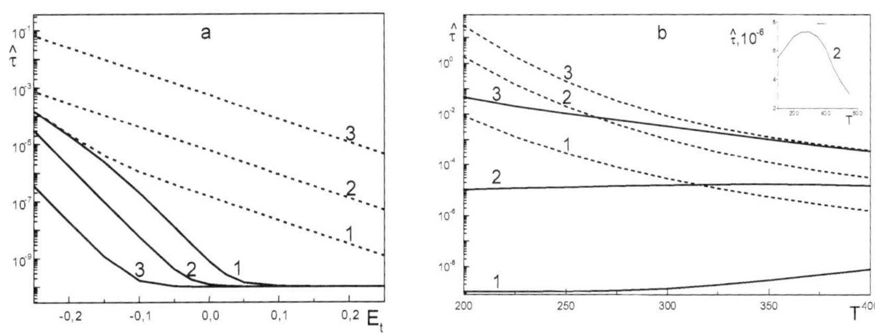


**Figure 4.** Dependences of maximal lifetime of holes $\hat{\tau}_p(N)$ (solid lines) and electrons $\hat{\tau}_n(N)$ (dashed lines) on recombination level energy $E_t$ (eV) for $Ge$ (1), $Si$(2), and $GaAs$ (3) at $T = 300$ K **(a)** and on temperature $T$ (K) for $Si$ at different values $E_t$ **(b)**: curve 1, 0; curve 2, (-101/152); curve 3, (-3/4). Recombination level energy is reckoned from the middle of forbidden gap. Physical parameters for $Ge$, $Si$, and $GaAs$ are obtained from monograph [3]. Adopted: $N_D$= $10^{15}$ cm$^{-3}$, $w_n = 10^{-8}$cm$^3$/s, $\theta = 10^2$

#### 4.3.1. *Hole lifetime*

Reciprocal hole lifetime

$$\tau_p^{-1} = \tau_{p1}^{-1} + \tau_{p2}^{-1} + \tau_{p3}^{-1} \tag{49}$$

consists of three partial components.

First component

$$\tau_{p1}^{-1} = w_p \times N_-^e = \frac{w_p \times n_t}{2} \times \frac{B + A \times \delta - \delta^2}{\delta} \tag{50}$$

corresponds to the change of capture rate of holes $\Delta p \times w_p \times N_-^e = \Delta p / \tau_{p1}$ caused only by deviation of hole concentration from its equilibrium value (capture of excess holes $\Delta p$ at equilibrium trapping centers $N_-^e$).

Second component

$$\tau_{p2}^{-1} = -w_p \times p_t \times \frac{2}{\delta} \times \frac{(\delta \times \theta - 1) \times (B + A \times \delta - \delta^2)}{A \times \delta + B \times [1 + \delta \times \theta \times (1 + \delta)] + \delta^3} \tag{51}$$

corresponds to the change of capture rate of holes $p_e \times w_p \times \Delta N_- = \Delta p / \tau_{p2}$ caused only by deviation of concentration of hole trapping centers from its equilibrium value (capture of equilibrium holes $p_e$ at nonequilibrium trapping centers $\Delta N_-$).

Third component

$$\tau_{p3}^{-1} = \delta \times \tau_{p2}^{-1} \tag{52}$$

corresponds to the change of thermal emission rate of holes from impurity level states into valence band $2 \times p_t \times w_p \times \Delta N_0 = \Delta p / \tau_{p3}$ caused by deviation of concentration of hole generation centers from its equilibrium value (thermal emission of holes from nonequilibrium centers $\Delta N_0$).

Lifetime $\tau_{p1}$ can be interpreted as capture time of excess holes by equilibrium traps, lifetime $\tau_{p2}$ can be interpreted as relaxation time of excess holes due to capture of equilibrium holes by nonequilibrium traps, and lifetime $\tau_{p3}$ can be interpreted as time of thermal emission of holes from nonequilibrium centers.

If conditions (27) are fulfilled and $N < N_D$, then recombination centers are almost completely filled with electrons ($\delta \equiv N_-^e / N_0^e > 1$). For this reason, even if $\theta \equiv w_p / w_n = 1$, capture time of hole $\tau_{p1}$ is much shorter than capture time of electron $\tau_{n1}$ for the relevant equilibrium trapping centers. In other words, equilibrium traps capture holes much more intensively than electrons. Therefore, the generation of excess free charge carriers initiates the formation of additional nonequilibrium centers of thermal generation of holes and, simultaneously, reducing concentration of trapping centers of electrons ($\Delta N_0 = -\Delta N_- > 0$). This change of charge state of recombination impurity atoms results in negative values of components $1/\tau_{p2}$ and $1/\tau_{p3}$ in expression (36); moreover $|\tau_{p3}| << |\tau_{p2}|$, because $\delta >> 1$ at $N < N_D$. This means that hole lifetime $\tau_p$ exceeds capture time of holes $\tau_{p1}$ at equilibrium traps due to dominating thermal emission of holes from relevant nonequilibrium centers. As long as $N < N_D$, concentration $N_-^e \cong N$ of equilibrium capture centers of holes grows with increased $N$, but concentration $N_0^e << N$ of capture centers of electrons still remains low. Therefore, concentration $\Delta N_0$ of nonequilibrium centers of hole thermal emission increases as well. This increase causes faster decreasing $|\tau_{p3}|$ than decreasing $\tau_{p1}$. As a result, starting with concentration $N = N_{1p}$, the rate of hole thermal emission from nonequilibrium centers and the capture rate of nonequilibrium holes at equilibrium traps become closer to each other. For this reason, $\tau_p$ starts to grow (Figure 1b).

When $N$ becomes larger than $N_D$, the concentration $N_-^e$ of equilibrium hole capture centers practically stabilizes, while concentration $N_0^e$ of equilibrium electron capture centers grows with the increase of recombination centers' concentration ($N_-^e \cong N_D, N_0^e \cong N - N_D$ at $n_t/2 << N - N_D << N_D^2/2p_t$). This means that the ratio $\tau_{p1}/\tau_{n1}$ increases with increasing $N$. For this reason, concentration of nonequilibrium hole thermal emission centers decreases, and concentration of hole traps $N_-$ increases. As a result, $\tau_p(N)$ decreases with increased $N$ (Figure 1b). When $N$ prevails $N_D^2/2p_t$, the concentration $N_-^e$ of equilibrium hole capture centers grows again with increased $N$ due to thermal emission of electrons from valence band to impurity level ($N_-^e \cong \sqrt{2N \times p_t}$). However, concentration of equilibrium capture centers of electrons grows much faster ($N_0^e \cong N$). Therefore, the decrease of $\tau_p(N)$ continues. As it is seen from (36)-(39), $\tau_p$ becomes less than $\tau_{p1}$, when product $\delta \times \theta$ becomes less than unity.

As shown above, minimum point $N = N_{1p}$ of dependence $\tau_p(N)$ shifts toward $N_D$ with growth $N_D$ (Figure 3a). The main reason is that equilibrium electrons are being captured at centers of hole thermal emission and decreased concentration $\Delta N_0$ of these centers. The higher the concentration of equilibrium electrons $n_e$, the more $\Delta N_0$ decreases. Concentration $n_e$ grows with increased $N_D$. When $N$ ascends, then $n_e$ descends and $N_-^e$ increases that causes increased $\Delta N_0$ at $N < N_D$. In other words, decreased $\Delta N_0$ with increased $N_D$ is compensated by increased $\Delta N_0$ with increased $N$. This is the reason why the greater the $N_D$, the closer the $N_{1p}$ to $N_D$.

For the same reasons, non-monotonic dependence $\tau_p$ on $N$ cancels out, as shown above (Figure 3b), at $N_D > 2 \times (\theta \times p_t)^2 / n_t$ (increased $\Delta N_0$ with increasing $N$ is not able to compensate decreasing $\Delta N_0$ with increasing $N_D$).

Non-monotonic character of dependence $\tau_p$ on $N$ does not occur and at low concentrations $N_D$ {see inequities (27), expressions (22), and Figure 3b}, when equilibrium electron population at recombination level is determined mostly by electron-hole transitions between that level and free bands. In this case, values $\delta$ cannot provide prevailing growth of hole thermal emission rate from nonequilibrium centers over the growth of capture rate of nonequilibrium holes at equilibrium hole traps with increasing $N$.

Maximal value of ratio $K_p \equiv \hat{\tau}_p / \check{\tau}_p$ is achieved at $N_D \cong 2 \times \theta \times p_t$ and equals to approximately

$$\left(K_p\right)_{\max} \cong \sqrt{\frac{\theta \times p_t}{n_t}}. \tag{53}$$

Note that with increasing energy $E_t$ of recombination level, non-monotonicity of dependence $\tau_p(N)$ fades out (Figure 3b) and then cancels out absolutely. This is caused by the increase in concentration $n_e$ of equilibrium electrons and decrease in value $\delta$ and fall of the probability of hole thermal emission from recombination level into valence band with increasing energy of recombination level referred to the top of valence band. For the same reason, value $\hat{\tau}_p$ decreases with increasing $E_t$ (Figure 4a).

The character of dependence $\hat{\tau}_p$ on temperature (Figure 4b) is determined by the following dependences on temperature:

$$\delta_{2p}(T) = \frac{1}{n_i(T)} \times \sqrt{2 \times N_D \times p_t(T) + 4 \times p_t^2(T)}, \tag{54}$$

$$p_t(T) = N_v \times \exp\left(-\frac{E_t}{kT}\right), n_e(T) = \frac{\delta_{2p}(T) \times n_i^2(T)}{2 \times p_t(T)}. \tag{55}$$

Values $p_t(T)$ and $n_e(T)$ increase always with temperature $T$ rise. Increased $p_t$ means increasing probability of thermal emission of hole from recombination center into valence band. Therefore, the above-mentioned process facilitates increasing $\hat{\tau}_p$ with $T$ rise. At the same time, growth $n_e(T)$ facilitates decreasing $\hat{\tau}_p$ with $T$ rise due to decreasing concentration of nonequilibrium centers $\Delta N_0$ of hole thermal emission.

Value $\delta_{2p}$ decreases with $T$ rise at $E_t \le E_g/3$ due to approaching $N_{2p}$ closer and closer to $N_D$ (see expression (36)). Value $\delta_{2p}$ decreases also at $E_t \ge E_g/2$ up to temperature at which non-monotonic dependences $\tau_p$ and $\tau_n$ on $N$ cancel out. Falling $\delta_{2p}$ decreases $\Delta N_0$ that facilitates decreasing $\hat{\tau}_p$ with $T$ rise. When $E_t \ge E_g/2$, then $p_t(T)$ increases faster and $\delta_{2p}(T)$ falls and $n_e(T)$ grows. As a result, $\hat{\tau}_p$ increases with temperature rise (Figure 4b). If $E_t \le E_g/3$, then increased $p_t$ with temperature rise cannot compensate decreased $\delta_{2p}(T)$ and growth $n_e(T)$. As a result, $\hat{\tau}_p$ decreases with temperature rise (Figure 4b). If $E_g/3 < E_t < E_g/2$, then at $T < \widetilde{T}$, dependence $\hat{\tau}_p(T)$ will be increasing, and at $T > \widetilde{T}$ dependence $\hat{\tau}_p(T)$ will be falling for the same reasons that in previous cases (see expression (39) and insert in Figure 4b).

*4.3.2. Electron lifetime*

By analogy with hole lifetime, reciprocal electron lifetime consists of three partial components:

$$\tau_n^{-1} = \tau_{n1}^{-1} + \tau_{n2}^{-1} + \tau_{n3}^{-1} \tag{56}$$

First component

$$\tau_{n1}^{-1} = w_n \times N_0^e = \frac{w_n \times n_t}{2} \times \frac{B + A \times \delta - \delta^2}{\delta^2} \tag{57}$$

corresponds to the change of electron capture rate $\Delta n \times w_p \times N_0^e = \Delta n / \tau_{n1}$ caused by deviation of electron concentration from equilibrium value (capture of excess electrons $\Delta n$ on equilibrium traps $N_0^e$).

Second component

$$\tau_{n2}^{-1} = \frac{w_n \times n_t}{2} \times \frac{(\delta \times \theta - 1) \times (B + A \times \delta - \delta^2)}{\theta \times (B + A \times \delta - \delta^2) + (1 + \delta) \times (\theta \times B + \delta)} \tag{58}$$

corresponds to the change of electron capture rate $n_e \times w_n \times \Delta N_0 = \Delta n / \tau_{n2}$ caused solely by deviation of concentration of electron capture centers from equilibrium value (capture of equilibrium electrons $n_e$ on nonequilibrium capture centers $\Delta N_0$).

Third component

$$\tau_{n3}^{-1} = (\delta \times \tau_{n2})^{-1} \tag{59}$$

corresponds to the change of thermal emission rate of electrons from impurity level into valence band $(1/2) \times n_t \times w_n \times \Delta N_- = -\Delta n / \tau_{n3}$ caused by deviation of concentration of electron thermal emission centers from equilibrium value (thermal emission of electrons from nonequilibrium centers $\Delta N_-$).

Times $\tau_{n1}$, $\tau_{n2}$ and $(-\tau_{n3})$ have physical sense similar to times $\tau_{p1}$, $\tau_{p2}$ and $(-\tau_{p3})$, respectively.

Value $\delta >> 1$ as long as $N < N_D$, and hence ratio $\tau_{p1} / \tau_{n1} << 1$. Therefore, the occurrence of excess free charge carriers leads to the formation of additional (nonequilibrium) capture centers of electrons and, at the same time, decrease in concentration ($\Delta N_0 = -\Delta N_- > 0$) of electrons' generation centers. Partial components $1/\tau_{n2}$ and $1/\tau_{n3}$ are positive values at such change of charge state of recombination centers; moreover $\tau_{n2} < \tau_{n3}$, because of $\delta > 1$ at $N < N_D$. It means that, due to preferable capture of equilibrium electrons on nonequilibrium traps, lifetime of electrons $\tau_n$ is shorter, than capture time of electrons on equilibrium traps. With further increased $N$, the number of equilibrium capture centers of holes grows, but the number of equilibrium capture centers of electrons remains still small. As a result, concentration $\Delta N_0$ of nonequilibrium capture centers of electrons increases. For this reason, starting from concentration $N = N_{1n}$, capture rate of equilibrium electrons on nonequilibrium traps becomes higher than capture rate of nonequilibrium electrons on equilibrium traps. In other words, partial component $1/\tau_{n2}$ becomes critical component defining reciprocal lifetime of electrons $1/\tau_n$.

Component $1/\tau_{n2}$ falls with growth $N$ due to the decrease in concentration $n_e$ of equilibrium electrons, so $\tau_n \approx \tau_{n2}$ and increases with growth $N$ (Figure 1b).

At values $N$ greater than $N_D$, the ratio $\tau_{p1}/\tau_{n1}$ increases with increasing $N$. This, again, leads to decreasing concentration of nonequilibrium capture centers of electrons with increasing $N$. Value $n_e$ continues to fall as well. As a result, partial component $1/\tau_{n1}$ becomes critical component defining reciprocal electron lifetime $1/\tau_n$, and therefore, $\tau_n(N)$ falls with increasing $N$ (Figure 1b).

As shown above, in contrast to dependence $\tau_p(N)$, ratio $N_{1n}/N_D$ decreases (Figure 3a) and ratio $K_n \equiv \hat{\tau}_n/\breve{\tau}_n$ always increases monotonically with increasing $N_D$ (Figure 3b). Such regularities are caused by increased $n_e$ with increasing $N_D$. Because of this, capture rate of equilibrium electrons at nonequilibrium traps becomes greater than capture rate of nonequilibrium electrons at equilibrium traps at lower concentrations $\Delta N_0$, i.e., at lower values of ratio $N/N_D$. In contrast to the situation with holes, here, decreasing $\Delta N_0$ with increased $N_D$ is compensated by increasing $n_e$.

Similar to the behavior of hole lifetime, non-monotonic dependence $\tau_n(N)$ fades gradually and then cancels out (Figure 3b) with decreasing $N_D$ or increasing $E_t$. First regularity is caused by decreased $n_e$ and $\delta = 2n_e/n_t$ with decreasing $N_D$. Second regularity is caused by decreasing $\delta$ and, hence, $\Delta N_0$, with increasing energy $E_t$ of recombination level. In this case, however, due to growth $n_e$, non-monotonicity of dependence $\tau_n(N)$ cancels out at larger values $E_t$ than in the case of holes.

Due to decreasing $\delta_{2n}$ with increasing $E_t$, value $\hat{\tau}_n$ decreases as well (Figure 4a). The type of dependence $\hat{\tau}_n$ on temperature (Figure 4b) is determined only by dependence $\delta_{2n}(T)$, because in maximum point $\tau_n = \tau_{n1}/2 \sim 1/N_0^e \sim \delta_{2n}$. In zeroth approximation, $\delta_{2n}(T)$ coincides with $\delta_{2p}(T)$, determined by expression (56). Therefore, $\hat{\tau}_n$ decreases always with temperature rise (Figure 4b).

## 5. Relation between concentrations of photoholes and photoelectrons outside approximation of quasi-neutrality

Note first, from formulas (10), (11), (20), and (23) follow

$$a_n = \frac{\varepsilon}{8\pi \times q} \times \frac{(1+\delta)\times f(\delta)\times w_p \times n_t}{\theta \times f(\delta) + (1+\delta)\times(\theta \times B + \delta)}, \tag{60}$$

$$a_p = \frac{\varepsilon}{2\pi \times q} \times \frac{(1+\delta^{-1}) \times f(\delta) \times w_p \times p_t}{B + (A + \theta \times B) \times \delta + (\theta \times B + \delta) \times \delta^2}. \tag{61}$$

Differentiating (12) with respect to $x$, we obtain

$$\frac{\partial \Delta E}{\partial x} = \frac{1}{q \times \mu_n \times n_e} \times \frac{\partial \Delta I_n}{\partial x} - \frac{E_0}{n_e} \times \frac{\partial \Delta n}{\partial x} - \frac{D_n}{\mu_n n_e} \times \frac{\partial^2 \Delta n}{\partial x^2}. \tag{62}$$

Recall that $\Delta E \equiv E(x) - E_0$ is the change of electric field intensity caused by deviation of concentrations of free charge carriers and their capture centers from equilibrium values by reason of band-to-band absorption of optical radiation: $E(x)$ and $E_0$ are electric field intensities with and without illumination.

From equation (16) and formula (5), we find

$$\frac{1}{q} \times \frac{\partial \Delta I_n}{\partial x} = \frac{\Delta n}{\tau_n} + a_n \times \frac{\partial \Delta E}{\partial x} - g. \tag{63}$$

Eliminating $\frac{1}{q} \times \frac{\partial \Delta I_n}{\partial x}$ from equations (62) and (63), we obtain that

$$\frac{\partial \Delta E}{\partial x} = \frac{1}{1-\xi_n} \times \left[ \frac{1}{\mu_n \times n_e} \times \left( \frac{\Delta n}{\tau_n} - g \right) - \frac{E_0}{n_e} \times \frac{\partial \Delta n}{\partial x} - \frac{D_n}{\mu_n \times n_e} \times \frac{\partial^2 \Delta n}{\partial x^2} \right], \tag{64}$$

where

$$\xi_n = \frac{a_n}{\mu_n \times n_e}. \tag{65}$$

Eliminating $\frac{\partial \Delta E}{\partial x}$ from equations (7) and (64) and taking into account expressions (23) and (65), we find that relation between concentrations of excess holes $\Delta p$ and electrons $\Delta n$ is determined by the following formula:

$$\Delta p = \frac{\tau_p}{\tau_n} \times \Delta n + \chi \times \left( \frac{\tau_p}{\tau_n} \times \Delta n - g \times \tau_p - \mu_n \times \tau_p \times E_0 \times \frac{\partial \Delta n}{\partial x} - D_n \times \tau_p \times \frac{\partial^2 \Delta n}{\partial x^2} \right), \tag{66}$$

where

$$\chi = \frac{a_n + a_p}{\mu_n \times n_t \times \frac{\delta}{2} - a_n}. \tag{67}$$

In quasi-neutrality approximation $a_n = a_p = 0$ ; therefore, in this case,

$$\Delta p = \frac{\tau_p}{\tau_n} \times \Delta n. \tag{68}$$

## 6. Derivation of equation for distribution function of photoexcited charge carriers' concentration outside quasi-neutrality

From expressions (12)–(14), it follows that

$$\Delta E = \frac{I_{ph} - q \times \left(\mu_n \times \Delta n + \mu_p \times \Delta p\right) \times E_0 + q \times D_p \times \frac{\partial \Delta p}{\partial x} - q \times D_n \times \frac{\partial \Delta n}{\partial x}}{q \times \left(\mu_n \times n_e + \mu_p \times p_e\right)}. \tag{69}$$

Plugging this expression for intensity $\Delta E$ of photoinduced electric field in (12) and taking into account (17), we obtain

$$\Delta I_n = \frac{b \times n_e \times I_{ph}}{b \times n_e + p_e} - \frac{n_e \times \Delta p - p_e \times \Delta n}{b \times n_e + p_e} \times q \times \mu_n \times E_0 + \left(n_e \times \frac{\partial \Delta p}{\partial x} + p_e \times \frac{\partial \Delta n}{\partial x}\right) \times \frac{q \times D_n}{b \times n_e + p_e}, \tag{70}$$

where $b = \mu_n / \mu_p = D_n / D_p$. Formulas (17) and (70) allow writing

$$\frac{1}{q} \times \frac{\partial \Delta I_n}{\partial x} = -\frac{n_e \times \frac{\partial \Delta p}{\partial x} - p_e \times \frac{\partial \Delta n}{\partial x}}{b \times n_e + p_e} \times \mu_n \times E_0 + \left(n_e \times \frac{\partial^2 \Delta p}{\partial x^2} + p_e \times \frac{\partial^2 \Delta n}{\partial x^2}\right) \times \frac{D_n}{b \times n_e + p_e}. \tag{71}$$

Plugging relation (66) between $\Delta p$ and $\Delta n$ in (71), we find that

$$\frac{1}{q}\times\frac{\partial\Delta I_n}{\partial x}=-\left(\mu_n^a+\frac{\xi}{1-\tilde{\xi}_n}\times\frac{\tau_p}{\tau_n}\times\mu_p\right)\times E_0\times\frac{\partial\Delta n}{\partial x}+$$
$$+\left(D_n^a+\frac{\xi}{1-\tilde{\xi}_n}\times\left(\tau_p\times\mu_p\times\mu_n\times E_0^2+\frac{\tau_p}{\tau_n}\times D_p\right)\right)\times\frac{\partial^2\Delta n}{\partial x^2}+$$
$$+\frac{\xi}{1-\tilde{\xi}_n}\left(-\tau_p\times\frac{\mu_p}{\mu_n}\times D_n^2\times\frac{\partial^4\Delta n}{\partial x^4}+\tau_p\times\mu_p\times E_0\times\frac{\partial g}{\partial x}-\tau_p\times D_p\times\frac{\partial^2 g}{\partial x^2}\right), \tag{72}$$

where

$$D_n^a=\frac{n_e\times\tau_p+p_e\times\tau_n}{(p_e+b\times n_e)\times\tau_n}\times D_n=\frac{\delta^2\times\tau_p+B\times\tau_n}{(B+b\times\delta^2)\times\tau_n}\times D_n, \tag{73}$$

$$\mu_n^a=\frac{n_e\times\tau_p-p_e\times\tau_n}{(p_e+b\times n_e)\times\tau_n}\times\mu_n=\frac{\delta^2\times\tau_p-B\times\tau_n}{(B+b\times\delta^2)\times\tau_n}\times\mu_n. \tag{74}$$

Formulas (73) and (74) are, none other than, well-known (in quasi-neutrality approximation) expressions for ambipolar diffusion constant $D_n^a$, and ambipolar mobility $\mu_n^a$ of electrons and dimensionless parameter

$$\xi=\frac{a_n+a_p}{\mu_n\times n_e+\mu_p\times p_e}=\frac{2\times(a_n+a_p)\times\delta}{(\delta^2\times\mu_n+B\times\mu_p)\times n_t} \tag{75}$$

is much less than unity (see below). On the other hand, as it follows from equations (63) and (64),

$$\frac{1}{q}\times\frac{\partial\Delta I_n}{\partial x}=\frac{1}{1-\tilde{\xi}_n}\times\left(\frac{\Delta n}{\tau_n}-g\right)-\frac{\tilde{\xi}_n}{1-\tilde{\xi}_n}\times\left(\mu_n\times E_0\times\frac{\partial\Delta n}{\partial x}+D_n\times\frac{\partial^2\Delta n}{\partial x^2}\right). \tag{76}$$

Equating (72) and (76), we obtain equation

$$\xi\times D_n\times D_p\times\tau_p\times\frac{\partial^4\Delta n}{\partial x^4}-\left[\left(1-\tilde{\xi}_n\right)\times D_n^a+\xi\times\tau_p\times\mu_p\times\mu_n\times E_0^2+\xi\times\frac{\tau_p}{\tau_n}\times D_p+\tilde{\xi}_n\times D_n\right]\times\frac{\partial^2\Delta n}{\partial x^2}+$$
$$+\left[\left(1-\tilde{\xi}_n\right)\times\mu_n^a+\xi\times\frac{\tau_p}{\tau_n}\times\mu_p-\tilde{\xi}_n\times\mu_n\right]\times E_0\times\frac{\partial\Delta n}{\partial x}+\frac{\Delta n}{\tau_n}==g+\xi\times\tau_p\times\left(\mu_p\times E_0\times\frac{\partial g}{\partial x}-D_p\times\frac{\partial^2 g}{\partial x^2}\right). \tag{77}$$

Because

$$D_n - D_n^a = \frac{b - \tau_p / \tau_n}{b \times n_e + p_e} \times n_e \times D_n, \mu_n^a + \mu_n = \frac{b + \tau_p / \tau_n}{b \times n_e + p_e} \times n_e \times \mu_n, \tag{78}$$

then

$$\left(D_n - D_n^a\right) \times \tilde{\xi}_n + \xi \times \frac{\tau_p}{\tau_n} \times D_p = \xi_n \times D_n + \xi_p \times \frac{\tau_p}{\tau_n} \times D_p,$$

$$\xi \times \frac{\tau_p}{\tau_n} \times \mu_p - \left(\mu_n^a + \mu_n\right) \times \tilde{\xi}_n = \xi_p \times \frac{\tau_p}{\tau_n} \times \mu_p - \xi_n \times \mu_n, \tag{79}$$

where

$$\xi_n = \frac{a_n}{\mu_n n_e + \mu_p p_e} << 1, \xi_p = \xi - \xi_n = \frac{a_p}{\mu_n n_e + \mu_p p_e} << 1. \tag{80}$$

Therefore, we can rewrite equation (77) as follows:

$$Q \times \frac{\partial^4 \Delta n}{\partial x^4} - D \times \frac{\partial^2 \Delta n}{\partial x^2} + \mu \times E_0 \times \frac{\partial \Delta n}{\partial x} + \frac{\Delta n}{\tau_n} = g + \xi \times \tau_p \times \left(\mu_p \times E_0 \times \frac{\partial g}{\partial x} - D_p \frac{\partial^2 g}{\partial x^2}\right), \tag{81}$$

where

$$D = D_n^a + D_E + D_\xi, \mu = \mu_n^a + \mu_\xi, \tag{82}$$

$$D_E = \xi \times \tau_p \times \mu_p \times \mu_n \times E_0^2, \tag{83}$$

$$Q = \xi \times D_n \times D_p \times \tau_p, D_\xi = \xi_p \times \frac{\tau_p}{\tau_n} \times D_p + \xi_n \times D_n, \mu_\xi = \xi_p \times \frac{\tau_p}{\tau_n} \times \mu_p - \xi_n \times \mu_n. \tag{84}$$

Equation (81) is the desired equation. It depicts adequately the continuity of electron, hole, and total currents [see (15)-(17)].

## 7. Solitary Illuminated Sample

In this section, we will consider opportunities for improving photoexcitation of charge carriers and photo-emf $V_{ph}$ by increasing concentration $N$ of recombination centers.

### 7.1. Preliminaries: Basic relations

We will call the sample as solitary, if it is not in external electric field and external electrical circuit is open.

For simplicity, we will characterize effectiveness of charge carriers' photoexcitation by light-propagation-length averaged concentrations of photoelectrons $<\Delta n>$ and photoholes $<\Delta p>$ (Figure 5).

Figures 6 and 7 show calculated dependences $<\Delta n>(N)$ and $V_{ph}(N)$.

We have not used in study quasi-neutrality approximation [2-9, 13, 18, 21, 22, 28-31] because it can lead to unacceptable errors in calculation of dependences $<\Delta n>(N)$, $<\Delta p>(N)$ (Figures 8 and 9), and $V_{ph}(N)$ (Figures 10 and 11) due to the fundamental contribution of photoexcited space charge into photoelectric effects in semiconductors. In other words, even in solitary sample, photoexcited electron-hole plasma in semiconductor may not always be quasi-neutral.

Let's consider a homogeneous semiconductor sample (Figure 5) with no voltage applied, i.e., in absence of illumination and intensity of electric field $E_0$=0. The density of photogeneration rate of charge carriers, in view of multiple internal reflections, is determined by the following expression:

$$g(x) = \gamma \times \left[ a_- \times \exp(-\gamma x) + a_+ \times \exp(\gamma x) \right], \tag{85}$$

in which

$$a_- = \frac{(1-R)\times F_0}{1-R^2 \times \exp(-2\gamma \tilde{W})}, a_+ = a_- \times R \times \exp(-2\gamma \tilde{W}), \tag{86}$$

where $R$ and $\gamma$ are coefficients of light reflection and absorption, $F_0$ is density of incident photon flux, and $\widetilde{W}$ is sample thickness along incident light direction (Figure 5a). As is clear from (66) and (81), in discussed conditions, relation between concentrations of excess holes $\Delta p$ and electrons $\Delta n$ is as follows:

$$\Delta p = \frac{\tau_p}{\tau_n} \times \Delta n + \chi \times \left( \frac{\tau_p}{\tau_n} \times \Delta n - g \times \tau_p - D_n \times \tau_p \times \frac{\partial^2 \Delta n}{\partial x^2} \right), \tag{87}$$

and function $\Delta n(x)$ obeys the equation

$$Q\times\frac{\partial^4\Delta n}{\partial x^4}-D\times\frac{\partial^2\Delta n}{\partial x^2}+\frac{\Delta n}{\tau_n}=g(x)-\xi\times\tau_p\times D_p\times\frac{\partial^2 g}{\partial x^2}, \tag{88}$$

where

$$D=D_n^a+D_\xi. \tag{89}$$

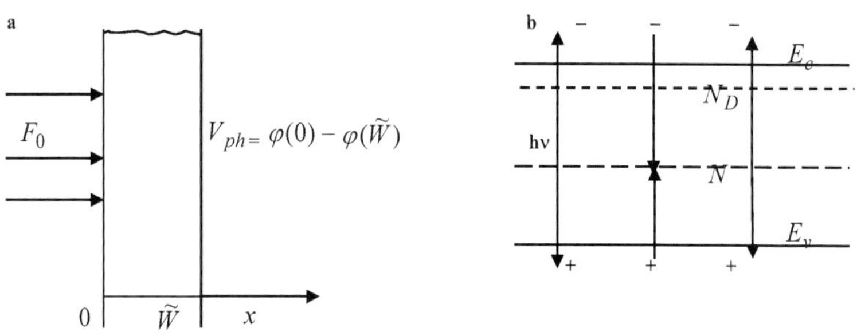


**Figure 5.** Layout of sample illumination (**a**) and generation-recombination processes in semiconductor (**b**). $F_0$, density of incident photon flux; $V_{ph}$, photo-emf; $\phi(x)$, electric potential; $\widetilde{W}$, thickness of sample along light propagation; $E_c$ and $E_v$, energy of conduction band bottom and valence band top

Denote: $\tau_n$ and $\tau_p$ are electron (24) and hole (25) lifetimes; $D_{n,p}$ are their diffusion constants and values $D_n^a$, $D_\xi$ and $Q$ and dimensionless parameters $\chi$ and $\xi << 1$ are determined by expressions (73), (84), (67), and (75), respectively.

Exact solution of equation (88) is

$$\Delta n(x)=\sum_{i=1}^{4}C_i\times\exp(k_i\times x)+T_n\times g(x), \tag{90}$$

where

$$k_{1,2}=\sqrt{\frac{L^2-\sqrt{L^4\mp4\times\xi\times L_n^2\times L_p^2}}{2\times\xi\times L_n^2\times L_p^2}},k_3=-k_1,k_4=-k_2, \tag{91}$$

$$T_n=\frac{\left(\frac{1}{\xi\times L_p^2}\right)-\gamma^2}{D_n\times(\gamma^2-k_1^2)\times(\gamma^2-k_2^2)}, \tag{92}$$

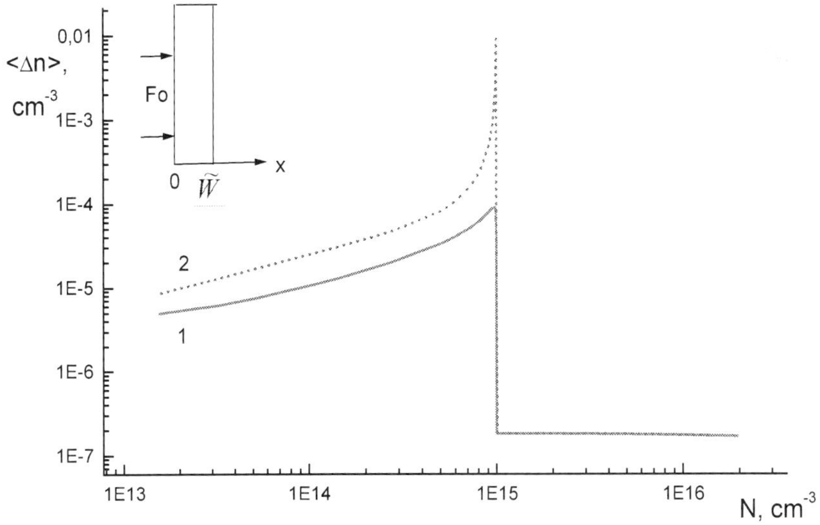


**Figure 6.** Dependences of mean concentration of photoelectrons $<\Delta n>$ in $GaAs$ for levels with energy (eV): $E_t = E_{t1}$ = –0.42 (curve 1) and $E_t = E_{t2} = -0.24$ (curve 2) on concentration of recombination impurity $N$ ; layout of sample illumination and axis $x$ are shown on insert. Adopted: light absorption coefficient $\gamma = 10^4$ cm$^{-1}$, diffusion constants of electrons $D_n = 221$ cm$^3$/s and holes $D_p$ =10.4 cm$^3$/s [2-9]; $\widetilde{W} = 10^{-3}$ cm; $F_0 = 1$ cm$^{-2}\times$s$^{-1}$; $T$ =300 K; concentration of shallow donors $N_D = 10^{15}$ cm$^{-3}$; $\theta \equiv w_p / w_n = 10^2$, $w_n = 10^{-8}$ cm$^3$/s [1-9], where $w_p$ and $w_n$ are capture probabilities of hole and electron

$$L^2 = L_a^{\ 2} + \xi_n \times L_n^{\ 2} + \xi_p \times L_p^{\ 2}, \tag{93}$$

$$L_a^{\ 2} = \frac{p_e \times L_n^2 + b \times n_e \times L_p^2}{p + b \times n_e} = \frac{B \times \tau_n + \delta^2 \times \tau_p}{B + b \times \delta^2} \times D_n \equiv D_n^a \times \tau_n = D_p^a \times \tau_p, \tag{94}$$

$L_n = \sqrt{D_n \times \tau_n}$ and $L_p = \sqrt{D_p \times \tau_p}$ are electron and hole diffusion lengths, $D_p^a$ is quasi-neutral ambipolar hole diffusion constant, and $L_a$ is quasi-neutral ambipolar diffusion length of charge carriers.

In quasi-neutrality approximation, parameters $\xi$, $\xi_n$ and $\xi_p$ are equal to zero; therefore, in this approximation, the distribution of excess electrons' concentration is determined by equation

$$D_n^a \times \frac{\partial^2 \Delta n}{\partial x^2} - \frac{\Delta n}{\tau_n} = -g(x). \tag{95}$$

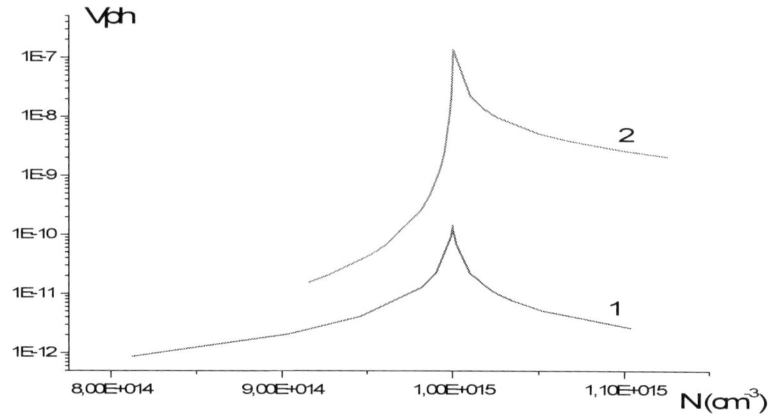


**Figure 7.** Dependence of photo-emf $V_{ph}$ (arbitrary units) in $GaAs$ for levels with energy $E_{t1}$ (curve 1) and $E_{t2}$ (curve 2) on concentration of recombination impurity $N$. Parameters are the same as in Figure 6

### 7.2. Effectiveness of charge carriers' photoexcitation

We define the mean value $<y>$ of variable $y(x)$ as

$$<y>=\frac{1}{\tilde{W}}\int_0^{\tilde{W}} y(x)dx. \tag{96}$$

Let's analyze the worst case, when recombination of excess charge carriers on illuminated ($x=0$) and unilluminated ($x=\widetilde{W}$) surfaces of the sample (Figure 5a) is so intensive that

$$\Delta n(0)=\Delta p(0)=\Delta n(\tilde{W})=\Delta p(\tilde{W})=0. \tag{97}$$

From (87), (90), and (97), it follows that

$$\Delta n(x)=\Delta n_1(x)+\Delta n_2(x)+T_n\times g(x), \tag{98}$$

$$\Delta n_1(x)=\frac{k_1^2\times(L_n^2\times k_2^2-1)}{D_n\times(k_2^2-k_1^2)\times(\gamma^2-K_1^2)}\times\frac{g(0)\times sh[k_1\times(\tilde{W}-x)]+g(\tilde{W})\times sh(k_1\times x)}{sh(k_1\times\tilde{W})}, \tag{99}$$

$$\Delta n_2(x)=-\frac{k_2^2\times(L_n^2\times k_1^2-1)}{D_n\times(k_2^2-k_1^2)\times(\gamma^2-k_2^2)}\times\frac{g(0)\times sh[k_2\times(\tilde{W}-x)]+g(\tilde{W})\times sh(k_2\times x)}{sh(k_2\times\tilde{W})}, \tag{100}$$

$$\Delta p(x)=\frac{\tau_p}{\tau_n}\times\Delta n(x)-\frac{\tau_p}{\tau_n}\times\chi\times\frac{(L_n^2\times k_1^2-1)\times(L_n^2\times k_2^2-1)}{D_n}\times \left\{\frac{A_1+A_2}{k_2^2-k_1^2}+\frac{\gamma^2\times g(x)}{(\gamma^2-k_1^2)\times(\gamma^2-k_2^2)}\right\} \tag{101}$$

where

$$A_1=\frac{k_1^2}{\gamma^2-k_1^2}\times\frac{g(o)\times sh[k_1\times(\tilde{W}-x)]+g(\tilde{W})\times sh(k_1\times x)}{sh(k_1\times\tilde{W})}, \tag{102}$$

$$A_2=-\frac{k_2^2}{\gamma^2-k_2^2}\times\frac{g(o)\times sh[k_2\times(\tilde{W}-x)]+g(\tilde{W})\times sh(k_2\times x)}{sh(k_2\times\tilde{W})}. \tag{103}$$

Thus, according to definition (96), we find

$$<\Delta n>=\eta_n\times<g>,<\Delta p>=\eta_p\times<g>, \tag{104}$$

where

$$\eta_n=T_n+\frac{\gamma\times cth(\gamma\times\tilde{W}/2)}{(k_2^2-k_1^2)\times D_n}\times\sum_{i=1}^{2}(-1)^{i-1}\times\frac{k_i\times\alpha_{3-i}}{\gamma^2-k_i^2}\times th\left(\frac{k_i\times\tilde{W}}{2}\right), \tag{105}$$

$$\alpha_{1,2}=k_{1,2}^2\times L_n^2-1, \tag{106}$$

$$<g>=\frac{(1-R)\times F_0}{W}\times\frac{1-\exp(-\gamma\times\tilde{W})}{1-R\times\exp(-\gamma\times\tilde{W})}, \tag{107}$$

$$\eta_p=\frac{4\pi}{\varepsilon}\times(a_n+a_p)\times\tau_p\times\eta_\rho+\frac{\tau_p}{\tau_n}\times\eta_n, \tag{108}$$

$$\eta_\rho=-\frac{\varepsilon}{4\pi}\times\frac{\gamma\times\alpha_1\times\alpha_2}{(\mu_n\times n_e-a_n)\times L_n^2}\times\left\{\Lambda^3-\frac{cth(\gamma\times\tilde{W}/2)}{k_2^2-k_1^2}\times\sum_{i=1}^{2}(-1)^i\times\frac{k_i}{\gamma^2-k_i^2}\times th\left(\frac{k_i\times\tilde{W}}{2}\right)\right\}\equiv\frac{<\Delta\rho>}{<g>}, \tag{109}$$

$$\Lambda^3 = \frac{\gamma}{(\gamma^2 - k_1^2)(\gamma^2 - k_2^2)}, \tag{110}$$

and $\Delta\rho$ is photoexcited space charge density.

Outside quasi-neutrality approximation, expressions (20)–(25), (60), (61), (75), (80), (91)–(94), and (104)–(110) determine, in parametric form (value $\delta = N_-^e / N_0^e$ is used as parameter), dependences $<\Delta n>(N)$ and $<\Delta p>(N)$. We will call found dependences (see Figures 6, 8, and 9) as exact, because, in linear approximation with respect to $F_0$, dependences are exact in contrast to quasi-neutrality approximation case.

Denote desired dependences in quasi-neutrality approximation as $<\Delta\tilde{n}(N)>$ and $<\Delta\tilde{p}(N)>$. In quasi-neutrality approximation,

$$\Delta n(x) = \Delta\tilde{n} \equiv \frac{\tau_n}{1 - \gamma^2 \times L_a^2} \times \left\{ g(x) - \frac{g(0) \times sh[(\tilde{W} - x) / L_a] + g(\tilde{W}) \times sh(x / L_a)}{sh(\tilde{W} / L_a)} \right\}, \tag{111}$$

$$< \Delta p > = < \Delta\tilde{p} > \equiv < \Delta\tilde{n} > \times \frac{\tau_p}{\tau_n}, \tag{112}$$

$$< \Delta n > = < \Delta\tilde{n} > \equiv \frac{(1 - R) \times F_0 \times \tau_n}{(1 - \gamma^2 \times L_a^2) \times \tilde{W}} \times \frac{1 - \exp(-\gamma\tilde{W}) - \gamma \times L_a \times \left[1 + \exp(-\gamma\tilde{W})\right] \times th(\tilde{W} / 2L_a)}{1 - R \times \exp(-\gamma\tilde{W})}. \tag{113}$$

Figure 6 shows that the effectiveness of charge carriers' photoexcitation may grow significantly with increasing $N$. Up to small corrections, dependences $<\Delta n>(N)$ and $<\Delta p>(N)$ reach maximums $<\Delta n>_{max}$ and $<\Delta p>_{max}$ at the same concentration $N = \hat{N} = N_D$ as for lifetimes (Figure 1b) and after that fall very strongly. Figures 8 and 9 illustrate the influence of photoexcited space charge in point $N = \hat{N}$ on the validity of results. It is clear from Figures 8 and 9 that with thinning $\tilde{W}$ of sample, using quasi-neutrality approximation leads to error up to several orders of magnitude. Let's clarify Figures 6, 8, and 9.

Lengths $L$ and $L_1 \equiv 1/k_1$, up to small corrections, equal to $L_a$; moreover

$$L_n > L_1 > L_2 \equiv \frac{1}{k_2}. \tag{114}$$

At $\tilde{W} << L_a$, we have

$$< \Delta\tilde{n} >= \frac{\psi(\tilde{W},\gamma)}{D_n^a}, < \Delta\tilde{p} >= \frac{\psi(\tilde{W},\gamma)}{D_p^a}, \tag{115}$$

where function $\psi(\widetilde{W}, \gamma)$ is independent on $\tau_n$ and $\tau_p$. On the other hand, if inequality $\widetilde{W} << L_2$ is sufficiently strong, and when $L_p >> \widetilde{W}$ as well, from expressions (104), (105), and (108)–(110), it follows that

$$< \Delta n >= \frac{\psi(\tilde{W},\gamma)}{D_n}, < \Delta p >= \frac{\psi(\tilde{W},\gamma)}{D_p}. \tag{116}$$

This means that diffusion of photoelectrons and photoholes is independent from each other.

Therefore, $L_2$ has physical meaning as shielding length of photoinduced space charge.

Length $L_2 << L_a$, $\widetilde{W}$ in the vicinity of the point $N = \hat{N}$, i.e. quasi-neutrality is valid. At values $N < \hat{N}$, length $L_a < \widetilde{W}$ due to small value $D_n^a$. When $N$ passing through the point $\hat{N}$ toward larger values $N$, then $D_n^a$ increases very strongly (by several orders of magnitude), and after that it is remaining substantially constant. As a result, length $L_a > \widetilde{W}$ at values $N$ higher $\hat{N}$. This explains the asymmetry of dependence $<\Delta n>(N)$ about point $N = \hat{N}$, and "plateau," when $N > \hat{N}$ as well (Figure 6).

At point $N = \hat{N}$ and about it, length $L_2$ is so long that even at $\widetilde{W} \sim 0.1$ cm, solution in quasi-neutrality approximation is unacceptable; moreover, with decreasing $\widetilde{W}$

$$\frac{< \Delta n >_{\max}}{< \Delta\tilde{n} >_{\max}} \to \frac{D_n^a}{D_n} << 1, \tag{117}$$

and

$$\frac{< \Delta p >_{\max}}{< \Delta n >_{\max}} \to \frac{D_n}{D_p} \cong 20, \tag{118}$$

despite the fact that $\tau_p << \tau_n$ (Figure 1b).

When $\gamma^{-1} << \widetilde{W}$ and inequity $\widetilde{W} << L_n$ are sufficiently strong, then the total quantity of photoelectrons is proportional to $\widetilde{W}$ due to reducing loss on unilluminated surface. Therefore, there is a "plateau" on dependence $<\Delta n>_{\max}$ on $\widetilde{W}$. For the same reason, at $\gamma^{-1} << \widetilde{W} << L_a$, dependence $<\Delta\tilde{n}>_{\max}$ on $\widetilde{W}$ has a "plateau" as well. However, in the last case, "plateau" height is much higher, and its width is much wider than true "plateau" (Figure 8, curve 1).

Moreover, shielding length of photoinduced space charge $L_2 \approx L_a$ when impurity level energy equals to $E_{t2}$. Therefore, solution obtained in quasi-neutrality approximation, even when $\widetilde{W} \to \infty$, differs from exact solution at least by several times (Figures 8 and 9, curve 2).

### 7.3. Effectiveness of photo-emf excitation

In view of the fact that under considered conditions

$$I_{ph} = 0, \tag{119}$$

from expressions (12)–(14) and (87), we obtain that photo-emf

$$V_{ph} = \int_0^{\tilde{W}} \Delta E(x)dx = \phi(0) - \phi(\tilde{W}), \tag{120}$$

i.e. potential $\phi(x)$ of illuminated surface with respect to unilluminated (Figure 5a) expressed by formula

$$V_{ph} = \frac{D_n \times \Delta_n - D_p \times \Delta_p}{\mu_n \times n_e + \mu_p \times n_e} = \frac{(L_n^2 - L_p^2) \times \Delta_n + \chi \times L_p^2 \times (\tau_n \times \Delta_g + L_n^2 \times \Delta_n'' - \Delta_n)}{(\mu_n \times n_e + \mu_p \times n_e) \times \tau_n}, \tag{121}$$

where

$$\Delta_n = \Delta n(0) - \Delta n(\tilde{W}), \Delta_p = \Delta p(0) - \Delta p(\tilde{W}), \tag{122}$$

$$\Delta_g = g(0) - g(\tilde{W}), \Delta_n'' = \left.\frac{\partial^2 \Delta n}{\partial x^2}\right|_{x=0} - \left.\frac{\partial^2 \Delta n}{\partial x^2}\right|_{x=\tilde{W}}. \tag{123}$$

It is clear from expression (121) that at infinite surface recombination rate, i.e., when conditions (97) are fulfilled, illumination produces no photo-emf. In this regard, assume that photoexcited charge carriers are not captured on surfaces and there is no charge on surfaces. In these conditions, photoexcited electric field intensity

$$\Delta E(0) = \Delta E(\tilde{W}) = 0 \tag{124}$$

And densities of electron $\Delta I_n$ and hole $\Delta I_p$ photocurrents

$$\Delta I_n(0) = \Delta I_p(0) = \Delta I_n(\tilde{W}) = \Delta I_p(\tilde{W}) = 0. \tag{125}$$

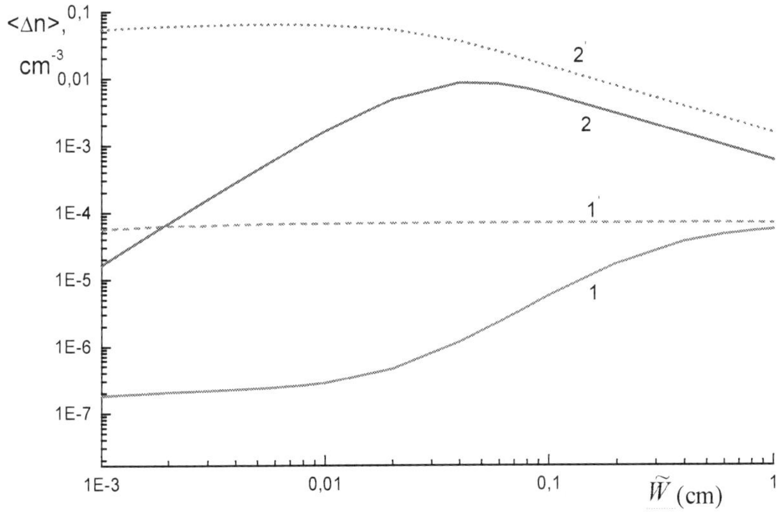


**Figure 8.** Dependences of $<\Delta n>_{\max}$ in the point of maximum functions $<\Delta n>(N)$ and $<\Delta p>(N)$ on thickness $\widetilde{W}$ for $GaAs$. Curves 1 and 2, recombination level energy $E_t$ equals to $E_{t1}$ and $E_{t2}$, respectively; solid lines - exact solutions; dashed lines - solutions in approximation of quasi-neutrality. Adopted parameters and other symbols are the same as in Figure 6

Therefore, we may write boundary conditions, in view of relation (87), as

$$\left.\frac{\partial \Delta n}{\partial x}\right|_{x=0} = \left.\frac{\partial \Delta n}{\partial x}\right|_{x=\tilde{W}} = 0, \tag{126}$$

$$D_n \times \left.\frac{\partial^3 \Delta n}{\partial x^3}\right|_{x=0} = -\left.\frac{\partial g}{\partial x}\right|_{x=0}, D_n \times \left.\frac{\partial^3 \Delta n}{\partial x^3}\right|_{x=\tilde{W}} = -\left.\frac{\partial g}{\partial x}\right|_{x=\tilde{W}}. \tag{127}$$

From (85), (86), (90), (126), and (127), it follows that

$$C_{1,2} = (-1)^{2,1} \times b_{1,2} \times \frac{g'_0 \times \exp(-k_{1,2} \times \tilde{W}) - g'_{\tilde{W}}}{2 \times sh(k_{1,2} \times \tilde{W})}, \tag{128}$$

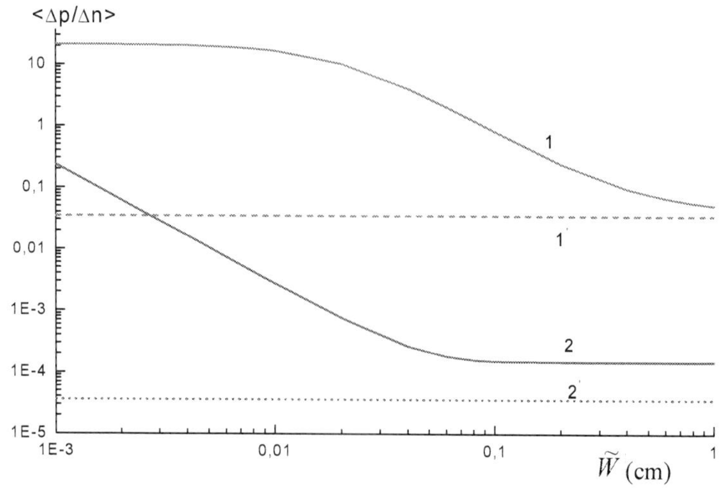


**Figure 9.** Dependences of ratio $<\Delta p>_{max} / <\Delta n>_{max}$ in the point of maximum functions $<\Delta n>(N)$ and $<\Delta p>(N)$ on thickness $\widetilde{W}$ for $GaAs$. Curves 1 and 2, recombination level energy $E_t$ equals to $E_{t1}$ and $E_{t2}$, respectively; solid lines - exact solutions; dashed lines - solutions in approximation of quasi-neutrality. Adopted parameters and other symbols are the same as in Figure 6

$$C_{3,4} = C_{1,2} \times \frac{g_0' \times \exp(k_{1,2} \times \tilde{W}) - g_{\tilde{W}}'}{g_0' \times \exp(-k_{1,2} \times \tilde{W}) - g_{\tilde{W}}'}, \tag{129}$$

where

$$b_{1,2} = \frac{k_{1,2} \times (k_{2,1}{}^2 \times L_n{}^2 - 1)}{D_n \times (k_2{}^2 - k_1{}^2) \times (k_{1,2}{}^2 - \gamma^2)}, \tag{130}$$

$$g_0' \equiv \left.\frac{\partial g}{\partial x}\right|_{x=0} = -\gamma^2 \times a_- \times [1 - R \times \exp(-2\gamma\tilde{W})], \tag{131}$$

$$g_{\tilde{W}}' \equiv \left.\frac{\partial g}{\partial x}\right|_{x=\tilde{W}} = \frac{(1-R) \times \exp(-\gamma\tilde{W})}{1 - R \times \exp(-2\gamma\tilde{W})} \times g_0'. \tag{132}$$

Find consistently

$$\Delta n(0) = T_n \times g_0 + \sum_{i=1}^{2}(-1)^{i+1} b_i \times [g'_0 \times cth(k_i \times \tilde{W}) - g'_{\tilde{W}} \times \cos ech(k_i \times \tilde{W})], \tag{133}$$

$$\Delta n(\tilde{W}) = T_n \times g_{\tilde{W}} + \sum_{i=1}^{2}(-1)^{i+1} b_i \times [g'_0 \times \cos ech(k_i \times \tilde{W}) - g'_{\tilde{W}} \times cth(k_i \times \tilde{W})], \tag{134}$$

where

$$g_0 \equiv g(0) = \gamma \times a_- \times [1 + R \times \exp(-2\gamma\tilde{W})], g_{\tilde{W}} \equiv g(\tilde{W}) = \frac{(1+R)\times\exp(-\gamma\tilde{W})}{1+R\times\exp(-2\gamma\tilde{W})} \times g_0. \tag{135}$$

$$\Delta_n = T_n \times \Delta_g + \sum_{i=1}^{2}(-1)^{i+1} \times b_i \times th\left(\frac{k_i \times \tilde{W}}{2}\right) \times \Sigma'_g, \tag{136}$$

$$\Delta''_n = \gamma^2 \times T_n \times \Delta_g + \sum_{i=1}^{2}(-1)^{i+1} \times k_i^{\,2} \times b_i \times th\left(\frac{k_i \times \tilde{W}}{2}\right) \times \Sigma'_g, \tag{137}$$

where

$$\Delta_g = g(0) - g(\tilde{W}) = \gamma \times F_0 \times (1-R) \times \frac{1-\exp(-\gamma\tilde{W})}{1+R\times\exp(-\gamma\tilde{W})}, \tag{138}$$

$$\Sigma'_g = g'_0 + g'_{\tilde{W}} = -\gamma^2 \times F_0 \times (1-R) \times \frac{1+\exp(-\gamma\tilde{W})}{1+R\times\exp(-\gamma\tilde{W})}. \tag{139}$$

If we utilize relation

$$Z \equiv L_n^2 - L_p^2 - \chi \times L_p^2 \times (k_1^2 \times L_n^2 - 1) \times (k_2^2 \times L_n^{\,2} - 1) = 0, \tag{140}$$

then expressions (121) and (136)–(139) allow us to obtain

$$V_{ph} = \frac{(1-R)\times\gamma\times F_0 \times (L_n^2 - L_p^2)}{\mu_n \times n_e + \mu_p \times p_e} \times M, \tag{141}$$

where

$$M = \frac{k_1^2 \times k_2^2 \times f_-(\gamma \tilde{W})}{(k_1^2 - \gamma^2) \times (k_2^2 - \gamma^2)} + \frac{\gamma \times k_1 \times k_2}{k_2^2 - k_1^2} \times \left[ \frac{k_1 \times th(k_2 \times \tilde{W} / 2)}{k_2^2 - \gamma^2} - \frac{k_2 \times th(k_1 \times \tilde{W} / 2)}{k_1^2 - \gamma^2} \right] \times f_+(\gamma \tilde{W}), \quad (142)$$

$$f_{-,+}(\gamma \tilde{W}) = \frac{1 \mp \exp(-\gamma \tilde{W})}{1 + R \times \exp(-\gamma \tilde{W})}. \quad (143)$$

In quasi-neutrality approximation, we get

$$V_{ph} = \tilde{V} \equiv \frac{(1-R) \times \gamma \times F_0 \times (L_n^2 - L_p^2)}{\mu_n \times n_e + \mu_p \times p_e} \times \tilde{M}, \quad (144)$$

where

$$\tilde{M} = \frac{\gamma \times L_a \times f_+(\gamma \tilde{W}) \times th(\tilde{W} / 2L_a) - f_-(\gamma \tilde{W})}{(\gamma \times L_a)^2 - 1}. \quad (145)$$

We refer to the dependences (141) of photo-emf $V_{ph}$ on $N$ and $\widetilde{W}$ as exact. The reason is that, in contrast to the case of quasi-neutrality approximation, said dependences are exact in linear approximation with respect to flux density $F_0$.

Figures 7, 10, and 11 show these dependences and solution in quasi-neutrality approximation as well. It is clear in Figures 10 and 11 that, in quasi-neutrality approximation, maximal value $V_{ph}(N)$ far exceeds the "true" value; moreover, exceedance may be several orders of magnitude.

We can explain results by the fact that very long shielding length (114) of photoexcited space charge $L_2$ in the point of maximum $N = \hat{N} \equiv N_D$ and nearby will cause diffusion of photoelectrons independently on photoholes (see Section 6.2) at $\widetilde{W} \sim 0.1$ cm. In other words, electron diffusion constant $D_n$ determines the distribution of photoelectron concentration $\Delta n(x)$. Aside from that, at $N \cong \hat{N}$, due to charge coupled to impurities, inequity $D_n >> D_n^a$ is fulfilled. It means that true effectiveness of photoelectrons' spreading is much higher, than that given by quasi-neutrality approximation. Therefore, ratio $(V_{ph})_{\max} / (\widetilde{V}_{ph})_{\max} << 1$, when $\widetilde{W} >> L_2$ (Figure 11).

Evidently, with thinning $\widetilde{W}$, spreading area of photoexcited charge carriers becomes wider, i.e., values $\Delta_n = \Delta n(0) - \Delta n(\widetilde{W})$ и $\Delta_p = \Delta p(0) - \Delta p(\widetilde{W})$ decrease. Therefore, values $(V_{ph})_{\max}$ and $(\widetilde{V}_{ph})_{\max}$ fall with thinning $\widetilde{W}$ (Figure 10).

Note that for impurity level energy equal to $E_{t2}$, the solution obtained in quasi-neutrality approximation, even when $\widetilde{W} \to \infty$, differs from the exact solution more than two times (Figure 10). This is due to the fact that in considered case the shielding length of photoexcited space charge $L_2 = 1/k_2 \approx L_a$, where $L_a$ is ambipolar diffusion length of charge carriers.

## 8. Photoelectric gain

We will consider uniform spatial distribution of density of charge carriers' photoexcitation rate $g$ and ignore surface recombination of photoexcited charge carriers.

We will facilitate mathematical description of photoelectric gain (see Figures 1a and 2)

$$G(N,V) \equiv \frac{I_{ph}}{q \times W \times g} \tag{146}$$

through utilizing small dimensionless parameter (75) $\xi << 1$, which will characterize the degree of deviation of semiconductor from local neutrality under illumination. Here $N$ is concentration of recombination centers; $I_{ph}$ is photocurrent density (14); $q$ is absolute value of electron charge; $V$ is bias voltage applied across the sample; $W$ is distance between current electrodes (see insert in Figure 1a).

Using linearized expressions for electron (12) and hole (13) components of photocurrent $I_{ph}$ and expressions (60) and (61), we may rewrite equations (5)–(7) as follows:

$$R_n = \frac{\Delta n}{\tau_n} - \xi_n \times \frac{\partial j}{\partial x}, R_p = \frac{\Delta p}{\tau_p} + \xi_p \times \frac{\partial j}{\partial x}, \Delta p = \frac{\tau_p}{\tau_n} \times \Delta n - \xi \times \tau_p \times \frac{\partial j}{\partial x}, \tag{147}$$

where

$$j = (\mu_n \times \Delta n + \mu_p \times \Delta p) \times E_0 + D_n \times \frac{\partial \Delta n}{\partial x} - D_p \times \frac{\partial \Delta p}{\partial x}, \tag{148}$$

$E_0 = V/W$, and dimensionless small parameters $\xi_n << 1$ and $\xi_p << 1$ as defined in (80). Relations (147) show that, by dimensionless small parameter $\xi$, we may really characterize the degree of deviation of semiconductor from local neutrality under illumination. If there is no external load (inset in Figure 1a), then illumination does not change voltage $V$ across the sample.

Therefore, expressions (4), (5), (8), and (9) with boundary conditions (1) allow us to write

$$I_{ph} = \left\{ \left( b + \frac{\tau_p}{\tau_n} \right) \times \langle \Delta n \rangle + \xi \times \frac{L_p^2}{W} \times \left( \frac{\partial \Delta p}{\partial x} - b \times \frac{\partial \Delta n}{\partial x} \right) \Bigg|_{x=0}^{x=W} \right\} \times q \times \mu_p \times E_0, \tag{149}$$

where $\langle n \rangle$ is arithmetic mean with respect to $x$ concentration of excess electrons (insert in Figure 1a)

$$\langle \Delta n \rangle = \frac{1}{W} \times \int_0^W \Delta n(x) dx, \tag{150}$$

$$b = \mu_n / \mu_p.$$

Actual distribution $\Delta n(x)$ at $g(x)$=$const$ is defined by equation

$$Q \times \frac{\partial^4 \Delta n}{\partial x^4} - (D_n^a + D_E + D_\xi) \times \frac{\partial^2 \Delta n}{\partial x^2} + (\mu_n^a + \mu_\xi) \times E_0 \times \frac{\partial \Delta n}{\partial x} + \frac{\Delta n}{\tau_n} = g. \tag{151}$$

Values of parameters $Q$, $D_\xi$, $\mu_\xi$, and $D_E$ are dictated by photoexcitation of space charge $\Delta\rho$ (4). Analysis shows that inequality $\xi << 1$, which is typically fulfilled with large margin, allowing to omit in equation (151) terms with parameters $Q$, $D_\xi$ and $\mu_\xi$. However, term with parameter $D_E$ must be retained, because, even at moderate electric fields, $D_E$ may exceed $D_n^a$ due to square-law dependence $D_E$ on $E_0$. For the same reason, we can omit in equation (149) terms including small parameter $\xi$. Thus, we arrive at relations

$$(D_n^a + D_E) \times \frac{\partial^2 \Delta n}{\partial x^2} - \mu_n^a \times E_0 \times \frac{\partial \Delta n}{\partial x} - \frac{\Delta n}{\tau_n} + g = 0, \tag{152}$$

$$I_{ph} = q \times \left( \mu_n + \frac{\tau_p}{\tau_n} \times \mu_p \right) \times \langle \Delta n \rangle \times E_0. \tag{153}$$

Equation (152), with boundary condition (1), and relation (153) allow us to obtain the formula for photoelectric gain:

$$G = \left( \mu_n \times \tau_n + \mu_p \times \tau_p \right) \times \frac{E_0}{W} \times \left\{ 1 - 4 \times \frac{L_{ef}}{W} \times \sqrt{\left( \frac{d_a}{2L_{ef}} \right)^2 + 1} \times \frac{sh\left( \frac{W}{2L_1} \right) \times sh\left( \frac{W}{2L_2} \right)}{sh\left( \frac{W}{2L_1} + \frac{W}{2L_2} \right)} \right\}, \tag{154}$$

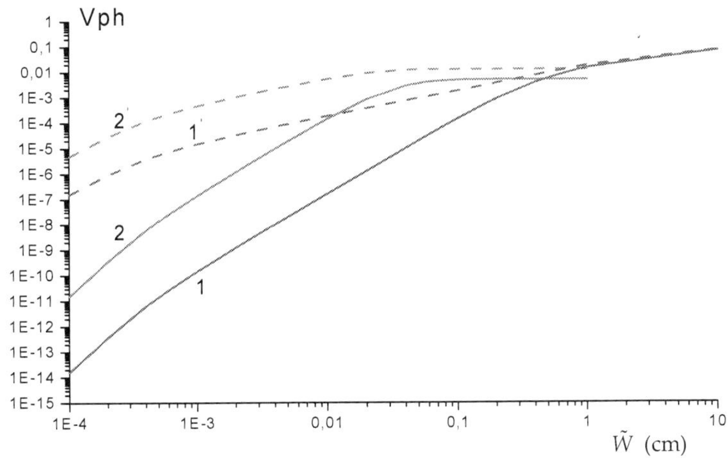


**Figure 10.** Dependences of maximal value $V_{ph}(N)$ in $GaAs$ on thickness $\widetilde{W}$. Curves 1 and 2, recombination level energy $E_t$ equals to $E_{t1}$ and $E_{t2}$, respectively; solid curves, exact solutions; dashed curves, solutions in approximation of quasi-neutrality. Adopted parameters and other symbols are the same as in Figure 6

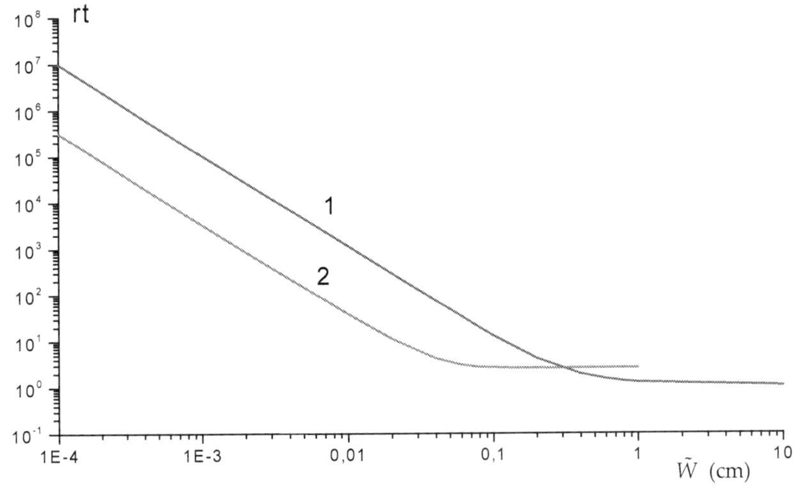


**Figure 11.** Dependence of ratio $rt = \widetilde{V}_{ph} / V_{ph}$ in the point of maximal value $V_{ph}(N)$ in $GaAs$ on thickness $\widetilde{W}$. Curves 1 and 2 – recombination level energy $E_t$ equals to $E_{t1}$ and $E_{t2}$. $V_{ph}$, exact solutions; $\widetilde{V}_{ph}$, solutions in approximation of quasi-neutrality. Adopted parameters and other symbols are the same as in Figure 6

where quasi-neutral ambipolar drift length of charge carriers

$$d_a = \mu_n^a \times \tau_n \times E_0, \tag{155}$$

effective diffusion length of charge carriers

$$L_{ef} = \sqrt{(D_n^a + D_E) \times \tau_n}, \tag{156}$$

and effective reciprocal diffusion-drift lengths $L_1$ and $L_2$ are defined by expressions

$$\frac{1}{L_{1,2}} = \pm \frac{d_a}{2L_{ef}^2} + \sqrt{\left(\frac{d_a}{2L_{ef}^2}\right)^2 + \frac{1}{L_{ef}^2}}. \tag{157}$$

Relations (18), (20)–(22), (24), (25), (73)–(75), (83), and (153)–(157) determine, in parametric form dependence $G(N, V)$ (see Figures 1a and 2a). Ratio $\delta = N_-^e / N_0^e$ is used as parameter in said relations.

It can be shown that equation $\mu_n^a(N)=0$, where $\mu_n^a$ is given by expression (74), has a solution when inequalities (27) and (41) are fulfilled. In zeroth-order with respect to small parameters (27) and (41), the root of this equation coincides with maximal extrema of functions $\tau_n(N)$ and $\tau_p(N)$ and equals to $N_D$ (Figure 1b and 1c).

Let's explain dependence $\mu_n^a$ on $N$ shown in Figure 1c.

Product

$$\mu_n^a \times \tau_n = \mu_p^a \times \tau_p \tag{158}$$

determines drift length and direction in electric field of concentrational perturbation – quasi-neutral cloud of positive and negative charges [2, 31], including bound at deep impurity (here $\mu_n^a$ and $\mu_p^a$ are electron and hole ambipolar mobility). Last mentioned bounding explains dependence $\mu_n^a$ on ratio $\tau_p(N)/\tau_n(N)$ in trap-assisted recombination. It would appear reasonable that charge carriers, which prevail in quantity, can easily shield photoexcited space charge, i.e., they are adjusted to drift of charge carriers of another type. That is why, in the case of band-to-band recombination ($\tau_n = \tau_p$), quasi-neutral cloud of positive and negative charges drifts in electric field with the same velocity and in the same direction as minority charge carriers, whereas in intrinsic material, cloud is out of control by electric field at all [2, 31] ($\mu_n^a = \mu_p^a \cong \mu_p > 0$ at $n_e >> p_e$, $\mu_n^a = \mu_p^a \cong -\mu_n < 0$ at $n_e << p_e$ and $\mu_n^a = \mu_p^a = 0$ at $n_e = p_e$). Similar situation, but not exactly identical, happens in the case of trap-assisted recombination.

In this case, due to the fact that $\tau_p < \tau_n$ (Figure 1c), vanishing $\mu_n^a$ occurs in n-type material (for specified parameters in Figure 1, at $n_e \cong 10 \times p_e$ in silicon and $n_e \cong 10^4 \times p_e$ in gallium arsenide). Positive sign of perturbation charge bound at deep impurities ($\Delta N_{-} < 0$) causes such behavior. Ratio $p_e / n_e$ begins to increase significantly, and very sharply, only when $N \cong N_D$. At the same time, ratio $\tau_p / \tau_n$ may not have so many orders of smallness as ratio $p_e / n_e$ may have. Therefore, $\mu_n^a$ vanishes when $N \cong N_D$, if deep level, according to conditions (27) and (41), lies in lower half of bandgap. If that level lies in upper half of bandgap, then, again, due to the fact that $\tau_p / \tau_n$ may not have so many orders of smallness as ratio $p_e / n_e$ may have, function $\mu_n^a(N)$ is always positive for actual values $N$, wherein lifetimes $\tau_p$ and $\tau_n$ decrease always with increasing $N$ (see Section 3). We denote solution based on relations (152) and (153) as approximate. Parameter $\xi(N)$, still remaining small, reaches its global maximum near point $N = \hat{N} = N_D$, where function $G(N)$ reaches maximal extremum $\hat{G}$ (Figure 1a). Deviation of approximate value $\hat{G} = \hat{G}_{appr}$ from exact value $\hat{G} = \hat{G}_{exact}$ {calculated with due regard for all terms in relation (149) and equation (151)} is shown in Figure 2b. Exact solution at $N = \hat{N}$ is not so difficult to find, because at this point $\mu_n^a = 0$. It is clear from Figure 2b that agreement $\hat{G}_{appr}$ with $\hat{G}_{exact}$ is quite good.

From expression (154), it follows that

$$\hat{G} = \left\{ 1 - 2 \times \frac{\hat{L}_{ef}}{W} \times th\left( \frac{W}{2\hat{L}_{ef}} \right) \right\} \times \left( \mu_n \times \hat{\tau}_n + \mu_p \times \hat{\tau}_p \right) \times \frac{E_0}{W}, \tag{159}$$

where

$$\hat{L}_{ef} = \sqrt{\hat{L}_a^2 + \hat{L}_E^2} \tag{160}$$

is effective ambipolar length (156) at $N = \hat{N}$ (i.e., at $\delta \cong \sqrt{A+B}$) and $\hat{\tau}_n$ and $\hat{\tau}_p$ are relevant electron (48) and hole (38) lifetimes. We can write that

$$\hat{\tau}_n = \frac{\sqrt{A+B}}{2 \times w_n \times N_D}, \hat{\tau}_p = \frac{2 \times A + \theta \times B \times \sqrt{A+B}}{2 \times w_p \times N_D \times (A+B)}, \tag{161}$$

$$\hat{L}_a^2 = \frac{(A/\theta) + B \times \sqrt{A+B}}{(A+B) \times D_n + B \times D_p} \times \frac{D_n \times D_p}{w_n \times N_D}, \tag{162}$$

$$\hat{L}_E^{\,2} = \frac{\varepsilon}{8\times\pi\times q} \times \frac{2\times A + (B+1)\times\theta\times\sqrt{A+B}}{w_p\times N_D\times n_t\times\sqrt{A+B}} \times \frac{\mu_n\times\mu_p\times E_0^2}{(A+B)\times\mu_n + B\times\mu_p}. \tag{163}$$

Value $\hat{L}_a$ is conventional ambipolar diffusion length at maximal extremum of function $G(N)$ (Figure 1a) calculated in quasi-neutrality approximation, i.e., when parameter $\xi$ is set to zero.

From expression (159), it follows that function $\hat{G}$ depends non-monotonically on applied bias voltage $V$ (Figure 2a). This is caused by increased $\hat{L}_{ef}$ with increasing $E_0=V/W$ that provides progressive loss of photoexcited charge carriers resulting from increasing diffusive inflow of photocarriers to current contacts' electrodes with follow-up recombination.

Increase in effective ambipolar diffusion constant $D_{ef}=D_n^a+D_E$ (coefficient before second derivative in equation (152)) causes elongation $\hat{L}_{ef}$ with increasing $E_0$. In turn, photoinduced space charge $\Delta\rho$ (4) causes monotonic increased $D_{ef}$ with increasing $E_0$. The analysis of expressions (159)–(163) shows that function $\hat{G}(V)$ reaches its maximum value (Figure 2c)

$$\hat{G}_{\max}(W) \cong \begin{cases} \dfrac{25}{72}\times\dfrac{W}{\hat{L}_a}\times\hat{G}_{\max}^{\max}, & when \quad W < 2\times\hat{L}_a \\ \hat{G}_{\max}^{\max}, & when \quad W > 2\times\hat{L}_a \end{cases}, \tag{164}$$

when bias voltage is applied across sample $V=V_{opt}(W)$, where optimal bias voltage (Figure 2d)

$$V_{opt}(W) \cong \begin{cases} \tilde{E}_0\times W, & when \quad W < 2\times\hat{L}_a \\ \dfrac{\tilde{E}_0\times W^2}{2\times\hat{L}_a}, & when \quad W > 2\times\hat{L}_a \end{cases} \tag{165}$$

Threshold value $G=\hat{G}_{\max}^{\max}$ (for given physical parameters of semiconductor) and electric field intensity $E_0=\tilde{E}_0$, at which $\hat{L}_{ef}=\sqrt{2}\times\hat{L}_a$, are defined by the following expressions:

$$\hat{G}_{\max}^{\max} = (\mu_n\times\hat{\tau}_n + \mu_p\times\hat{\tau}_p)\times\frac{3\times\tilde{E}_0}{25\times\hat{L}_a}, \tag{166}$$

$$\tilde{E} = \sqrt{\frac{8\pi\times kT\times n_i}{\varepsilon}}\times\left(1+\frac{A}{B}\right)^{1/4}. \tag{167}$$

## 8. Summary

Capture rate of excess charge carriers increases with increasing concentration $N$ of deep impurity levels, i.e., recombination centers (traps). However, as shown in this chapter on the example of single-level acceptor, this increase does not lead to unavoidable decrease in lifetime of excess electrons $\tau_n$ and holes $\tau_p$, when nonequilibrium filling of recombination level states is very low. The matter is that lifetimes are determined not only by capture of excess charge carriers at equilibrium traps but also by bound-to-free transitions of electrons and holes from nonequilibrium capture centers due to thermal emission and by capture of equilibrium charge carriers at nonequilibrium traps as well. Therefore, lifetimes of excess charge carriers can be either more or less than the time of their capture at equilibrium traps and can be strongly non-monotonic functions of recombination center concentration (Figure 1b). In the case of acceptor recombination level, it can happen if recombination level is located in lower half of forbidden gap. In the case of donor recombination level, it must be located in upper half of forbidden gap. It is essential that the ratio of lifetimes in maximum and minimum of functions $\tau_n(N)$ and $\tau_p(N)$ can be several orders of magnitude (Figures 1b and 3b).

It seems, authors of article [25] have reported first about the availability of minimum and portion of weak growth (up to 24 %) on experimental dependence of excess charge carriers' lifetime on recombination center concentration, which increased because of bombarding sample by high-energy electrons. Many years later, increase in lifetime, presumably, caused by increasing $N$, but already gained in several times, was observed experimentally [32].

The main reason for giant splash of photoresponse in semiconductors with increasing recombination center concentration $N$ (Figures 1a, 6, and 7) is the growth of charge carriers' lifetime in orders of magnitude.

This reason is also sufficient to provide increase, in order of magnitude and more, in efficiency of charge carriers' photoexcitation with increasing $N$ (Figure 6).

At and about point $N = \hat{N} \cong N_D$, where charge carriers' lifetime is maximal, equilibrium concentration of charge carriers becomes small, where $N_D$ is shallow dopant concentration. Therefore, increase in Dember photo-emf $V_{ph}$ in several orders of magnitude (Figure 7) is caused by both strongly non-monotonic dependences $\tau_n(N)$ and $\tau_p(N)$ (Figure 1b) and highly non-monotonic dependence of sample dark resistance on concentration $N$ [1-3, 8, 30, 31].

However, increase in orders of magnitude in charge carriers' lifetime with increasing $N$ (Figure 1b) is not a good reason for the development of giant splash in photoelectric gain $G$ with increasing $N$ (Figure 1a). As follows from [18, 19], $G$ increases with increasing charge carriers' lifetime, if ambipolar mobility $\mu_a$ (see (74), (158), [2]) is equal to zero, or if there is no recombination on current contact electrodes ($x = 0$ and $x = W$ ; see inset in Figure 1a).

In reality, recombination on contact electrodes occurs always to more or less extent [5, 9]. Therefore, under normal conditions ($\mu_a \neq 0$), increase in lifetimes, beginning from some lifetime values, does not increase in photocurrent density $I_{ph}$ [5, 18, 19].

Saturation in $I_{ph}$ is most clear in the case of high-rate recombination at contact electrodes (sweep-out effect on contacts [5, 18, 19]), when there are no photocarriers at contacts, i.e., conditions (1) are fulfilled. At trap-assisted recombination, function $\mu(N)$, under the same conditions (27) and (41), when there are non-monotonic dependences $\tau_n(N)$ and $\tau_p(N)$, vanishes at the same, up to small, correction value $N \cong \hat{N}$, at which functions $\tau_n(N)$ and $\tau_p(N)$ reach their maximal extrema $\hat{\tau}_n$ and $\hat{\tau}_p$ (Figure 2b and 2c). Therefore, $I_{ph}$ and, consequently, $G$ increase to the extent of increasing $\hat{\tau}_n$ and $\hat{\tau}_p$. These are physical grounds of giant splash in photoelectric gain $G$ with increasing $N$ (Figure 2a). Above mentioned results of strict analytical calculations (i.e., outside commonly used local approximation of quasi-neutrality) show that photoinduced local space charge affects substantially on giant splash of semiconductor photoelectric response with increasing concentration of recombination centers.

Strict solutions of problems concerning the quantity of photoexcited electrons $N_{ph}$ and holes $P_{ph}$ and Dember photo-emf $V_{ph}$ may be fundamentally different from solutions obtained in approximation of quasi-neutrality $\widetilde{N}_{ph}$, $\widetilde{P}_{ph}$, and $\widetilde{V}_{ph}$, respectively.

It may be that $P_{ph} / N_{ph} >> 1$ even if hole lifetime $\tau_p$ is much less than electron lifetime $\tau_n$ (Figure 2b). At the same time, in approximation of quasi-neutrality, $\widetilde{P}_{ph} / \widetilde{N}_{ph} = \tau_p / \tau_n << 1$ (Figure 2b). At point $N = \hat{N}$, at which functions $N_{ph}(N)$, $P_{ph}(N)$, and $V_{ph}(N)$ reach maximum values, and for thin samples (with thickness along light propagation $\widetilde{W} < 0.1$ cm), in surroundings of point $N = \hat{N}$, solutions obtained in quasi-neutrality approximation may differ from solution outside quasi-neutrality in several orders of magnitude (Figures 8–11). Moreover, even $\widetilde{W} \to \infty$, neglecting by photoinduced space charge is not always possible, i.e., it is not always possible to solve problem in quasi-neutrality approximation. The reason is that when recombination level is deep enough, then shielding length of photoexcited space charge may be of the order of ambipolar diffusion length of charge carriers.

At sweep-out effect on contact electrodes, splash of $G(N)$ with increasing $N$ depends non-monotonically on applied voltage $V$ across the sample (Figure 2a). That non-monotonic behavior is not related to heating of charge carriers or lattice and charge carriers injecting contacts. The reason is the increase in effective ambipolar diffusion constant $D$ (coefficient before second derivative in equation (81) determining the distribution of photocarriers) with increasing $V$, leading to huge loss of photocarriers due to faster diffusion to contacts and subsequent recombination. In turn, increase in $D \sim V^2$ is caused by photoinduced local space charge. What is important is that, at optimum voltage $V_{op}$ (Figure 2d), value $G$ can have several orders of magnitude (Figure 2c) at high concentrations of recombination centers.

As shown in [33], when recombination impurity $N$ has three charged states (two-level recombination center), then, again, strong increase in $\tau_n(N)$ and $\tau_p(N)$ with increasing $N$ may occur; moreover dependences $\tau_n(N)$ and $\tau_p(N)$ may have two charged states and two minimums and maximums. Opposite to single-level recombination center, in the case of two-level recombination center, maximum $G(N)$, as shown in [34], can be reached at lower concentrations $N$ and have greater peak value. Photoelectric gain $G$, to the left of maximum value $G(N)$, is larger in the case of two-level recombination center, than in single level. The reason is the low ambipolar mobility of charge carriers in the case of two-level recombination center.

As shown in [35], significant growth of charge carriers' lifetimes with increase in concentration of recombination impurity in certain range could be stimulated by uncontrolled (background) doping by deep impurities. Even two maximums can occur.

Above-mentioned regularities occur at arbitrarily low-level photoexcitation and they become the more evident, the wider the semiconductor bandgap.

The theory of giant splash of photoresponse in semiconductors at low-level illumination with increasing concentration of recombination centers could develop further through the generalization of boundary conditions on semiconductor surfaces and current contact electrodes, accounting for nonuniformity of charge carriers' photoexcitation along the line of current flow and fluctuation processes. The study of nonstationary (frequency domain and transient) characteristics is of particular interest.

From physical essence of considered effects, it follows that similar effects can occur in other mediums with recombination of dissociative or ion-ion type, for example, in multicomponent plasma [36]. More details about topic are given in [37-46].

## Acknowledgements

Authors are grateful to Prof. Sergey A. Nikitov for valuable support of publication.

## Author details

Viacheslav A. Kholodnov[1,2] and Mikhail S. Nikitin[3*]

*Address all correspondence to: nikiboxm@yandex.ru

1 V.A. Kotelnikov Institute of Radio Engineering and Electronics Russian Academy of Sciences, Moscow, Russia

2 Moscow Institute of Physics and Technology (MIPT), Dolgoprudny town, Moscow Region, Russia

3 JSC Shvabe-Photodevice, Moscow, Russia

## References

[1] Milns A. Deep Impurities in Semiconductors. New York: John Wiley and Sons; 1973.

[2] Smith R. Semiconductors. London: Cambridge University Press; 1978.

[3] Sze S. Physics of Semiconductor Devices. New York: John Wiley and Sons; 1981.

[4] Bube R. Photoelectronic Properties of Semiconductors. Cambridge: Cambridge University Press; 1992.

[5] Rogalski A., editor. Infrared Photon Detectors. Bellingham: SPIE Press; 1995.

[6] Colinge J.-P. & Colinge C. Physics of Semiconductor Devices. New York: Springer; 2002.

[7] Lutz G. Semiconductor Radiation Detectors. Berlin: Springer; 2007.

[8] Sze S. & Ng Kwok K. Physics of Semiconductors Devices. New York: John Wiley and Sons; 2007.

[9] Rogalski A. Infrared Detectors. New York: CRC Press, Taylor & Francis Group; 2011.

[10] Hall R. Recombination of electrons and holes in germanium. Phys. Rev. 1952; 87(2) 387.

[11] Shockley W. & Read W. Statistics of the recombination of holes and electrons. Phys. Rev. 1952; 87(5) 835-842.

[12] Blackmore J. Semiconductor Statistics. Oxford: Pergamon Press; 1962.

[13] Ryvkin S. Photoelectric Effects in Semiconductors. New York: Consultants Bureau; 1964.

[14] Blackmore J. Solid State Physics. London: Cambridge University Press; 1985.

[15] Downey P.; Martin R.; Nahory R. & Lorimor O. High speed, ion bombarded InGaAs photoconductors. Appl. Phys. Lett. 1985; 46(4) 396.

[16] Schaelin A.; Loepfe R; Melchior H.; Suter M. & Woelfli W. Beryllium-bombarded $In_{0.53}Ga_{0.47}As$ and InP photoconductors with response times below 3 ps. Mater. Sci. Eng. B 1989; 2(1-3) 81-85.

[17] Balberg I. & Naidis R. Sensitization of the minority carriers lifetime in a photoconductor. Phys. Rev. B 1998; 57(12) R6783-6787.

[18] Rittner E. Electron processes in photoconductors. Photoconductivity Conference at Atlantic City, November 1954, R.G. Breckenbridge, B. Russel & E. Hahn (eds.), 215-268, New York: Wiley & Sons; 1956.

[19] Shacham-Diamand J. & Kidron I. Contact and bulk effects in intrinsic photoconductive infrared detectors. Infrared Phys. 1981; 21(2) 105-115.

[20] Drugova A. & Kholodnov V. Concept of the realization of the ultimate possible photogain in intrinsic photoresistors under given concentration of the impurity recombination centres. Proceedings of International Semiconductor Device Research Symposium, v. 1, 197-200, December 1995, Charlottesville, Virginia, USA.

[21] Keyes R., editor. Optical and infrared detectors. Berlin: Springer; 1977.

[22] Beneking H. Gain and bandwidth of fast near-infrared photodetectors: A comparison of diodes, phototransistors, and photoconductive devices. IEEE Trans. Electron. Dev. 1982; ED-29(9) 1420-1430.

[23] Kholodnov V. Giant burst of photoconductivity in semiconductors upon an increase in the concentration of recombination centers. JETP Lett. 1998; 67(9) 685-691.

[24] Osipov V. & Kholodnov V. Theory of the diodes with radiative and unradiative impurity recombination. Sov. Phys. Semicond. 1970; 4(12) 1932-1937.

[25] Wertheim G. Transient recombination of excess carriers in semiconductors. Phys. Rev. 1958; 109(4) 1086-1091.

[26] Tsang W., editor. Lightwave Communication Technology: Photodetectors. Semiconductors and Semimetals v. 22, Part D, Willardson R. & Beer A. (eds.), New York: Academic Press; 1985.

[27] Kireev P. Physics of Semiconductors (in Russian). Moscow: Higher School; 1969.

[28] Moss T. Optical Properties of Semiconductors. London: Butterworth; 1959.

[29] Pikus G. Foundations of the Semiconductor Devices Theory (in Russian). Moscow: Nauka; 1965.

[30] Auth J.; Genzow D. & Herrmann K. Photoelektrische Erscheinungen (in German). Berlin: Akademie-Verlag 1977.

[31] Bonch-Bruevich V. & Kalashnikov S. Semiconductor Physics (in Russian). Moscow: Nauka; 1990.

[32] Basheleishvili Z.; Garnyk V.; Gorin S. & Pagaeva T. Change of the minority carrier lifetime under irradiation of high-ohmic p-type silicon. Sov. Phys. Semicond. 1984; 18(9) 1074-1075.

[33] Kholodnov V. & Serebrennikov P. Critical importance of the single-level approximation to account for the highly non-monotonic dependences of carrier lifetimes on recombination impurity concentration. Tech. Phys. Lett. 1997; 23(4) 268-270.

[34] Kholodnov V. & Serebrennikov P. Effect of photo-induced space charge on dependence of intrinsic threshold photoresistor amplification on two-level recombination impurity concentration. Proceedings SPIE, v. 5126, 352-356, Bellingham: SPIE Press; 2003.

[35] Kholodnov V. & Serebrennikov P. Dependence of electron and hole lifetimes on the majority recombination impurity concentration in the presence of auxiliary carrier trapping centers. Tech. Phys. Lett. 1997; 23(12) 969-971.

[36] Rayzer Yu. Physics of Gas Discharge (in Russian). Moscow: Intellect; 2009.

[37] Drugova A. & Kholodnov V. Character of the dependences of non-equilibrium electron and hole lifetimes on the concentration of recombination centers in impurity-type recombination semiconductors. Solid-St. Electron. 1995; 38(6) 1247-1252.

[38] Kholodnov V. Contribution to the Hall-Shockley-Read theory of recombination. Semiconductors 1996; 30(6) 538-544.

[39] Kholodnov V. Giant splash of photoelectric amplification in photoconductors with increasing of recombination centers' concentration under weak optical radiation from the region of intrinsic absorption (in Russian). XV International Scientific and Engineering Conference on Photoelectronics and Night Vision Devices, Abstracts, p. 14, October 1998, Moscow, Russia.

[40] Kholodnov V. Super non-monotonic dependence of gain in threshold intrinsic photoresistors on the concentration of recombination centers (in Russian). International Conference "Applied Optics-98", Abstracts, p. 88, December 1998, Sankt-Petersburg, Russia.

[41] Kholodnov V. Gigantic splash of the weak optical radiation gain in intrinsic threshold photoconductive devices (photoresistors) upon an increase in the concentration of recombination centers. Proceedings of SPIE v. 3819, 98-115, Bellingham: SPIE Press; 1999.

[42] Kholodnov V. & Drugova A. On impact of the photo-induced space charge upon semiconductor photoresponse dependence on the concentration of recombination centers under weak optical radiation. Proceedings of SPIE v. 4340, 186-212, Bellingham: SPIE Press; 2000.

[43] Kholodnov V. & Drugova A. The influence of a photoinduced volume charge on the weak-radiation-induced interband photoproduction of carriers in semiconductors with impurity recombination centers. Tech. Phys. Lett. 2002; 28(9) 714-718.

[44] Kholodnov V. & Drugova A. The principal role of non-quasi-neutrality of semiconductor photoexcited plasma in effect a giant splash the photoresponse of homogeneous semiconductors with increasing of recombination centers concentration (in Russian). VI Russian Conference on Semiconductor Physics, Abstracts, p. 204-205, October 2003, Sankt-Petersburg, Russia.

[45] Kholodnov V. & Drugova A. On the degree to which the increase in the concentration of the recombination centers raises the efficiency of inter-band photoexcitation of carriers under weak optical radiation. Proceedings of SPIE v. 5126, 367-380, Bellingham: SPIE Press; 2003.

[46] Kholodnov V. & Drugova A. The degree of efficiency enhancement of interband photoexcitation of the electromotive force of Dember at the cost of increasing the recombination centers concentration at weak optical radiation. XIX International Scientific and Engineering Conference on Photoelectronics and Night Vision Devices. Abstracts, p. 190, May 2006, Moscow, Russia.

Chapter 13

# Optoelectronic Biomedical Systems for Noninvasive Treatment and Control with Informated Support in Solutions

Barylo Hryhoriy, Gotra Zenon, Ivakh Mariya, Kozhukhar Oleksandr, Makara Ivanna and Virt Volodymyr

Additional information is available at the end of the chapter

http://dx.doi.org/10.5772/61120

**Abstract**

The chapter presents the development of an optical-electronic system with informated support in doctor solutions for its application in hematology, otorhinolaryngology, and dermatology. With the purpose of improving the quality of treatment by the perfection of the technology, the introduction of new contact lenses controls the experimental researches of spectrums of different objects and uses an optical-electronic system for the control of blood during and after the process of photopheresis. Optoelectronic systems for the treatment of ear noise by method of sedation with the use of optical bio-impact on the human psycho-physical condition through visual receptors with a special light medical information program. The proposed specialized optical-electronic system for dermatology allowed to significantly reduce the length of time on the process of diagnosis, provide a rapid process and expand diagnostic capabilities for the identification of detected pathologies.

**Keywords:** optical-electronic system, informated support, contact lenses controls, hematology, otorhinolaryngology, dermatology

## 1. Introduction

The development of optical and optoelectronic non-invasive means of diagnosis and research of opportunities for their effective use, in particular, to assess the structural condition of the tissue area in normal and pathologically is an actual problem of modern biomedicine [1, 2]. A variety of pathological tissue area neoplasms forced to seek new approaches to conditioning, differentiation, and diagnosis. Such possibilities in dermatology, otolaryngology, hematology, and other fields of medicine have become a reality with the beginning of the application of the new principles of optoelectronic systems, highly efficient miniature light sources, and the design of new technologies using promising microcontrollers and flexible algorithms based on probabilistic methods approach (Bayesian method) and consistent statistical analysis (Wald method). The functional, pathophysiological, and psychophysical conditions of the patient, as well as changes in the conditions during illness, treatment, and individual treatments affect the optical properties of tissues. In particular, the express diagnosis of surface and near-surface layers of the skin on the principle analysis of signals reflected traversed, and its own electromagnetic radiation output information of the simulated process is to analyze the spectra of the light reflection of the scanned tissue.

For example, there have been new specialized optoelectronic and optical-electronic systems developed for [3]:

- automating the analysis of peripheral parameters and psychophysical conditions in the patient;
- activating the control blood system during photopheresis treatment procedure and (in the research study) optoelectronic active control of micro- and nano structures of biological fluids for new bacterial implantens technologies;
- physiotherapeutic procedures on the photo treatment of ear noises, tonsillitis, and pharyngitis; and
- express diagnostics and authentication of skin illnesses.

## 2. Optical-electronic system for the photopheresis treatment in hematology

### 2.1. Photospheresis

Photospheresis is a modern nanotechnology of blood, which is used for the treatment of heavily cured illnesses such as a T-cells lymph adenoma of the skin and psoriasis, considered as the precancerous stages. Due to the division of aluminous factions of blood with the introduction of a sensitizer under the action of electromagnetic radiation of area of a fellow creature of ultraviolet range and the electronic absorption of radiation of certain wavelengths, it is possible to carry out the treatment of the higher mentioned illnesses. Thus, the indexes of the key-in of optical radiation blood change in the different areas of spectrum [4].

### 2.2. Experiment and results

With the purpose of improving the quality of treatment by the perfection of technology due to the introduction of new contact lens controls in the experimental researches of spectrums of different objects that touch control of blood after and in the process of photopheresis were conducted. Conducting of these researches is extremely a necessity for the determination of prospective and possibility of method. The capacity of the structure consonant with the existing technology of photopheresis and the subsequent development of the checking system are offered below [5, 6].

The structure of the optoelectronic element of the research (Fig. 1) consists of the radiate 2 and sensory parts 5. The radiate part shows, by itself, the matrix of varicolored light-emitting diodes 1 with the chart of the electric feed. The light streams are from each of which, passing through the contactor from blood, is disposed between the radiate and sensory parts 3, gets on a photo detector sensory part 4. The sensory part includes itself, except for a photo detector or photo detectors electronic charts of the previous strengthening, feed, and tuning of the intensity of radiation. Such structure enables the comparison of the noted optical descriptions of blood with a marked higher resolution photo in the medical technology.

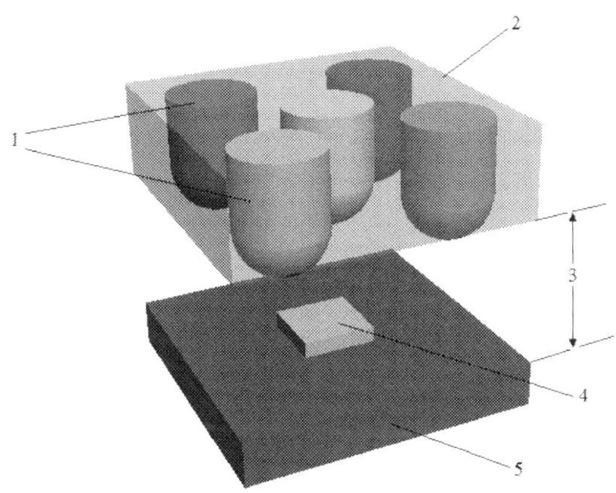


**Figure 1.** Structure of optoelectronic element. 1 - LED, 2 - radiative part, 3. Place of location of contactor, 4 – photo detector of sensory part, 5 - sensory part.

On the basis that we got and published earlier, recommendations in relation to principles and constituents of the checking system of the chart of experiment (Fig. 2) were developed, and the ground of choice of its components is carried out. A chart allowed to conduct the research of the dependencies of the intensity of electric signals, which were derived from the sensory part by the passing of a light stream from every light-emitting diode through the objects set in an experiment. By them were standard capacity with the bio-pierce of blood, empty capacity, layer of air with the sizes of capacity for blood and others like that.

For the choice of the constituents of an irradiate part - LED block 1, experimental researches that are spectral were conducted with the descriptions of light-emitting diodes in different areas within a visible range. The results of the researches of the light-emitting diodes are shown in Fig. 3. These results enabled us to choose certain types of light-emitting diodes according to some previous works pertaining to spectrum bars. There are also foreseen changes of light transmittance as a result of photopheresis. For the exception of the unforeseen influencing of

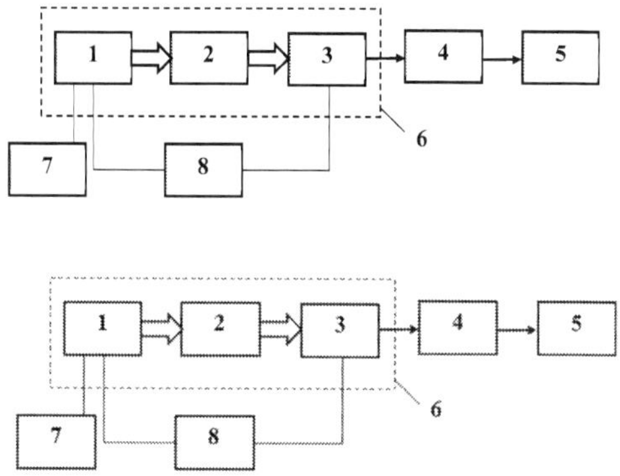


**Figure 2.** Chart of experiment: 1 – LED block, 2 – bio-sample, 3 – photo detector block, 4 – strengthener of photo-stream, 5 – measuring device of photo-stream, 6 – optoelectronic element, 7 – switchboard, 8 – feed.

extraneous radiates optoelectronic element 6 disposed in a light protective shell. For the influence on the results of the spectral descriptions of light transmittance and light reflection cuvette, the probed blood took place. And air around a cuvette is observed for the research of the passing of radiation of the chosen light-emitting diodes through this material, and the layer of air is chosen after its sizes. The intensity of radiation on spectral descriptions (Fig. 3) is resulted through the values of intensities and the maximal values for this description:

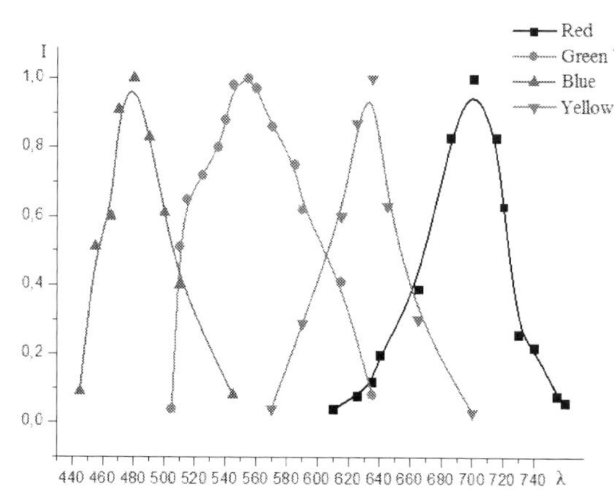


**Figure 3.** Spectral descriptions of separate light-emitting diodes I(λ).

On the basis of the derived results, the coefficients expected higher admissions of the objects chosen, the value of which is shown in Table 1. It is possible to correct the spectral description, which allows the exactness and authenticity of experiment.

| Color of LED<br>Object | Yellow | Blue | Green | Red |
|---|---|---|---|---|
| Air | 1,7 | 5,8 | 8,7 | 10 |
| Cuvette | 1,4 | 5,4 | 6 | 6 |
| Cuvette with blood | 1,2 | 2,2 | 4,3 | 5,8 |

**Table 1.** Expected coefficients of admission.

To the sensory part, silicious photodiodes were applied as FD-24Ê with a spectral sensitiveness in the area of wave lengths of 590-980nm. The type and location of the photodiodes in a photo detector block enabled to get signals about the changes of radiation intensity after its transmission through the chosen object from each of the noted light-emitting diodes.

Both parts of the flat, thin-walled cuvette, which was filled the bio-pierce of blood, were set. The standard medical cuvette or, as it is named in hematological technologies, the contactor, was made from a nontoxic optically transparent polyethylene with a 2mm thickness of walls. The volume of blood simultaneously contained in the cuvette is 140x23x1mm.

After the known values of the volume of blood in a cuvette and its ultraviolet display in accordance with technology of photopheresis, the energy of excitation that blood gets during irradiation is estimated.

The radiation of the visible range of wave lengths, which is fixed with the basis of the control of changes in the spectral descriptions of blood as a result of OOH displays, must go through the walls of the cuvette twice. The developments of the controls and needs of the experimental researches with its own spectral descriptions must be taken into account. Given all these, the research for the different areas of the spectrum is done for every bio-sample and its absence on the basis of the value of the admission coefficient of optical radiation in the different areas of the spectrum.

The light-emitting diodes feed was carried out through the managed ballast resistors with the consistently included current measuring devices. For the switching of light-emitting diodes, an electro-mechanic switchboard was applied.

For the control of diets, light-emitting diodes and photo detectors were foreseen as the proper measuring devices of current and tension. The signals of the photo detector after strengthening by an electronic chart was seen in the measuring device through photo stream 5.

A research method included light-emitting diodes and measuring each of them to the photo stream. Results achieved with the use of a radiate block on the basis of 4 light-emitting diodes for different objects, which apply in photopheresis, are presented in Fig. 4.

Through the relation of the photo stream attained for the different light-emitting diodes in the different areas of spectrum for the objects chosen, the value of coefficients of light transmittance was higher. With the account of changes in the spectrum as a result of the cooperation of light with the cuvette and air, the expected values of the coefficients for the chosen light-emitting diodes are presented in the light transmittance of blood shown in Table 2.

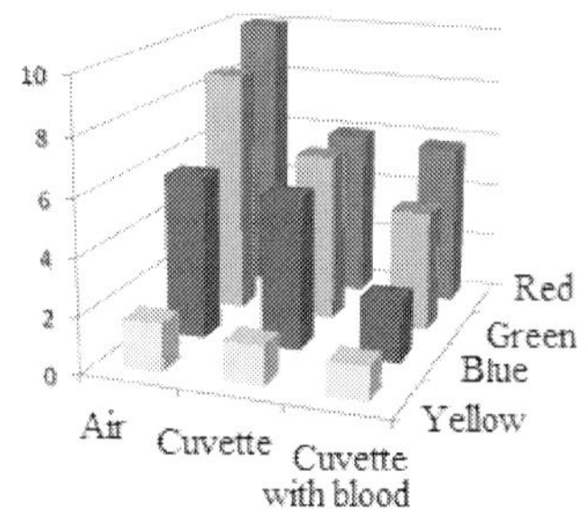


**Figure 4.** Dependencies of photo streams are for the chosen light-emitting diodes and different objects which apply in photopheresis.

| Colour of LED | Blue | Green | Yellow | Red |
|---|---|---|---|---|
| Coefficients of transmittance | 0,4 | 0,72 | 0,86 | 0,97 |

**Table 2.** The value of coefficients of light transmittance for blood.

These results are illustrated below in Fig. 5.

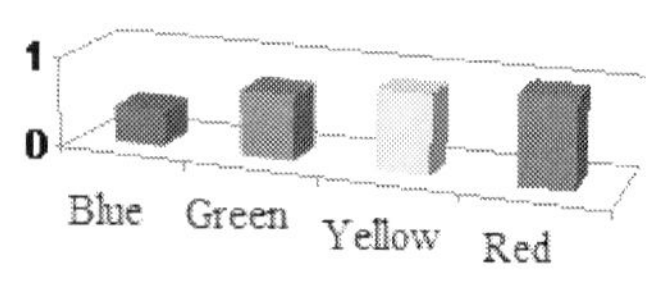


**Figure 5.** Coefficients of light transmittance for blood and the radiative part from four chosen light-emitting diodes.

Such set of bars can be applied to assess the facilities of the electro-optical research of blood as a result of its photo technological treatment.

On the developed operating optoelectronic prototype of the research of blood system during photopheresis or other photo-medic technologies, experimentally probed changes of coefficient admission of exposed to the rays at photopheresis are seen in four basic areas of spectrum.

It is shown that most spectral changes from photopheresis are tested by red, green, and dark blue areas of spectrum, which can be applied in the development of the optoelectronic checking system as a constituent of equipment for conducting of photopheresis.

After the results are acquired, it is possible to consider that the application of the offered parameters of the optoelectronic checking system will enable the development not only to control the technology of photopheresis but also for its perfection.

http://dx.doi.org/10.5772/61120

### 2.3. Model device for the study of optical parameters of blood during photopheresis

There is an urgent need to improve the efficiency of photomedical technology (PMT) for many medical areas such as hematology, otorhinolaryngology, dermatology, surgery, etc. Also, there is the problem of creating devices on the principles of programmable photo stimulation and information feedback from bio-objects (BO), minimal interference to BO, and no damage to the patient. One of prospective ways of solving this problem is to formulate the basic medical and technical requirements and the principles of development and application in medical practice and photo medicines, the basis of which would be assigned to new advances in research in the direction of optoelectronics and the design specialized for specific PMT optoelectronic systems (OES). Such thermal systems, discharge plasma, and semiconductor sources aiming incoherent radiation to allow the creation of BO stimulating irradiation with software controllable (according to the spatial frequency characteristics of the irradiated BO) dynamics and continuous noninvasive research of medicinal sessions for the analysis of changes in the optical parameters of the BO. This analysis can be based on the comparison of optical characteristics that are relevant to a particular medical technology (photo or any other technology) because, like the patient's blood, blood-filled organs, such as healthy, infected, or operated tissues, etc., due to differences in their BO radiation, reflection, and transmission BO test light streams [7-9].

Thus, a specialized OES has been developed for hematological PMT, including photopheresis technology as one of the most promising treatment for difficult and incurable illnesses such as psoriasis and T-cell lymphoma skin. The results of theoretical and experimental research regularities are also revealed. A clinical research device was created and implemented to provide a certain intensity and wavelengths of certain parts of the spectrum of ultraviolet and visible medical test irradiation taken from the patient's blood samples. To get the best stimulation effect of therapeutic irradiation flux, frequencies must be modulated, which corresponds to the frequency processes in irradiated samples of blood or blood fractions. Considering that there are only some accurately known ranges of frequencies, as they may be different for various patients, conditions should be provided for resonance frequency scanning. Contactless sensors are used as the correction modes for session therapeutic irradiation and for further treatment, based on the feedback provided, which is informative about the progress and effects of the treatment session. With the device, the volume of samples taken from the patients irradiated the treatment stream in the UV-A (320-400 nm) with specific power of 0.1... 0.4 mW/cm$^2$.

Upon the completion of the session, which had a duration of 5-10 minutes, the irradiated sample is returned to the patient's bloodstream. Information support based on continuous control changes, with the help of the optical characteristics of a subsidiary doctor who conducted the photopheresis, may be an important condition to contribute to the success of the treatment components of the information.

A similar treatment on the basis of a developed specialized OES with a dynamic radiation capacity of 10-10$^2$ mW (Fig. 6) allows therapeutic irradiation, control, and correction [10, 11].

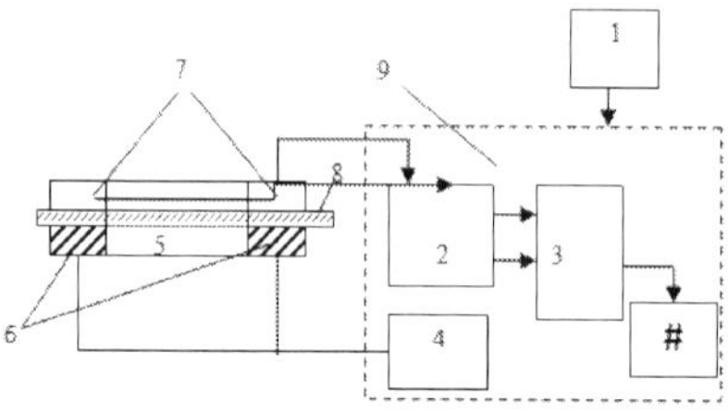


**Figure 6.** Block diagram of the online device for the analytical research of photopheresis efficiency: 1 - power supply 2 - the Amplifier, 3 - comparator, 4 - switch 5 – UV, 6 – LED, 7 - photodetector, 8 - sample BO, 9 – measurement unit, # – display.

OES is guided by a microprocessor, which laid irradiation modes and continuous control that are transmitted to a separate reflector and to a computer, via a USB-port, processing and display.

Before the start of the treatment, the doctor calibrates the OES, selects the mode of irradiation, puts the cell with the biosample in the device, and turns on the device. He watches the changes in the optical parameters of the biological test on the display. Upon reaching changes that correspond to those obtained in the clinical experiments' generalized value (Table 3), the doctor makes decisions on the sufficiency of the session or if a repeat session is necessary. At the same time, the doctor considers receiving understated indicators, the inexpediency for the further treatment, and in some cases, about unacceptable treatments for photopheresis of the patient.

Based on our research, the proposed model of the device is made for substantiated test light streams in spectral wavelengths of 642nm, 590nm, 505nm, and 465nm, which are also based on the changes in the intensity (Fig. 7) and colour temperature (Table 3) of the light streams through their passage in the BO.

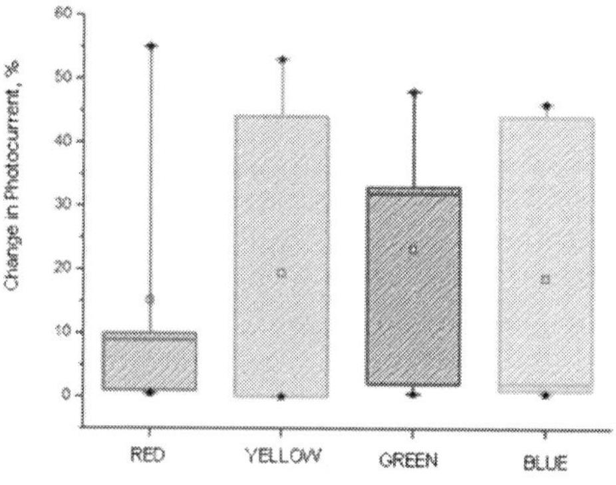


**Figure 7.** Changes in the photocurrents test light streams passed through the sample BO in a given spectral region.

| Group of patients | Befor irradiation | After irradiation |
|---|---|---|
| | $T_K$, [K] | $T_K$, [K] |
| 1 | 5300 | 4900 |
| 2 | 5550 | 4900 |
| 3 | 5350 | 5000 |

**Table 3.** Generalized changes in the optical parameters of biosample color temperature of test light flux after the irradiation and treatment process.

## 3. Optical-electronic system in otorhinolaryngology

### 3.1. Photomedical technology with light stimulation

On the basis of scientific and practical researches established, rhythmic light stimulation (light stimuli) can influence brain activity, and the dependence of the activity of the cerebral cortex on the stimulation frequency is associated with frequencies of biorhythms [12-13]. Thus, β-waves - over 13 Hz - support and activate the wakeful state, whereas α-waves - 8... 13 Hz debilitate the wakeful state, and τ and δ-waves - less than 8 Hz cause a slight or profound trance, or relaxation. Light-stimulation modes including those with light pulse repetition frequencies in accordance with the frequencies of space-time processes in biomedical objects (BO) can cause the bio-resonance effect in these objects. One can reach the effect of profound relaxation using an optical programmable trigger for the brain waves in the direction of low-frequency τ and δ-waves.

The obtained reactions can be explained by the changes in the activity of the somatosensory areas of the cerebral hemispheres' cortex under the influence of the photostimulation program. The moment of stabilization of these changes indicates that the cerebral cortex centers that are responsible for the relaxation are stimulated sufficiently. These changes are reflected by influencing stem vegetative and somatic nerve centers, and by reducing the tone of skeletal muscles and peripheral vessels. This results to the registered optical and thermal effects, and this fact agrees with physiology of processes in a human body during the stimulation of photoreceptors. It is proposed to use the appropriate contactless sensors for getting information about the achievement of the required state of the patient's organism.

Visual receptors that perceive a given light information program on the communication channels are excited centers of the cerebral cortex responsible for the psycho-emotional state of a person. As a result of such exposure, the most favorable conditions for treatment is when it is administered in the patient's condition atony (relaxation). It was determined that such influence on the patient's condition is accompanied by increased heat production, which caused an increase in the blood flow of small vessels and capillaries. In response, the system activates the regulation of temperature homeostasis.

These processes change the optical properties of tissue blood filled authority (BFA) [14], in particular, the detected variation of transmittance and reflectance flux at its interaction with a certain connective tissue, especially tissue BFA. Change in the first parameter causes, above all, an increase of the blood flow to the cells, and the second has an impact to the heat production on the surface of epithelial tissue changes (cf. deformation, change shape, sweating, etc.). Moment stabilizing changes in these parameters over time indicates the achievement of the required excitation centers of the cerebral cortex responsible for relaxing the body.

The optoelectronic system for the treatment of ear noise by method of sedation with the use optical bio-impact on the human psychophysical condition through visual receptors (Fig. 8) was equipped with a special light medical information program (Fig. 9). The program provides the appropriate frequency to change photos' stimulus on visual receptors of patients within 30 minutes of treatment [15].

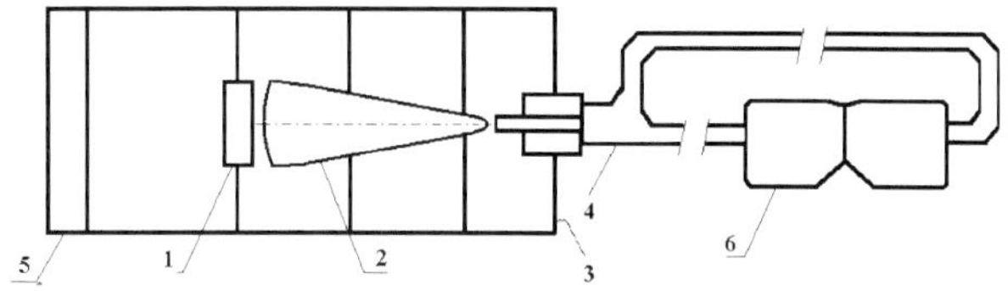


**Figure 8.** Model of photo stimulation device: 1 - LED terminal, 2 - focusing cone, 3 - case, 4 - lightguide, 5 - control unit, 6 - dimming glasses patient.

The measurement of change in the optical and thermal parameters of blood-filled object (BFO) in this experiment was carried out during the term of the medical program photo stimulation of bio influence.

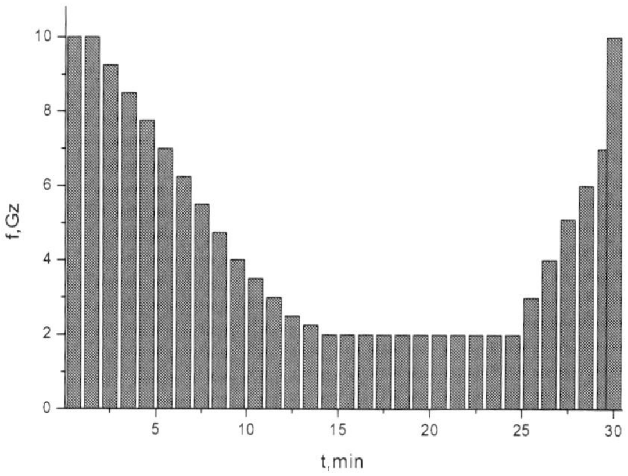


**Figure 9.** Special medical light information program of dependence time of the frequency photo stimulus for ear noise treatment by sedation [13].

In response, the system activates the regulation of the temperature of homeostasis. These processes change the optical properties of the blood-filled authority tissue. In particular, it is possible to fix the variation of transmittance and reflectance of the light flux in the interaction with certain connective tissues. The change of the first parameter causes, above all, an increase in the blood flow to the cells, while the second cause an impact on heat generation on the surface changes of the epithelial tissue (pores deformation, change shape, sweating, etc.). Moment stabilizing changes in these parameters achieves the required excitation centers of the cerebral cortex responsible for relaxing the body over time.

Based on these physiological processes developed and implemented for monitoring (changes of optical parameters of peripheral organ), the effectiveness of the procedures of the photo treatment of people who suffer from the disease otosclerotic is identified. The assessment of the impact on subjective indicators requires highly skilled medical staff to work during a treatment session and the use of complex and costly diagnostic equipment.

The achievement of an atony condition that is fixed by assessing changes in the optical parameters and temperature of the BFO can be used to signal the beginning of medical technologies by other factors, for example, the color information program on the specified area of organ and acupressure points. In order to obtain information of the necessary conditions of the patient organism at the start of the next medical technology, a contactless sensory actuator element, as an example, should be provided. Its objective is to obtain temperature and changes in the optical characteristics of BFO, and furthermore, to use hardware to determine the original set of changes. This task can be implemented, for example, by means of LED and photodiode optoelectronic components receiver and thermal radiation, which allows determining its temperature settings at the time. Information on optical performance and the temperature of the BFO is also assessed at the time when the original state of these changes is close to zero. This means that the system signals the biological response. The advantage of the proposed approach over the other is to carry out remedial action, capturing the individual signals of the bioresonance response to the impact. However, information on experimental research in this direction is not enough. Therefore, there is an urgent need for further studies on the implementation of modern control systems for medical purposes.

The measurement of change in the optical and thermal parameters of blood-filled object in this experiment were carried out during the term of the medical program photo stimulation bioeffects suggested using the exercise shown in the overarching framework of Fig. 10 and Fig. 11 [16-18].

The optical radiation from LED elements of the device (Fig. 11) after the interaction with the soft tissue of the BFO is situated on the input window element photo diode device that is located on the axis of the reflected and passed rays through the BFO. At the same time, the actual radiation of the BFO focuses on the reflex element and is directed to the photo convertor. These elements are partly measuring-controlling units whose task is to obtain optical signals after the interaction with the CA that are carriers of the reviews on the body healing effect. Subsequently processed opto-electronic units are converted from optical to electrical signals and after further analysis and comparison of test signals, the medical personnel are informed about the results on a computer monitor. The results of the described device provide a whole

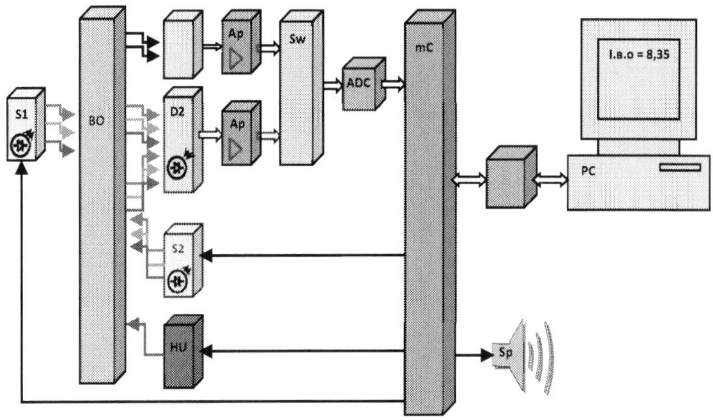


**Figure 10.** Generalized structure of optoelectronic photo diagnostic systems: BO - biological object, S1, S2 - light source, D1, D2 - transducers, Ap - signal amplifier, Sw - switch, ADC - analog - digital converter, mC - microcontroller, HU - medical device, TR - circuit connection, Sp - signaling device, PC - personal computer.

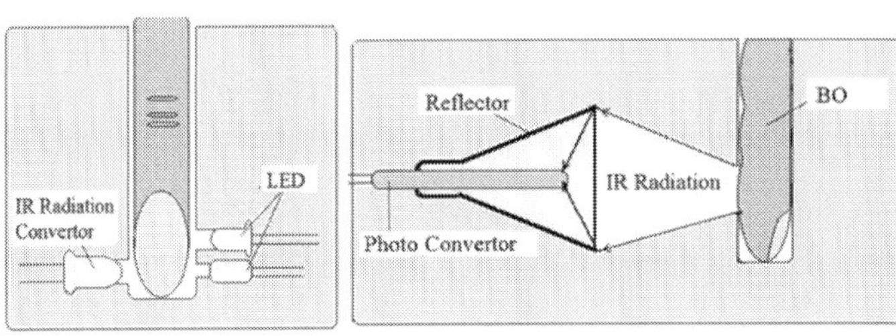


**Figure 11.** Converter circuit of the opto-electronic information system of evaluating the effectiveness of treatment procedures.

body response to the therapeutic effect of photostimul devices and systems, used for future decisions.

At the time of achieving biological resonance, the proposed fixed equipment during the photostimul program, the system allows to apply the proposed method of OES treatment. With the use of computer technology or online help from a physician-operator, the method for further treatment is determined by choosing the light program information and referral optical radiation with the settings specified in the scope.

The experimental studies applied some research, which are shown in Figs. 12, 13, 14. The transducer (TD) is placed on the optical axis, which focuses radiation on the TD BO, two LEDs, and photodetectors. The non-contact measurement of temperature and changes in the optical parameters of the experiments were carried out with the BO. The TD was placed on the focal length of the direct and reflected heat rays. The LED emitters are placed so that the fixed photodetector passed through the BFO and reflected from the surface of the light output.

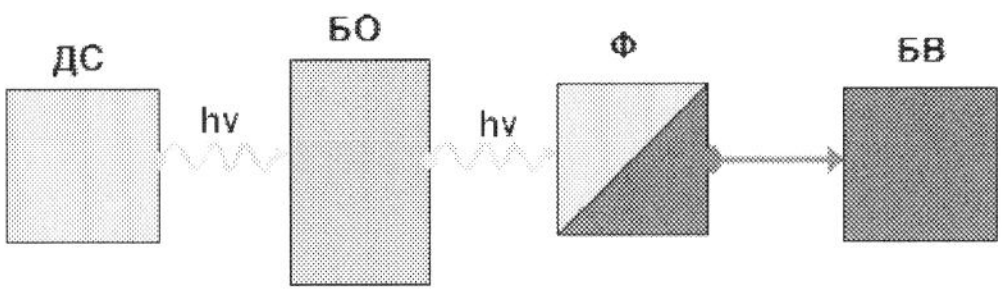


**Figure 12.** Experimental research of the transmission coefficients: ДС - light source, БО - biological object, Ф - hotodetector, БВ - unit measurement.

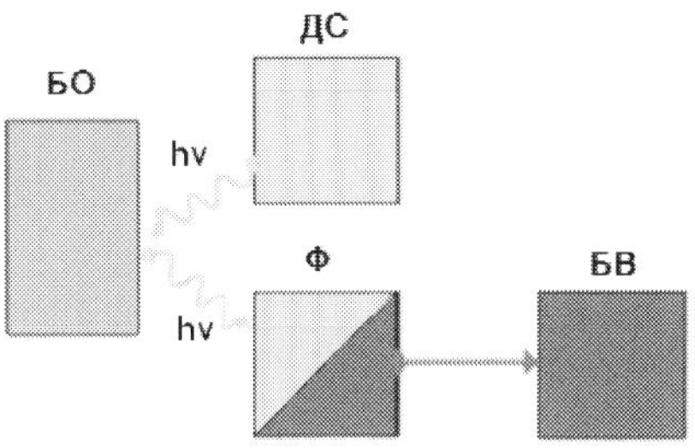


**Figure 13.** Experimental research reflecting coefficients: ДС - light source, БО - biological object, Ф - hotodetector, БВ - unit measurement.

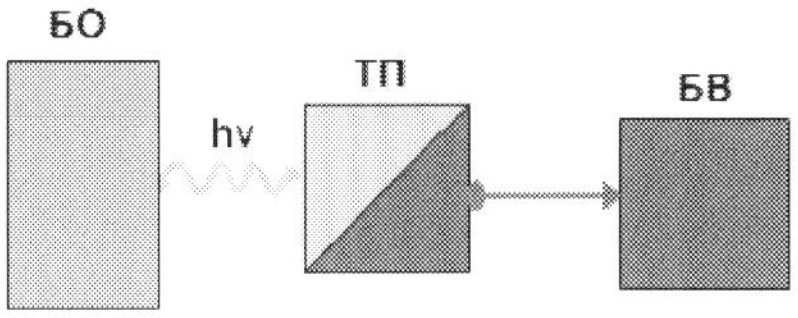


**Figure 14.** Experimental research showing temperature change: БО - biological object, ТП- thermo receiver, БВ- unit measurement.

The choice of LEDs to conduct research according to Fig. 12 and Fig. 13 was carried out based on the obtained spectral characteristics of the BO, which in this case is the phalanx of the finger (Fig.15).

Thus, the choice of the optical fiber array to conduct research to determine the coefficient of light transmission is made within the spectral characteristics of the BO. To determine the coefficient of the reflection elected LED emission spectrum, which does not fall within a specified range transmission BO.

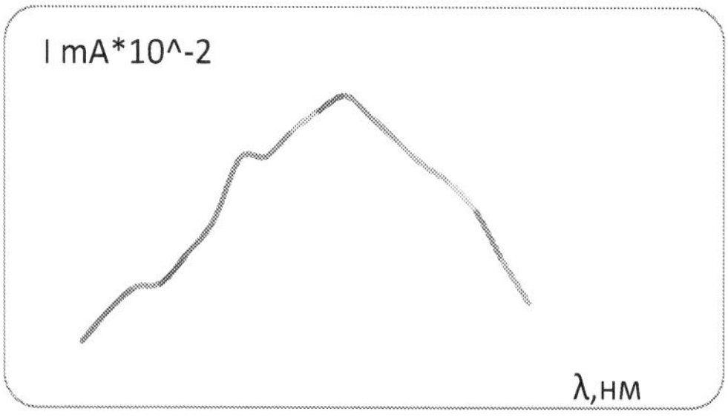


**Figure 15.** Spectral characteristics of BO.

For research to determine the dynamics of change in the transmission coefficient, the BO used the LED red area of the spectrum (Fig. 12). For research according to Fig. 13, the white LED spectral characteristics were used.

The results of the experimental research for one of the 20 patients are shown in Fig. 16 and Fig. 17.

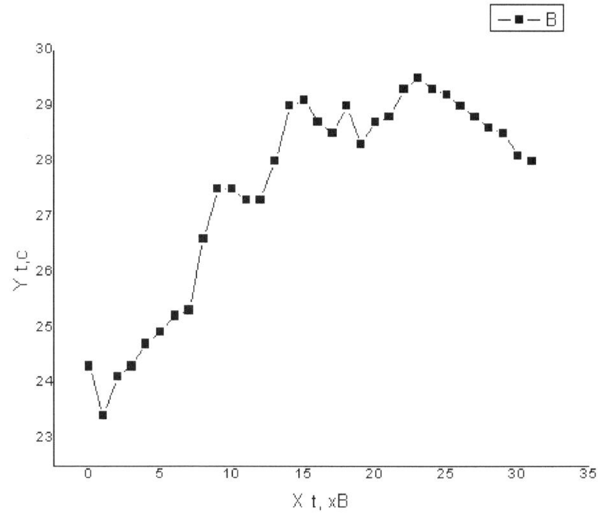


**Figure 16.** The change in temperature over time photo stimulation.

However, in all cases for the patients studied, the recurrence of dynamic changes of the measured parameters was confirmed. The difference in the time change of these settings for each patient was within 10-15 minutes of the actual photostimul medical applications. This is

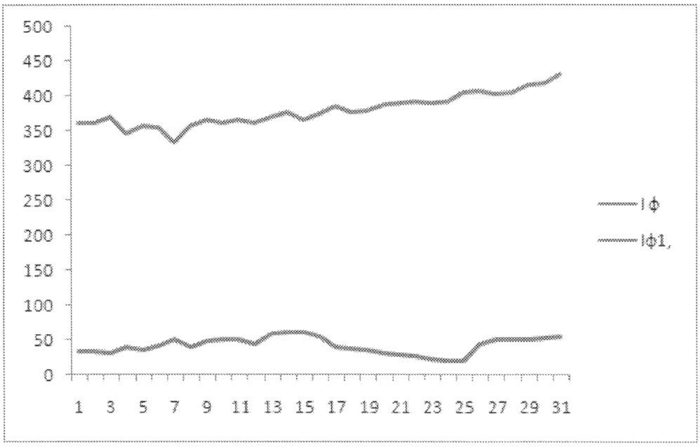


**Figure 17.** Change in photocurrent for the reflected (ϕ2) and passed through the BO (ϕ1) luminous flux over time.

explained by the individual characteristics of bioresonance formed as the body responds to the impact of a given medical photostimul program. Sharp changes in temperature and optical (traversed and reflected light beam) parameters in the first 10-20 minutes of the medical applications further institutionalize these changes in the future. This is due to the introduction of the body into a state of atony 10 minutes after the start of the medical program, which was recorded in the experiments. The results obtained are consistent with the literature described in the physiology processes of the human body during the stimulation of photoreceptors.

The results regarding changes in the temperature of the coefficient of reflection and light transmission scheme of the BO confirm the possibility of using this method and being a recommended optoelectronic element for implementation.

### 3.2. Problem of wide using light stimulation

The main problem of wide using light stimulation and the development of a controlled relaxation effect based on light stimulation consists of the absence of the principles of developing light stimulation tools with controlling light stimulation processes and providing their interactive conducting.

One promising solution to this problem is to formulate the basic medical and technical requirements and principles of development and the application in medical practice photos medicines, the basis of which would assign new advances in the field of research in optoelectronics and the design of specialized OES for specific FT. Such OES based on the thermal, plasma-discharge, and semiconductor sources of directional incoherent radiation can be created on the BO stimulation exposure with programmably controlled dynamics (according to the space-frequency characteristics of the irradiated BO) and continuous non-destructive control of a session for the analysis of change from the optical parameters of the BO.

Special attention should be paid to light stimulation through visual analyzers. They perceive a given light stimulation program of the communication channels of the cerebral cortex' excited centers, which is introduced because of the state of relaxation. Particularly, providing guidelines of photo-pulse to the very retina causes induced electrical potentials. It is clear from these reactions that retinal neurons convert photo-stimuli benched by force to electrical signals benched by amplitude. This is a relatively low-frequency modulation of the resting potential leading to relaxation.

This effect is accompanied by the increased thermogenesis caused by the increase of the blood flow in small vessels and capillaries. As a result, the optical properties of tissues, namely the coefficients of transmission and reflection of the light beam in the interaction with certain connective tissues, change. This confirmed the results of medical research conducted (Figs. 18, 19) [19-21].

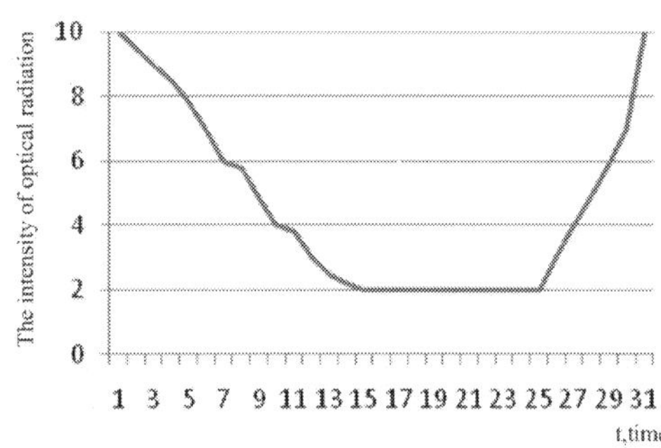


**Figure 18.** Change in the intensity of the optical radiation device during the treatment session.

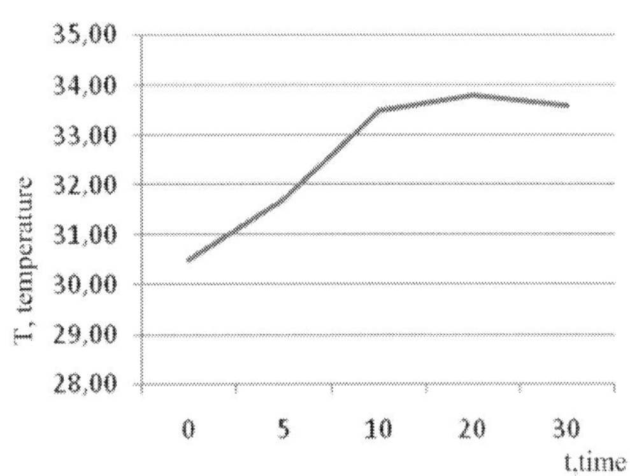


**Figure 19.** The temperature variation of tissue (skin) of the patient during the treatment session.

Taking the studies of the impact of stimulating light on the human body into account, the opto-electronic system (IPS) is proposed based on incoherent optical emitters, which holds a stimulating light through visual analyzers. The system consists of a generator, a generator's frequency control unit, a power supply, an output stage amplification, a LED matrix, a LED matrix control unit, and a light guide. A change of modes is provided by the generator, controlling the frequency and output power that can be set to choose medical program exposure modes (Fig. 20).

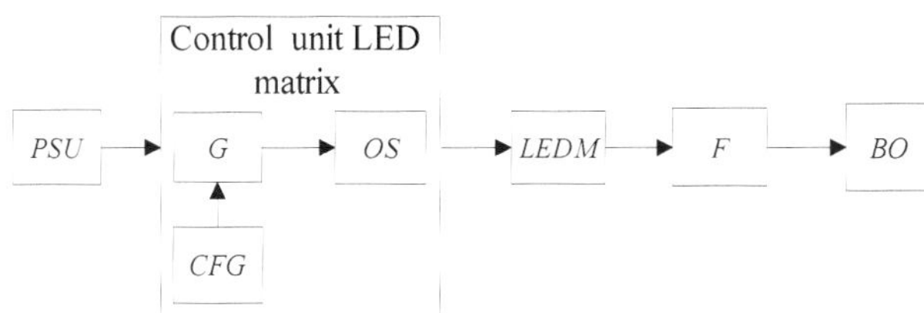


**Figure 20.** Block diagram of the opto-electronic system for radiation.

In order to provide information density and interactive conduction of photo-stimulation, we propose an OES, being noninvasive and harmless, for recording psychophysical state changes in the BO based on three converters using the method of simultaneous comparison of changes in optical properties and heat flows caused by changes in the temperature of peripheral BO's (Fig. 21).

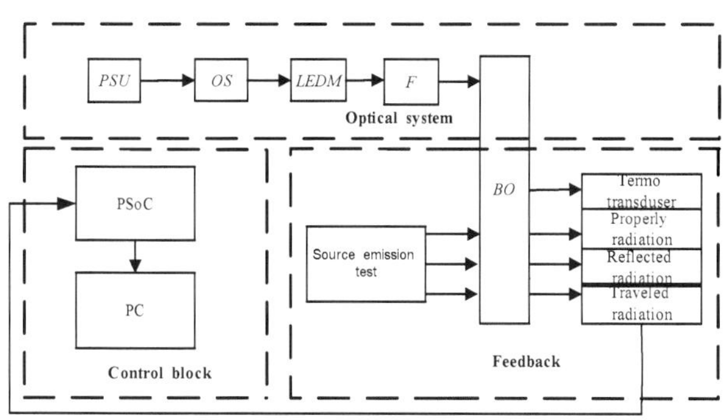


**Figure 21.** Block diagram of stimulating light.

The system consists of a block of optical radiation, the control block, and the control unit. Block optical radiation consists of the following components: a power supply (PSU), the output stage (OS), and the radiation source (LEDM) with a light guide (F) and creates the required for the irradiation luminous flux in the BO in the process of the treatment session. Under the influence of this radiation, changes in the optical and thermal parameters of BO, which is fixed by the circle feedback control unit, is implemented. Controlled optical signals in the state of the BO registered respective transducers and processed microcontroller PSoC. Based on the results of the analysis, a control output stage radiation source was carried out. Additionally, obtained results are transmitted via a serial interface to a PC for software processing and displayed on the screen in interactive mode.

Upon experimental investigations, it was detected that the BO's measured signals change during 30–40 min of a treatment session. The obtained dependencies underlie the algorithm of the BO's state monitoring system function. The limits of the controlled intensity parameters of the BO's own radiation, the radiation reflected, and radiation that passed through a BO were determined. On the basis of these three parameters, one can evaluate changes in the physiological state of the total organism.

In the process of the monitoring session, the dominant parameter of the most expressed changes of the dynamics was selected. The remaining parameters are additionally used. One of the three software options were selected by means of determining the value of maximum change (Fig.22).

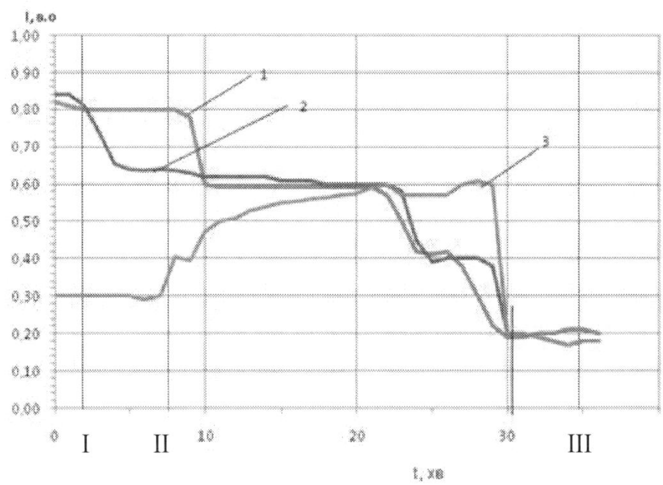


**Figure 22.** Experimental results dependencies of the intensity controlled light streams: 1 - passed through BO luminous flux test, 2 - test the reflected light beam, 3 - own stream of optical radiation.

As seen from the experimental results at phase I, which is the beginning of the treatment session, the dynamics of change between the absorbed and reflected radiation tissue does not. After exposing to stimulating light for 5-7 min (phase II), there is a change in the dynamics maximum values at $\Delta I = 0.15$ r.u., therefore, the patient is in a state of relaxation. In the next 15 minutes, time stabilization occurs, achieving the desired excitement centers of the brain that shows an effect of relaxation. In the last 2 min of the treatment session, the dynamics observed at ΔI = 0.1 r.u. This shows the state of the vigil centers of the brain, therefore, the patient is in a state of relaxation. As seen from the graph, the intensity of the absorbed and reflected radiation at the third stage, after a 30 min session is almost identical and very low at ≈ 0,2 r.u. and is in a condition for the termination of the treatment session.

In order to implement the interactive operation mode, the following algorithm for analysis of the parameters of the controlled optical was proposed (Fig. 23).

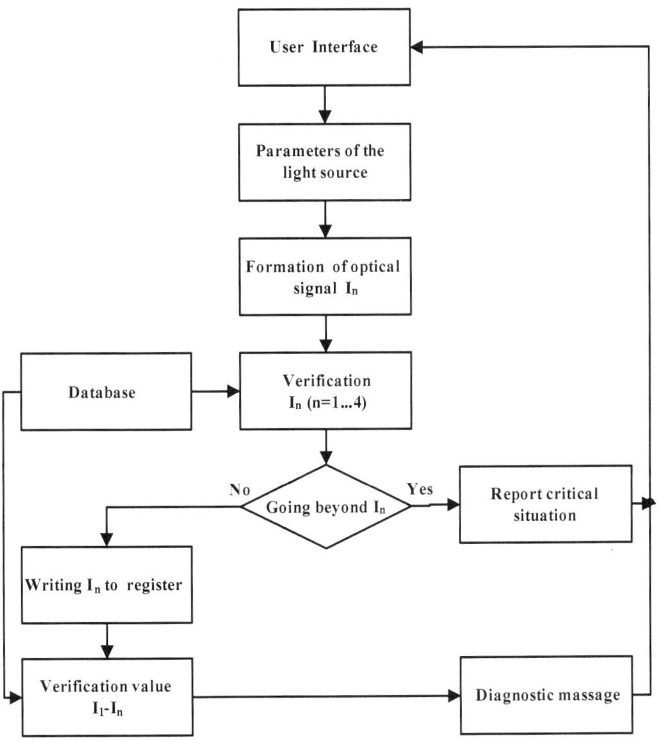


**Figure 23.** The algorithm interactive mode of a light stimulated session.

Process management is carried out using a PC through an informative user interface. Before starting the work in an automatic or manual mode, one has to set the parameters of the LED matrix: duration of light signals, amplitude, and limits of the spectrum of test signals. After starting the process, each of the input parameters I1 - I4 is controlled continuously. At the same time, each parameter is checked for compliance with the value ranges defined in the database. In the case when the controlled parameter In exceeds the allowable limits, a message about the achievement of the relaxation effect is formed, and the light stimulation session is automatically terminated. Checking the values of all the parameters In remained continuous throughout the session. After the session, on the basis of the results, a diagnostic message is formed according to a combination of optical parameters In set in the database of the system.

The view of the test sample of the controlled light stimulation device is shown in Fig. 24. Channel 1 provides programmable light pulses to a visual analyzer via a light guide cable to dimming glasses. Channel 2 provides a programmable photo-stimulation through other receptor areas [22, 23].

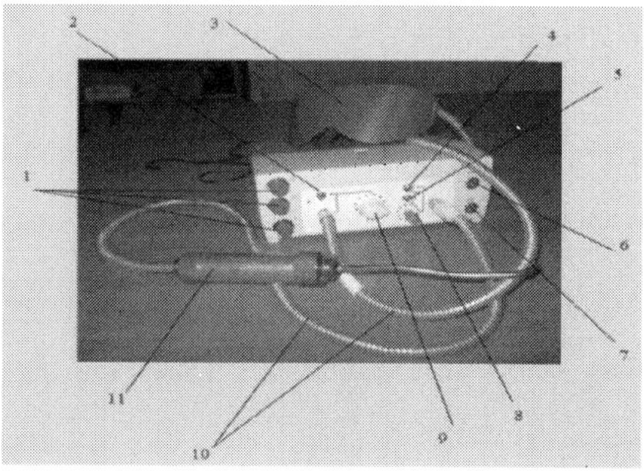


**Figure 24.** on / off RGB 1st – channel, 2 – mode switching manual / automatic 1-st channel, 3 – glasses for a patient, 4 – mode switching manual / automatic 2-channel, 5 – choice programs in incorporating LEDs 2nd channel, 6 –RESET reset switch program of the 2nd channel, 7 – STOP program stop switch of the 2nd channel, 8 – switching frequency control of the 2nd channel in manual mode, 9 – frequency control of the 1st channel, 10 – fiber optics cables, 11 fiber-tip – instrument.

In the developed OES, a hardware and software system was implemented that allows contactless obtaining of effective results of the current changes in the the psychophysical state of the organism during the session of the light-stimulation radiation program.

The software and hardware of the developed OES with the current values of change in indicators reflect the results of the analysis and provide a notification of the preschedule termination of the current session or unnecessary further sessions.

### 3.3. Application of the principle of comparing the importance of signals for the multi-parameter diagnostic technology

The doctor who conducts the medical session observes the changes in the optical parameters of the BFO on display. The display shown on a computer monitor are the results of the analysis of at least three optical parameters of the BFO. Having to work interactively, the doctor decides on the adequacy of medical procedures and the need for its repetition or appropriate further treatment of the chosen scheme. Subject to the attainment of atony, the doctor may use the second channel with a dynamic radiation exposure for the chosen color therapeutic program as the treatment for other diseases.

The results given were of the clinical research in the Lviv National Medical Institute. Daniel Galician indicates the functional ability of the proposed device and the prospects of its application in otorhinolaryngology.

Recently, in the treatment of skin diseases, there is an urgent need for express diagnosis. This necessitates the development of new methods and creating comfortable, lightweight, compact, simple, and inexpensive electronic means. This means that for the detection and identification of skin diseases by survey, the skin surface of the patient were developed on the principles of contactless electromagnetic radiation of the OES. The method is based on a comparison of the spectra in the reflected light skin problematic plot flows with test spectra accumulated in the database for the external manifestations of common dermatological diseases. This enables the physician, during the examination of the patient, to make a quick diagnosis of skin diseases such as eczema, seborrhea, surface manifestations of psoriasis, and others.

At the end or during the technology session, based on the results of the tested information, messages on a technology action result to an accurate medical decision.

What is important here is that the information content of the results of the medical procedures ensures the timely adoption of the medical decision. It could be more efficient to reduce the technology process in general, making possible cases which result to the redundancy of selective exposure to unwanted side effects.

If we consider the traditional methods of the testing processes, optical signals, and decision-making, it should be noted that until recently, licenses for the most widely-called tabular algorithm is based on the calculation with using tables. Due to the lack of sensitivity to the situation, some values (sometimes the most informative) is not taken into consideration for whatever reason. These drawbacks can be eliminated when using flexible algorithms based on a probabilistic approach. A Bayesian method or the method of sequential statistical analysis method (Wald method) is used [24].

The decision to this problem must be based on a probabilistic approach using a sequential statistical analysis of the results of the continuous automatic testing of the patient and peer reviews for developing the variants of solution by the doctor. The use of logic functions that form the basis of artificial intelligence elements has a huge importance.

At the first approach, logic functions can be viewed as a process of manipulating information and could present certain information signals, such as X1,..., XK. In this approach, the input signals are primary and logical reasoning is secondary. According to Bayes' formula, the measure of the reliability of conclusions about the effectiveness of either efficiency or inefficiency of medical procedures is the probability P (Yj / Xi) in a recurrent form:

$$P\left(Y_j / X_1,\ldots,X_K\right) = P\left(Y_j / X_1,\ldots,X_{K-1}\right)\frac{P\left(X_K / Y_j\right)}{P\left(X_K\right)} \tag{1}$$

If the analysis of one of the two available options, and provided that P (Y1) = P (Y2), the following are valid for statistically independent features:

$$\frac{P(Y_1 / X_1,...,X_K)}{P(Y_2 / X_1,...,X_K)} = \prod_{i=1}^{K} \frac{P(X_i / Y_1)}{P(X_i / Y_2)} \tag{2}$$

Or after the logarithm in a recurrent form:

$$u_K = u_{K-1} + \ln z_K \tag{3}$$

The decision rules in this case as follows:

$$\begin{aligned} u_K &\geq 0 \rightarrow X \in Y_1, \\ u_K &< 0 \rightarrow X \in Y_2. \end{aligned} \tag{4}$$

Together with equation (3) and isolating rule (4), the formula is interpreted as follows: if, after the consideration of another parameter the signal of sign magnitude $U_K$ has not changed, there are grounds for the termination of the analysis.

On the basis of the implementation of a flexible algorithm (Fig. 25).

For each time interval, the test values and the dynamics of change exceeded the critical standards of the emerging reports critical condition - $Y_K$ on the input signal $X_I$. In addition, for each time interval, the entire procedure is checked. In the case of exceeding the present time and the absence of the dynamics parameters of the dominant signals, the message YNE forms the effectiveness of the treatment procedure in the case of emergencies, which formed a disturbing message about the need for an immediate cessation in the treatments. The resulting data set is stored in a database that is used in the design of similar devices.

According to the proposed analysis algorithm for the test signal, the designed structural diagram of a medical tested device is based on the microcontroller PSoC CY8C24x94 family with built-in full-speed USB [25-27].

The basis of the microcontroller is a powerful microprocessor M8C with Harvard architecture, which controls the operation of the digital and analog blocks and surrounding devices. The digital part consists of four 8-bit internal blocks PSoC and interface modules of the user, which are used for connecting external devices: keyboard, display, power audio signal, and PC. The analog part consists of six indoor units, which allows the use of the following features: analog-to-digital and digital-to-analog conversion, amplification, programmable gain, pulse width modulation, detection, comparison, multiplexing, and correlation. The processing of input information carried out under the control of the embedded software, which is used to store energy-volatile memory (EEPROM and Flash). To store intermediate results during processing, an internal random access memory (RAM) is used.

According to the proposed analysis algorithm for the test signal, a structural diagram of a medical tested device based on the microcontroller PSoC CY8C24x94 family with built-in full-speed USB (Fig.26) was designed.

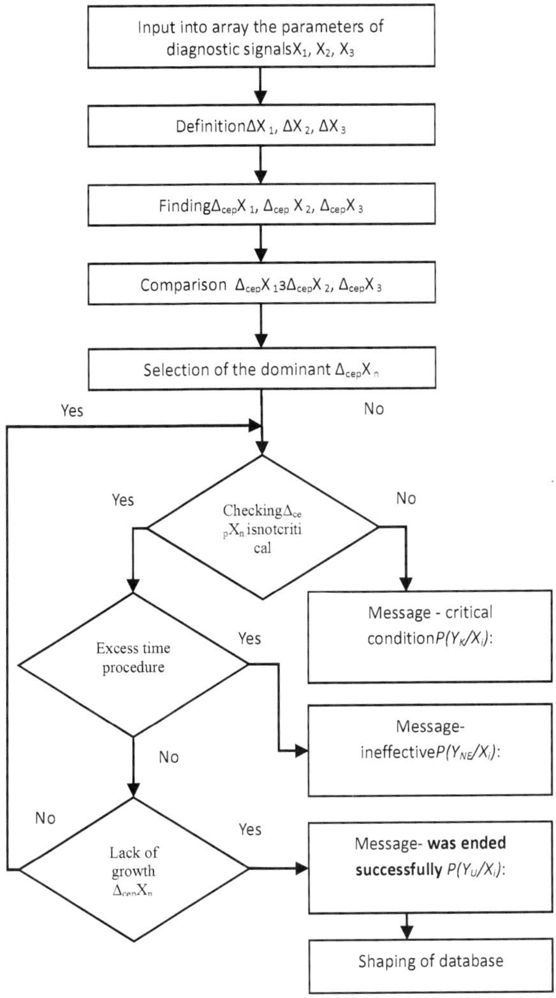


**Figure 25.** The flexible processing of the algorithm.

For the practical implementation of the device, which uses a PSoC Designer software by which the internal configuration of the microcontroller PSoC CY8C24x94 is debugged according to the functional diagram. Input signals X1-X3 after amplifiers AMR1-AMR3 filtered SF1-SF3 and through the multiplexer MUX to the ADC. Information obtained during the conversion process code that corresponds to the value of one parameter in a fixed point in time is stored in the RAM, the main memory uses for analysis, according to the developed algorithm. For the

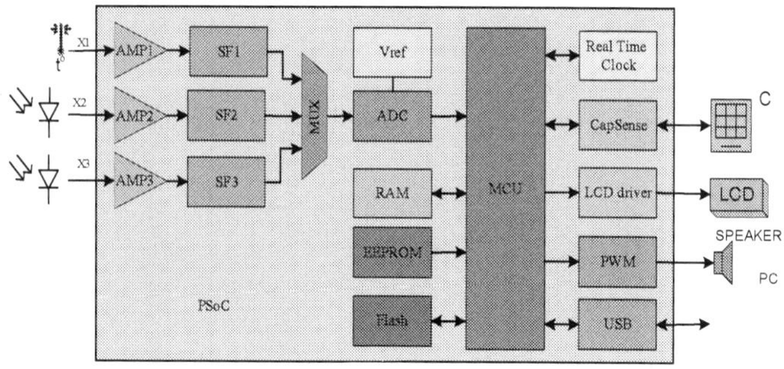


Figure 26. Functional diagram of implementation in the medical device.

formation of the timeframe during which the registration of input signals is carried, a Real-Time Clock is used. The managing works of the device is performed by using the keyboard via the corresponding interface module CapSense. The modes shown on the displayed operation is determined by the internal driver of the LCD module. In the event of a critical situation, additionally formed sound signals, its frequency and duration is given by module PWM. For a detailed analysis and the formation of the database of input parameters and corresponding test states, the results are transferred via USB to a PC.

It is proposed to improve the interactive physician format of the flexible new algorithm based on the probabilistic approach using sequential statistical analysis of the results of the continuous automatic testing of the patient for medical technologies efficacy of the procedure with the adoption of the basis of user solution.

The proposed method uses the elements of artificial intelligence and algorithm analysis with sufficient accuracy.

The results allow the creation of a new class of devices for medical technologies with continuous automatic evaluation of the effectiveness of a technological action results for an accurate medical decision-making and the creation of reports on previous technological strategies.

## 4. Research on an optoelectronic system for dermatology

Modern medical diagnosis requires the solution of problems related to the study and comparison of healthy (normal) biological tissues with corresponding areas of certain systemic diseases (e.g., the diagnosis and stage presence of diseases such as psoriasis, eczema, seborrhea, dermatitis, etc.). One of the achievements of recent years can be called transillumination method that is widely used, particularly in dermatology for the noninvasive diagnosis and identification of disease [19].

The necessary noninvasive diagnosis can be done by measuring and comparing the spectrophotometric parameters of normal and pathological areas. For a simplified version, transillumination, in the express diagnosis of skin pathologies, makes it feasible to create an inexpensive device that has an optical irradiator visible as colored fields and many diode matrix for light fluxes of different wavelengths on the surface of skin pathologies.

In the creation of a model of such a device, the LED, which studied the spectral characteristics of multi-colored LEDs that have been selected for optoelectronic matrix, is selected. The scheme of the experiment is shown on Fig. 27.

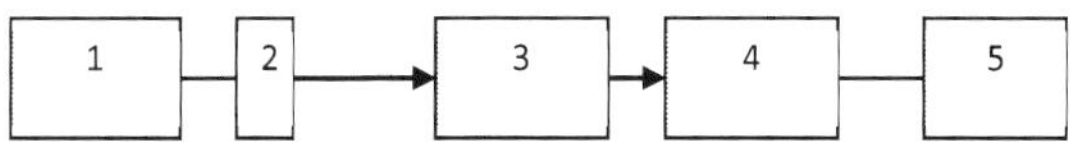


**Figure 27.** Block diagram investigation of the spectral characteristics of LED: 1 - LED power supply, 2 – LED, 3 – monochromator, 4 – photodetector, 5 - power meter radiation.

The research results presented in Table 4 are based on the spectral characteristics of present values of intensity (Figs. 28, 29) selected for the model of LEDs.

| red | | orange | | yellow | | green | | azure | | blue | | violet | |
|---|---|---|---|---|---|---|---|---|---|---|---|---|---|
| λ, nm | I, a.u. | λ, nm | I, a.u. | λ, nm | I, a.u. | λ, nm | I, a.u. | λ, nm | I, a.u. | λ, nm | I, a.u. | λ, nm | I, a.u. |
| 610 | 0 | 633 | 0 | 568 | 0 | 508 | 0 | 400 | 0 | 416 | 0 | 420 | 0 |
| 633 | 0.57 | 639 | 1 | 586 | 0.35 | 530 | 0.26 | 420 | 1 | 429 | 0.8 | 426 | 0.16 |
| 639 | 0.69 | 666 | 0.77 | 590 | 1 | 541 | 1 | 442 | 0.5 | 435 | 1 | 435 | 0.25 |
| 666 | 1 | 680 | 0.35 | 599 | 0.31 | 565 | 0.52 | 469 | 0 | 442 | 0.6 | 443 | 0.12 |
| 680 | 0.35 | 710 | 0 | 610 | 0 | 592 | 0 | | | 444 | 0 | 452 | 0 |
| 709 | 0 | | | | | | | | | | | | |

**Table 4.** The dependence of the radiation intensity of LED I on the wavelength λ.

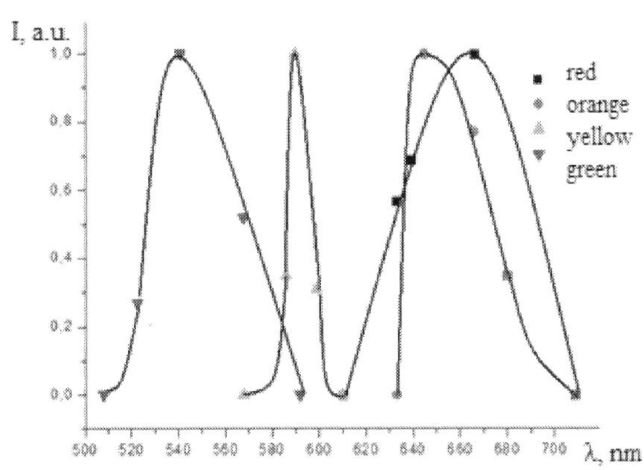


**Figure 28.** Spectral characteristics of longwave LEDs chosen for the developed model.

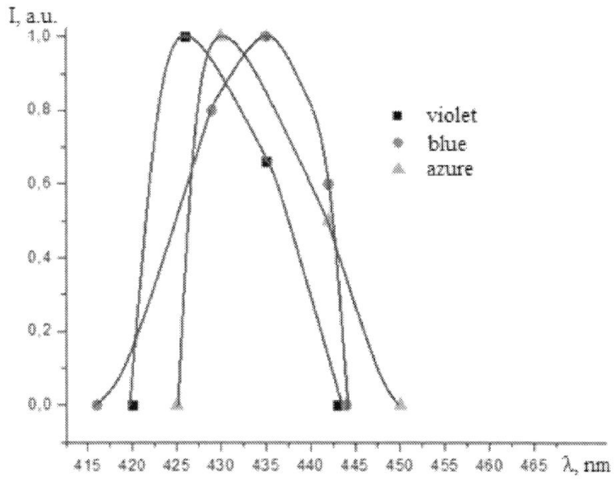


**Figure 29.** Spectral characteristics of shortwave LEDs chosen for the developed model.

From the graphs (Figs. 28, 29), the full range of radiation selected of LEDs completely covers the need for a radiation area. Thus, the radiation that is created by the LED strip covers the following wavelengths: blue - 508 to 592 nm, yellow - 568 to 610 nm, red - 610 to 709 nm, orange - 633 to 709 nm, azure - 429 to 450 nm, blue - 416 to 444 nm, and purple - 420 to 443 nm. The maximum radiation intensity of LED corresponds to these wavelengths: green - $\lambda$ = 541 nm, yellow - $\lambda$ = 590 nm, red - $\lambda$ = 666 nm, orange - $\lambda$ = 639 nm, blue - $\lambda$ = 430 nm, blue - $\lambda$ = 435 nm, and violet - $\lambda$ = 426 nm.

Based on the selected LEDs set, the LED matrix optoelectronic element mock (Fig. 30).

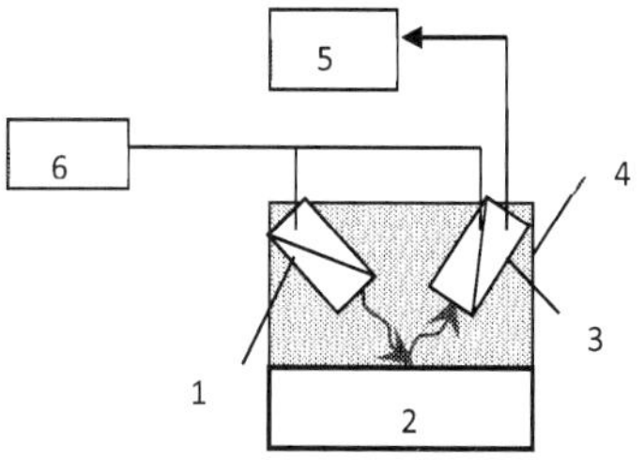


**Figure 30.** Scheme of researches: 1 – photodetector, 2 - color test sample, 3 - LED 4, - dimming building, 5 - radiation power meter, 6 - power supply.

The LED and the photodetector are located on the reflecting surface (sample) at angles close to 45º so that the radiation coming from the LED is reflected and reaches the photodetector.

The LED and photodetector were in the dimming housing for protection against visible light when they were measured.

The obtained results are reduced to a common matrix (Table 5).

| Color sample | red | orange | yellow | green | azure | blue | violet |
|---|---|---|---|---|---|---|---|
| LED | | | | | | | |
| red | 8.26 | 8.25 | 8.26 | 4.92 | 7.47 | 4.05 | 6.58 |
| orange | 5.6 | 7.04 | 7.64 | 3.76 | 5.22 | 2.63 | 4.38 |
| yellow | 1.47 | 5.12 | 5.48 | 1.47 | 2.74 | 1.44 | 1.47 |
| green | 3.75 | 4.72 | 8.23 | 6.13 | 8.24 | 4.03 | 4.12 |
| azure | 6.48 | 5.14 | 7.59 | 7.08 | 3.36 | 8.25 | 7.84 |
| blue | 5.44 | 6.17 | 7.85 | 3.87 | 8.17 | 3.78 | 3.84 |
| violet | 6.22 | 5.63 | 6.21 | 3.02 | 4.78 | 2.29 | 3.64 |

**Table 5.** Matrix intensity of the reflection surface of the skin for different parts of the visible spectrum.

Based on the data obtained, the field developed a policolor probe as part of the dermatological OES, allowing the identification of the individual colors of the reflected radiation intensity for the matrices in seven major areas of the visible spectrum. It is aimed at the examined surface radiation source that constitutes LEDs that turn on their radiation spectra to cover the entire visible wavelength range, located at an angle to the optical axis and coaxially around the sensor mounted on. With the possibility to be away from the examined surface, the reflected light beams from each of the LEDs are enable alternate or simultaneous switching.

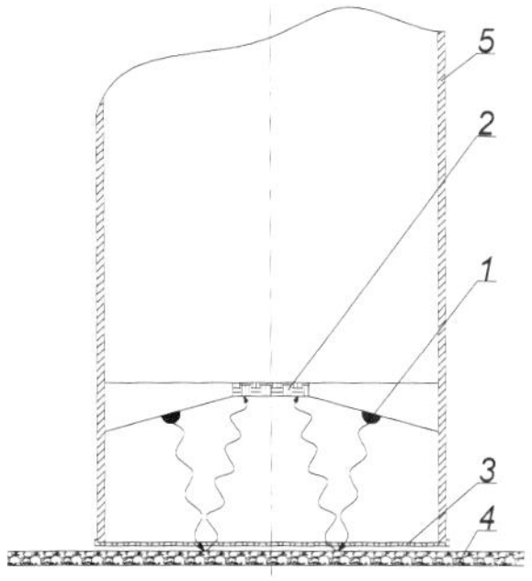


**Figure 31.** Opticoelectronic probe: 1 – LED matrixs, 2 – photodetector, 3 - Protective optically transparent element, 4 – biomedical object, 5 – hull.

The optoelectronic probe system consists of LED matrix 1 consisting of LED type MUV9C, photodetector 2, which serves as a photodiode PD-24, and an optically transparent protective element 3, which is a translucent polystyrene plate (Fig. 31). The falling light output of the LED matrix 1 after passing through the optically transparent protective element 3 creates a reflection from the surface of the investigated object. The biomedical luminous flux again passes through an optically transparent protective element 3 and reaches the photodetector 2. The signal from the sensor enters the amplifier and then measures the photocurrent contained in the measuring-controlling unit.

On the basis of experimental and computational studies created a working model (Fig. 32).

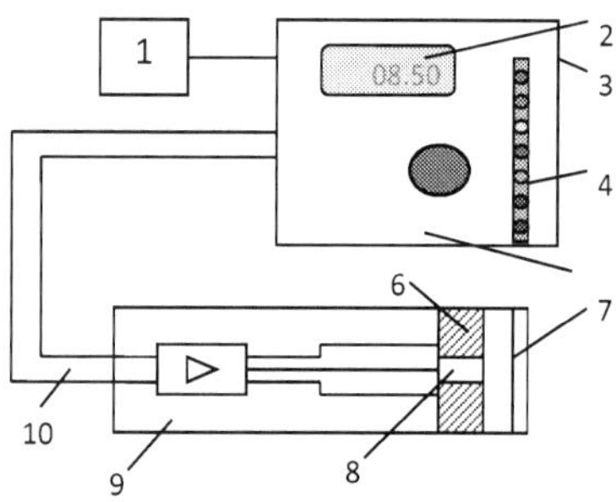


**Figure 32.** The structure of the model: 1 - power supply, 2 - value of photocurrent, 3 - hull, 4 – indicator, 5 - measurement and a control unit, 6 - LED matrix, 7 - protective cover, 8 – photodetector, 9 - optoelectronic unit, 10 - cable.

The dimensional control block is designed for the control of optoelectronic measuring unit and its signal. By pressing the "Start", the LEDs are alternately turned on in the LED matrix composed of red, orange, yellow, green, blue, violet, and blue LEDs. The indicators reflect a powerful monitor LED. The luminous flux of the LED matrix reflected by the research object reaches the photodetector. The signal from the photodetector reaches the amplifier and gets the value of the photocurrent.

In medical practice, for the detection and identification of disease by the spectral characteristics of skin manifestations and with appropriate spectral characteristics, it is convenient to use the tabulated values of the intensity of the reflected light from surfaces of such flows. For each of these tables, there are LED test matrices to identify the disease in the course of its treatment (Table 6).

| LED | | red | orange | yellow | green | azure | blue | violet |
|---|---|---|---|---|---|---|---|---|
| Healthy skin | I, в.о. | 8.14 | 5.5 | 3.07 | 5.7 | 8 | 5.86 | 4.55 |
| Psoriasis | | 7.82 | 5.72 | 2.66 | 5.17 | 7.26 | 5.24 | 4.53 |
| Seborrhea | | 6.74 | 4.6 | 2.09 | 5.23 | 6.78 | 4.67 | 3.78 |
| Chronic dermatitis | | 8.16 | 4.96 | 2.3 | 4.98 | 6.98 | 5.32 | 4.58 |

**Table 6.** The test matrix of skin surface.

The resulting matrix development can be used in clinical trials to further establish the basis of the reflective optical imaging device.

On the basis of what has been detected during the mathematical and physical modeling regularities in the interaction of light streams with pathological skin areas, optimized selection and relative positions photodetectors, and irradiative components of the proposed device to hospitals on the basis of a specialized OES (optical-electronic probe), a display and access to a personal computer is included in each visit (Fig. 33) [28].

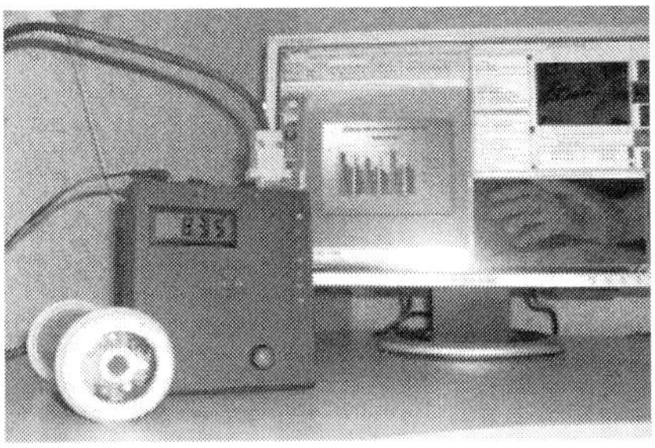

**Figure 33.** OEC components and probe express diagnostics of skin diseases (from left to right - probe measurement unit, PC). 1 - LED matrix, 2 – photodetector, 3 - protective optically transparent element, 4 – BO, 5 – enclosure.

The methodology for light diagnostic research by means of the developed device is an external surface diagnosing skin with a suspected pathology of supply optoelectronic probe to the problem area. A new application of the specialized OES allows the doctor to make long-term monitoring of the changes in cell behavior, unaccompanied by heated skin, and the different effects on biochemical, histological, and functional levels and replenishment on this bases of biological tissues in diseases including digital pictures of the cutaneous manifestations of disease. Clinical studies have shown that the application of the proposed specialized OES significantly reduced the process of diagnosing the level with a rapid process and expanded the diagnostic capabilities for the identification of detected pathologies.

## Author details

Barylo Hryhoriy[1], Gotra Zenon[1], Ivakh Mariya[1], Kozhukhar Oleksandr[1*], Makara Ivanna[2] and Virt Volodymyr[1]

*Address all correspondence to: akozhukha@ukr.net

1 The ED Department, Lviv Polytechnic National University, Ukraine

2 The IM Department, Lutsk National Technical University, Ukraine

## References

[1] Tuchin V.V. "Handbook of Optical Biomedical Diagnostics", SPIE Press, 2002 - 1093 p.

[2] Rodney Cotterill "Biophysics: an introduction", 2002, 408 p.

[3] Ivakh Mariya, Kozhukhar Alexander, Zazuljak Andriy, Kosyi Evgen "Expansion of therapeutic possibilities based on the developed optronic systems monitoring progress of medical treatment", Biomedical Engineering and Electronics, №1, 2012, http://www.es.rae.ru:8888/biofbe/.

[4] Kozhukhar O.T., Kurgan M.G., Skira M.S., Kurgan D.M. "Problem device for illumination in treatment by photopheresis"in 4th International Forum on Innovative Technologies for Medicine (ITMED 2010), Bialystok, 2010, P. 32.

[5] Oleksandr Kozhukhar, Mariya Skira, Yurij Kuzio "Descriptions of optoelectronic elements are for control of changes of optical properties of blood after the photopheresis" in Xth International Conference "Modern problems of radio engineering, telecommunications and computer science", 23–27 Feb. 2010, P. 122.

[6] Alternative approaches to improving ultraviolet actuators photopheresis (in Ukrainian) / Z. Gotra, M Ivakh, A. Kozhukhar, K. Voytsihovska / Systems of information processing, № 2 (92), 2011, pp. 239-243.

[7] Gotra Zenon, Mariya Ivakh, Kozhukhar Alexander "System of testing the patient's blood advisability photopheresis treatment" in 4th International Forum Science and Technology Days Poland –East, 12-14 April, 2011,Bialowieza, 2011, P. 31.

[8] Z.Y. Hotra, A.T. Kozhukhar, A.M. Zazuljak, M.S. Ivakh "Development of noncoherent optical-electronic systems in modern photo medical technology" Proc. SPIE 8698, Optical Fibers and Their Applications 2012, 86980J (January 11, 2013); doi:10.1117/12.2019268

[9] Zenon Gotra, Alexander Kozhukhar, Igor Didych, Andrij Zazuljak, Marija Skira, Olena Tkachenko. The universal sensory optronic system for estimation of blood technology quality, The conference catalogue chronicles theproceedings of the 3rd International Forum on Innovative Technologies for Medicine, Białystok, 2009, p. 39.

[10] Device for photostimulation therapy in hematology (in Ukrainian): patent Ukraine 62114 / Hotra Zenon; Kozhukhar Oleksandr; Ivakh Mariia (2011)

[11] Optical electronic control of activity of treatment procedures / Kozhukhar Alexander, Andriy Zazuljak, Ivakh Mariya // 4th International Forum Science and Technology Days Poland –East: Forum procedeengs. – Bialowieza: Innovative Eastern Poland Asociation, 2012, pp. 47-52.

[12] Grishchenko G. Bioresonance stimulation of the human organism as a substitute for application of chemical medicines (in Ukrainian), Proceedings [electronic resource], 2010, Mode of access:http://bibl.kma.mk.ua/index.php?m=10&s=3&t=6.

[13] *S.Tonnies "Entspannung fur Tinnitusbetrofenne durch Photostimylation" Springer Medizin Verlag, № 54, 2006, pp.481-486.*

[14] Kozhukhar O.T., Zazuliak A.M., Kuchak E.V. "The influence fotostymuliv with frequencies on human biorhythms body" Electronics" Bulletin ou "LP", № 680, 2010, pp. 225-230.

[15] I Didych, A. Zazulyak, O. Kozhukhar "The programable system is for photomedic technologies". In XIII Miedzynarod. Szkoly Komputerowego wspomaganja projektovanja. Wytwarzania I eksploatacji. Jurata. 11-15 05.2009.

[16] A. Kozhukhar, M. Kostiv "Photomedic light sources with controlled dynamics radiation" (in Ukrainian) in First All-Ukrainian Congress of "Medical and Biological Informatics and Cybernetics" with international participation. - Kyiv. 23-26 June 2010, p. 254.

[17] Hotra Z., Zazuliak A., Kozhukhar A. "Otorhinolaryngological photo therapeutics programmatically controlled complex with a continuous the assessment of efficiency procedure" Perspective technologies and devices, № 2, 2012, p. 30-36.

[18] *Zazuljak, A. Kozhuкhar, O.Tkachenko "Therapeutic and diagnostic devices based on encephalographic frequencies programmed low intensity photostimulation on visual Receptors" in 2nd Forum Science&Nechnology Days Poland., 2009, p. 42.*

[19] Tkachuk R., Makara I., Kozhukhar O., Tkachenko O. "Exspress diagnostics and authentication of illnesse is with application of the new optical-electronic systems" in 6th International Forum on Innovative Technologies for Medicine (ITMED 2012), 21–23 November 2012, P. 24-25.

[20] Z. Hotra, A. Zazuljak, M. Ivakh "New approaches to continuos information support and evaluation effectiveness of physiotherapeutic procedures" in 12-th International conference "The experience of desigining and application of cad systems in microelectronics", 19-23 Feb. 2013, Polyana-Svalyava (Zakarpattya), 2013, pp. 152-153.

[21] Zenon Gotra, Hryhoriy Barylo, Aleksandr Kozhukhar, Mariya Ivakh, "Implementation of flexible algorithms for continuous monitoring and operative correction of medical technologies Innovative eastern poland association" in 7th International forum on innovative technologies for medicine, 5-7 decem. 2013, Bialystok, 2013, p. 10.

[22] Device for phototherapy via visual receptor (in Ukrainian): patent Ukraine 42525: Kozhukhar Oleksandr; Skunts Nazar; Zazuliak Andrii (2009)

[23] Noninvasive sensor device for registration of changes in psycho-physical state of human person (in Ukrainian): patent Ukraine 60600, Hotra Zenovii; Kozhukhar Oleksandr; Zazuliak Andrii; Kuchak Yevhen (2010)

[24] Guosheng Yin Bayesian Generalized Method of Moments, Bayesian Analysis (2009) Number 2, pp. 191-208

[25] Robert Ashby. Designer's Guide to the Cypress PSoC (Embedded Technology).

[26] D. Kilochek "Design on programmable systems-on-chip PSoC Cypress" Components and technology, № 4, 2006.

[27] G.I. Barylo, M.S. Ivakh, A.M. Zazulyak, O.T. Kozhukhar "Information support of medical procedures based on automated analysis perephericperformance" in the twentieth international scientific-practical. conf. "Information Technology: Science, Technology, education, health: CH.III (May 15-17, 2012, Kharkiv), 2012. - P. 96

[28] Light-diagnostic survey device: patent Ukraine 53154 / Melnyk Ivanna; Kozhukhar Oleksandr.

# Chapter 14

# A New FPN Cancellation Circuit for Time-Domain CMOS Image Sensors

Fernando de S. Campos, José Alfredo C. Ulson,
José Eduardo C. Castanho,
Bruno Albuquerque de Castro and Rudolf Riehl

Additional information is available at the end of the chapter

http://dx.doi.org/10.5772/60756

**Abstract**

A fixed-pattern noise correction technique for time-domain CMOS imagers with high dynamic range is presented in this chapter. Analytical derivations are presented showing how the circuit variations affect the time measured. The error in the time measured can be reduced by using lower reference voltages achieving values smaller than 4%. The fixed-pattern noise correction technique proposed is based on a new readout method for time-domain imagers employing two reference voltages for the discharge time measurement. This new technique is non-sensitive to circuit parameter variations that contribute to fixed-pattern noise such as hold voltages of transistors. A simple electronic circuit is proposed to implement the technique. Circuit and simulation results are presented to demonstrate the feasibility of the proposed technique.

**Keywords:** CMOS image sensors, Active Pixel Sensor, fixed-pattern noise, photodiode

## 1. Introduction

CMOS image sensors have been an important source of research and industrial development in the area of image sensors due to the present high level of integration and low power consumption compared to charge-coupled devices (CCDs). The technology advances of CMOS image sensor results in image sensors with high dynamic range (>=100dB). Higher dynamic range become other important advantage of CMOS image sensors over CCDs. However, most of CMOS imager architecture with high dynamic range neglect the fixed-pattern noise (FPN) and do not present any kind of cancelling circuit or impact on such architectures. Therefore, although conventional linear CMOS sensors incorporate FPN cancellation techniques as correlated-double sampling (CDS), in general, it is still missing in the literature of FPN cancellation techniques for high dynamic range CMOS imagers.

Several techniques were proposed to achieve high dynamic range using CMOS image sensors [1-6]. In [2], by integrating a comparator and a dynamic flip-flop in each pixel, a multi-sampling technique was proposed. In [3], a self-reset technique using one comparator per pixel was described. In general, these techniques are effective as they achieved a dynamic range >100dB, but, none of them show the impact of FPN or present FPN cancellation techniques.

Recently, high dynamic range CMOS imagers using analog-to-digital converters have been proposed and they can incorporate CDS techniques for FPN cancellation although they do not present any FPN impact analysis [7-12]. However, these approaches achieve dynamic ranges of only 70–80dB, which is too low for some applications that typically demand dynamic range higher than 100dB.

In this chapter we present a FPN impact analysis and a technique of FPN cancellation for time-domain CMOS imagers with high dynamic range.

## 2. Principle of operation of time-domain CMOS imagers

Figure 1a shows the typical time-domain CMOS pixel. It comprises a photodiode, a reset transistor and a comparator. The pixel operation has two stages. The first stage is the reset period. During the reset period the reset transistor operating as switch is closed and the photodiode voltage ($V_{pd}$) is charged to an initial voltage $V_{reset}$ close to $V_{dd}$. At the beginning of the second stage, the integration period, the reset transistor is turned off and the photodiode at high impedance starts to discharge. The discharge speed during discharge is according to the photocurrent intensity (see Figure 1b). The discharge time is smaller for higher photocurrent values.

The time-domain CMOS sensor operation consists of measuring the discharge time of the photodiode during the integration period by comparison of photodiode voltage with a reference voltage (see Figure 1b). This comparison is made by using a comparator as seen in Figure 1a. The discharge time is defined as the instant in which the photodiode voltage drops below (ou to) the reference voltage and the output comparator goes high. According to [2], the discharge time as function of the photocurrent for a reference voltage $V_{ref}$ is given by:

$$t_d = \frac{C_{pd}}{\left(I_{dark} + I_{ph}\right)}\left(V_{reset} - V_{ref}\right) \tag{1}$$

http://dx.doi.org/10.5772/60756

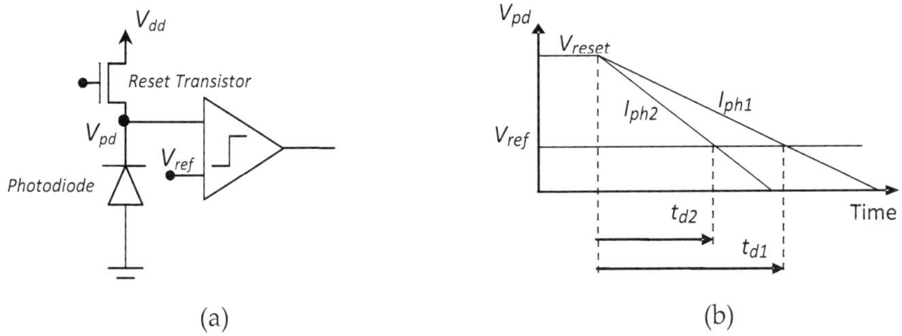


**Figure 1.** (a) Time-domain CMOS pixel; (b) main waveforms.

where $C_{pd}$ is the photodiode's capacitance, $I_{dark}$ is the dark current, $I_{ph}$ is the photocurrent, $V_{reset}$ is the initial photodiode's voltage and $V_{ref}$ is the reference voltage.

Figure 2 shows the discharge time for different values of reference voltage assuming $C_{pd}$=30fF and $I_{dark}$=100fA. As one can see, the discharge time is smaller for higher reference voltages. For photocurrent smaller than 100fA, the curve saturates due to the presence of the dark current.

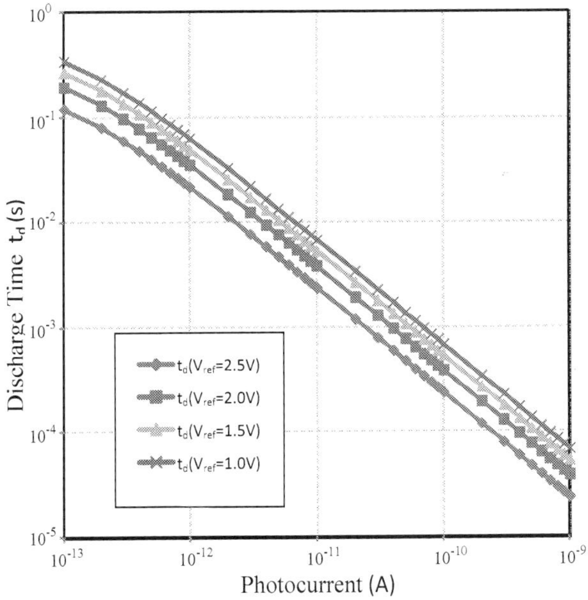


**Figure 2.** Discharge time of a time-domain CMOS sensor at different values of reference values.

# 3. Fixed-pattern noise in time-domain CMOS imagers

The fixed-pattern noise is defined as the variation of pixel's photoresponse in a matrix when a uniform light intensity focuses the matrix. Time-domain CMOS imagers have two main sources of variation that contributes to fixed pattern noise; (1) the variation of the reset voltage ($\Delta V_{reset}$) due to threshold voltage variation of the reset transistor and (2) the offset voltage of voltage comparator ($\Delta V_{offset}$). The threshold voltage variation causes a variation on the discharge time that can be written as

$$t_d + \Delta t_d = \frac{C_{pd}}{(I_{dark} + I_{ph})}\left(V_{reset} \pm \Delta V_{th} - V_{ref} \pm \Delta V_{offset}\right) \qquad (2)$$

where $\Delta t_d$ is the error reflected in the discharge time measured, $\Delta V_{th}$ is the variation introduced by the threshold voltage variation of reset transistor and $\Delta V_{offset}$ is the voltage variation introduced by the offset voltage of the voltage comparator. It is possible to demonstrate from the previous equation that the error reflected in the discharge time ($\Delta t_d$) can be written as

$$\Delta t_d = \frac{C_{pd}}{\left(I_{dark} + I_{ph}\right)}\left(\Delta V\right) \qquad (3)$$

where, $\Delta V = \Delta V_{th} + \Delta V_{offset}$.

According equation (3), the discharge time error is proportional to voltage error introduced ($\Delta V$), and inversely proportional to photocurrent and. therefore the discharge time error varies according light intensities changes. Figure 3 shows the discharge time error as function of photocurrent assuming $C_{pd}$=30pF, $I_{dark}$=100fA for three different values of voltage errors.

Manipulating equation (3), it can be demonstrated that the percentage of relative error is given by

$$\frac{\Delta t_d}{t_d}\% = \frac{\Delta V}{\left(V_{reset} - V_{ref}\right)}100\% \qquad (4)$$

According to equation (4) the relative error of discharge time is independent of light intensity (photocurrent) and inversely proportional to reference voltage. Figure 4 shows the relative discharge time error for three different values of voltage error. As one can see, the relative discharge time error is smaller for small reference voltage values. It is smaller than 4% at $V_{ref}$=0.5V.

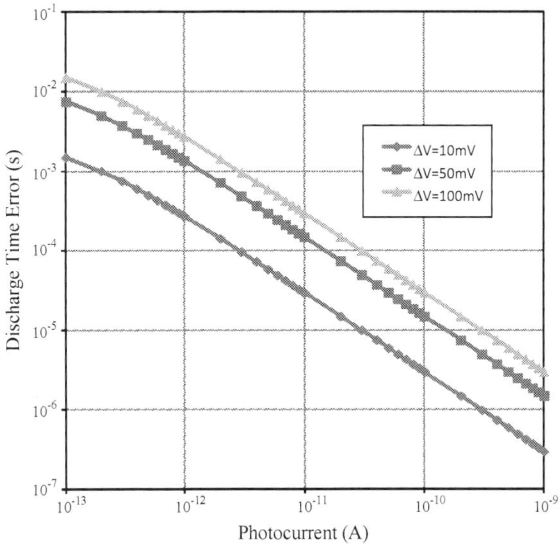


**Figure 3.** Discharge time error.

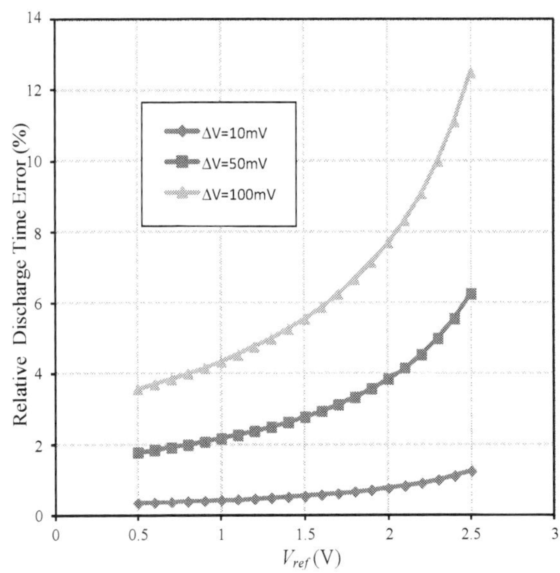


**Figure 4.** Relative discharge time error.

## 4. New readout method for time-domain with FPN reduction

In order to cancel the voltage error introduced (see equation 2), we propose to measure the discharge time between the comparisons with two reference voltages as shown in Figure 5.

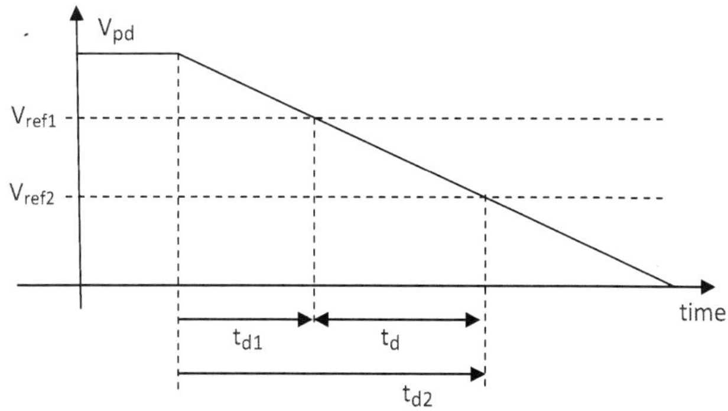


**Figure 5.** New readout method main waveforms.

The instant of comparison are two, $t_{d1}$ and $t_{d2}$. For $V_{ref1}$ the comparison time is given by

$$t_{d1} = \frac{C_{pd}}{\left(I_{dark} + I_{ph}\right)}\left(V_{reset} \pm \Delta V_{th} - V_{ref1} \pm \Delta V_{offset}\right) \tag{5}$$

For $V_{ref2}$ the comparison time is given by

$$t_{d2} = \frac{C_{pd}}{\left(I_{dark} + I_{ph}\right)}\left(V_{reset} \pm \Delta V_{th} - V_{ref2} \pm \Delta V_{offset}\right) \tag{6}$$

The discharge time is now given as the interval time between $t_{d1}$ and $t_{d2}$. Subtrating equations (5) of (6), it is possible to demonstrate the discharge time given by

$$t_d = t_{d2} - t_{d1} = \frac{C_{pd}}{\left(I_{dark} + I_{ph}\right)}\left(V_{ref1} - V_{ref2}\right) \tag{7}$$

According to equation (7) the new discharge time is independent of the source of fixed-pattern noise because it is independent of reset voltage and the offset voltage of voltage comparator. However, the same comparator must be used in order to achieve the offset voltage cancelling.

Figure 6 shows the transfer function graph of $t_d$ for $V_{ref}1$-$V_{ref2}$=2V, $C_{pd}$=30fF and $I_{dark}$=100fA. The characteristic of time versus photocurrent is the same of conventional time-domain CMOS imager as can be seen in Figure 3.

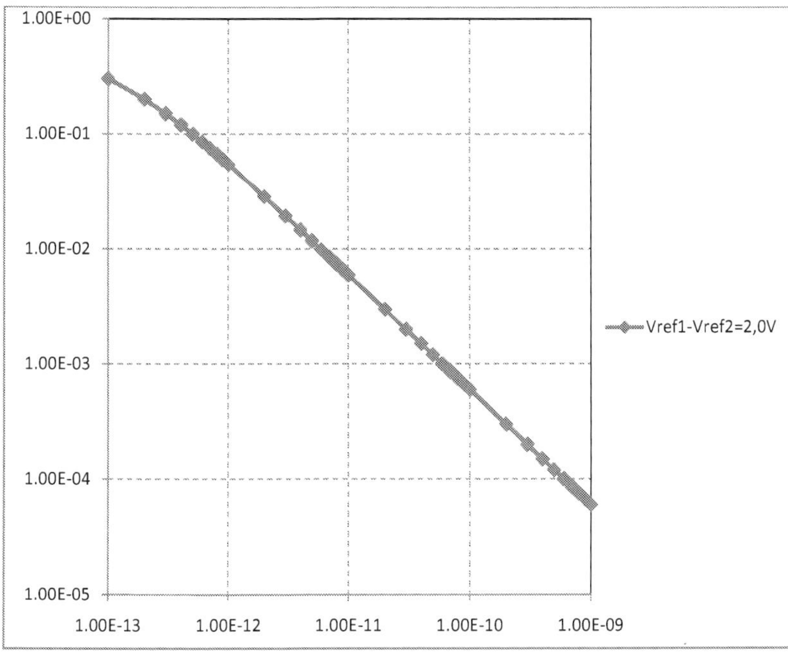


**Figure 6.** Discharge time (s) versus photocurrent (A).

## 5. Simulation results

### 5.1. FPN time-domain characteristic without FPN cancellation

Figure 7 shows the equivalent circuit of the CMOS sensor image used for simulations. The simulated pixel is comprised of a reset transistor and capacitance and a source current as equivalent circuit of photodiode. We assume that the photocurrent is constant and independent of voltage as usual in the literature. For simulations we used a design kit of a 0.35μm 3.3V CMOS technology.

In order to achieve the transfer characteristic time versus photocurrent simulated, we use a constant reference voltage of 1.5V, reset voltage of 3.3V and a typical capacitance of 30fF. By varying the photocurrent different comparison times for reference voltage of 1.5V were found

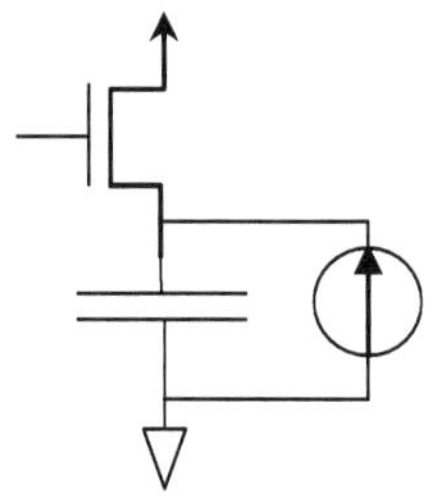

**Figure 7.** Discharge time (s) versus photocurrent (A).

and the obtained transfer characteristic is shown in Figure 8. Our analysis indicates that the simulation results are in agreement with equation (1).

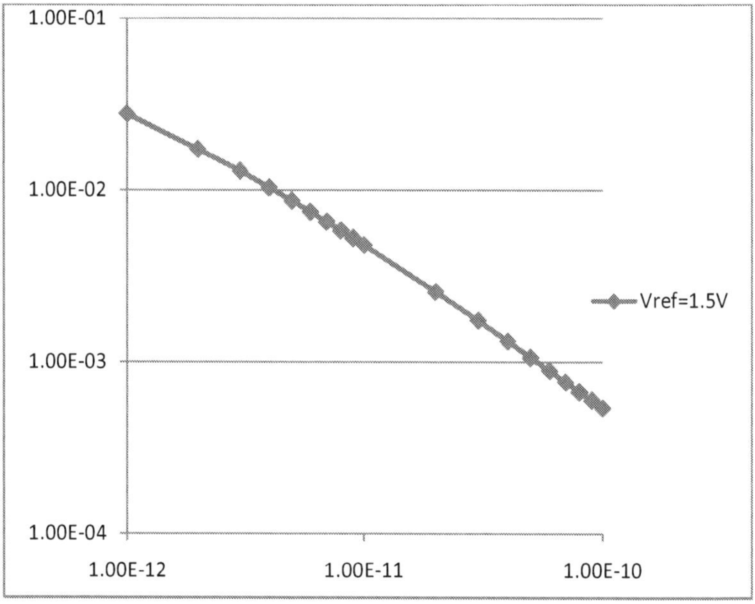


**Figure 8.** Discharge time (s) versus photocurrent (A).

For simulating the error voltage introduced by circuits, we choose to change the reset voltage (initial voltage) from 3.3V to 3.2V. Therefore, we simulate a error voltage of 100mV caused by the variation of the threshold voltage of the reset transistor and/or by the offset voltage of the comparator. Using a reference voltage of 1.5V and changing the initial voltage in the simulations we found a comparison time error from 6% to 10% as shown in Figure 9.

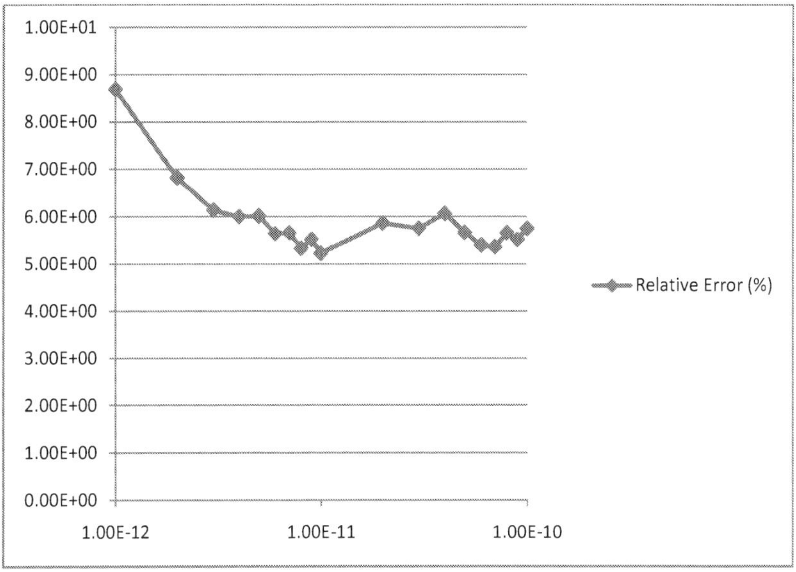


**Figure 9.** Relative error (%) versus photocurrent (a) for 100mV of variation in reset voltage.

This result indicates that the FPN in time domain for an entirely pixel matrix is less than 10%. The error is calculated as

$$error(\%) = 100 * \frac{t_d(3.3V) - t_d(3.2V)}{t_d(3.3V)} \tag{8}$$

### 5.2. FPN time-domain characteristic with FPN cancellation

Two references voltages are used for FPN cancellation. We use $V_{ref1}$=2.5V and $V_{ref2}$=2.0V as reference voltages. For this case, the transfer characteristic discharge time versus photocurrent is shown in Figure 10. The discharge time $t_d$ was measured as the difference between the discharge time to V1 and the discharge time to V2 ($t_d = t_d(2.5V) - t_d(2.0V)$). Our analysis indicates that the transfer characteristic indicated in Figure 10 is in agreement with equation (7).

A simulation with 100mV of error introduced again was performed and the error was calculated as equation (8). For this case, the relative error found is shown in Figure 11. This simulation result indicates that the relative error is less than 2% percent. Therefore, the simulation result shows that this technique can reduce the FPN in time-domain imagers.

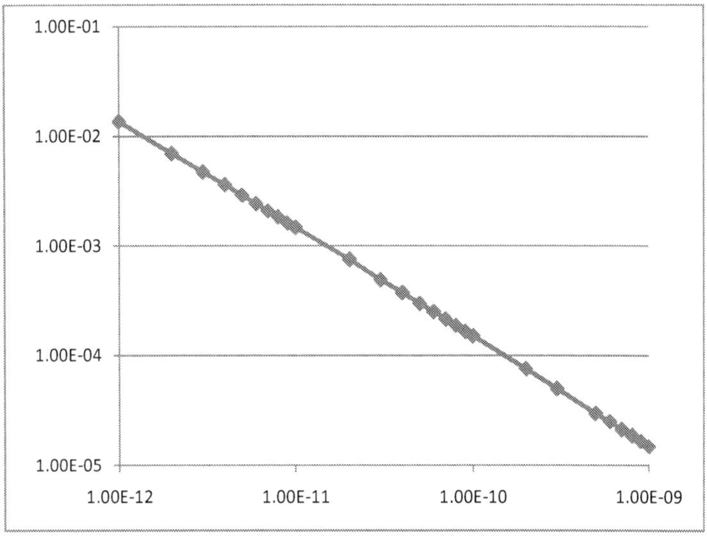


**Figure 10.** Discharge time (s) versus photocurrent (A) for Vref1=2.5V and Vref2=2.0V.

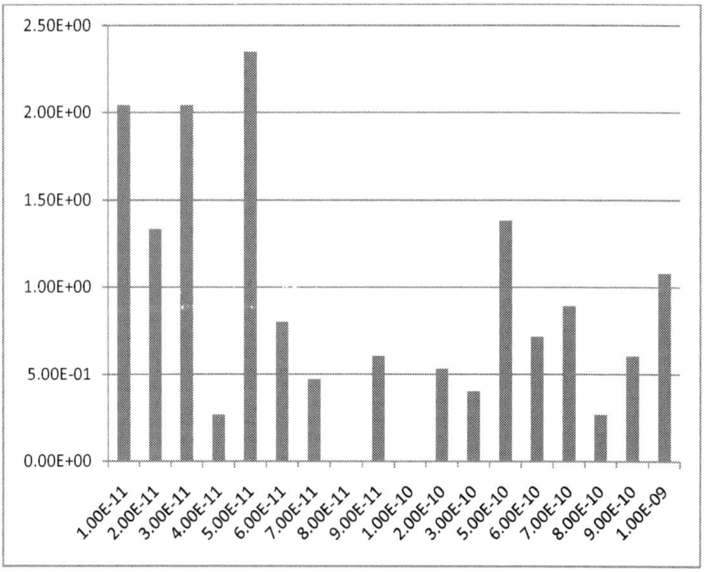


**Figure 11.** Relative error (%) versus photocurrent (A).

http://dx.doi.org/10.5772/60756

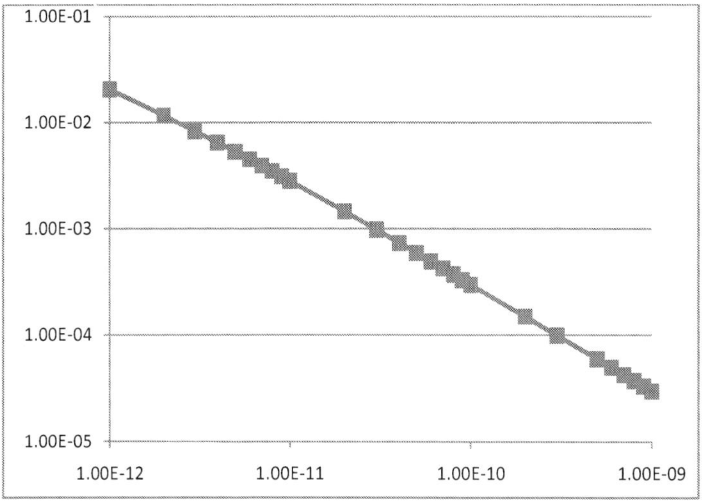


**Figure 12.** Discharge time (s) versus photocurrent (A) for Vref1=2.5V and Vref2=1.5V.

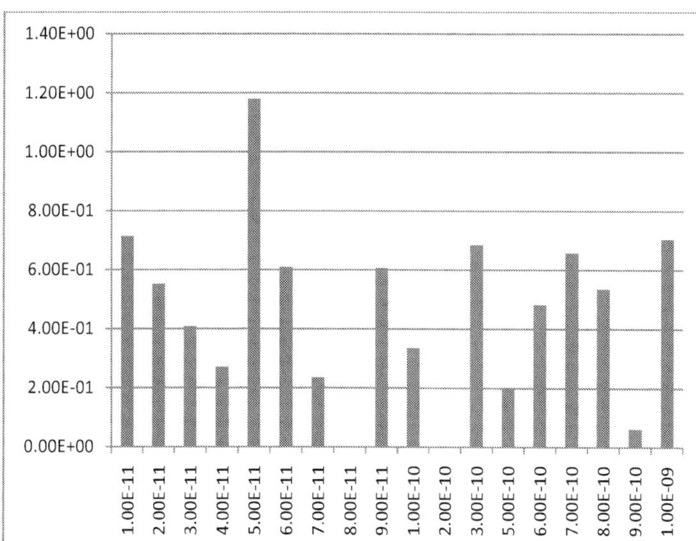


**Figure 13.** Relative error (%) versus photocurrent (A).

In order to confirm the results, we decide to simulate using reference voltages of V1=2.5V and V2=1.5V. As one can see in Figure 12, for this case the transfer characteristic is in agreement with equation (7) and the transfer characteristic in Figure 10.

For this case, an error of 100mV was introduced in the circuit simulated as stated before. The relative error found in this case is shown in Figure 13. Again the relative error is less than 2% showing that the technique can reduce the FPN.

## 6. Electronic circuit for time domain FPN cancellation

The challenge in this technique implementation is to develop a circuit that uses the same comparator (same offset voltage) to compare two reference voltages. For this purpose we propose the circuit shown in Figure 14.

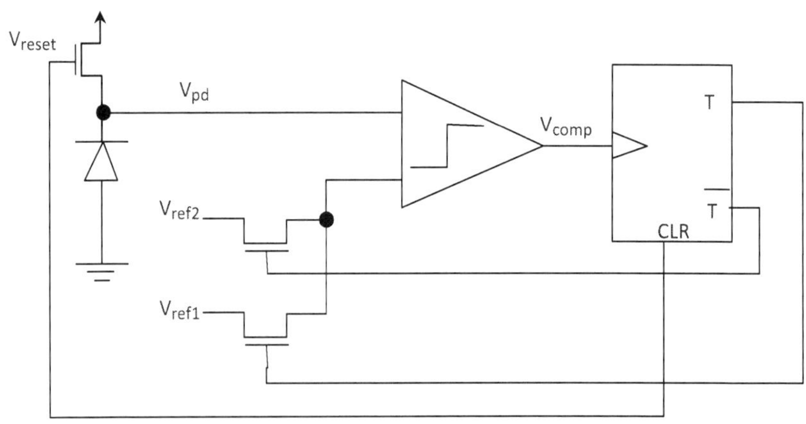


**Figure 14.** Electronic circuit of new FPN cancelling for time-domain imagers.

The circuit is composed of a comparator, two selector transistors (switches), and a T Flip-Flop. The main waveforms of signal are shown in Figure 15. First, the reset signal used to reset the pixel is also used to reset the flip-flop and the counter. Initially at low, the output of flip-flop control the two switches selecting only one of the reference voltages ($V_{ref1}$). After reset, and during the beginning of integration time, the $V_{ref1}$ reference voltage is selected for comparison. At instant of comparison, when the output pixel is $V_{ref1}$, the output comparator triggers the flip-flop that changes to high. At this moment, the switches are biased disconnecting $V_{ref1}$ and connecting $V_{ref2}$. The output of the comparator goes low again but the output of the flip-flop remains high. When the output pixel voltage reaches $V_{ref2}$ the comparator output pulses again trigger the flip-flop output to low. At this moment the counter stops and the reference voltage $V_{ref1}$ returns to be connected to the comparator input.

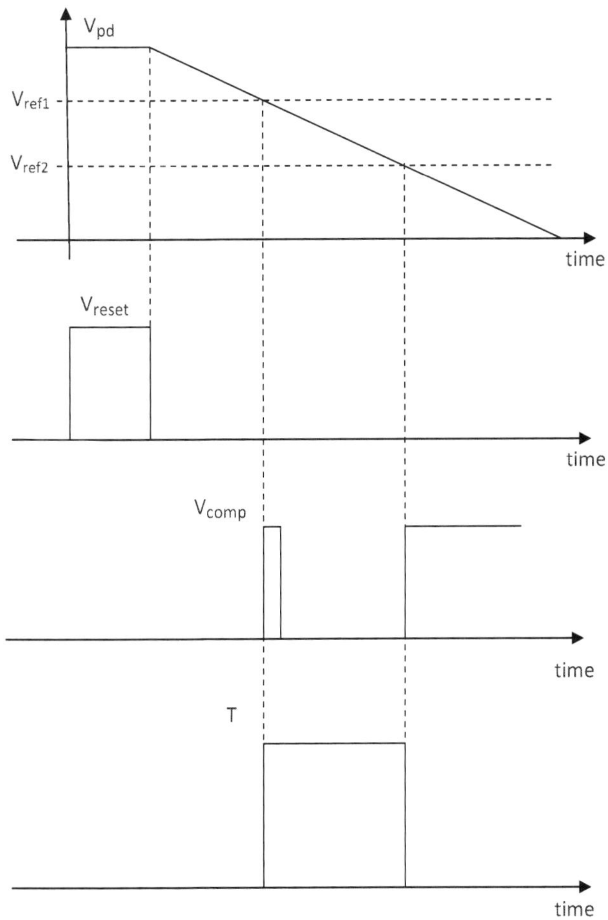


**Figure 15.** Electronic circuit main waveforms.

## 7. Conclusions

High dynamic range imager CMOS image sensors are one of the most important applications of CMOS image sensors. Most of the high dynamic range CMOS image sensors proposed in the literature neglect the fixed pattern noise, which is one of the major drawbacks of CMOS image sensors compared to CCDs. In this work, a fixed-patter noise reduction technique is presented for high dynamic range CMOS image sensors operating in time domain. According to simulation results, the new technique can reduce the fixed patter noise down to 2%.

## Author details

Fernando de S. Campos*, José Alfredo C. Ulson, José Eduardo C. Castanho, Bruno Albuquerque de Castro and Rudolf Riehl

*Address all correspondence to: fcampos@feb.unesp.br

Univ. Estadual Paulista "Júlio de Mesquita Filho" (UNESP) – Bauru campus, Brazil

## References

[1] A. Belenky, A. Fish, A. Spivak, and O. Yadid-Pecht, "A snapshot CMOS image sensor with extended dynamic range," *IEEE Sensors Journal*, vol. 9, pp. 103-111, 2009.

[2] F.S. Campos, O. Marinov, N. Faramarzpour, F. Saffih, M.J. Deen, J.W. Swart, "A multisampling time-domain CMOS imager with synchronous readout circuit," *Analog Int. Cir Signal Proc*, vol. 57, pp. 151-159, 2008.

[3] D. Park, J. Rhee, and Y. Joo, "A wide dynamic-range CMOS image sensor using self-reset technique," *IEEE El Dev Lett*, vol. 28, no. 10, pp. 890-892, 2007.

[4] L.G. McIlrath, "A low-power low-noise ultrawide-dynamic range CMOS imager with pixel-parallel A/D conversion," *IEEE J. Solid-State Circuits*, vol. 36, pp. 846-853, 2001.

[5] D. Yang, A. El Gamal, B. Fowler, and H. Tian, "A 640x512 CMOS image sensor with ultra wide dynamic range floating point pixel level ADC," *IEEE J. Solid-State Circuits*, vol. 34, pp. 1821-1834, 1999.

[6] S. Decker, R.D. McGrath, K. Brehmer, and G. Sodini, "A 256x256 CMOS imaging array with wide dynamic range pixels and column parallel digital output," *IEEE J. Sol-St Cir*, vol. 33, pp. 2081-2091, 1998.

[7] M.-W. Seo, T. Sawamoto, T. Akahori, Z. Liu, T. Iida, T. Takasawa, T. Kosugi, T. Watanabe, K. Isobe, and S. Kawahito, "A low-noise high-dynamic-range 17b 1.3-Megapixel 30-fps CMOS image sensor with column-parallel two-stage folding-integration/cyclic ADC," *IEEE Trans. Electron Devices*, vol. 59, no. 12, Dec. 2012.

[8] Z. Ignjatovic, D. Maricic, M.F. Bocko, "Low power, high dynamic range CMOS image sensor employing pixel-level oversampling ΣΔ analog-to-digital conversion, " *IEEE Sensors Journal*, vol. 12, no. 4, pp. 737-746, April 2012.

[9] M.W. Seo, et al., "A low-noise high intrascene dynamic range CMOS image sensor with a 13 to 19b variable-resolution column-parallel folding-integration/cyclic ADC", *IEEE J Sol-St Cir*, vol. 47, no. 1, pp. 272-283, Jan 2012.

[10] Y.C. Chae, J.M. Cheon, S.H. Lim, M.H. Kwon, K.S. Yoo, W.K. Jung, D.H. Lee, S.H. Ham, and G.H. Han, "A 2.1 M pixels, 120frames/s CMOS image sensor with column-parallel ΔΣ ADC architecture," *IEEE J. Solid-State Circuits*, vol. 46, no.1, pp. 236-247, Jan. 2011.

[11] J. H. Park, S. Ayoama, T. Watanabe, K. Isobe, and S. Kawahito, "A high-speed low noise CMOS image sensor with 13-b column-parallel single-ended cyclic ADCs," *IEEE Trans. Electron Devices*, vol. 56, no. 11, pp. 2414-2422, Nov. 2009.

[12] M. Furuta, Y. Nishikawa, and S. Kawahito, "A high-speed, high-sensitivity digital CMOS image sensor with a global shutter and 12-bit column-parallel cyclic A/D converter," *IEEE J. Solid-State Circuits*, vol. 42, no. 4, pp. 766-774, Apr. 2007.

Chapter 15

# All-Optical Waveguide-Type Switch Using Saturable Absorption in Graphene

Misaki Takahashi, Hiroki Kishikawa,
Nobuo Goto and Shin-ichiro Yanagiya

Additional information is available at the end of the chapter

http://dx.doi.org/10.5772/60437

**Abstract**

An all-optical switch using saturable absorption that accompanies refractive-index change such as in graphene films is proposed. The switching conditions are theoretically derived for arbitrary values of absorption and refractive-index change. It is found that switching can be performed by weaker control light when the refractive-index change is accompanied in saturable absorption. Switching conditions for various combinations of absorption and refractive-index change are confirmed by FD-BPM simulation. As an experimental demonstration of saturable absorption in graphene, we discuss measured nonlinear absorption in a graphene-loaded waveguide. The saturable absorption and refractive-index change through vertically placed multilayered graphene sheets are also discussed based on reported experimental results.

**Keywords:** Graphene, integrated optics, optical switch, saturable absorption, nonlinear refractive-index change

## 1. Introduction

Fast-response optical switches are one of the key devices in photonic routers and in optical signal processors. The switching response depends on the control mechanism of the switch. All-optical switching using optical nonlinear effects has a potential for picosecond-order response. Optically controlled switches include devices using phase shift induced by optical nonlinear Kerr effect [1] and phase shift in semiconductor optical amplifiers (SOAs) [2] and in quantum dots [3].

We have proposed an all-optical switch controlled by Raman amplification [4,5] or by saturable absorption [6-8], where optical signal amplitude is controlled instead of the signal phase. In the previously proposed switches, refractive-index change induced by all optical nonlinear interactions was assumed to be small enough to be negligible [4,6] or to be adjusted at an integer multiple of $2\pi$ [7]. When we assume graphene sheets as the saturable absorbing material, the induced refractive-index change cannot be ignored [9]. In this book chapter, we theoretically analyze the proposed switch to derive switching conditions in general when both the absorption and the index change are induced by nonlinear interaction [8]. It is found that, with arbitrary combination of absorption and induced phase shift, optical switching can be performed by employing appropriate fixed attenuator and fixed phase-shifter in the switching device. We confirm the switching performance by finite-difference beam-propagation-method (FD-BPM) simulation. As an experimental demonstration of saturable absorption in graphene, we also discuss measured nonlinear absorption in a graphene-loaded waveguide [7]. Nonlinear refractive-index change in graphene is also discussed to evaluate optical nonlinear phase change through vertically placed multilayer graphene based on the reported experimental results [9].

## 2. Graphene for optical processing

Graphene is a two-dimensional (2D) monolayer of $sp^2$ -bonded carbon atoms and has a dense honeycomb crystal structure. This carbon material has been attracting increasing interest, in particular, since the Nobel Prize was awarded to Novoselov and Geim for their significant contribution in graphene researches [10]. The application areas of graphene are shown in Fig. 1 [11]. Unique characteristics of graphene are driven by its linear massless band structure. In electronic devices, electrons in graphene move at ultrafast speed, and various applications have been investigated. In the field of optics, a monolayer graphene shows large broadband optical absorption of 2.3%, and excellent saturable absorption phenomena having ultrafast carrier dynamics. The saturable absorption has been applied to mode-locked fiber lasers [12-14].

In our proposed optical switch, the saturable absorption characteristics are employed to control optical signal light. In such optical waveguide devices, two schemes, loaded type and vertically inserted type, to introduce graphene in the optical waveguide can be considered as shown in Fig. 2, where graphene is introduced in one of the two arms of a Mach-Zehnder interferometer (MZI) as an example. The former is easy to fabricate in integrated-optic device structure. On the other hand, the latter is easy to evaluate the effect of graphene, whereas optical scattering at the inserted plane of the waveguide may induce insertion loss. We consider all-optical control of signal light by intense control light through saturable absorption.

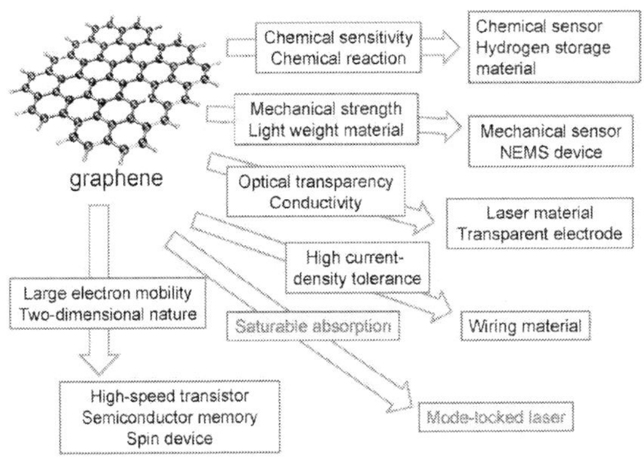


**Figure 1.** Application area of graphene.

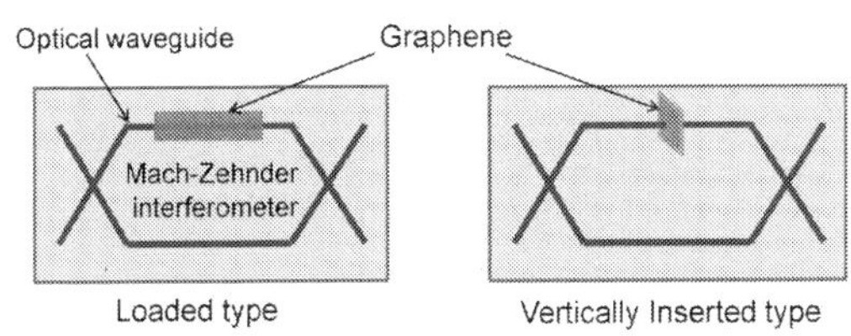


**Figure 2.** Introduction of graphene in optical waveguide.

We consider a model for controlling the transmittance through a graphene-introduced waveguide. The signal light with electrical-field amplitude $E_{s,in}$ attenuates to be $E_{s,out}$ through the vertically inserted graphene sheets or the graphene-loaded waveguide as shown in Fig. 3(a). The intense control light with electrical-field amplitude $E_{c,in}$ also attenuates to be $E_{c,out}$. The attenuation of the signal light can be decreased by the intense control light due to saturable absorption of graphene. This control of absorption can be equivalently represented by a cascade connection of a fixed attenuator with amplitude attenuation coefficient $t_{low}$ and an amplifier controlled by the control light whose amplitude amplification coefficient is $\alpha$ as shown in Fig. 3(b). The amplification corresponds to decrease of the absorption due to the saturable absorption induced by the intense control light. An induced nonlinear phase shift through saturable absorption is denoted by $\phi$. We denote the coefficient including amplification and phase shift as $\alpha^{'}=\alpha \exp(j\ \phi)$. These coefficients are defined by

$$t_{low} = \frac{E_{s,out}(E_{c,in}=0)}{E_{s,in}}, \tag{1}$$

and

$$\alpha' = \alpha e^{j\phi} = \frac{E_{\mathrm{s,out}}(E_{\mathrm{c,in}} \neq 0)}{E_{\mathrm{s,out}}(E_{\mathrm{c,in}} = 0)}. \tag{2}$$

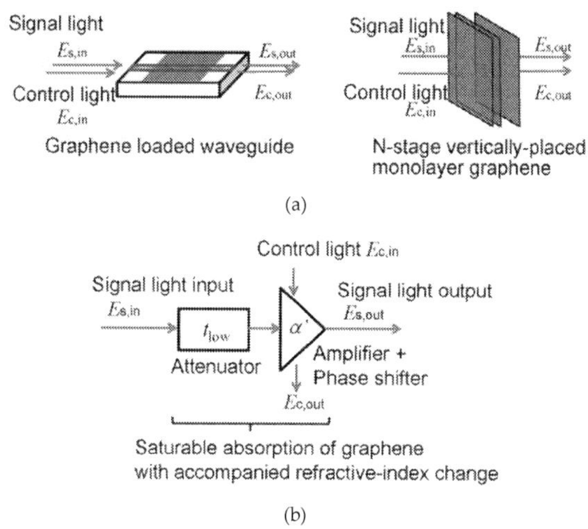


**Figure 3.** Modeling of light control by graphene; (a) signal and control lights in graphene, and (b) a model with fixed attenuator and variable amplifier and phase-shifter controlled by control light.

# 3. Optical switch configuration

The proposed switch consists of two cascaded MZIs connected with asymmetric X-junction couplers, as shown in Fig. 4 [8]. We consider two types of configuration, Type A and B, depending on whether a fixed attenuator is employed in the upper arm of the second MZI or in the lower one, respectively. It is noted that a conventional switch using phase control can be formed with a single MZI. Optical signals in the first MZI are controlled by vertically inserted or overlaid graphene. Control light at a different wavelength or in an orthogonally different polarization from the signal light is coupled in both arms of the first MZI. By inserting the control light, the absorption in the graphene film is reduced due to saturable absorption, which is accompanied by the refractive-index change in the graphene film. Thus, the attenuation and the phase of the signal light along the first MZI arm are controlled by the control light. The transmittance for optical field amplitude in each arm is given by $t_{\mathrm{low}}\ \alpha_i \exp(j\phi_i) = t_{\mathrm{low}}\ \alpha_i'$, $i$ = A and B. In either arm of the second MZI, a fixed attenuator with a fixed phase-shifter has the transmittance given by $\beta_{\mathrm{fix}} = |\beta_{\mathrm{fix}}| \exp(j\ \arg(\beta_{\mathrm{fix}}))$.

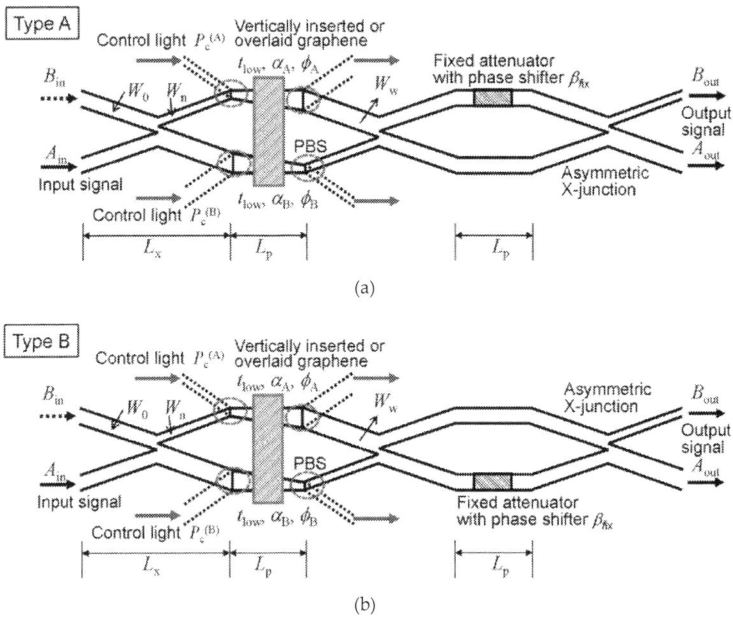


**Figure 4.** Switch structure; (a) Type A and (b) Type B.

Switching response time is mainly limited by the response of the saturable absorption in graphene. It is reported that, from pump-probe experiments, the carrier relaxation time and carrier–carrier intraband scattering time-constant are less than 1 ps [13]. A response time in a 20-μm-long graphene-coated microfiber was also measured to be 2.2 ps [15]. Since the switching is kept as on-state during the control signal existence, optical control pulse with rise and fall time less than 1 ps can result in picosecond switching.

## 4. Switching conditions

The input-to-output relation for optical field amplitudes in case of Type A shown in Fig. 4(a) is given by

$$\begin{aligned}\begin{bmatrix} A_{out} \\ B_{out} \end{bmatrix} &= \left(\frac{1}{\sqrt{2}}\right)^3 \begin{bmatrix} 1 & 1 \\ -1 & 1 \end{bmatrix} \begin{bmatrix} 1 & 0 \\ 0 & \beta_{fix} \end{bmatrix} \begin{bmatrix} 1 & 1 \\ -1 & 1 \end{bmatrix} \begin{bmatrix} t_{low}\alpha_B' & 0 \\ 0 & t_{low}\alpha_A' \end{bmatrix} \begin{bmatrix} 1 & 1 \\ -1 & 1 \end{bmatrix} \begin{bmatrix} A_{in} \\ B_{in} \end{bmatrix} \\ &= \frac{t_{low}}{2\sqrt{2}} \begin{bmatrix} a_{11} & a_{12} \\ a_{21} & a_{22} \end{bmatrix} \begin{bmatrix} A_{in} \\ B_{in} \end{bmatrix}\end{aligned} \tag{3}$$

where

$$\begin{cases} a_{11} = \alpha_A^{'} - \alpha_B^{'} - \beta_{\text{fix}}(\alpha_A^{'} + \alpha_B^{'}) \\ a_{12} = \alpha_A^{'} + \alpha_B^{'} + \beta_{\text{fix}}(-\alpha_A^{'} + \alpha_B^{'}) \\ a_{21} = -\alpha_A^{'} + \alpha_B^{'} - \beta_{\text{fix}}(\alpha_A^{'} + \alpha_B^{'}) \\ a_{22} = -\alpha_A^{'} - \alpha_B^{'} + \beta_{\text{fix}}(-\alpha_A^{'} + \alpha_B^{'}) \end{cases}. \tag{4}$$

Similar equations for input-to-output relation in Type B shown in Fig. 4(b) are derived as

$$\begin{aligned} \begin{bmatrix} A_{\text{out}} \\ B_{\text{out}} \end{bmatrix} &= \left(\frac{1}{\sqrt{2}}\right)^3 \begin{bmatrix} 1 & 1 \\ -1 & 1 \end{bmatrix} \begin{bmatrix} \beta_{\text{fix}} & 0 \\ 0 & 1 \end{bmatrix} \begin{bmatrix} 1 & 1 \\ -1 & 1 \end{bmatrix} \begin{bmatrix} t_{\text{low}}\alpha_B' & 0 \\ 0 & t_{\text{low}}\alpha_A' \end{bmatrix} \begin{bmatrix} 1 & 1 \\ -1 & 1 \end{bmatrix} \begin{bmatrix} A_{\text{in}} \\ B_{\text{in}} \end{bmatrix} \\ &= \frac{t_{\text{low}}}{2\sqrt{2}} \begin{bmatrix} a_{11} & a_{12} \\ a_{21} & a_{22} \end{bmatrix} \begin{bmatrix} A_{\text{in}} \\ B_{\text{in}} \end{bmatrix} \end{aligned} \tag{5}$$

where

$$\begin{cases} a_{11} = -\alpha_A^{'} - \alpha_B^{'} + \beta_{\text{fix}}(\alpha_A^{'} - \alpha_B^{'}) \\ a_{12} = -\alpha_A^{'} + \alpha_B^{'} + \beta_{\text{fix}}(\alpha_A^{'} + \alpha_B^{'}) \\ a_{21} = -\alpha_A^{'} - \alpha_B^{'} - \beta_{\text{fix}}(\alpha_A^{'} - \alpha_B^{'}) \\ a_{22} = -\alpha_A^{'} + \alpha_B^{'} - \beta_{\text{fix}}(\alpha_A^{'} + \alpha_B^{'}) \end{cases}. \tag{6}$$

We consider the case that an optical signal is incident only at the lower port, that is, $A_{\text{in}}=E_{\text{in}}$ and $B_{\text{in}}=0$. Then, eqs. (3) and (5) are simplified to be

$$\begin{bmatrix} A_{\text{out}} \\ B_{\text{out}} \end{bmatrix} = \frac{t_{\text{low}}E_{\text{in}}}{2\sqrt{2}} \begin{bmatrix} a_{11} \\ a_{21} \end{bmatrix}. \tag{7}$$

We now derive the conditions required for complete switching. We assume that switching is operated by feeding a control light with power $P_c^{(A)}$ in the upper arm of the first MZI to switch to the lower output port as $A_{\text{out}}$, and by feeding a control light $P_c^{(B)}$ in the lower arm to switch to the upper output port as $B_{\text{out}}$. Therefore, the parameters $\alpha_A^{'}$ and $\alpha_B^{'}$ due to saturable absorption are to be set as $\alpha_A^{'}=1$ to switch to the upper output port as $B_{\text{out}}$ and $\alpha_B^{'}=1$ to switch to the lower output port as $A_{\text{out}}$ because no control light is fed in these arms.

Switching conditions for $\beta_{\text{fix}}$ and the output field $B_{\text{out}}$ in the case of Type A are obtained from eqs. (4) and (7) as follows:

1. Switching to $B_{\text{out}}$ :

$$\beta_{\text{fix}} = \frac{\alpha_B^{'} - 1}{\alpha_B^{'} + 1} \text{ and } B_{\text{out}} = \frac{t_{\text{low}} E_{\text{in}} (1 - \alpha_B^{'})}{\sqrt{2}}. \tag{8}$$

2. Switching to $A_{\text{out}}$ :

$$\beta_{\text{fix}} = \frac{\alpha_A^{'} - 1}{\alpha_A^{'} + 1} \text{ and } A_{\text{out}} = \frac{t_{\text{low}} E_{\text{in}} (1 - \alpha_A^{'})}{\sqrt{2}}. \tag{9}$$

It is found from these conditions that, when $\alpha_A^{'}$ for switching to $A_{\text{out}}$ is equal to $\alpha_B^{'}$ for switching to $B_{\text{out}}$, switching can be operated with a fixed $\beta_{\text{fix}}$.

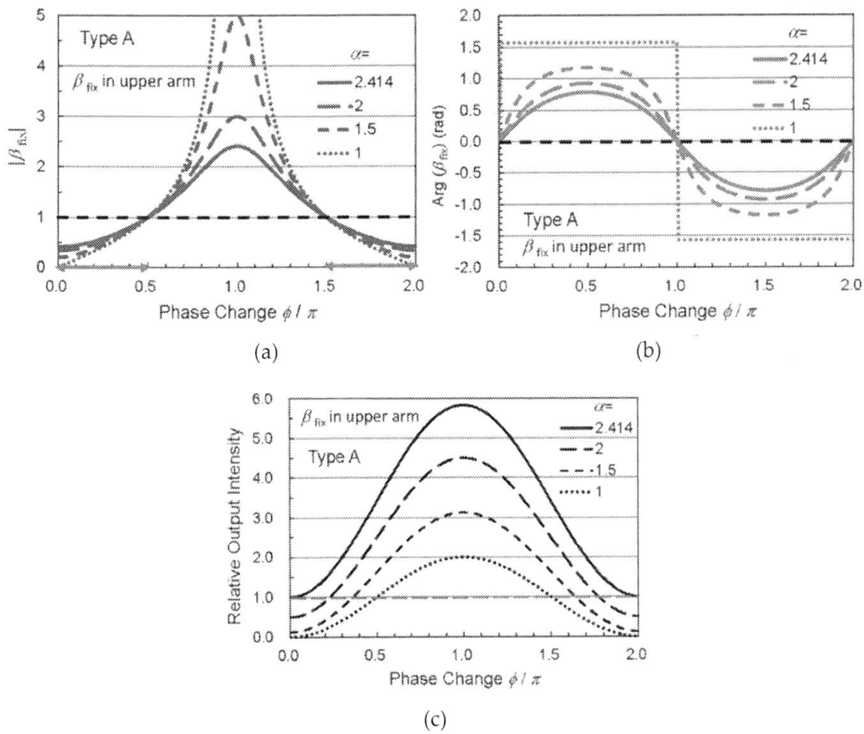


**Figure 5.** Switching conditions in switch Type A; (a) the fixed attenuator and (b) phase-shifter in the second MZI required for switching and (c) the output intensity.

Similarly, switching conditions in the case of Type B are obtained from eqs. (6) and (7) as follows:

**1.** Switching to $B_{\mathrm{out}}$ :

$$\beta_{\mathrm{fix}} = \frac{\alpha_B^{'} + 1}{\alpha_B^{'} - 1} \text{ and } B_{\mathrm{out}} = -\frac{t_{\mathrm{low}} E_{\mathrm{in}} (1 + \alpha_B^{'})}{\sqrt{2}}. \tag{10}$$

**2.** Switching to $A_{\mathrm{out}}$ :

$$\beta_{\mathrm{fix}} = \frac{\alpha_A^{'} + 1}{\alpha_A^{'} - 1} \text{ and } A_{\mathrm{out}} = \frac{t_{\mathrm{low}} E_{\mathrm{in}} (1 + \alpha_A^{'})}{\sqrt{2}}. \tag{11}$$

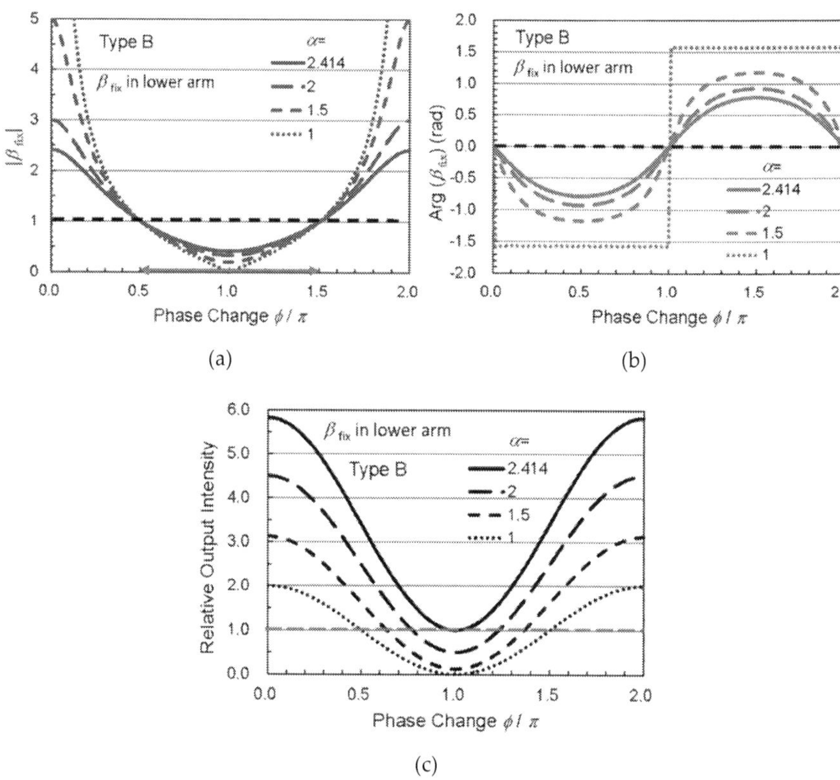


**Figure 6.** Switching conditions in switch Type B; (a) the fixed attenuator and (b) phase-shifter in the second MZI required for switching and (c) the output intensity.

From these equations, switching conditions required for $\beta_{fix}$ and the relative output intensities $|A_{out}/(t_{low}E_{in})|^2$ and $|B_{out}/(t_{low}E_{in})|^2$ are calculated as a function of phase change $\phi_A$ or $\phi_B$ as shown in Figs. 5 and 6 for Type A and B, respectively, where $\alpha_i=1+\sqrt{2}$, 2.0, 1.5, and 1.0 are assumed as a parameter. Since $|\beta_{fix}|\leq 1$ can be realized by an attenuator, Type A should be selected for $|\phi_i|\leq\pi/2$ and $3\pi/2\leq|\phi_i|\leq 2\pi$ and Type B for $\pi/2\leq|\phi_i|\leq 3\pi/2$, respectively, in the range of $0\leq|\phi_i|\leq 2\pi$. An arbitrary combination of $\alpha_i$ and $\phi_i$ can be used to operate switching by adjusting $|\beta_{fix}|$ and arg( $\beta_{fix}$).

We consider three cases as examples of switching. As a first case, case (1), we assume that no phase shift is accompanied in saturable absorption, that is, $\phi_A=\phi_B=0$, and no phase shift is set in $\beta_{fix}$. This corresponds to the case discussed in ref. [7]. The switching conditions in Type A are derived from eqs. (8) and (9) as follows:

1. Switching to $B_{out}$ $(\alpha_A'=1,\ \alpha_B'=1+\sqrt{2})$ :

$$\beta_{fix}=\frac{1}{1+\sqrt{2}}\left(\rightarrow \arg\left(\beta_{fix}\right)=0\right) \text{ and } B_{out}=-t_{low}E_{in}. \tag{12}$$

2. Switching to $A_{out}$ $(\alpha_A'=1+\sqrt{2},\ \alpha_B'=1)$ :

$$\beta_{fix}=\frac{1}{1+\sqrt{2}}\left(\rightarrow \arg\left(\beta_{fix}\right)=0\right) \text{ and } A_{out}=-t_{low}E_{in}. \tag{13}$$

This case corresponds to $\alpha_i=1+\sqrt{2}\cong 2.414$ and $\phi_i=0$ in Fig. 5(a). The operating point for $\beta_{fix}$ and the output intensity are indicated by red circles on green curves in Fig. 7(a).

As a next case, case (2a), we assume the same $\alpha_i=1+\sqrt{2}$ as case (1), whereas a phase shift $\phi_i=\pi/2$ is accompanied. The switching conditions in Type A are derived from eqs. (8) and (9) as follows:

1. Switching to $B_{out}$ $(\alpha_A'=1,\ \alpha_B=1+\sqrt{2},\ \phi_B=\pi/2)$ :

$$\beta_{fix}=\frac{j\left(1+\sqrt{2}\right)-1}{j\left(1+\sqrt{2}\right)+1}\left(\rightarrow\left|\beta_{fix}\right|=1,\arg\left(\beta_{fix}\right)=\frac{\pi}{4}\right) \text{ and } B_{out}=\frac{t_{low}E_{in}[1-j\left(1+\sqrt{2}\right)]}{\sqrt{2}}. \tag{14}$$

2. Switching to $A_{out}$ $(\alpha_A=1+\sqrt{2},\ \phi_A=\pi/2,\ \alpha_B'=1)$ :

$$\beta_{fix}=\frac{j\left(1+\sqrt{2}\right)-1}{j\left(1+\sqrt{2}\right)+1}\left(\rightarrow\left|\beta_{fix}\right|=1,\arg\left(\beta_{fix}\right)=\frac{\pi}{4}\right) \text{ and } A_{out}=\frac{t_{low}E_{in}[1-j\left(1+\sqrt{2}\right)]}{\sqrt{2}}. \tag{15}$$

It is noted that, since $\{[1-j(1+\sqrt{2})]/\sqrt{2}\}^2=2+\sqrt{2}$, the output intensity is increased from 1.0 to 3.41 compared with case(1). Thus, when $\phi_i=\pi/2$ is accompanied in saturable absorption, the excess loss through the switch is decreased by $10\log_{10}(2+\sqrt{2})=5.33$ (dB). The operating point for $\beta_{fix}$ and the output intensity are indicated in Fig. 7(b).

When $\phi_i$ increases further to $\pi/2\le\phi_i\le 3\pi/2$, switching in type A cannot be realized with a fixed attenuator because $|\beta_{fix}|\ge 1$, and the switch structure of Type B has to be used to realize with a fixed attenuator. In this case, $|\beta_{fix}|$, $\arg(\beta_{fix})$ and the normalized output are found in Fig. 6. We consider a case of $\phi_i=\pi$ as case (2b). The switching conditions in Type B are derived from eqs. (10) and (11) as follows:

1. Switching to $B_{out}$ $(\alpha_A^{'}=1,\ \alpha_B=1+\sqrt{2},\ \phi_B=\pi)$:

$$\beta_{fix}=\frac{1}{1+\sqrt{2}}\left(\rightarrow \arg\left(\beta_{fix}\right)=0\right) \text{ and } B_{out}=t_{low}E_{in}. \tag{16}$$

2. Switching to $A_{out}$ $(\alpha_A=1+\sqrt{2},\ \phi_A=\pi,\ \alpha_B^{'}=1)$:

$$\beta_{fix}=\frac{1}{1+\sqrt{2}}\left(\rightarrow \arg\left(\beta_{fix}\right)=0\right) \text{ and } A_{out}=t_{low}E_{in}. \tag{17}$$

The same switching conditions are found for $\phi_i=(2n-1)\pi,\ n=1,\ 2,\ ....$ It is noted that, for $\phi_i=2n\pi$, the same switching conditions as for $\phi_i=0$ in Type A are applied. The operating point for $\beta_{fix}$ and the output intensity are indicated in Fig. 7(c).

We consider improvement of control efficiency by taking the phase shift $\phi_i$ and $\arg(\beta_{fix})$ into account. Although $\alpha_i=1+\sqrt{2}$ is required with a fixed $\beta_{fix}$ with $\arg(\beta_{fix})=0$ in the second MZI, switching can be performed even with $\alpha_i<1+\sqrt{2}$ as shown in Figs. 5 and 6. We consider a case of $\phi_i=0$. By decreasing $\alpha_i$ from $1+\sqrt{2}$ to 2.0, the required $\beta_{fix}$ decreases to $1/3$, resulting in decrease of the normalized output from 1 to 1/2 as shown in Fig. 5. Thus, by decreasing $\alpha_i$ from $1+\sqrt{2}$ to 2.0, insertion loss of the switch increases by 3 dB. Next, we consider a case of $\phi_i=\pi/2$ and $\alpha_i=2.0$ as case(3). The switching conditions in Type A are derived from eqs. (8) and (9) as follows:

1. Switching to $B_{out}$ $(\alpha_A^{'}=1,\ \alpha_B=2,\ \phi_B=\pi/2)$:

$$\beta_{fix}=\frac{3+j4}{5}\left(\rightarrow|\beta_{fix}|=1,\ \arg\left(\beta_{fix}\right)=0.927\right) \text{ and } B_{out}=\frac{t_{low}E_{in}(1-j2)}{\sqrt{2}}. \tag{18}$$

2. Switching to $A_{out}$ $(\alpha_A=2,\ \phi_A=\pi/2,\ \alpha_B^{'}=1)$:

$$\beta_{fix}=\frac{3+j4}{5}\left(\rightarrow\left|\beta_{fix}\right|=1,\arg\left(\beta_{fix}\right)=0.927\right) \text{ and } A_{out}=\frac{t_{low}E_{in}(1-j2)}{\sqrt{2}}. \tag{19}$$

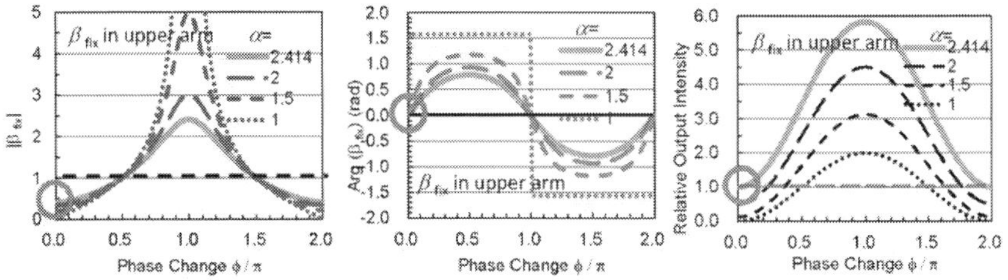


(a)

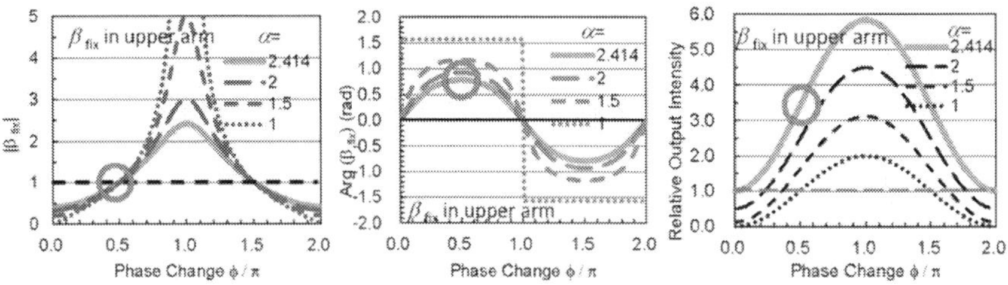


(b)

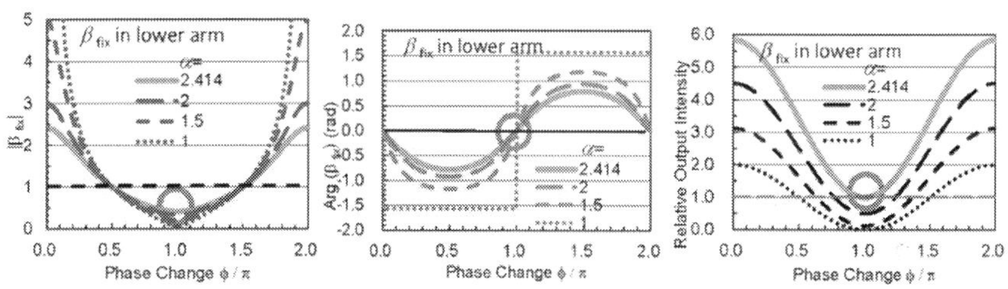


(c)

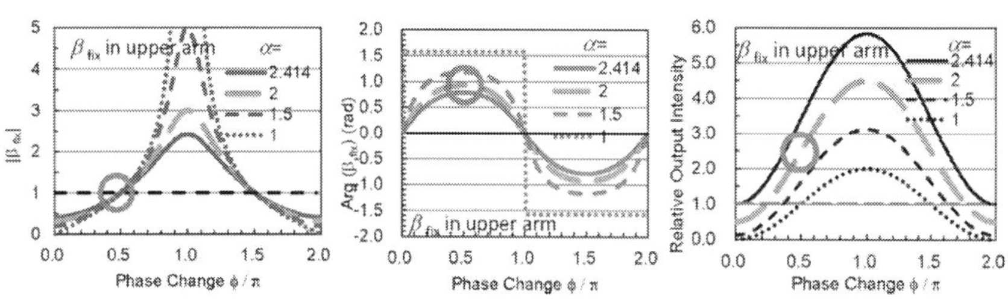


(d)

**Figure 7.** Operation points in the switching conditions for four kinds of switches: (a) case(1), (b) case(2a), (c) case(2b), and (d) case(3).

Since $\left[(1-j2)/\sqrt{2}\right]^2$=2.5, the output intensity is increased by 2.5 times compared with case(1). The operating point for $\beta_{\text{fix}}$ and the output intensity are indicated in Fig. 7(d). In the case with a further reduced $\alpha_i$, switching can be operated by employing appropriate fixed $|\beta_{\text{fix}}|$ and $\arg(\beta_{\text{fix}})$ according to the phase shift $\phi_i$.

Finally, we consider a limited case with $\alpha_i=1.0$, that is, no saturable absorption is used but the phase shift $\phi_i$ alone is used in switching. When $\phi_i$ is increased to $\pi/2$, $\beta_{\text{fix}}$ is derived to be $\beta_{\text{fix}}=(-1+j)/(1+j)=e^{j\pi/2}$, that is, a $\pi/2$ phase-shifter is used as $\beta_{\text{fix}}$. The normalized output intensity is 1.0. When $\phi_i$ is further increased to $\pi$, $|\beta_{\text{fix}}|$ reduced to 0 in Type B. No output is obtained in this case. This switching condition of $\phi_i=\pi$ is used in a conventional switch consisting of a single MZI. That is, the switching is performed at the output of the first MZI. Since $\beta_{\text{fix}}$ is designed to equalize the optical intensity in the two arms of the second MZI, switching cannot be performed with the proposed two-stage MZI switch, for $\alpha_i=1.0$ and $\phi_i=\pi$.

## 5. FD-BPM simulation

In order to confirm the switching operation with the proposed switch, FD-BPM simulation was performed under various switching conditions. We consider a 2D slab waveguide model. The core and cladding regions have refractive indices of $n_c$ = 1.461 and $n_s$ = 1.45, respectively. The waveguide widths of the fundamental, narrow, and wide waveguides of the asymmetric X-junctions are $W_1$=3.0 $\mu$m, $W_n$=2.6 $\mu$m, and $W_w$=3.4 $\mu$m, respectively. Optical waves are assumed to be TE mode. The lengths of the X-junction coupler and the parallel waveguides in arms are $L_x$=16 mm and $L_p$=1 mm, respectively. The total length is 50 mm. The distance of the two input ports is 23 $\mu$m. The optical wavelength is 1550 nm.

In the simulation, $\alpha_i$ and $\phi_i$ caused by saturable absorption are equivalently simulated just by multiplying the optical electric field by the coefficient $\alpha_i e^{j\phi_i}$ at a plane located at the end of the graphene-loaded or graphene-inserted waveguide. The waveguides for coupling the control light are not modeled in this simulation. In a similar manner, the attenuation corresponding to $t_{\text{low}}$, the fixed attenuator and the phase-shifter $\beta_{\text{fix}}$ are modeled by multiplying the optical electric field by $t_{\text{low}}$ and $|\beta_{\text{fix}}|\exp[j\ \arg(\beta_{\text{fix}})]$, respectively. The value of $t_{\text{low}}$ is assumed to be an ideal value of $1/(1+\sqrt{2})$, where the optical signal does not attenuate through graphene with intense control light.

Fig. 8(a) shows optical intensities along the switch in case (1) obtained by FD-BPM simulation, where $\alpha_i=1+\sqrt{2}$, $\phi_i=0$, and $\beta_{fix}=1/(1+\sqrt{2})$ are assumed. Similarly, optical intensities along the switch in case (2a), case (2b), and case (3) are plotted in Fig. 8(b), (c), and (d), respectively. The switched output intensities are summarized in Table 1. The simulated outputs show good agreement with theoretical results with error less than 3.5%.

## 6. Measurement of saturable absorption through graphene-loaded waveguide

In this section, we show an experimental result of optical saturable absorption in a monolayer-graphene-loaded waveguide. The optical waveguide was fabricated by $K^+$ ion exchange on a soda-lime glass substrate. Optical waveguide was formed by $K^+$ ion exchange through a 30 μm opening in an aluminum mask. Potassium nitrate was used as the ion source and ion exchange was carried out at 370 degrees Celsius for 4 hours. After polishing the end facets of the waveguide, a sheet of monolayer graphene was transferred on the glass substrate surface. The monolayer graphene was formed by CVD on copper foil (iTRIX Corporation). The length of the graphene sheet is about 7 mm. The waveguide structure is shown in Fig. 9(a). A microscopic top-view of the waveguide and a near-field output intensity profile from the waveguide output end facet are also shown in Fig. 9(b) and (c), respectively. Raman spectrum of the transferred graphene on the glass substrate was measured as shown in Fig. 10. In the Raman spectrum, the D peak at 1350 $cm^{-1}$ is correlated with the disorder of the graphene lattice; the G peak at 1580 $cm^{-1}$ and the 2D peak around 2700 $cm^{-1}$ correspond to the phonon excitation at the Brillouin zone center and the second-order of zone-boundary phonons, respectively [16].

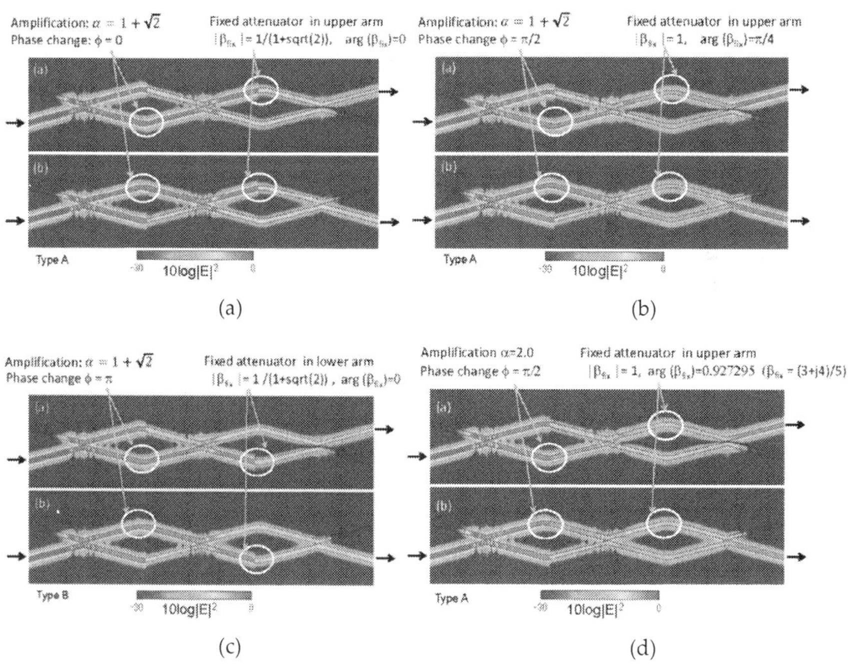


**Figure 8.** Optical signal intensities along the switches obtained by FD-BPM simulation for (a) case(1), (b) case(2a), (c) case(2b), and (d) case(3).

| Case | $\alpha_i$ | $\phi_i$ | $\lvert\beta_{fix}\rvert$ | $\arg(\beta_{fix})$ | Control arm | Theory | | Simulation | |
|---|---|---|---|---|---|---|---|---|---|
| | | | | | | $\lvert A_{out}\rvert^2$ | $\lvert B_{out}\rvert^2$ | $\lvert A_{out}\rvert^2$ | $\lvert B_{out}\rvert^2$ |
| (1) | $1+\sqrt{2}$ | 0 | $\frac{1}{1+\sqrt{2}}$ | 0 | B | 0 | 0.1716 | $6.281\times10^{-4}$ | $1.653\times10^{-1}$ |
| | | | | | A | 0.1716 | 0 | $1.689\times10^{-1}$ | $6.195\times10^{-4}$ |
| (2a) | $1+\sqrt{2}$ | $\frac{\pi}{2}$ | 1 | $\frac{\pi}{4}$ | B | 0 | 0.5858 | $1.197\times10^{-3}$ | $5.647\times10^{-1}$ |
| | | | | | A | 0.5858 | 0 | $5.704\times10^{-1}$ | $1.168\times10^{-3}$ |
| (2b) | $1+\sqrt{2}$ | $\pi$ | $\frac{1}{1+\sqrt{2}}$ | 0 | B | 0 | 0.1716 | $6.158\times10^{-4}$ | $1.665\times10^{-1}$ |
| | | | | | A | 0.1716 | 0 | $1.680\times10^{-1}$ | $6.060\times10^{-4}$ |
| (3) | 2 | $\frac{\pi}{2}$ | 1 | $\arg\left(\frac{3+j4}{5}\right)$ | B | 0 | 0.4289 | $1.101\times10^{-3}$ | $4.135\times10^{-1}$ |
| | | | | | A | 0.4289 | 0 | $4.173\times10^{-1}$ | $1.070\times10^{-3}$ |

**Table 1.** Theoretical and simulated output intensities.

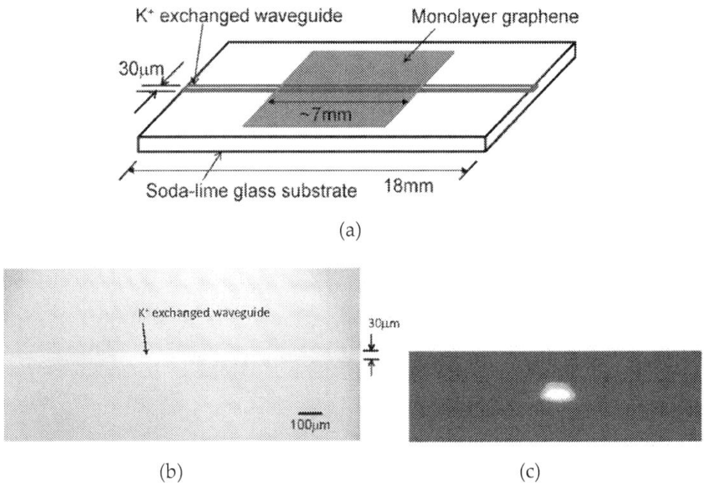


**Figure 9.** Graphene-loaded optical waveguide; (a) the structure, (b) the microscopic top-view, and (c) the near field pattern from the waveguide.

The experimental setup for measuring saturable absorption is illustrated in Fig. 11. A 1.56 μm femtosecond laser with width of 0.4 ps at repetition rate of 41.96 MHz was used as the laser source. The average power is 4.4 mW. The polarization-controlled laser light was coupled into the waveguide by ×20 objective lens.

Figure 12 shows the measured insertion loss for TE and TM modes as a function of the incident average power at the waveguide input. The measured loss includes coupling loss, waveguide-propagation loss and attenuation loss due to graphene. The insertion loss reduces by more than 10 dB as the incident power increases. This is considered to be caused by saturable

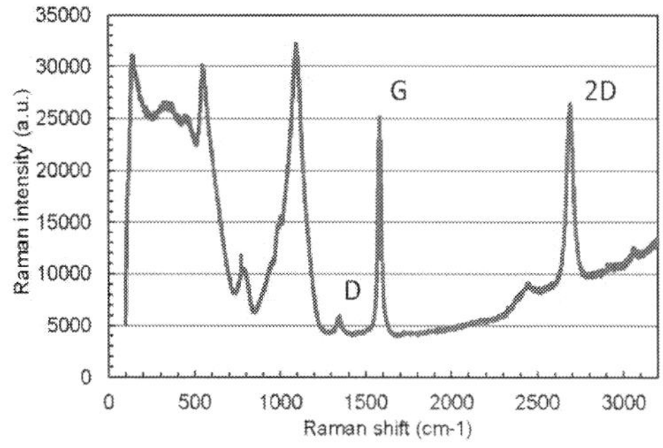


Figure 10. Raman spectrum from the graphene-loaded waveguide.

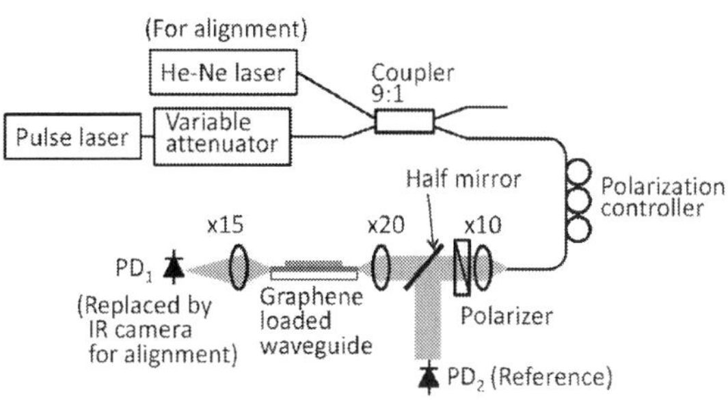


Figure 11. Experimental setup for measurement of nonlinear insertion loss.

absorption of graphene. This modulation depth of the transmittance through the waveguide satisfies the requirement for optical switching. The insertion loss in TM mode is larger than that in TE mode by about 10 dB at low optical power. The difference reduces to about 3dB at high optical power.

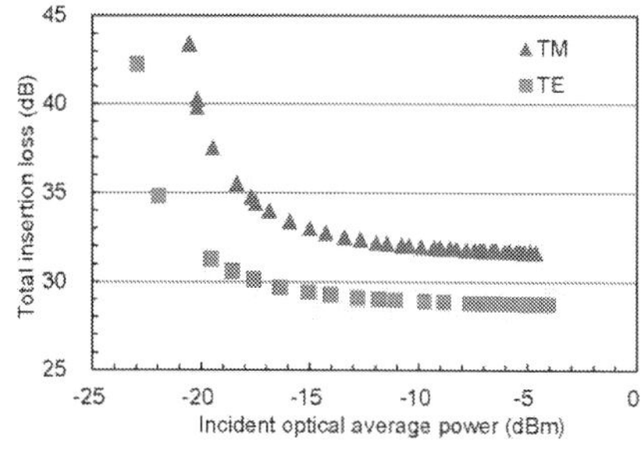


Figure 12. Nonlinear insertion loss through a graphene-loaded optical waveguide.

## 7. Evaluation of transmittance and phase shift in loosely multilayered graphene

It is found from the discussion in the previous sections that the switch can be operated by adjusting the fixed $\beta_{\text{fix}}$ in the second MZI even if the phase shift $\phi_i$ is unknown. In this section, we discuss the transmittance and the phase shift in vertically inserted multilayered graphene based on a reported experimental result by H. Zhang et al. [9]. The transmittance in loosely stacked three-layer graphene was measured and was reported to be fitted by the following experimental formula:

$$T(I) = T(0) + \Delta T_0 - \frac{\Delta T_0}{1 + I / I_{\text{sat}}} \tag{20}$$

where $I$ is the incident optical peak power density, $T(0)$ is the transmittance at weak optical power density. The parameters fitted with the experimental data are $\Delta T_0$=0.051, $I_{\text{sat}}$=0.074 GW/cm$^2$. When $T(0)$ is assumed as $T(0)$=0.977$^3$=0.9326, the transmittance $T(I)$ is plotted as shown in Fig. 13(a).

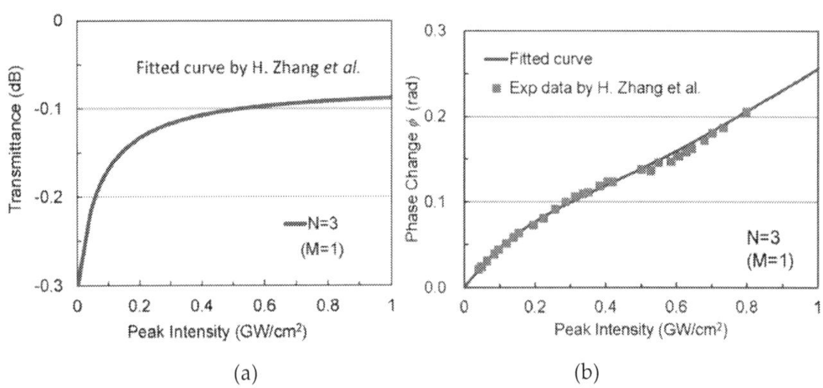


**Figure 13.** Transmittance and phase change through loosely three-layered graphene.

Reported experimental data of the accompanied phase change are shown in Fig. 13(b) [9]. We find an experimental formula from these data as follows:

$$\phi(I) = k_0 L I \left[ a_1 \exp\left(-a_2 I^{a_3}\right) + n_{2,c} \right] \times 10^9 \tag{21}$$

where $k_0$=$2\pi/\lambda$ with $\lambda$=1.55 μm, $L$ =1 nm, and the fitted parameters are $n_{2,c}$=6.3×10$^{-8}$ cm$^2$/W, $a_1$=6.75×10$^{-8}$, $a_2$=6.49, and $a_3$=1.34. The optical incident peak power density is in unit of GW/cm$^2$. The fitted curve is also shown in Fig. 13(b). This phase change was derived from nonlinear refractive index which might include parametric process such as four-wave mixing and strong nonparametric process [9].

We now estimate transmittance $T_{3M}$ and phase change $\phi_{3M}$ in $3M$-layer graphene. Using $T(I)$ and $\phi(I)$ for three-layer graphene, we can derive $T_{3M}(I)$ by

$$\begin{cases} T_{3M}\left(I_{\text{in}}\right)=T(I_{\text{in}})\prod_{k=1}^{M-1}T(I_{\text{out},k}) \\ I_{\text{out},k}=I_{\text{out},k-1}T\left(I_{\text{out},k-1}\right) \text{ and } I_{\text{out},0}=I_{\text{in}} \end{cases} \tag{22}$$

where $I_{\text{in}}$ is the incident optical peak power density. The phase change $\phi_{3M}(I)$ from the phase at $I_{\text{in}}=0$ is given by

$$\phi_{3M}\left(I_{\text{in}}\right)=\sum_{k=0}^{M-1}\phi(I_{\text{out},k}) \tag{23}$$

Fig. 14 shows the transmittance and the phase change, where $N=3M$, It is approximately estimated that the transmittance can be changed by $2K$ dB with $N=30K$, where $K$ = 1,..., 4. Phase change of $K$ rad is accompanied by increasing the peak power density to 0.4~0.5 GW/cm$^2$. From these results, we can estimate the required layer number to satisfy the switching conditions when the graphene is vertically introduced in the waveguide. However, it is noted that the nonlinear phase shift discussed in this section is the phase shift of the incident intense light, that is, not for weak signal light whose wavelength or polarization is different from that of the control light. Therefore, nonlinear phase shift in the proposed switch structure has to be investigated in detail.

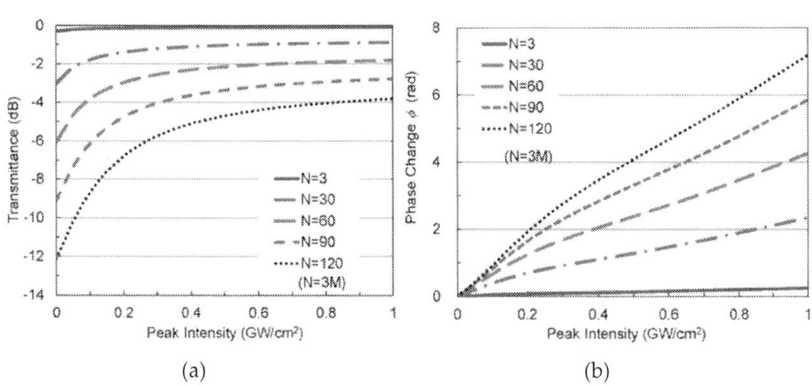


**Figure 14.** Transmittance and phase change through loosely multi-layered graphene.

## 8. Conclusions

Generalized switching conditions for the proposed switch were derived and confirmed by FD-BPM. It is found that the associated refractive-index change can reduce the required absorption decrease in saturable absorption. As a measured saturable absorption, nonlinear insertion loss along graphene-loaded waveguide was discussed. Also, evaluation of nonlinear transmittance and phase shift in multilayered graphene was described. Since conventional switches utilize only phase change to operate switching, the proposed switch, which uses both phase change and absorption change will give a new switching scheme.

## Acknowledgements

This work was supported in part by JSPS KAKENHI (23656243 and 24360150).

## Author details

Misaki Takahashi, Hiroki Kishikawa, Nobuo Goto * and Shin-ichiro Yanagiya

*Address all correspondence to: goto.nobuo@tokushima-u.ac.jp

Department of Optical Science and Technology, Tokushima University, Tokushima, Japan

## References

[1] Doran N. J. and Wood D. Nonlinear-Optic Loop Mirror. *Optics Lett*. 1988, 13(1), 56-58.

[2] Nakamura S., Ueno Y., Tajima K., Sasaki J., Sugimoto T., Kato T., Shimoda T., Itoh M., Hatakeyama H., Tamanuki T., and Sasaki T. Demultiplexing of 168-Gb/s Data Pulses with a Hybrid-Integrated Symmetric Mach-Zehnder All-Optical Switch. *IEEE Photon Tech Lett*. 2000, 12(4), 425-427.

[3] Kitagawa Y., Ozaki N., Takata Y., Ikeda N., Watanabe Y., Sugimoto Y., and Asakawa K. Sequential Operations of Quantum Dot/Photonic Crystal All-Optical Switch With High Repetitive Frequency Pumping. IEEE/OSA *J Lightwave Technol*. 2009, 27(10), 1241-1247.

[4] Kishikawa H., Kimiya K., Goto N., and Yanagiya S. All-Optical Wavelength-Selective Switch Consisting of Asymmetric X-junction Couplers and Raman Amplifiers for Wide Wavelength Range. IEEE/OSA *J of Lightwave Technol*. 2010, 28(1), 172-180.

[5] Kishikawa H., Goto N., and Kimiya K. All-Optical Wavelength-Selective Switch by Intensity Control in Cascaded Interferometers. In *Frontiers in Guided Wave Optics and Optoelectronics*: Bishnu Pal (ed.), Intech; 2010. pp. 257-268.

[6] Oya M., Kishikawa H., Goto N., and Yanagiya S. All-Optical Switch Consisting of Two-Stage Interferometers Controlled by Using Saturable Absorption of Monolayer Graphene. *Opt Express*. 2012, 20(24), 27322-27330.

[7] Takahashi M., Ueda W., Goto N., and Yanagiya S. Saturable Absorption by Vertically Inserted or Overlaid Monolayer Graphene in Optical Waveguide for All-Optical Switching Circuit. *IEEE Photon J*. 2013, 5(5), 6602109.

[8] Takahashi M., Kishikawa H., Goto N., and Yanagiya S. All-Optical Switch with Cascaded Two-Stage Mach-Zehnder Interferometers Using Saturable Absorption Accompanied by Refractive-Index Change in Graphene. *IEEE/OSA J of Lightwave Technol*. 2014, 32(21), 3624-3630.

[9] Zhang H., Virally S., Bao Q., Ping L. K., Massar S., Godbout N., and Kockaert P. Z-scan Measurement of the Nonlinear Refractive Index of Graphene. *Opt Lett*. 2012, 37(11), 1856-1858.

[10] Geim A. K. and Novoselov K. S. The Rise of Graphene. *Natur Mater*. 2007, 6, 183-191.

[11] Iechika, Y. Application of Graphene to High-Speed Transistors: Expectations and Challenges. *Sci Technol Trends Quarter Rev* 2010, 37, 76-92.

[12] Sun Z., Hasan T., Torrisi F., Popa D., Privitera G., Wang F., Bonaccorso F., Basko D. M., and Ferrari A. C. Graphene mode-locked ultrafast laser. *ACS Nano* 2010, 4(2), 803-810.

[13] Bao Q., Zhang H., Ni Z., Wang Y., Polavarapu L., Shen Z., Xu Q.-H., Tang D. and Loh K. P. Monolayer Graphene as a Saturable Absorber in a Mode-Locked Laser. *Nano Res*. 2011, 4(3), 297-307.

[14] Yamashita S. A Tutorial on Nonlinear Photonics Applications of Carbon Nanotube and Graphene. *IEEE/OSA J Lightwave Technol*. 2012, 30(4), 427-447.

[15] Li W., Chen B., Meng C., Fang W., Xiao Y., Li X., Hu Z., Xu Y., Tong L., Wang H., Liu W., Bao J. and Shen Y. R. Ultrafast All-Optical Graphene Modulator. *Nano Lett*. 2014, 14(2), 955-959.

[16] Ferrari A. C., Meyer J. C., Scardaci V., Casiraghi C., Lazzeri M., Mauri F., Piscanec S., Jiang D., Novoselov K. S., Roth S., and Geim A. K. Raman Spectrum of Graphene and Graphene Layers. *Phys Rev* Lett. 2006, 97, 187401-1-4.

Chapter 16

# Novel Floating and Auto-stereoscopic Display with IRLED Sensors Interactive Virtual Touch System

Jian-Chiun Liou

Additional information is available at the end of the chapter

http://dx.doi.org/10.5772/61113

**Abstract**

A wide range of the types of interactive virtual touch system have been in research and development. As displayed by the trends, users do not need any special equipment, can interact with the images, and under normal circumstances of the interactive nature, saves trouble. We have been studying relevant video interactive systems in which a virtual image, like in the real world, exists to display objects. We developed a floating display and the principle, which is based on an interactive video system, to enable more realistic auto-stereoscopic images.

We developed and applied a multi-view auto-stereoscopic concept, and this newly developed system contains static and moving images. To prove that this concept can be applied, we have developed a series of infrared light-emitting diode sensors character interaction system. In the study, the system is presented in any virtual objects displayed on the screen. These virtual objects move around and interact with the user. The user need not key in any input mode of operation and can use body language to achieve interactive function, as if the fictitious image exists in the ambient space. Here, we use infrared LED array sensors to sense human body motion and finger movements, and we use LED modules to transmit and receive arrays. The systems uses a three-dimensional arrangement of the LED array architecture in order to effectively and accurately detect the user's finger movements and body motions.

The purpose of the entire framework is to have the principle of an actual image go through the optical system of the lens from the external surface of the system in 5 cm floating images. The optical lens system frontage of 50 cm can be seen as a pop-up image, and the image shows a 15° angle. People use their forefinger in a non-contact space virtual touch system to control the floating image menu. Touching virtual objects in virtual space will become the next generation human-machine operations. Touching virtual objects architecture will have an effect on other devices. This fictitious space inter-osculate virtual keyboard idea will have a benefit on people's health (not infected) and can be applied to other public equipment like the ATM, hospital elevator (buttons), and so on.

**Keywords:** Virtual, Interactive, Floating image, Auto-stereoscopic

## 1. Introduction

This study shows that the interface does not require any keyboard input and/or a high precision interactive device operation. The purpose of the study is for the actual object image through the optical lens system design to go beyond 5 cm from the system of the drift image. Outside the optical system, in about 50 cm in front of the viewer, a pop-up image can be observed at a viewing angle of 15°. Viewers can use a non-contact touch through IR LED arrays shown in Figure 1 A.

The general contact system with touch-sensitive keys may cause contamination, and germs can potentially spread and reproduce through contact. The development of a touch sensing technology and multi-touch operation technology has gradually been used in all kinds of electrical products for everyday use. Because of privacy requirements, research and development staffs have developed personal information security needs and different ways to deal with the different system architectures [1-6].

As a subsequent introduction, we mainly focus on the common use of electronic equipment in public places: ATM buttons, health and other equipment, and hospital elevator buttons; the system does not touch the electronic products since the operation principle of the function keys can be positioned. Different aspects that are addressed further include a non-touch system to explore the given potential. There are now a popular variety of touch devices, which are necessary to control systems through non-touch input texts, which may be useful in their daily lives in the future.

In the current non-contact control system, the system is set up as a matrix and sent by the infrared - receiver modules. The infrared detector arrays are generally formed by interrupting, and the infrared ray is used to deposit the position of the object. However, in the infrared transmitter, detecting the receiver module current means the infrared transmitting and receiving modules continue to be touched because the object interrupts the corresponding position in the infrared[7-9].

The LCD flat panel display has become the world's leading product. In view of this phenomenon, the trend that dominated the stereo system is the LCD type, giving the viewer a disparity in Figure 1B and the final 3D effect. The 3D effect is assigned a different image, a different view of the area showing the direction of the optical device.

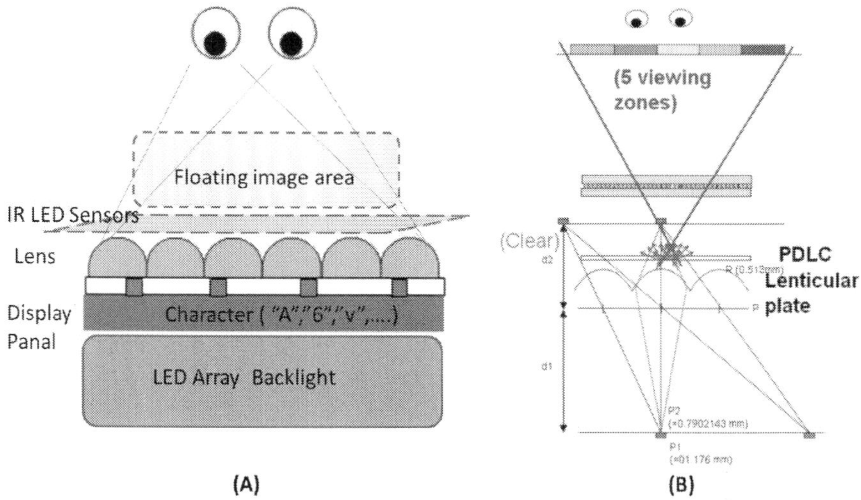


**Figure 1.** (A). Floating image with interactive touch system. Figure 1(B). Multi-view auto-stereoscopic system.

Some studies on multi-view 3D displays have been developed a while back. Multi-view stereo displays are usually followed up either by the integral imaging method [10-12] or the parallax barrier method [13]. All of these existing methods display a 3D image on a flat screen, and the image cannot be superimposed in real space.

Our interactive imaging studies may be used to pop up in real space through a floating 3D image, and the user can see the floating images without using glasses. Superimposing the projection methods using a real space image projection display has been proposed [14-16]. Based on a non-planar image, users can actually touch the projected image device using these methods, thereby allowing the user objects to experience the real projection surface interaction. However, such a 2D image projection of nature cannot produce multi-view parallax images. In this study, our goal is to create a real image of the touch system and air-infrared, interactive, virtual image with LED sensor.

## 2. Architecture

Since the fictitious appearance is formed in air, the user can locate the three-dimensional position through the direct contact with the air touch virtual panel. Infrared LED illumination is made in a cylindrical lens mounted in front of a fictitious image as shown in Figure 2.

The fictitious appearance is located in a plurality of sub-regions, which belongs to the operator interface. The object is touched in a sub-region that corresponds by send - receiving. For the corresponding operation function block to perform, the sub-region must be touched. A synchronous processing system is electrically connected to the corresponding transmission - reception. The system is arranged in the same direction of the object sensing delay enabled devices, and the synchronous processing system is configured to synchronize the control settings for each unit.

To solve mishandling, which continuously trigger a current with the same key issues, the operator was asked to correct the perceived image on the virtual panel in the right viewing area. Out of the region, observers perceive only a partial view of the blind or image map. We use the delay function and synchronization to control the virtual panel of the system.

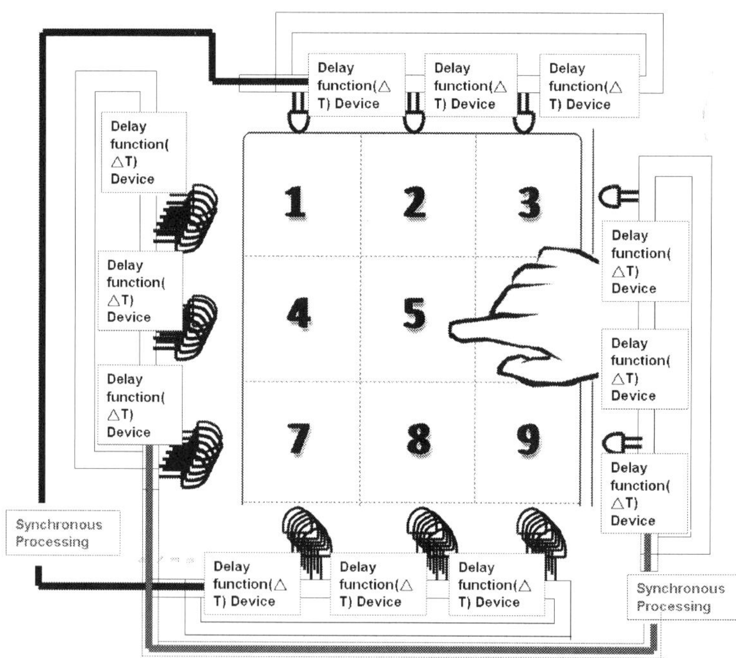


**Figure 2.** Virtual Touch system with IR LED device array

Novel floating and auto-stereoscopic display with IR LED sensors interactive virtual touch system include two major parts. The first part is the floating image and auto-stereoscopic display. The second part is the design and fabrication of the IR LED sensors multiplexer(mux) system.

## 2.1. Floating image and auto-stereoscopic display

In recent years, the three-dimensional television market grew quickly. To add, it is included the other three-dimensional floating image systems. The stereo image is shown in Figure 3.

http://dx.doi.org/10.5772/61113

For this reason, the system constructed a floating color image projecting device. According to principles of geometric optics, through a set of lenses or concave mirror and 3D objects that project true 3D images, we can achieve the principles of purpose. Therefore, it uses a display method to replace so that the image is changing the wishes of necessary real 3D objects. However, the two-dimensional image floating image, rather than the three-dimensional real image, has been achieved. Therefore, we propose the free space of the three strategic projects with a wide-angle auto-stereoscopic floating picture below.

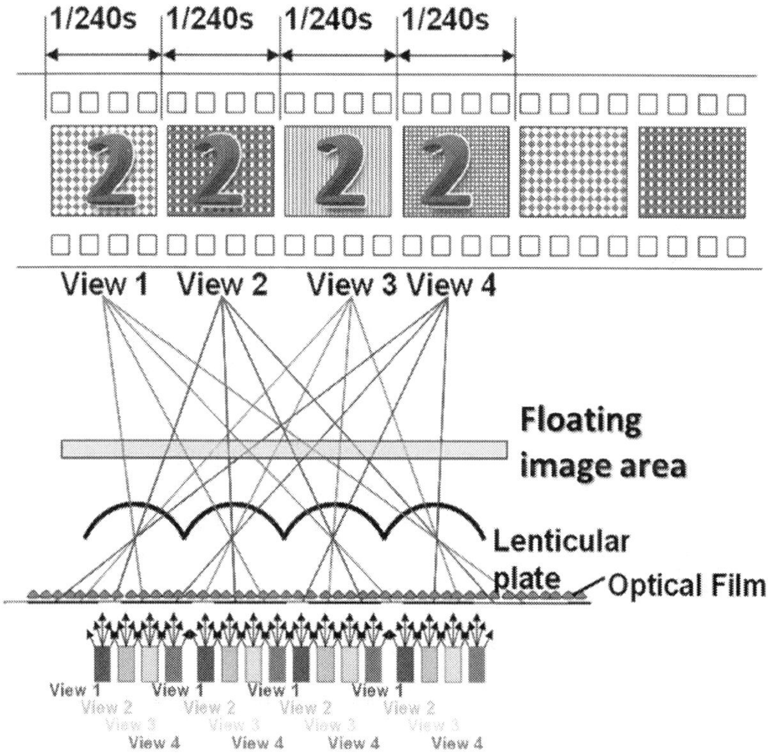


**Figure 3.** The floating image products with a virtual image.

We propose a floating image display apparatus in Figure 4 that has two main parts: (1) an image appearance system, and (2) an auto-stereoscopic display.

An optical projection system of this type comprises a projection lens, illumination optics in front of the projection lens, and one light modulator, such as a digital mirror device (DMD). The light modulator allows the reflection of the light in a controllable manner to different directions, such as when the ON state light cones, in which the light is projected onto the

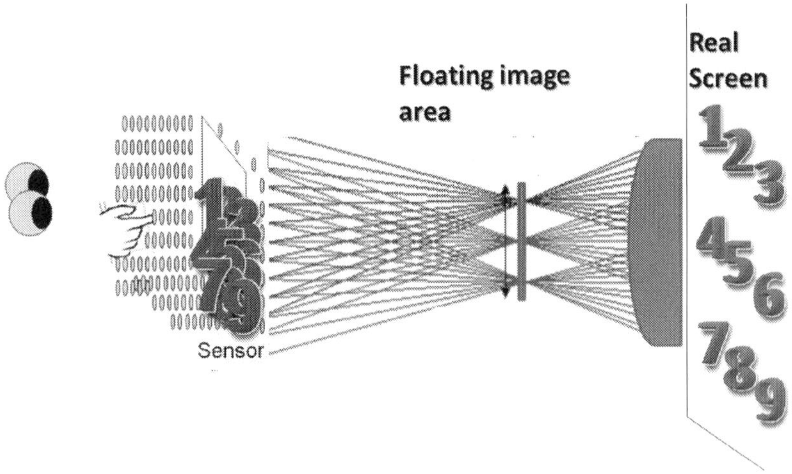


Figure 4. Sensors array in the appearance image system.

projection lens, and the OFF state light cones are formed. In this research, the optical projection system is shown in Figure 5.

Figure 5. The projection lens and one light modulator.

The designed microlens is shown in Figure 6 such that the width of the region through the microlens at a certain viewing zone may be observed as equal to or smaller than the LCD pixel array in a sub-pixel. You can predict high resolution image when viewed from a viewing area to the center of the area when the rays pass through the lenticular center.

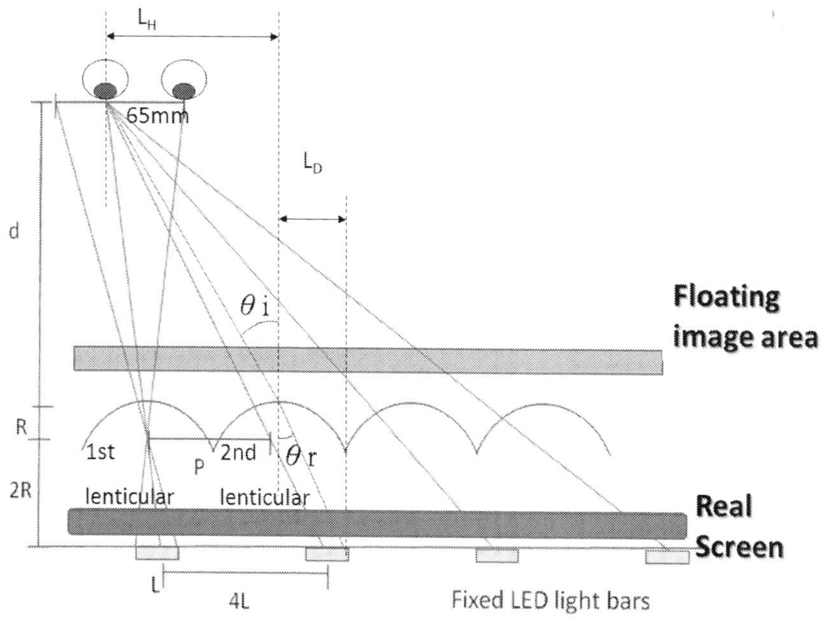


**Figure 6.** Designed microlens.

The image generating auto-stereoscopic display and its parameters include the viewing distance and the viewing area. In general, the pitch of the viewport to match the eyes and the viewing distances of the monitor determines the purpose. In this apparatus, in particular, the stereoscopic image is viewed by the projection lens shown in Figures 7, 8, 9, and 10. Thus, it illustrates the details of the theory and methods.

A floating real image between the projection lens and the user was set. Moreover, by using the depth detecting module to identify the user position and the control unit electrically connected to the image generator, the projection lens is set. And the depth detecting module, the image generator, and the projection lens set can adjust the position of the floating real image according to the user position.

Regarding the keypad module, it should be noted that by utilizing the control unit and the depth detecting module, the display apparatus in the disclosure can provide a user-friendly operation and a realistic interaction for the user as shown in Figure 10. Specifically, the control unit steers the movement of the image generator according to the position of the user detected by the depth detecting module, so as to adjust the relative positions of the image generator and the projection lens set, the position of the floating real image, and the size of the floating real image. The depth detecting module detects the position of the user, and may also detect the position of the body of the user, or the position of the fingers of the user touching the floating real image.

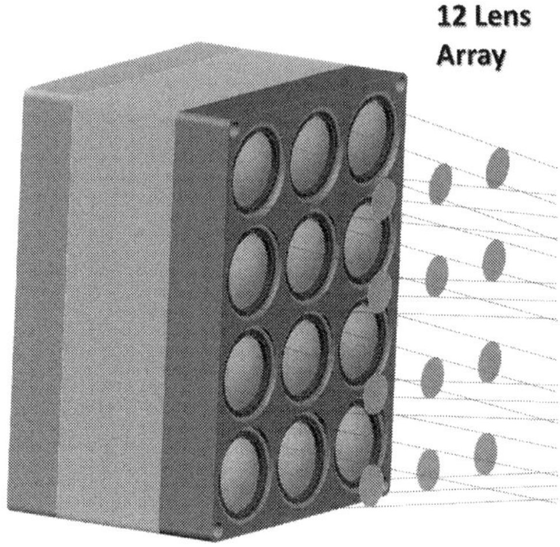


Figure 7. Twelve lens array.

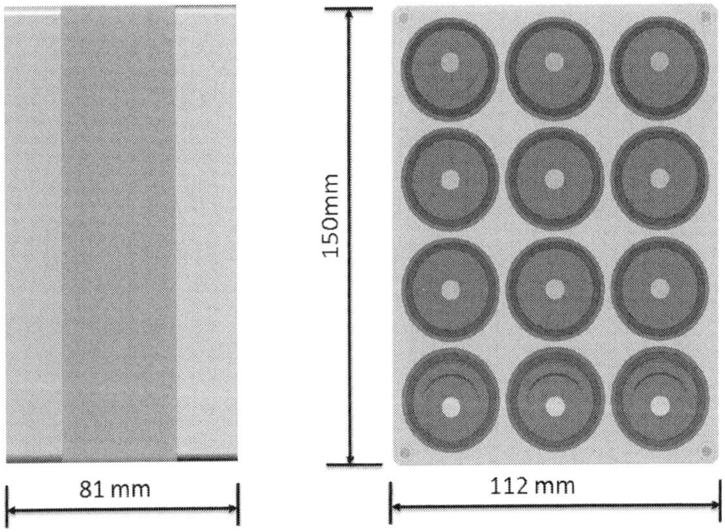


Figure 8. Top view of twelve lens array.

The depth detecting module feeds the detected position information of the user. The control unit can calculate the position of the user, the position of the floating real image, and the required size of the floating real image. Accordingly, the image generator and/or the projection lens set are shifted by corresponding distances to achieve the needed image variation effects.

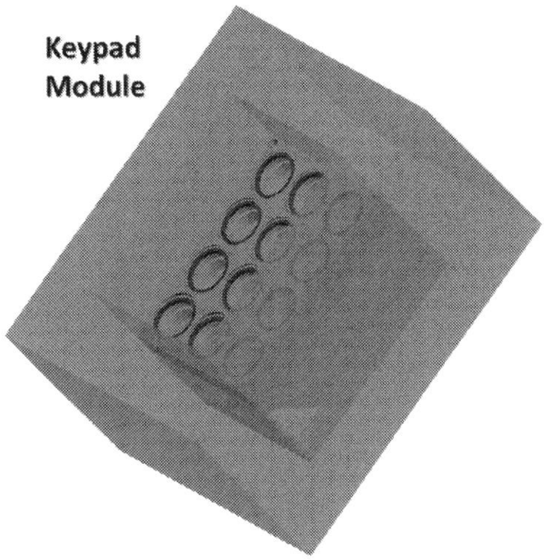


**Figure 9.** Keypad module.

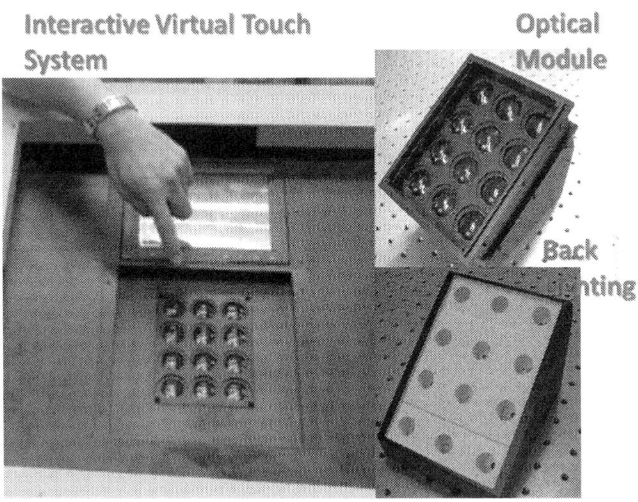


**Figure 10.** The stereoscopic images through the projected lenses.

## 2.2. Design and fabrication of IR LED sensors multiplexer (mux) system

We propose an auto-stereoscopic display and an interactive technology locker door floating image device as shown in Figure 8. The pop-up appearance system has two main parts: (1) an image pop-up system, and (2) two sights of auto-stereoscopic display. In these two parts, the audiences can observe the appearance of 3D images in the air with a disparity from a wide viewing angle. The purpose of the image is to float in a free space that is accessible to the viewer's hand. Accordingly, the touch and interactive technologies can include free space to enhance system capabilities. The new device can have many applications, especially for products and information kiosks in public places.

The IR LED sensors system is shown in Figure 11, with the non-touch positioning sensing devices using infrared (IR) LED and proximity sensor. The signals, from a set of proximity sensors, are used to determine the position of the hand through the trilateration method. The distances between the hand and sensors can be obtained from proximity sensors. The positioning results of the hand are verified over mobile devices. Therefore, the system allows a user to intuitively interact with the mobile device real time. This design can waive the necessity of touching the panel or screen of the mobile device.

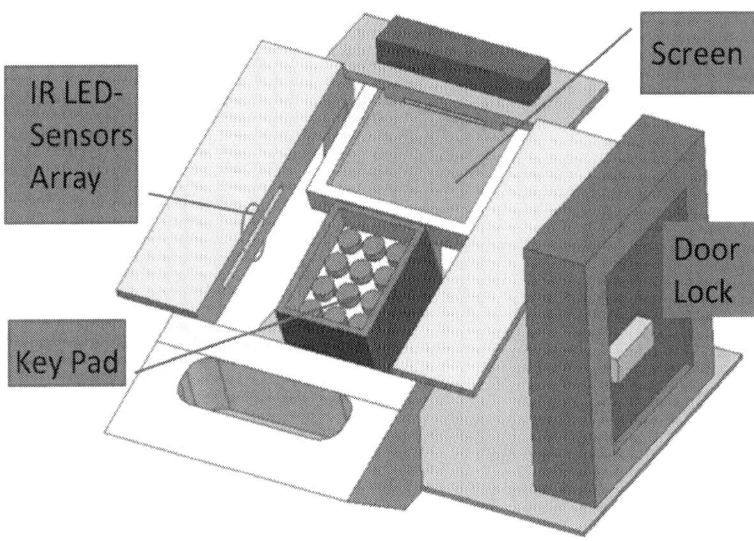


**Figure 11.** IR LED sensors system.

The multiplexer data control system of the shift register provided a sequence signal. All of the IR LED sensors architecture methods are shown in Figure 12. The floating image is a pop-up space located between the IR LEDs array and sensors array.

Given several mode designs, we have to solve the noise cross-talk between channels and the energy consumption. On the first mode, we designed an "even group and odd group" circuit architecture that is shown in Figure 13. Additionally, the waveform is shown in Figure 14.

The second addressable type of the design is an interval of two elements on different time sequence driving architecture. The architecture shows a sequence that involves the 1st, 4th, 7th;

2nd, 5th, 8th; 3rd, 6th, 9th cyclical signal output controlled elements as shown in Figure 15. The waveform is shown in Figure 16.

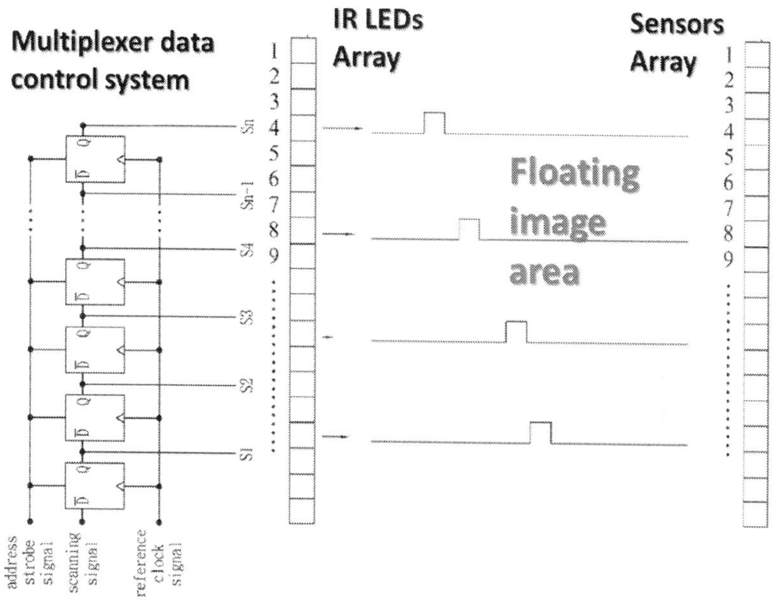


**Figure 12.** IR LED Sensors multiplexer (mux) system.

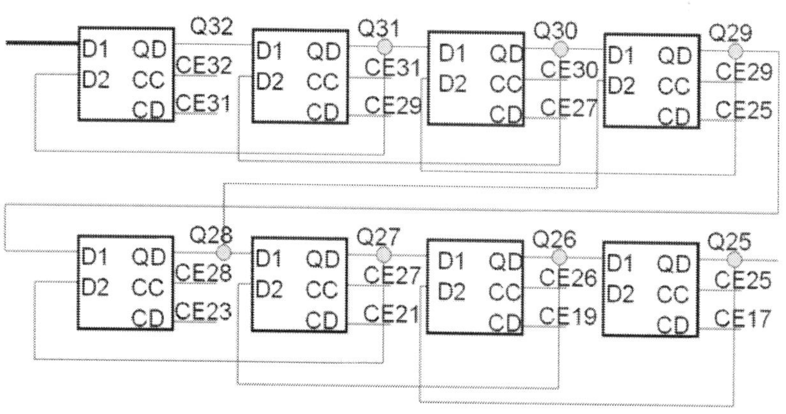


**Figure 13.** Even and odd group output in different time cycle.

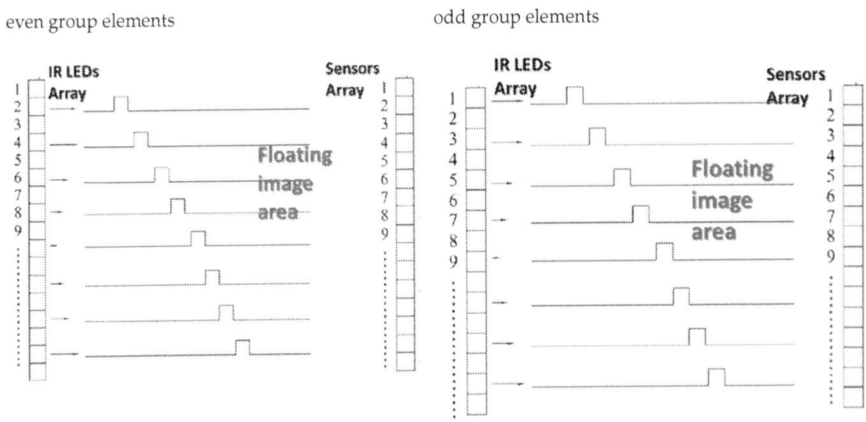


**Figure 14.** The waveform of even and odd elements.

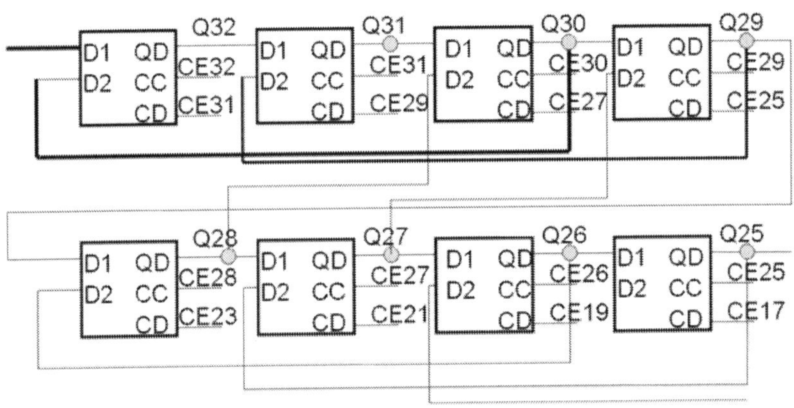


**Figure 15.** An interval of two elements on a different time driving.

Thirdly, there is an interval of three elements on a different time driving. The architecture shows a sequence that involves the 1st, 5th, 9th; 2nd, 6th, 10th; 3rd, 7th, 11th; 4th, 8th, 12th cyclical signal output controlled elements as shown in Figure 17. The waveform is shown in Figure 18.

Every shift register is used to receive part of the timing counting signal and combined with an odd-even number selection mechanism or a specific number sequence (e.g. 1, 4, 7) element mechanism to generate a set of enabling signals as shown in Figure 19. In this manner, the driving control of the present element circuits can be reached via any random combination of the address signal and each set of enabling signals, i.e., the driving of each element circuit is

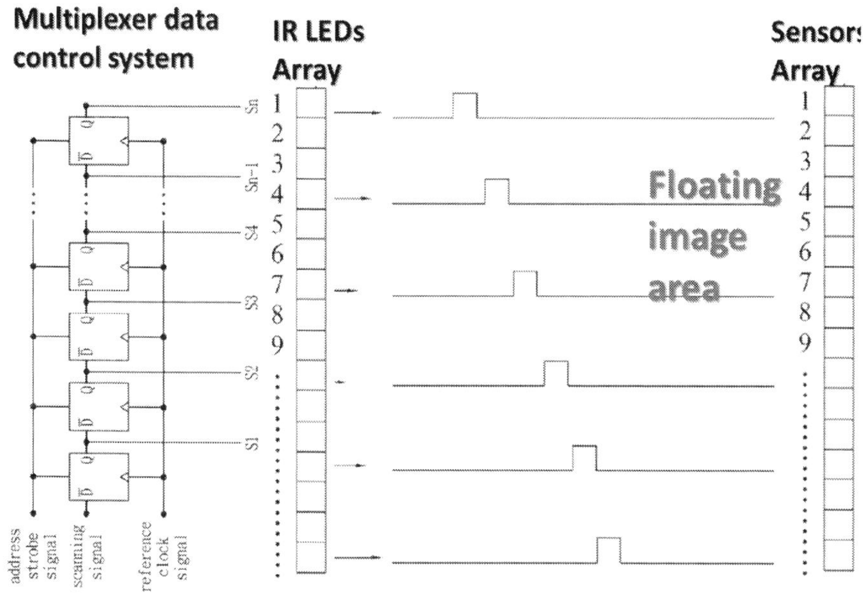


Figure 16. The waveform of interval of two elements.

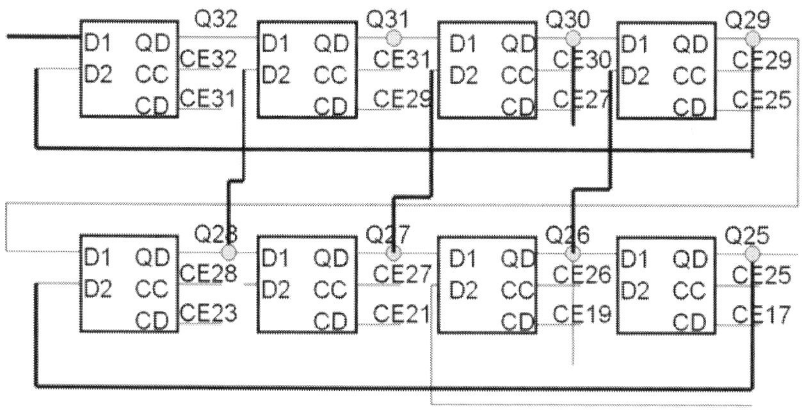


Figure 17. An interval of three elements on a different time driving.

fulfilled via the control of an address signal and one set of the enabling signals. This arrangement is able to prevent generating erroneously triggering signals in the original due to the mutual interference of the IR LED sensors circuits.

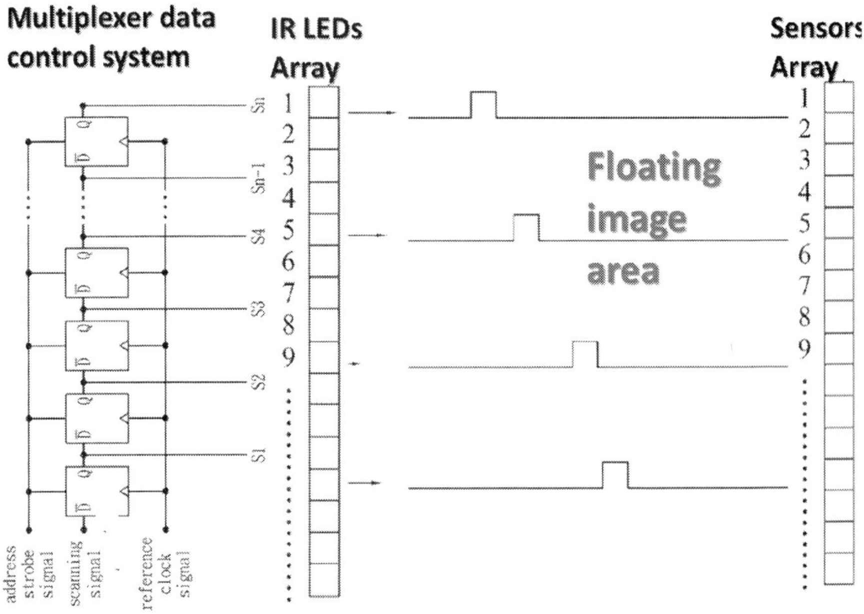


Figure 18. The waveform of the interval of three elements.

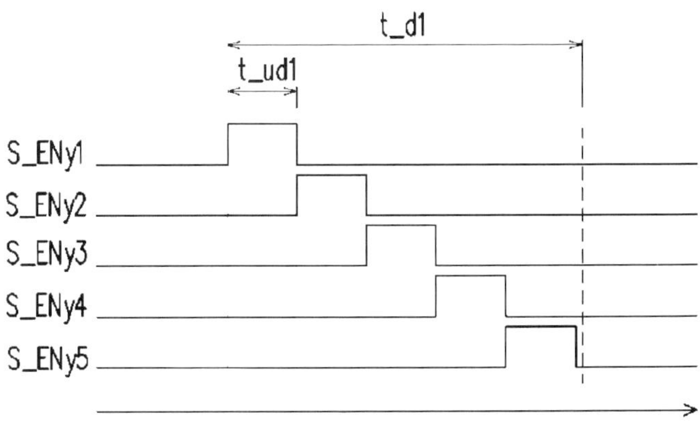


Figure 19. A set of enabling signals.

http://dx.doi.org/10.5772/61113

## 3. Simulation results

The scanning sequence of the multiplexer is shown in Figure 20. An interleaving data encoding sequence is labeled in red and blue arrows respectively. The inference between neighboring elements is also simulated in Figure 20. The output voltage noise of the third element is lower than half of the level of the second element. The error signal excited by the first elements is significantly reduced.

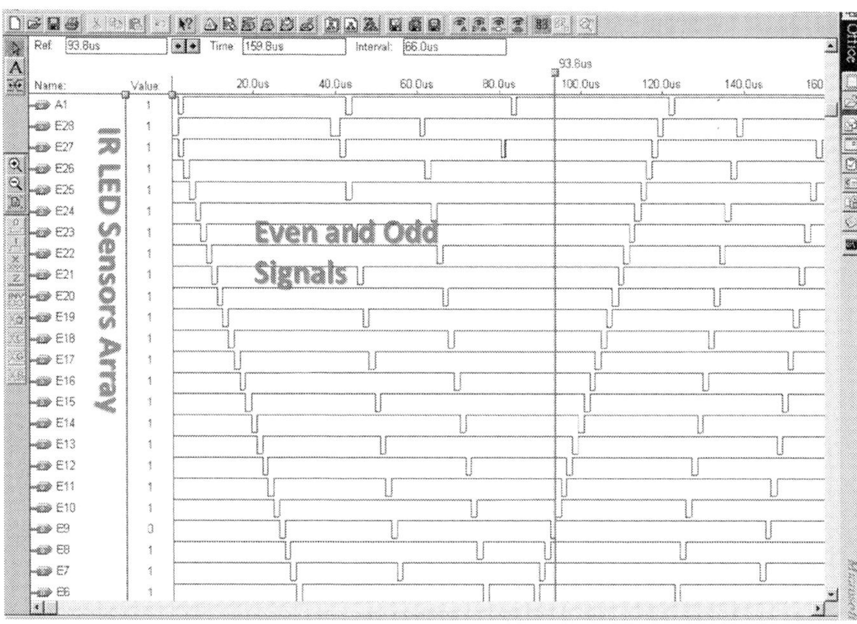


**Figure 20.** Simulated data encoding sequence for medical ultrasound multiplexer.

The control circuit on IR LED sensor elements verifies addressing elements by a multiplexer logic simulation. There are four kinds of printing tape: type 1 is a sequence of data input for IR LED sensor elements 1, 2, 3,...up to IR LED sensor elements n; type 2 is a sequence of data encoding for IR LED sensor elements n, (n-2), (n-4),...down to final IR LED sensor elements 2; type 3 is a sequence for IR LED sensor elements 1, 3, 5,...up to final IR LED sensor elements; and type 4 is a sequence for IR LED sensor elements (n-1), (n-3), (n-5),....down to final IR LED sensor elements. The driver system design has to simulate the function work, and we used FPGA to verify this.

A novel IR LED sensor elements multiplexer data registration system has been designed and simulated. The controller can be easily adapted to different sizes of the IR LED sensor elements without much hardware arrangement. The simulated results show that the proposed multiplexer controller can reduce the noise interference caused by the excitation of neighboring channels as shown in Figure 21.

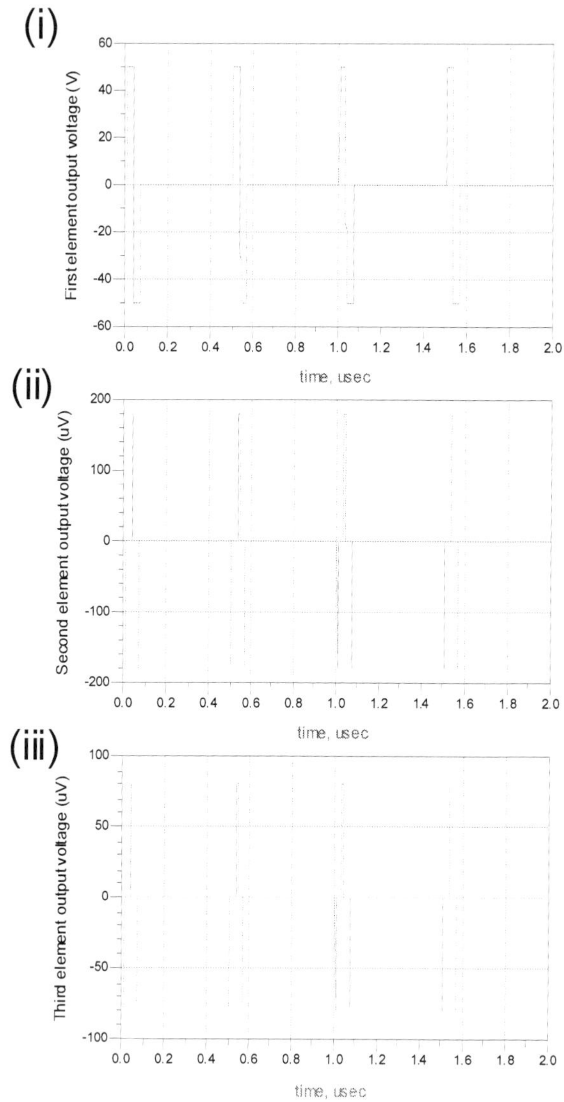


**Figure 21.** Simulated voltage inferences among the IR LED sensor elements.

A two-dimensional data control system is shown in Figure 22. Under the two-dimensional circuit configurations, a shifting sub-circuit uses a special control method to generate the enabling signals in a serial-in and parallel-out manner. The special control method connects an output of a n-th D flip-flop from the output end to a trigger end of a (n+2)-th D flip-flop.

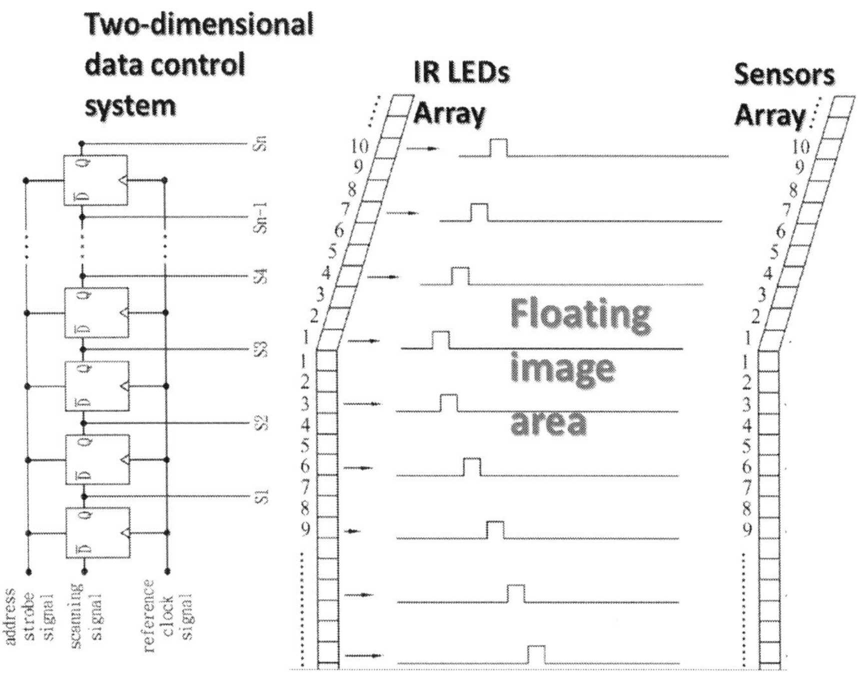


Figure 22. Two-dimensional data control system.

The multi-dimensional data control system is shown in Figure 23. For the same architecture, the three-dimensional data control system method of the shifting circuit connects an output of the n-th D flip-flop from the output end to a trigger end of a (k+3)-th D flip-flop.

## 4. Integrated system

Figure 24 shows a floating image of the virtual touch system photo. It is the top view of the floating image system. Figure 25 is a cross view of the floating image system. The fictitious image is formed by a focus lens.

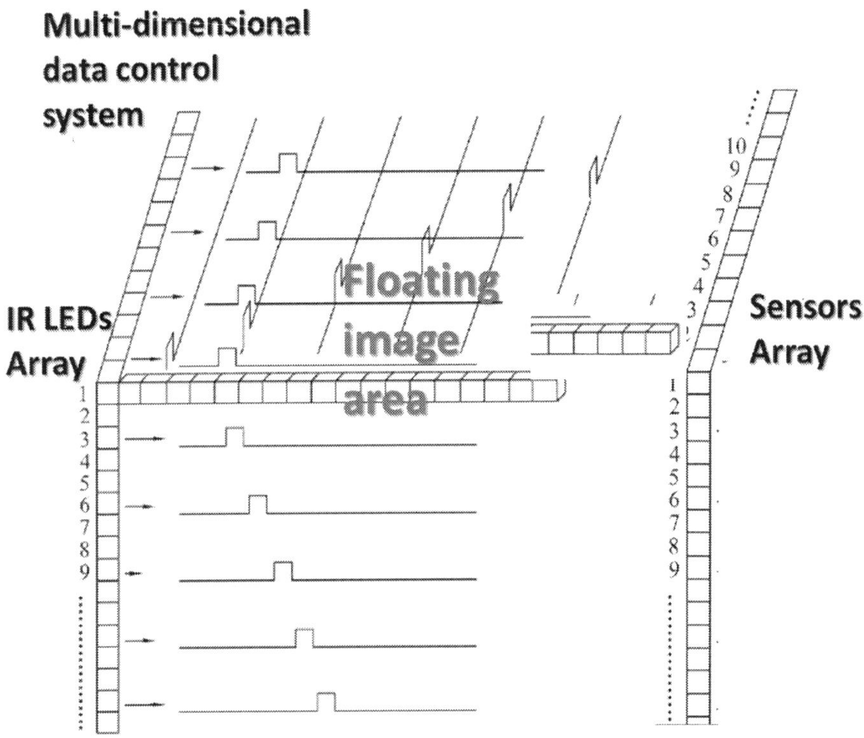


Figure 23. Multi-dimensional data control system.

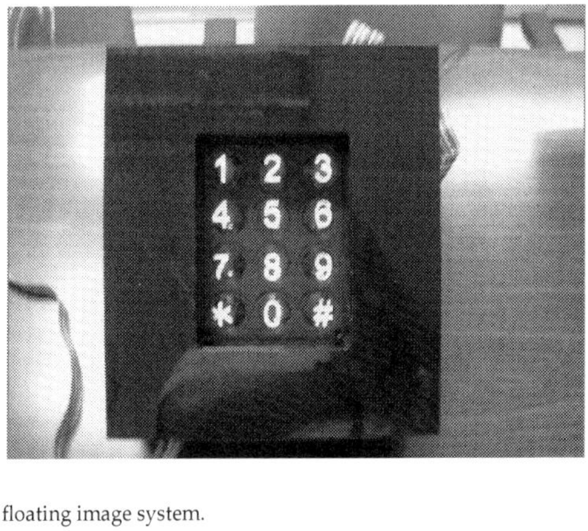

Figure 24. Top view of the floating image system.

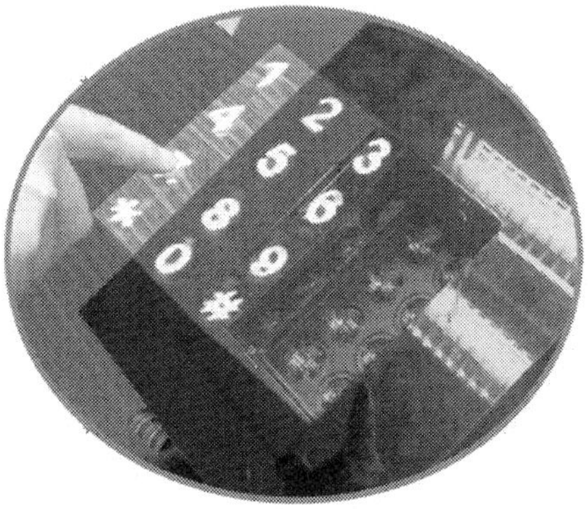

Figure 25. Cross view of the floating image system.

## 5. Discussion and result

The integrated infrared light sensor module, a floating 3D image architecture and image display is shown in Figures 26 and 27. The image is 20 cm away from the system, and viewers could observe the image to be about 50 cm within the best viewing distance. In addition to displaying three-dimensional images, the application is also a virtual keyboard for the telephone, as shown in Figure 28. Thus, the device can be used to interact with the user interface function. The virtual system displays its application to the locker room door as shown in Figure 29.

Figure 26. IR LED sensors module.

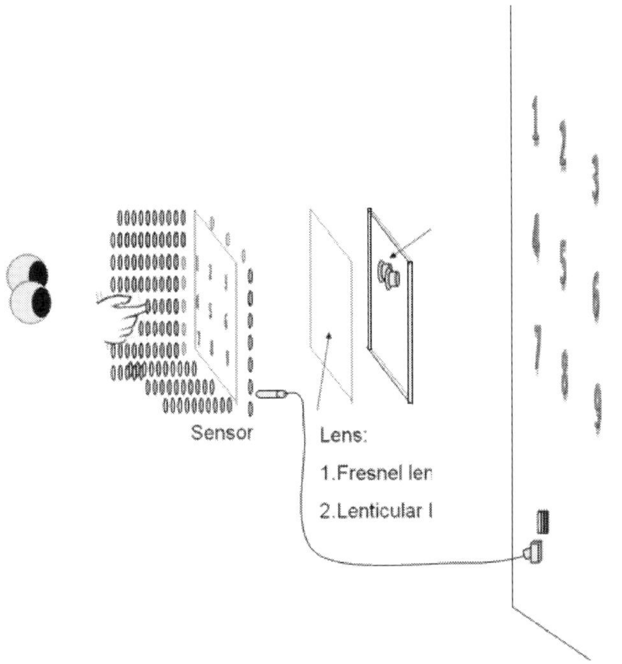


**Figure 27.** Floating 3D image architecture.

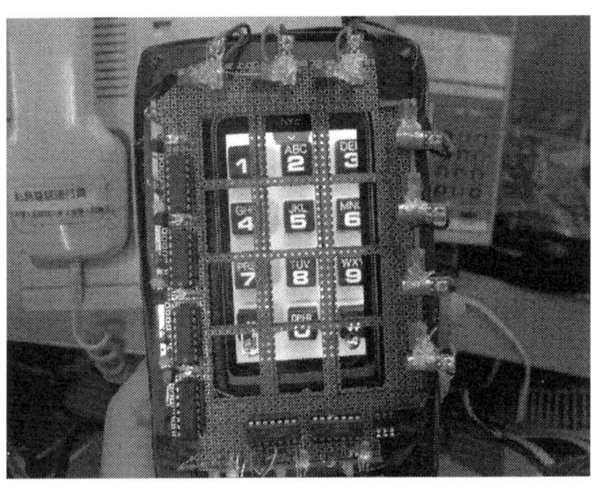

**Figure 28.** Virtual keypad for telephone.

http://dx.doi.org/10.5772/61113

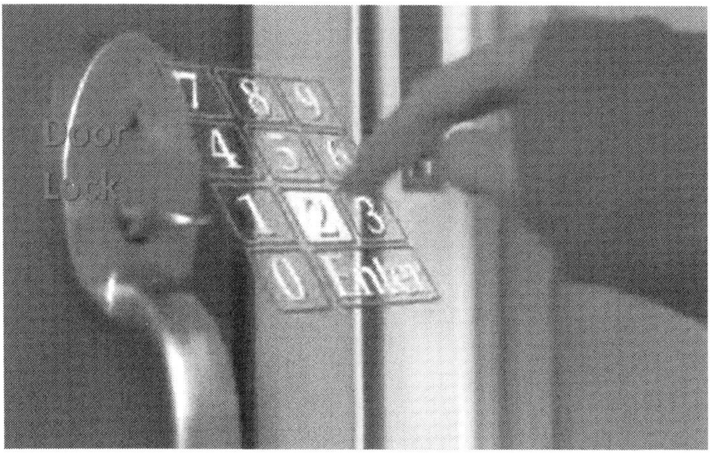


**Figure 29.** Virtual systems for the door locker.

## Author details

Jian-Chiun Liou

Address all correspondence to: jcliou@kuas.edu.tw

National Kaohsiung University of Applied Sciences (KUAS), The Department of Electronic Engineering, Taiwan

## References

[1] Han, J.Y., Low-cost multi-touch sensing through frustrated total internal reflection. ACM UIST, pp. 115-118,(2005).

[2] Kitamura, Y., Nakayama, T., Nakashima, T., and Yamamoto, S., The illusionhole with polarization filters. ACM VRST, p. 244-251, (2006).

[3] Chan, L.W., Chuang, Y.F., Chia, Y.W., Hung, Y.P., and Jane Hsu., A new method for multi-finger detection using a regular diffuser. International Conference on Human-Computer Interaction, (2007).

[4] Chan, L.W., Chuang, Y.F., Yu, M.C., Chao, Y.L., Lee, M.S., Hung, Y.P. and Jane Hsu, Gesture-based interaction for a magic crystal ball. ACM VRST, pp. 157-164(2007).

[5] Wilson, A. TouchLight: An Imaging Touch Screen and Display for Gesture-Based Interaction. International Conference on Multimodal Interfaces, pp. 69-76, (2004).

[6] Morris, M.R., Morris, D., and Winograd, T., Individual audio channels with single display groupware: effects on ommunication and task strategy. ACM CSCW, p. 242-251,(2004).

[7] Sugimoto, M., Hosoi, K., and Hashizume, H., Caretta: system for supporting face-to-face collaboration by integrating personal and shared spaces ACM CHI, p. 41-48, (2004).

[8] Kitamura, Y., Osawa, W., Yamaguchi, T., Takemura, H., and Kishino, F., A display table for strategic collaboration preserving private and public information. International Conference on Entertainment Computing, pp. 167-179,(2005).

[9] Kakehi, T., Iida, M., Naemura, T., Shirai, Y., and Matsushita, M., Lumisight table: an interactive view-dependent tabletop display. IEEE Computer Graphics and Applications, January/February, 25(1), p. 48-53,(2005).

[10] G. Lippmann: La photographie integrale, Comptes-Rendus 146, pp. 446.451, (1908).

[11] Roberto Lopez-Gulliver, Shunsuke Yoshida, Sumio Yano, and Naomi Inoue: gCubik : Real-time Integral Image Rendering for a Cubic 3D Display, ACM SIGGRAPH 2009 Emerging Technologies, New Orleans, USA, August 3-7 (2009).

[12] Koike, T., Sakai, H., Shibahara, T., Oikawa, M., Yamasaki, M., and Utsugi, K., Light field copy machine. In ACM SIGGRAPH ASIA 2009 Emerging Technologies: Adaptation (Yokohama, Japan, December 16 – 19, SIGGRAPH ASIA '09,(2009).

[13] Sakamoto, K., Kimura, R., and Takaki, M.: Parallax polarizer barrier stereoscopic 3D display systems, Active Media Technology, 2005. (AMT 2005). Proceedings of the 2005 International Conference on, pp.469- 474, 19-21 May (2005).

[14] Daisuke Kondo, Toshiyuki Goto, Makoto Kouno, Ryugo Kijima, and Yuzo Takahashi: A Virtual Anatomical Torso for Medical Education using Free From Image Projection, Proc. of the 10th International Conference on Virtual Systems and Multimedia, pp678-685, (2004).

[15] A. Cassinelli and M. Ishikawa: Khronos projector. In SIGGRAPH 2005, Emerging Technologies, (2005).

[16] Y. Kawaguchi: The Art of Gemotion in Space, Information Visualization, 2006. IV 2006. Tenth International Conference on, pp.658-663, 5-7 July (2006).

Chapter 17

# OFDM and SC-FDMA over Fiber Using Directly Modulated VCSELs

Henrique M. Salgado, Rúben E. Neto,
Luís M. Pessoa and Pedro J. Batista

Additional information is available at the end of the chapter

http://dx.doi.org/10.5772/61118

**Abstract**

Radio-over-fiber technology, used in the transport of radio signals over optical fiber by means of an optical carrier between a remote site and a central node of a cellular network, is an attractive solution for backhauling of a large number of remote antennas, enabling the shifting of the hardware complexity from base stations to a central station.

Integration of both optical and wireless broadband infrastructures into the same backhaul network leads to significant simplification and cost reduction of base stations permitting equipment sharing and dynamic allocation of resources, which in turn leads to simplified system operation and maintenance.

Wireless systems on the other hand are evolving rapidly and new standards are appearing, such as the Long-Term Evolution aiming at satisfying the required need for increasing bandwidth. Radio-over-fiber systems are known to be susceptible to noise and non-linear distortion in particular to the large peak-to-average power ratio of orthogonal frequency division multiplexing signals employed in these standards. In this work we compare, experimentally and through simulation, the performance of orthogonal frequency division multiplexing and single carrier frequency division multiple access signals, in radio-over-fiber applications, using directly modulated VCSELs.

**Keywords:** RoF, VCSEL, OFDM, SC-FDMA

## 1. Introduction

Radio-over-fiber (RoF) technology is used in transporting radio signals over optical fiber by means of an optical carrier between a remote site and a central node of a cellular network. RoF techniques are increasingly seen as a promising solution to facilitate the backhauling of a large number of remote antennas, enabling the shifting of the hardware complexity from base stations to the central station [1]. RoF allows for the combination of the advantages of optical systems such as their high bandwidth and low power consumption, with the advantages of wireless systems, namely the flexibility to use multiple standards such as Long-Term Evolution (LTE) or other upcoming standards.

The number of LTE users has been increasing due to advantages such as the ability to reach a peak throughput of 300 Mbps on the downlink and 75 Mbps on the uplink [2]. To achieve high radio spectral efficiency as well as enable efficient scheduling in time and frequency domains, a multicarrier approach for multiple access was chosen by the 3rd Generation Partnership Project. Orthogonal frequency division multiple access (OFDMA) and single carrier frequency division multiple access (SC-FDMA) were selected for the downlink and uplink, respectively, the latter also known as discrete Fourier transform–spread OFDMA.

## 2. Motivation

When OFDMA or SC-FDMA type of signals are transmitted through directly modulated vertical-cavity surface emitting lasers (VCSELs), they suffer from intermodulation distortion, due to the large number of electrical subcarrier combination in the laser cavity which degrades system performance in addition to relative intensity noise (RIN), clipping noise at the transmitter as well as shot noise and thermal noise at the receiver.

This phenomenon is well known in the literature [3], and is a result of the interaction between the electrons and photons in the active region, which is generally well described by the rate equations [4]. As a motivation for this problem, Figure 1 and Figure 2 show the spectrum of an OFDM electrical signal at the laser output (after conversion by means of a photodiode), for two values of the laser bias current, 3 and 6 mA, respectively. The original OFDM signal is centred at 1 GHz, and directly modulates the VCSEL. The high nonlinear distortion of the signal (with centre at 1 GHz) can be clearly seen, which is caused by intermodulation distortion. Third order intermodulation products (IMPs) of the type, $f_i + f_j - f_k$ or $2f_i - f_j$, where $f_{i,j,k}$, represents the frequency of the OFDM subcarriers, coincide with the transmission band of the signal and severely limit system performance. The number of intermodulation terms falling on channel $r$, $IM_{111}^{N}(r)$, $IM_{21}^{N}(r)$, of type $f_i + f_j - f_k$ and $2f_i - f_j$, respectively, where $N$ is the number of subcarriers, is given as follows [5]:

$$IM_{21}^{N} = \frac{1}{2}\left\{N - 2 - \frac{1}{2}\left[1 - \left(-1\right)^{N}\right]\left(-1\right)^{r}\right\} \tag{1}$$

http://dx.doi.org/10.5772/61118

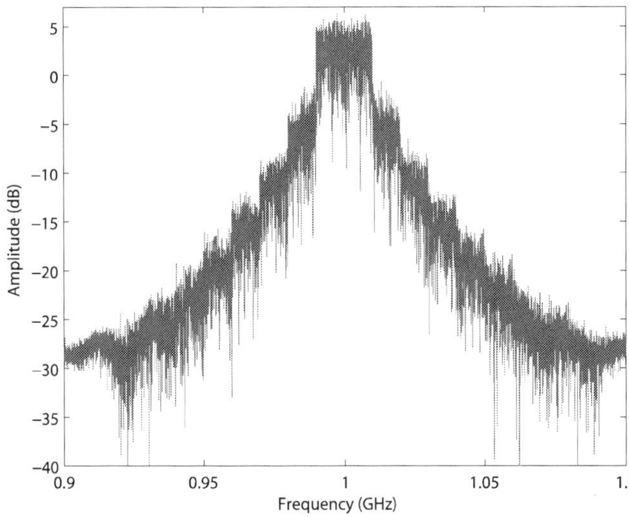


**Figure 1.** OFDM spectrum at the laser output for 3 mA bias current.

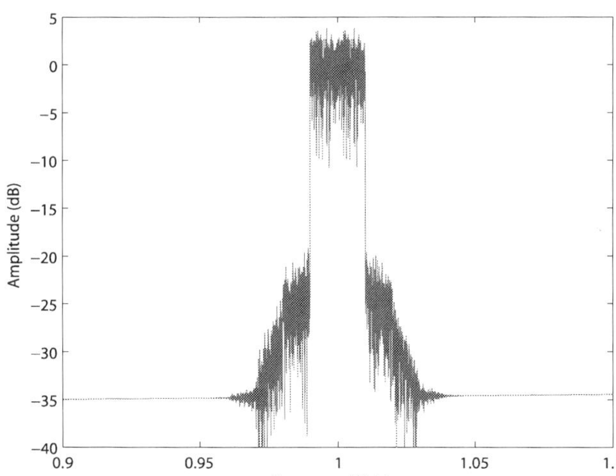


**Figure 2.** OFDM spectrum at the laser output for 6 mA bias current.

$$IM_{11\bar{1}}^{N} = \frac{r}{2}\left(N - r + 1\right) + \frac{1}{4}\left[\left(N-3\right)^{2} - 5\right] - \frac{1}{8}\left[1 - \left(-1\right)^{N}\right]\left(-1\right)^{N+r} \quad (2)$$

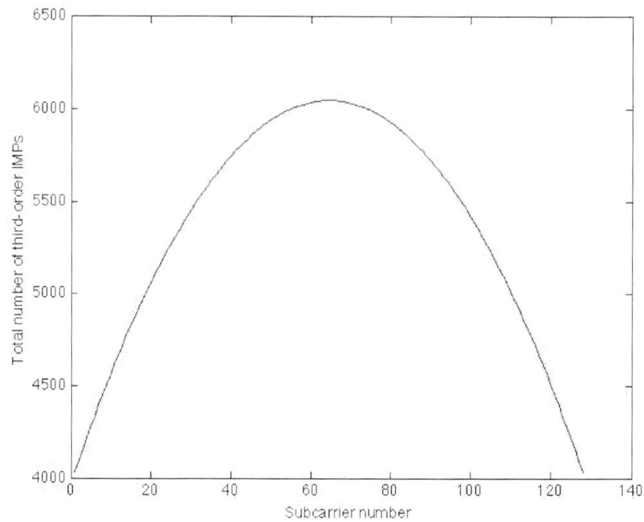


**Figure 3.** Total number of third-order intermodulation products as a function of subcarrier number for a 128 subcarriers signal.

Figure 3 shows the total number of third-order IMPs as a function of channel number, for 128 subcarriers. For a large number of subcarriers $IM_{111}^{N}$ and $IM_{21}^{N}$, approach the asymptotic value of $3N^{2}/8$ and $N/2$, respectively. On the other hand, the output spectrum is highly dependent on the operating point of the laser and on the allocation of the channels relative to the resonance frequency of the laser. On increasing the bias current to 6 mA, the resonance frequency, which previously was centred at 1 GHz, moves away to 2.5 GHz; therefore, the output signal presents a lower distortion as depicted in Figure 2. The interplay between the biasing of the laser, subcarriers frequency operation and noise, be it shot noise or RIN of the laser, is a complex one that needs to be modeled accurately for an adequate assessment of system performance. Hence, it is important to obtain a realistic model of the VCSEL device, including the electrical circuit associated with the parasitic elements and for that effect to extract the relevant parameters from experimental measurements.

## 3. Vertical-cavity surface-emitting laser

The VCSEL has emerged as an important class of semiconductor lasers in recent years. Its main characteristics, associated with the vertical-cavity geometry, are light emission perpendicular to the surface of the wafer and single longitudinal mode due to its short cavity length. The VCSEL is a microcavity laser consisting of a thin active region (< 1 μm) sandwiched between epitaxially grown distributed Bragg reflectors (DBRs). Since the first demonstration of the

VCSEL laser in Tokyo Institute of Technology in 1979 and after three decades of research, many potential applications have emerged. As a result several VCSEL manufacturers turned up in the market.

Due to their geometry, VCSELs offer a number of significant advantages over edge-emitting lasers listed below:

- Low threshold currents: the volume of the active region is relatively small, hence they possess low threshold currents and therefore consume less power than edge-emitting lasers.
- Circularly shaped beam: the symmetry in the wafer plane means the laser output is a narrow, low divergence circular beam, permitting high coupling efficiency to optical fibers with relaxed alignment tolerances.
- Single-mode operation: due to the microscopic cavity length, VCSELs inherently operate in a single-longitudinal mode which makes them suitable for high-bit-rate fiber optic communications.
- Low-cost-wafer fabrication: VCSELs allow for a high packing density in the form of two-dimensional arrays and cost-effective fabrication and testing at the wafer level.
- High modulation speed: experimental studies [3] indicate that the VCSELs have very fast intrinsic dynamic properties with relaxation oscillation frequencies as high as 71 GHz.

These advantages make the VCSEL device a suitable candidate for the applications targeted in ROF scenarios. However, the VCSEL as any semiconductor laser exhibits a nonlinear dynamic behavior that needs to be modeled accurately for the assessment of the impact of distortion on system performance. In the next section, the model of the VCSEL based on a set of rate equations for the carrier and photon density is presented. This model will be the basis for the investigation of the laser nonlinear distortion.

### 3.1. VCSEL modeling

The basic structure of the VCSEL is shown in Figure 4. The laser output is taken vertically through one of the mirrors, in contrast to conventional edge-emitting laser which emits light in the plane of the wafer surface. The conventional structure employs an active region consisting of multiple quantum wells between a n-type and a p-type DBR mirrors. Because of the short cavity length ($\approx 1\ \mu m$) and thickness of the active region, the mirror reflectivities ($R$) of the DBRs must be greater than 99%. To achieve these high values of reflectivity, the DBR mirrors are made up of 20 to 40 alternating quarter-wavelength thick layers of high and low refractive indices made of semiconductors with different compositions, typically $Al_xGa_{(1-x)}As$.

The operation of a VCSEL can be understood by accounting for the rate of recombination of carriers in the active region and the rate of generation and loss of photons. For laser emission to occur, stimulated emission should be the dominant recombination mechanism. The threshold gain is defined as the gain required to sustain the optical field after travelling one round trip in the cavity. Assuming the optical gain is constant over the whole length of the laser, this leads to the condition

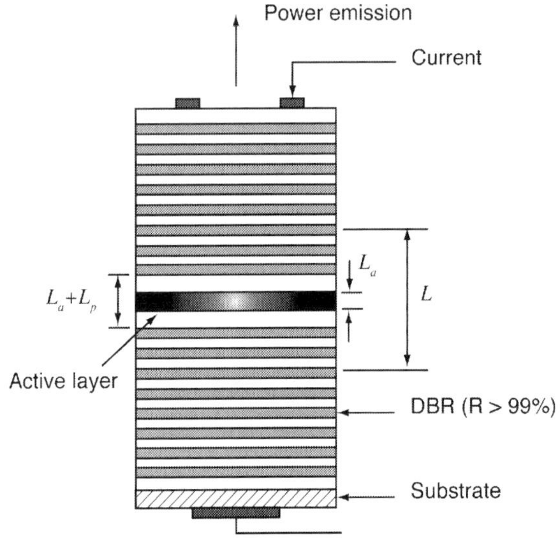


**Figure 4.** Schematic diagram of the laser structure, indicating active region thickness *La*, and effective cavity length *L*.

$$\Gamma g_{th} = \alpha_i + \frac{1}{L}\ln\left(\frac{1}{R}\right) \quad (3)$$

where $\Gamma$ is the optical confinement factor, $\alpha_i$ is the average internal loss and the second term is defined as the mirror loss $\alpha_m$. This equation shows that the gain per unit length must be sufficient to cancel out the optical losses and the losses due to light emission. Since the contribution of spontaneous emission in this simple analysis has not been considered, the actual gain will be slightly lower than the threshold gain. The description of laser operation is complete once the carrier density, $N$, is related to the injected current, $I$. This is accomplished through a rate equation that incorporates all the mechanisms by which the carriers are generated or lost inside the active region. The continuity equation which describes the rate of change of carriers in its general form is

$$\frac{dN}{dt} = \frac{\eta_i I}{qV} - \gamma_e(N)N - R_{st}P \quad (4)$$

The first term governs the rate at which the carriers are injected into the active layer due to external pumping; $q$ is the value of the electron charge, $\eta_i$ is the injection efficiency and $V$ is the volume of the active region. The second term takes into account the carrier loss owing to various recombination processes: spontaneous emission and non-radiation. The last term of equation (4) is due to stimulated emission recombination that leads to coherent emission of light.

A suitable form for the carrier recombination rate $\gamma_e(N)$ and the corresponding carrier lifetime $\tau$, for lightly doped material is

$$\gamma_e(N) = \frac{1}{\tau} = A_{nr} + BN + CN^2 \tag{5}$$

where the terms with the coefficients $A_{nr}$, $B$, and $C$ represent defect, bimolecular recombination and Auger recombination, respectively.

A corresponding rate equation for the photon density can be obtained from the Maxwell's equations using a classical approach [4, 6]. By a simple bookkeeping of the supply, annihilation, and creation of carriers and photons inside the laser cavity, we get

$$\frac{dP}{dt} = P\left(\Gamma R_{st} - \frac{1}{\tau_p}\right) + \beta\Gamma R_{sp} \tag{6}$$

In which the photon lifetime is defined by

$$\tau_p^{-1} = v_g\left[\alpha_i + \frac{1}{L}\ln\left(\frac{1}{R}\right)\right] \tag{7}$$

Equation (6) states that the rate of increase in photon density is equal to the photon generation by stimulated emission $\Gamma R_{st}P$ less the loss rate of photons $-P/\tau_p$ (as characterized by the photon lifetime $\tau_p$), plus the rate of spontaneous emission into the photon mode $\beta R_{sp}$, where $\beta$ is the fraction of the total spontaneous emission coupled into the laser mode.

The net stimulated rate which tells us how many photons are generated per unit of time per existing photon, yields a generation rate of new photons $dP/dt$ according to the following equation

$$\frac{dP}{dt} = R_{st}P \tag{8}$$

The resulting stimulated gain coefficient relates to the stimulated emission coefficient, $R_{st}$, by

$$g = \frac{\text{power emitted per unit volume}}{\text{power crossing a unit area}} = \frac{\mathrm{R_{st}}}{v_g} = \frac{R_{st}\bar{\mu}}{c} \tag{9}$$

where $v_g$ is the group velocity, $c$ is the free-space velocity of light, and $\bar{\mu}$ is the group refractive index of the material taking dispersion into account: $[\bar{\mu} = (\mu + v d\mu / dv)]$. For a multiple quantum well laser, a logarithmic function of the carrier density fits the gain well over a wide range of $N$, (see [7]),

$$g(N) = g_c \log\left(\frac{N}{N_{0m}}\right) \tag{10}$$

where $N_{0m}$ is the carrier density for transparency (zero gain) and $g_c$ is the gain coefficient. The gain function may be linearized about the carrier density at transparency yielding [8],

$$g(N) \approx a(N - N_{0m})(1 - \varepsilon P) \tag{11}$$

where $a$ is the differential gain, $\partial g / \partial N = g_c \cdot m / N_{0m}$ and $m$ is a linearization parameter obtained so that $g_{log}(N) = g_{lin}(N)$ at threshold, that is for $N = N_{th}$. Gain compression is also accounted phenomenologically through the term $(1 - \varepsilon P)$, $\varepsilon$ being the gain compression factor expressed in cubic meters.

Thus, we can rewrite the carrier and photon rate equations as

$$\frac{dN}{dt} = \frac{\eta_i I}{qV} - \frac{N}{\tau} - v_g a(N - N_{0m})(1 - \varepsilon P)P \tag{12}$$

$$\frac{dP}{dt} = \Gamma v_g a(N - N_{0m})(1 - \varepsilon P)P - \frac{1}{\tau_p}P + \beta \Gamma R_{sp} \tag{13}$$

The evolution of the signal transmitted over the optical fiber requires knowledge of the phase of the electric field. To account for dispersion effects, these equations may be complemented by an additional equation for the phase:

$$\frac{d\phi}{dt} = \frac{\alpha}{2}\left[\Gamma a v_g (N - N_{0m}) - \frac{1}{\tau_p}\right] \tag{14}$$

where $\alpha$ is the linewidth enhancement factor. Equations (11-13) represent the basic relations for describing the dynamic characteristics of laser diodes, as long as the noise sources may be omitted.

The first-order transfer function of the intrinsic laser can be obtained by linearization of the previous equations and is given by the following equation

$$\frac{H_1(\omega)}{H_1(0)} = \frac{\gamma(1-\varepsilon p_0)p_0}{(j\omega)^2 + j\omega(\gamma\Gamma\beta j_{th}/p_0 + \gamma\varepsilon p_0) + \gamma(1-\varepsilon p_0)p_0} \tag{15}$$

$$\gamma = \frac{\tau}{\tau_p} \tag{16}$$

where for the ease of numerical calculation, the physical quantities have been normalized according to [9] and $p_0$, $n_0$ correspond to the steady-state values of the photon and carrier density within the active region, associated with the bias current, $j_{th}$

$$n_0 = g_0 \tau_p N_0 \tag{17}$$

$$p_0 = \frac{\beta n_0}{1/\Gamma + n_{0m} - n_0} \tag{18}$$

$$j_{th} \cong n_{th} = 1/\Gamma + n_{0m} \tag{19}$$

### 3.2. Package and chip parasitic elements

When dealing with high-frequency electronics, the frequency limits are usually established by the parasitic elements. It is then required to know whether the laser modulation characteristics are due to the laser alone or due to the parasitic elements. To this aim, one must treat the laser as an electrical element and establish an equivalent circuit that includes the parasitic elements. Characterization of an electrical network at high frequencies is usually done using the scattering parameters.

The elements of the laser-equivalent circuit are derived from the rate equations augmented by the heterojunction voltage-current and space-charge characteristics. The resulting equivalent circuit is a parallel RLC resonant circuit [9, 10]. The carrier density and quasi-Fermi levels are clamped above threshold which manifests in the equivalent circuit as an "ac" short and no voltage can develop. The magnitude of the impedance of the entire circuit $|Z(\omega)|$ is therefore essentially zero at all frequencies except near the relaxation oscillation resonance, where its value does not exceed ≈1 Ω. For deriving the relation between the total external current $I_s$ and the current through the active region $I_a$ and in comparison to the relatively large external elements, the intrinsic laser diode can be regarded as a short circuit at all frequencies. Under zero bias, the intrinsic laser can be modeled by the active layer space-charge capacitance [11].

Chip parasitic elements vary widely among different laser structures. In practice, they take the form of a resistance in series with the intrinsic device combined with a shunt capacitance. An equivalent circuit model of the package and chip parasitic elements is shown in Figure 5; it includes a series inductor representing the wirebond, a shunting capacitor representing the contact capacitance, and a series resistor representing the contact resistance and the Bragg mirror stacks.

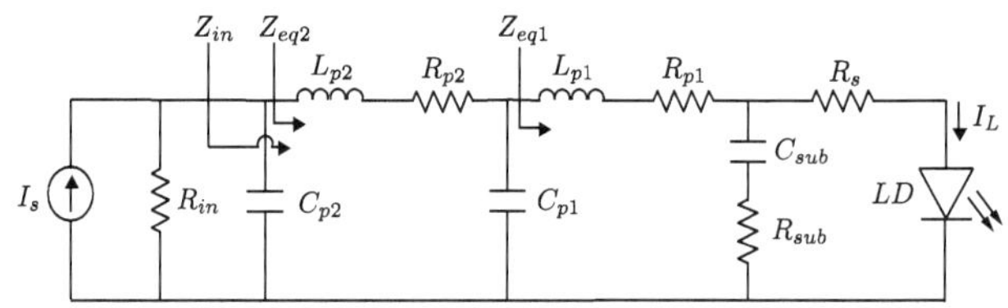


**Figure 5.** Equivalent circuit model for the laser parasitic elements.

We now define the following impedances:

$$Z_{p2} = R_{p2} + j\omega L_{p2} \tag{20}$$

$$Z_{p1} = R_{p1} + j\omega L_{p1} \tag{21}$$

$$Z_{sub} = R_{sub} + \frac{1}{j\omega C_{sub}} \tag{22}$$

The transfer function of the laser parasitic elements corresponding to ratio of the current flowing through the intrinsic laser, $I_L$ , and the source current, $I_S$, is given by:

$$Z_{in} = \frac{R_{in}}{1 + j\omega C_{p2} R_{in}} \tag{23}$$

$$Z_{eq1} = \frac{Z_{in} + Z_{p2}}{j\omega C_{p1}\left(Z_{in} + Z_{p2}\right)} \tag{24}$$

$$Z_{eq2} = \frac{Z_{eq1} + Z_{p1}}{Z_{sub} + Z_{eq1} + Z_{p1}} \tag{25}$$

$$\frac{I_L}{I_S} = \frac{Z_{eq1}}{Z_{eq1} + R_s} \frac{Z_{eq2}}{Z_{eq2} + Z_{p1}} \frac{Z_{in}}{Z_{in} + Z_{p2}} \tag{26}$$

The full laser transfer function is then the product of the intrinsic laser [equation (14)] and parasitic transfer functions (equation (26)).

### 3.3. Laser characterization

In modeling the VCSEL for simulation purposes, it is important to obtain a model as faithful as possible to the real device. This is achieved by extracting the laser parameters from experimental data. The $S_{21}$ and $S_{11}$ parameters are measured using a vectorial network analyzer (Lightwave Component Analyzer which characterizes devices in the electric and optical domains), a current source (Laser Diode Controller LDC-3700B), a bias-T (allows the continuous current injection in the laser), and a test fixture, as shown in Figure 6. The test fixture allows the connection between the laser and the SMA (Subminiature Version A) connector. The latter was designed in Advanced Design System considering Rogers 4000 series as a substrate. It should be noted that it is necessary to subtract the test fixture impact on the measurements using the de-embedding technique [12, 13].

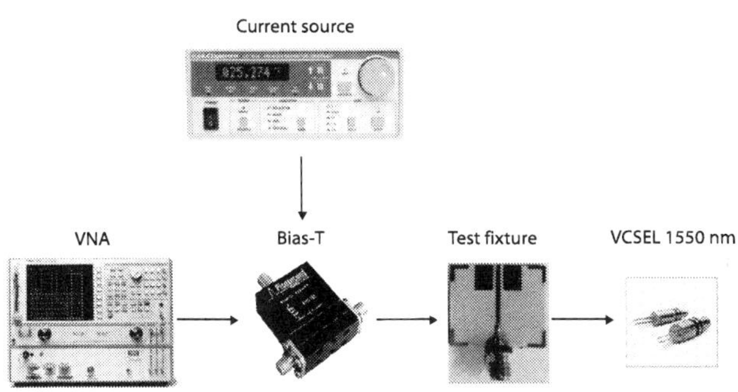


**Figure 6.** Setup for the measurement of the 1550 nm VCSEL.

### 3.4. Extraction of laser parasitic elements

It is possible to determine the input impedance $Z_{in}$ using the experimentally measured $S_{11}$ parameter:

$$Z_{in} = 50 \times \frac{1 + S_{11}}{1 - S_{11}} \tag{27}$$

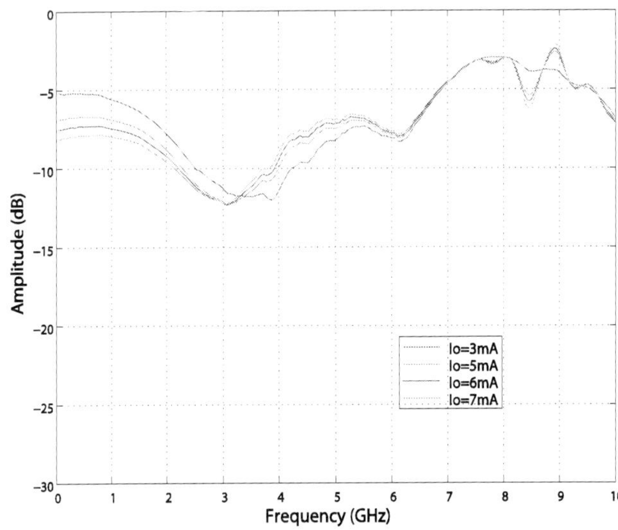


Figure 7. $S_{11}$ parameter for the 1550 nm VCSEL.

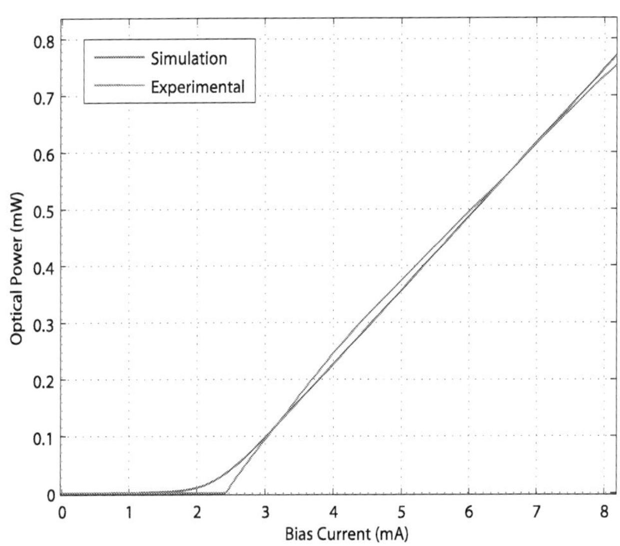


Figure 8. Characteristic curve for the 1550 nm VCSEL.

The measured $S_{11}$ parameter for the 1550 nm VCSEL for different bias currents above threshold is represented in Figure 7. The threshold current (2.14 mA) was obtained by inspection of the optical power versus current (*P-I*) characteristic curve shown in Figure 8.

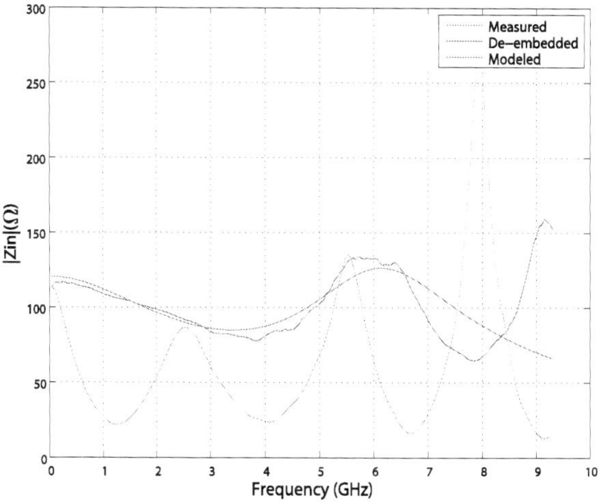


**Figure 9.** Magnitude of the 1550 nm VCSEL input impedance.

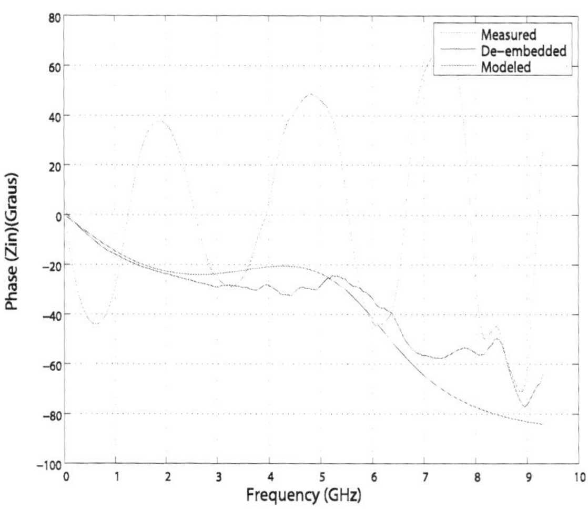


**Figure 10.** Phase of the 1550 nm VCSEL input impedance.

The laser parasitic elements are then obtained by means of an optimization process that fits the magnitude and phase of the input impedance of the equivalent circuit of the parasitics to the corresponding experimental result after the de-embedding procedure. From the results represented in Figure 9 and Figure 10, it is possible to verify the good approximation between the theoretical model (modeled) obtained using equation (23) and the experimental measurement (de-embedded), up to 7 GHz. Moreover, it is clear the importance of the de-embedding operation to obtain a good estimate of the laser parasitic elements.

The parasitic elements obtained before the optimization process are represented in Table 1:

| Element | Value | Element | Value |
|---|---|---|---|
| $R_{in}$ | 50 Ω | $L_{p1}$ | 0.68 nH |
| $C_{p2}$ | 0.3498 pF | $R_{p1}$ | 0.5 Ω |
| $L_{p2}$ | 2.828 nH | $C_{sub}$ | 0.04 pF |
| $R_{p2}$ | 52.387 Ω | $R_{sub}$ | 0 Ω |
| $C_{p1}$ | 0.6696 pF | $R_s$ | 113.02 Ω |

**Table 1.** Parasitic elements of the equivalent circuit

The frequency response of the circuit, as defined in equation (26), is shown in Figure 11, where the lowpass characteristic is observed.

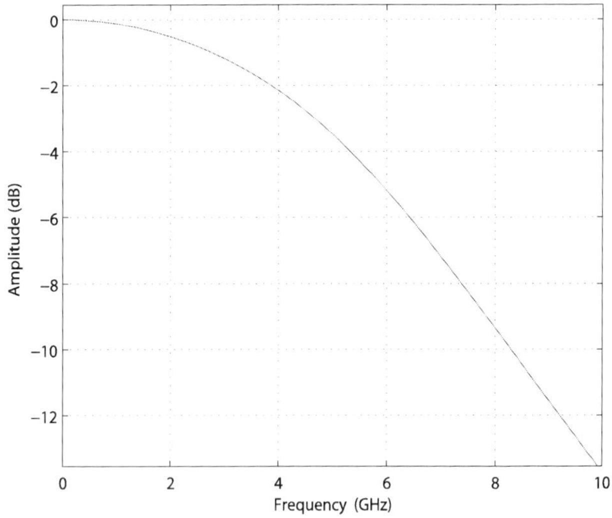


**Figure 11.** Normalized frequency response of the parasitic elements circuit.

### 3.5. Extraction of laser intrinsic parameters

The method employed for the extraction of the laser intrinsic parameters follows the frequency subtraction method described in reference [14]. To that purpose the laser transfer functions at different bias currents are obtained, using the $S_{21}$ parameter.

The experimental results, for different bias currents (above the threshold current), are represented as dashed lines in Figure 12. These results were used to extract the laser intrinsic parameters $[H_{ILD}(f)]$, by dividing each curve by the reference transfer function (measured for a bias current of 3 mA). The resulting function does not depend on the parasitic circuit $[H_{PC}(f)]$ or the test fixture $[H_{TC}(f)]$, as shown in equation (28) [15]:

$$\frac{H_{Global}\left(f,I_{Bias}\right)}{H_{Global}\left(f,I_{ref}\right)}=\frac{H_{ILD}\left(f,I_{Bias}\right)H_{PC}\left(f\right)H_{TC}\left(f\right)}{H_{ILD}\left(f,I_{ref}\right)H_{PC}\left(f\right)H_{TC}\left(f\right)}=\frac{H_{ILD}\left(f,I_{Bias}\right)}{H_{ILD}\left(f,I_{ref}\right)} \tag{28}$$

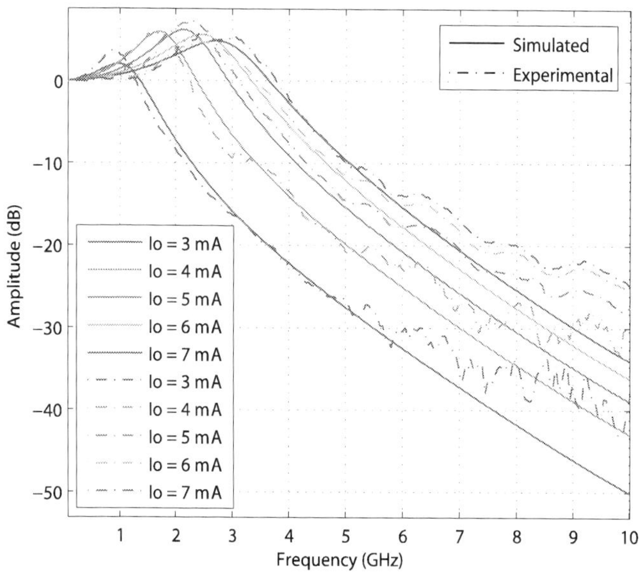


**Figure 12.** Frequency response of the 1550 nm VCSEL for different bias currents.

It is then possible to fit the corresponding theoretical response to the measured data through an optimization process of the laser parameters. That procedure was applied to the 1550 nm VCSEL, RC33xxx1-F from *RayCan*, using the Optimization Toolbox of MATLAB; the final result is shown in Table 2.

| Parameter | Value |
|---|---|
| $V$ , active region volume | $4.93 \times 10^{-18}$ m$^3$ |
| $g_{0,}$ gain slope constant | $2.50 \times 10^{12}$ m$^3$s$^{-1}$ |
| $N_{0m}$ , electron density at transparency | $2.71 \times 10^{24}$ m$^{-3}$ |
| $\beta$ , spontaneous emission factor | $6.5 \times 10^{-3}$ |
| $\Gamma$ , optical confinement factor | $3 \times 10^{-2}$ |
| $\tau_{s,}$ electron lifetime | 2.6 ns |
| $\tau_p$ , photon lifetime | 4.0 ps |
| $\varepsilon$ , gain compression factor | $5.0 \times 10^{-23}$ m$^3$ |
| $\eta_i$ , internal quantum efficiency | 0.8 |

**Table 2.** Extracted intrinsic parameters of the 1550 nm VCSEL

The frequency response of the VCSEL is represented in Figure 12, for different bias currents, including the simulated and experimental results. A good approximation is obtained between the simulated and experimental results.

# 4. OFDM and SC-FDMA over fiber applications

The VCSEL can be directly modulated using signals such as OFDM or SC-FDMA to encode digital data on multiple subcarrier frequencies. OFDM and SC-FDMA are used in applications such as wireless networks and LTE mobile communications. In order to avoid the high peak-to-average-power ratio (PAPR) inherent to OFDM modulation, the LTE standard employs SC-FDMA [16-18], an alternative modulation technique for the uplink with a similar low-complexity. Additionally, OFDM requires highly linear power amplifiers operating with a large backoff from their peak power, which results in low power efficiency [19]. In this context, it becomes pertinent to study the impact of employing SC-FDMA modulation within a RoF system based on directly modulated VCSELs.

## 4.1. Method and setup

The OFDM and SC-FDMA signal generation and demodulation is carried out in MATLAB environment. The experimental RoF setup illustrated in Figure 13 includes the Vector Signal Generator for the generation of the radio frequency (RF) signal that directly modulates the VCSEL, the optical fiber, the optical attenuator, and the optical receiver. The received signal is then sampled with a digital sampling oscilloscope at 20 Gsamples/s for offline demodulation in MATLAB. The validation of the transmitter and receiver blocks was performed in back-to-back configuration, by comparison with the theoretical PAPR results from the literature.

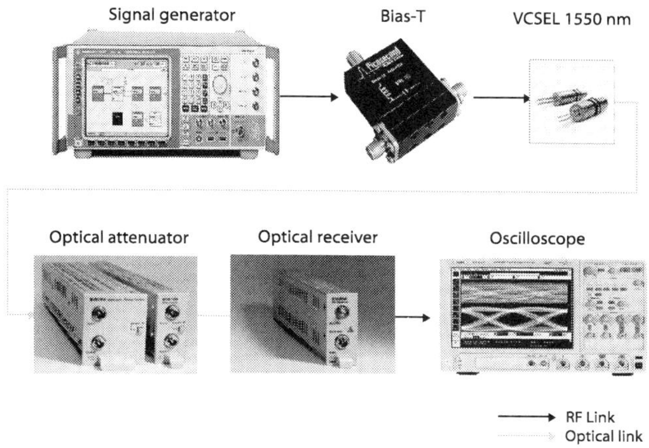


**Figure 13.** Experimental setup.

### 4.2. Transmitter

Both OFDM and SC-FDMA signals were modulated with a bit sequence, using three different possible modulation formats: QPSK (Quadrature Phase Shift Keying), 16-QAM (Quadrature Amplitude Modulation) and 64-QAM. Pilot subcarriers were added to the resulting symbols to estimate the effect of channel propagation. A diagram of both transmitters is shown in Figure 14.

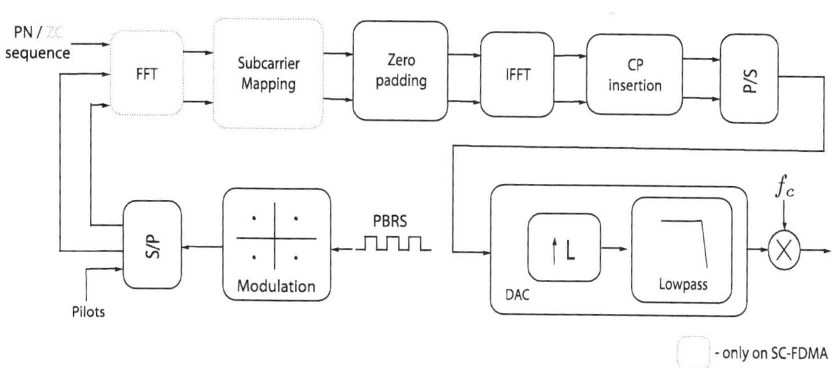


**Figure 14.** OFDM and SC-FDMA transmitters.

In the case of OFDM, the first symbol is used to facilitate time synchronization (finding the first symbol of the signal) at the receiver. This symbol is obtained using the inverse fast Fourier transform (IFFT) of a sequence formed by the real part of a pseudorandom noise sequence at even frequencies and by zeros at odd frequencies, as proposed by Park [20]. After applying zero padding (adding zeros on both sides of the signal spectrum) in order to ease the filtering operation, a N-point IFFT is applied to convert the signal to the time domain. Before the digital-to-analog converter (DAC) where the signal is upsampled, the cyclic prefix (CP) (copy of the last part of the signal) is added to prevent multipath delay. Finally, the signal is upconverted to a RF carrier.

In the case of SC-FDMA, a Zadoff-Chu sequence [21] is used in the LTE standard, which functions as the first SC-FDMA symbol for synchronization purposes at the receiver. After this, a N-point fast Fourier transform (FFT) is performed and the resulting N subcarriers are mapped into M subcarriers using one of two different mapping methods: the interleaved mapping also known as interleaved frequency division multiple access, where the subcarriers are equidistantly distributed over the entire spectrum; and the localized mapping, also called localized frequency division multiple access, where the subcarriers are confined to a fraction of the spectrum. Thereafter, the zero padding and M-point IFFT are applied. Lastly, the CP is added and, as in the case of OFDM, the upsampling and the upconversion operations are performed.

In order to allow the performance comparison between the two modulation formats, only one user is considered.

The In-phase (*I*) and Quadrature (*Q*) components of both modulations formats were loaded to the signal generator, represented in Figure 13, where the frequency and the power of the RF-transmitted signal was specified.

### 4.3. Receiver

As shown in Figure 15, the received OFDM signal is first baseband filtered, to eliminate the noise outside the band, and then downconverted at the same frequency specified on the generator in order to obtain the baseband signal, followed by the lowpass filtering to remove the harmonics generated. Then the downsampling and quantization operations are applied to the signal followed by a temporal synchronization using Park's method. Then, the inverse of the operations carried out in the OFDM transmitter are performed at the receiver, namely the channel estimation and corresponding equalization in the frequency domain, followed by the removal of both the pilot subcarriers after the FFT and zero padding operations.

In the case of the SC-FDMA receiver, the operations are as follows: first baseband filtering, then temporal synchronization using the cross-correlation between the received signal and the reference signal (Zadoff-Chu sequence), and finally downconversion and downsampling. In order to recover the transmitted symbols, the inverse of the operations carried out at the SC-FDMA transmitter are performed at the receiver, including the channel estimation and the frequency domain equalization, similarly to the case of the OFDM receiver, after the IFFT and the subcarrier demapping.

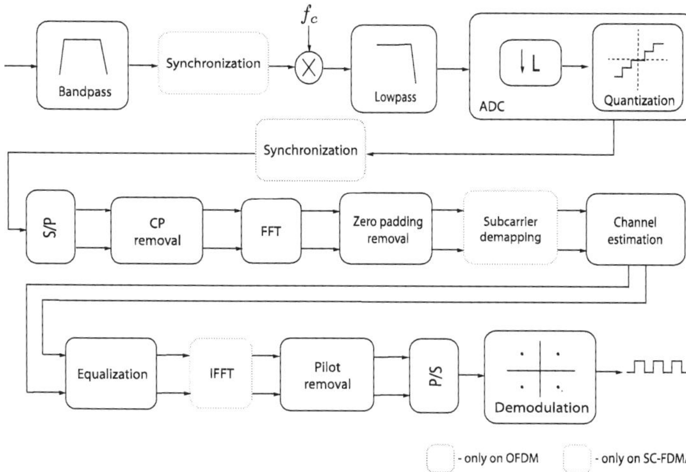


**Figure 15.** OFDM and SC-FDMA receivers.

## 5. System performance assessment

In this section, we present and discuss the relevant results using the signal-to-noise ratio (SNR) as a figure of merit to compare the OFDM and SC-FDMA signals with the same signal parameters.

### 5.1. Signals parameters

The signals generated are characterized by the following parameters: 16-QAM modulation, 10 Mbps, 1 user, 128 subcarriers, 8 pilot subcarriers for channel estimation, $4\times \text{FFT}_{size}=512$ of zero padding, CP of $\text{FFT}_{size}/4=32$, 11 transmitted symbols (1st symbol for synchronization and the others for data) and RF carrier located at 2.4 GHz. On the receiver side, the zero forcing equalization is used after the least squares channel estimation.

### 5.2. Analysis

We now assess the performance of a RoF system and compare the performance of OFDM and SC-FDMA wireless signals over fiber. Theoretically, the optical SNR is given by the following equation [22]:

$$\text{SNR} = \frac{I_{RX}^2}{I_{RIN}^2 + I_{SN}^2 + I_{TH}^2 + I_{IMI}^2} \tag{29}$$

where the four noise currents are: $I_{RIN}^2$ is the RIN noise current, $I_{SN}^2$ is the shot noise current, $I_{TH}^2$ is the thermal noise current due to equivalent load resistance and pre-amplifier noise, and $I_{IMI}^2$ is the noise current due the intermodulation distortion. For lower modulation indices, the RIN is dominant in comparison to the thermal and the quantum noises [23].

The performance of the system is assessed on the basis of SNR or the error vector magnitude (EVM) figures of merit. The EVM expresses the quality of a digital modulated signal and is defined as the difference vector between the measured and the reference signals. Then the SNR can be calculated from the EVM. These figures of merit are calculated as follows:

$$\text{EVM} = \frac{\frac{1}{N_S}\sum_{p=0}^{N_S}\left|S_{p,i} - S_{p,m}\right|^2}{\frac{1}{N_S}\sum_{p=0}^{N_S}\left|S_{p,i}\right|^2} \tag{30}$$

$$\text{SNR} = \left(\frac{1}{\text{EVM}}\right)^2 \tag{31}$$

where $S_i$ and $S_m$ are the ideal and measured constellations, respectively, $p$ is the constellation symbol index, and $N_S$ is the number of constellation symbols.

In Figures 16 to 18, the SNR results for OFDM and SC-FDMA transmissions are represented for three different laser bias currents ($I_0$= 4, 5, and 6 mA) as a function of the RF signal power, which is defined as:

$$P_{RF} = 10\log_{10}\left(\tau_{i,signal} Z_0\right) + 30\left[\text{dBm}\right] \tag{32}$$

where $\tau_{i,signal}$ is the signal variance and $Z_0$ is equal to 50 Ω.

From the results presented, it is possible to conclude that there is a good matching between the simulation and the experimental results. It is clear from Figures 16 to 18 that for lower RF power, noise is dominant whereas for higher RF power, the intermodulation distortion, introduced by the VCSEL, becomes the limiting performance factor. The SC-FDMA signal is more sensitive to noise then the OFDM signal for lower RF power, while for higher RF power, the SC-FDMA signal is more robust to the intermodulation distortion. Despite this fact, the maximum SNR values attained are identical in both cases, albeit at a higher RF power for the SC-FDMA case, with no clear performance improvement of the SC-FDMA with respect to the OFDM.

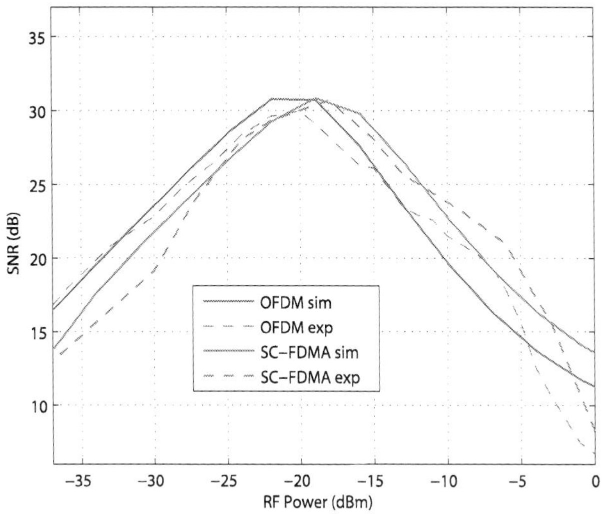


**Figure 16.** Simulated and experimental SNR for both OFDM and SC-FDMA as a function of the RF power for *Io* = 4mA.

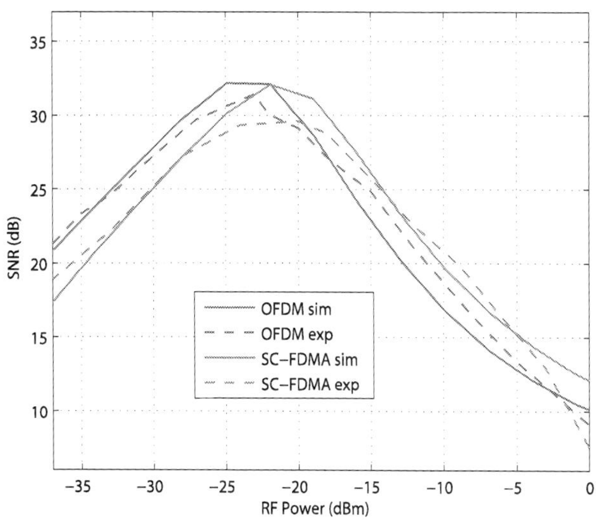


**Figure 17.** Simulated and experimental SNR for both OFDM and SC-FDMA as a function of the RF power for *Io* = 5mA.

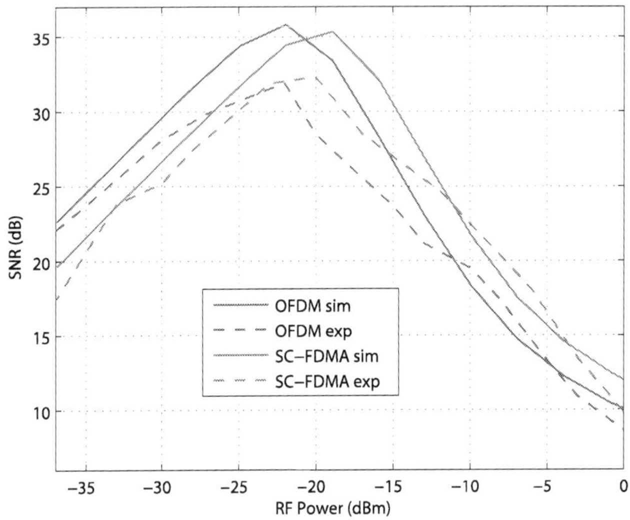


**Figure 18.** Simulated and experimental SNR for both OFDM and SC-FDMA as a function of the RF power for *Io* = 6mA.

# 6. Conclusion

We have assessed the performance of OFDM and SC-FDMA in the context of a RoF scenario, based on a directly modulated VCSEL operating at 1550 nm, and direct detection. To that purpose an accurate theoretical model of the laser was presented, for which device parameters were extracted based on fitting the model to experimental data of frequency response and input impedance. The simulation model fits well the experimental results, and we conclude that the SC-FDMA modulation presents a lower PAPR than the OFDM modulation as expected. Moreover, it is observed that the SC-FDMA is more susceptible to noise, yet it is more immune to intermodulation distortion than the OFDM modulation.

# Acknowledgements

We acknowledge support from Project "NORTE-07-0124-FEDER-000058" financed by the North Portugal Regional Operational Programme (ON.2 - *O Novo Norte*), under the National Strategic Reference Framework (NSRF), through the European Regional Development Fund (ERDF), and by national funds, through the Portuguese funding agency, *Fundação para a Ciência e a Tecnologia* (FCT). This work was carried out with the support of the TEC4SEA research infrastructure (www.tec4sea.com).

http://dx.doi.org/10.5772/61118

## Author details

Henrique M. Salgado[1,2*], Rúben E. Neto[2], Luís M. Pessoa[2] and Pedro J. Batista[1,2]

*Address all correspondence to: hsalgado@inesctec.pt

1 Faculdade de Engenharia da Universidade do Porto, Portugal

2 Instituto de Engenharia de Sistemas e Computadores – Tecnologia e Ciência - INESC TEC, Porto, Portugal

## References

[1] Chun-Ting Lin, Chen J, Peng-Chun Peng, et al. Hybrid optical access network integrating fiber-to-the-home and radio-over-fiber systems. IEEE Photonics Technology Letters. 2007;19(8):610-612.

[2] 3GPP. LTE [Internet]. Available from: http://www.3gpp.org/technologies/keywords-acronyms/98-lte [Accessed: January 6, 2014].

[3] Tauber D, Wang G, Geels RS, Bowers JE, Coldren LA. Large and small signal dynamics of vertical cavity surface emitting lasers. Applied Physics Letters. 1993;62(4):325

[4] Marcuse D. Classical derivation of the rate equation. IEEE Journal of Quantum Electron. 1983;19(8):1228-1231.

[5] Westcott J. Investigation of multiple f.m./f.d.m. carriers through a satellite t.w.t. operating near to saturation. Proceedings of the Institution of Electrical Engineers. 1967;114(6):726-740.

[6] Petermann K. Laser Diode Modulation and Noise. Advances in Opto-Electronics (ed.). Netherlands: Springer; 1988, p. 315. DOI: 10.1007/9789400929074

[7] Coldren LA, Corzine SW, Mashonavitch ML. Diode Lasers and Photonic Integrated Circuits (2nd ed.). John Wiley and Sons, Inc.; 2012, p. 744. DOI: 10.1002/9781118148167

[8] Bruensteiner M, Papen GC. Extraction of VCSEL rate-equation parameters for low-bias system simulation. IEEE Journal of Selected Topics in Quantum Electronics. 1999;5(3):487-494.

[9] Salgado H. Performance Assessment of Subcarrier Multiplexed Optical Systems: Implications of Laser Nonlinearities. Ph. D. dissertation. School of Electronic Engineering and Computer Systems, 1993.

[10] Tucker RS, Pope DJ. Microwave circuits models of semiconductor injection lasers. IEEE Transactions on Microwave Theory and Techniques. 1983;31(3):289-294

[11] Tucker RS, Kaminow IP. High-frequency characteristics of directly modulated InGaAsP ridge waveguide and buried heterostructures. Journal of Lightwave Technology. 1984;2(4):385-393.

[12] Agilent Technologies. De-embedding and embedding S-parameters networks using a vector network analyzer: Application Note 1364-1.2004.

[13] Agilent Technologies. In-fixture measurements using vector network analyzers: Application Note AN1287-9. 2006.

[14] Morton PA, Tanbun-Ek T, Logan RA, et al. Frequency response subtraction for simple measurement of intrinsic laser dynamic properties. IEEE Photonics Technology Letters. 1992;4(2):133-136.

[15] Silva S, Salgado HM. VCSEL Laser characterization and modelling for future optical transceiver at the super Large Hadron Collider. In: Transparent Optical Networks, 11th International Conference held on June 28, 2009--July 2, 2009. Azores: IEEE; 2009, p. 1-5.

[16] Ciochina C, Sari H. A review of OFDMA and single-carrier FDMA. In: Wireless Conference (EW), European held from 12-15 April 2010. Lucca: IEEE; 2010, p. 706-710.

[17] Myung HG. Introduction to single carrier FDMA. In: 15th European Signal Processing Conference held from 3-7 September 2007. Poznań, Poland: EURASIP, 2007.

[18] Myung HG, Junsung Lim, Goodman D. Peak-to-average power ratio of single carrier FDMA signals with pulse shaping. In: Personal, Indoor and Mobile Radio Communications, 2006 IEEE 17th International Symposium held from 11-14 September 2006. Helsinki: IEEE; 2006, p. 1-5.

[19] Myung HG, Junsung L, Goodman D. Single carrier FDMA for uplink wireless transmission. IEEE Vehicular Technology Magazine. 2007;1(3):30-38.

[20] Byungjoon Park, Cheon H, Kang C, Daesik Hong. Novel timing estimation method for OFDM systems. IEEE Communications Letters. 2003;7(5):239-241.

[21] Chu D. Polyphase codes with good periodic correlation properties. IEEE Transactions on Information Theory. 1972;18(4):531-532.

[22] Keiser G. Optical Fiber Communication (4th ed.). New York: McGraw-Hill Science/Engineering/Math; 2010, p. 688.

[23] Coelho D, Oliveira JMB, Pessoa LM, Salgado M, Castro JCS. Performance analysis of WDM-PON architecture for UWB distribution in aircraft networks. i-ETC:ISEL Academic Journal of Electronics, Telecommunications and Computers. 2013;2(1):ID16.

Chapter 18

# The Digital Silicon Photomultiplier

N. D'Ascenzo , V. Saveliev, Q. Xie and L. Wang

Additional information is available at the end of the chapter

http://dx.doi.org/10.5772/61127

**Abstract**

The Silicon Photomultipliers (SiPMs) are the new step in the development of the modern detection structures in the area of low photon flux detection with a unique capability of detection up to the single photons. The Silicon Photomultiplier intrinsically represents a digital signal source on the elementary cell level. The materials and technology of SiPMs are consistent with the modern electronics technology. We present the realization and implementation of a fully digital Silicon Photomultiplier Imager with an enclosed readout and processing on the basis of modern 3D technology.

**Keywords:** Silicon Photomu/ltiplier, Low photon flux, single photon detection, digital detection structures

## 1. Introduction

The conventional Silicon Photomultipliers (SiPMs) have a great progress and success in the past ten years[1,2]. The excellent performance of SiPMs, which is up to single photon detection capability, changed the understanding of the fundamental principle of light detection and gave the possibility to reach the limit of the detection of the photon flux. The detection of the single photons opens new methods, instrumentations, and analysis of the information, and it is important for the investigation of the most fundamental physics processes, such as the nature of the electromagnetic processes and interactions, the quantum nature of the electromagnetic processes, and the photons' nature.

Many successful applications of the conventional SiPMs could be mentioned, particularly in High Energy Physics and Medical Imaging Systems[3].

The structure and technology of the Silicon Photomultipliers are based on the modern challenges in physics and technology, which are now much richer and more flexible for the creation of the new generation intelligent photon detection structures. It is important to mention that the sensor technology of the Silicon Photomultiplier is fully compatible with the modern semiconductor technology such as CMOS technology, and it could be produced at standard Silicon mass production technology facilities[4]. This gives a huge perspective in the creation of the new generation of SiPMs, which combines the sensors and front-end and processing electronics with a wide range of variation of the performance and compatibility to connection with the modern data processing systems.

One of the first approaches to develop the structure and technology, with implementation of the readout and processing electronics, is already done with great success through the Digital Counting Silicon Photomultiplier (dSiPM)[5]. The Digital Counting Silicon Photomultiplier solved the problems of early digitization of the breakdown avalanche microcell signal and implementation of the integrated electronics (cell signal digitalization, trigger network, time-to-digital converter (TDC), and control logic on chip). In contrast to conventional SiPMs, the Counting Digital SiPM is, therefore, an all-digital (digital-in/digital-out) device and provides the digital counting of the number of photons and precise detection time. As a result, it produces faster and more accurate photon counts with extremely well-defined timing of the first photon detection. This realization is fully functional for the medical imaging systems, particularly for the Positron Emission Tomography (PET) diagnostic systems. The use of the Counting Digital Silicon Photomultipliers gives the possibility of the implementation of novel methods such as individual readout of the small size of scintillating crystals, the method of point of interaction for the improving the spatial resolution, and the time of flight method for the reduction of the background. Finally, it significantly reduces the electronic readout chain.

The new approach of the advanced digital Silicon Photomultiplier will provide full, available information of the detected low photon flux. Generally, the SiPM structure is a fine space distributed elementary sensors structure with a single photon detection capability on the single microcell level and high time resolution. The full information that is available from the SiPM structure includes the single photon detection i.e., the number of the detected photons, space distribution of the detected photons on the detected area, and detection time. The number of photons and the space distribution of the photons over the detected area could be very important for many applications and are called digital imaging.

The next step is the development of the new generation of the digital Silicon Photomultipliers, which will cover very important areas of the Digital Imaging or Real-Time Digital Imaging for the very low photon flux of the many important areas of applications such as homeland security, medical, space and astrophysics, and others.

Another area of the application of Silicon Photomultiplier-like structures is the digital coordinate sensors for the detection and the precision tracking of the charged particles in high energy physics – Avalanche Pixel Sensor (APiX)[6]. One of the important advantages of the tracking detectors on the basis of Silicon Photomultiplier-like structures is the significant reduction of the material budget, which is currently the main fundamental limitation of the improvement in precision tracking.

In the future, the development of the advanced Digital Silicon Photomultiplier structures for different applications will rapidly grow due to complete consistency with the modern semiconductor technology.

http://dx.doi.org/10.5772/61127

## 2. Physical principles of photons detection

The basic feature of the Silicon Photomultiplier as the detection structure for the low photon flux up to a single photon is the provision of internal amplification, which allows for overcoming the noise level of the front-end electronics to recognize the extremely low charge signal and to perform the processing. The internal amplification must be quite substantial, on the level of $10^6$, and can be reached in the specific conditions and mode operation. The conditions for the creation of the special type of avalanche process, which allows such internal amplification, could be reached by specific semiconductor structures. The basic physics principle is based on the creation of the breakdown avalanche process due to the intensive secondary ionization of the charged carriers of both types in the strong electric field formed in the special type of semiconductor structures.

### 2.1. Physics of photon detection and the internal amplification

The process of detection of the low photon flux or single photon is defined by the nature of photons, the physics of the photon interaction with matter (detection media), and the formation of the electric signal, i.e., the conversion process of the photons energy into the electric signal, which is used in the measurement systems.

The fundamental problem in the detection of the photons is that the energy of the photons (especially visible light photons) is almost the minimal possible energy value in nature. The energy of photons could be estimated, as an example for the 500 *nm* visible light photons, as 2.2 *eV*. This is one of the smallest energy values, which could be found in nature, and its detection is a great challenge under many aspects. Furthermore, the detection of a single photon within a single photon distribution is a special interesting task involving the study of the fundamental quantum nature of light.

The photoelectric effect is the main physics process of photons' interaction with matter for the visible range of light. It provides the conversion of the energy of photons to the electric signal in a particular charge. The interaction media that are most suitable for the present time are semiconductor materials such as Silicon, which gives direct conversion of the energy of the light photon to the electric charge. For the visible light and semiconductor material (as detection media), this process gives the converting ratio 1:1 – one photon correspondent energy creates one electron-hole pair, and this amount of electric charge should be transferred to and measured by the electronic system.

The basic principle of the detecting structure on the basis of semiconductor materials allows to use the result of the photoelectric interaction in creating the semiconductor structure within the depleted region (free from free charge carriers), e.g., special type of the *pn*-junction under reverse bias.

By applying the reverse bias to the structure, between two regions with different types of conductivity, the depleted area is formed with a low concentration of minor carriers and in-build electric fields.

The process of the creation of the electron-hole pairs due to the photoelectric interaction of photons in semiconductor structures and transport of the charge to the output is shown schematically in Fig. 1.

Photons with energy higher than the band gap of the semiconductor material are absorbed in the depleted area, thus creating electron-hole pairs inside. Carriers are generated during this process and are separated by in-build electric field: electrons drift to positive enhanced *n*-region and holes to negative enhanced *p*-region.

The charge carriers are then collected on the electrodes and pass through the external electric circuit generating the output electric signal. As mentioned before, for the single photon, the value of the signal created inside the detection volume is extremely low. In terms of measurement, the energy is equivalent to the charge level approximately $10^{-19}$ C. The registration of such signals is a very complicated task and the noise of the electronic measurement system is the main problem. The electronic noise of the measurement system could be estimated in terms of the equivalent noise charge for comparison to the charge signal convertion from the photon energy and for the modern electronic readout to be estimated at room temperature conditions as few hundreds of electrons or ~ $10^{-17}$ C. It means that the minimal signal that could be measured with modern electronic channels should be a few hundreds of electron hole pairs or photons higher. This is still far from the goal of measuring the signal correspondent to a single photon.

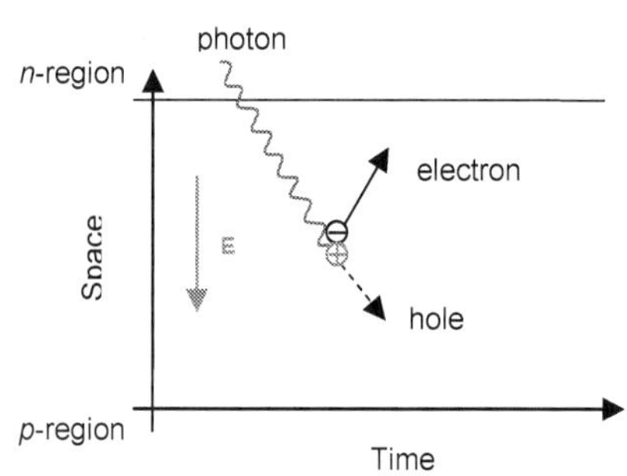


**Figure 1.** The creation of an electron-hole pair due to photon absorption in the semiconductor material.

The way to overcome this problem is to provide the internal amplification of the signal inside the detection media before the electronic signal processing. The value of the amplification should be at a range of $10^4$-$10^6$. This is the main conceptual idea of detecting the low photon flux or single photon by the semiconductor structure such as Silicon Photomultipliers.

The avalanche processes in the semiconductor media due to secondary ionization processes could realize the intrinsic amplification in semiconductor structures. In the high electric field, usually higher than $10^5$ V/cm$^2$, free carriers are accelerated and could reach energy higher than the ionization energy of valence electrons.

http://dx.doi.org/10.5772/61127

The minimal energy required for the impact of the ionization is called threshold ionization energy. This value is one of the main parameters of the theory of avalanche multiplication in semiconductor materials. The impact ionization parameters of the electrons and holes in the semiconductor materials are used to characterize the dynamics of the avalanche processes. They are defined as the inverse value of the average distance (along the electric field), which is necessary for electrons or holes to produce a secondary ionization and create an electron-hole pair.

The consequence of the secondary impact ionization interaction is the avalanche multiplication of the electron-hole pairs and the increase in value of the electric charge correspondent to the initial charge created by the interaction of photons.

Two types of avalanche processes could be realized in the semiconductor structures: the single carrier avalanche process and the breakdown avalanche process. These strongly depend on the value and ratio of impact ionization coefficients of the electrons and holes in Silicon. The values of the impact ionization coefficients strongly depend on the electric field. For the low electric field ~$10^{-4}$, shown in Fig 2 (left), the impact ionization coefficient of holes is much lower and the avalanche process is created only by one type of carrier – the electrons.

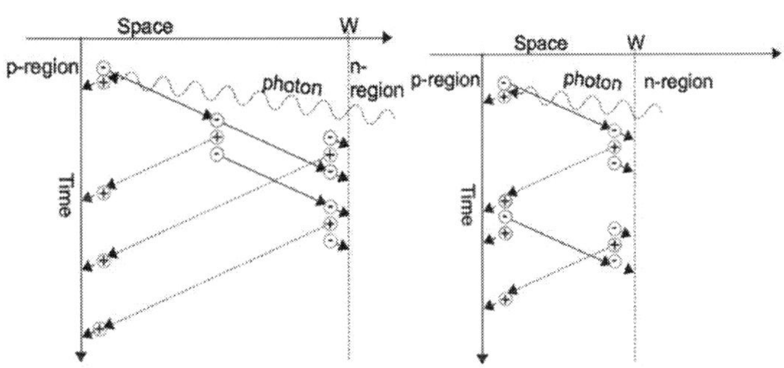


**Figure 2.** Two types of avalanche processes in the Silicon structures: (left) avalanche process with one type of carrier (electrons), (right) breakdown avalanche process with two types of carriers (electrons and holes)

The avalanche process, in this case, is one directional in space and self-quenched when the carriers reached the border of the depleted area in the Silicon. This type of avalanche process is used in the conventional avalanche photo detector (APD). In this case, the amplification gain is strongly dependent on the thickness of the depletion region and on the point of the interaction of the photon inside this area.

For the high level of electric field in the Silicon structure, the process in principle is different and is shown in Fig. 2 (right). The impact ionization coefficients of electrons and holes are close to each other and both types of carriers could participate in the avalanche process. As a result, the avalanche process is self-sustaining, and the number of carriers rises exponentially with time. The avalanche process develops more in time in the thin Silicon depleted thickness, which is comparable with the ionization path of the carriers. Moreover, the amplification gain does not depend on the point of interaction of the photons with the Silicon and practically does not depend on the thickness of the depleted region.

In the first case of the avalanche process of only one type of carriers (electrons), the gain of multiplication is limited by the thickness of depleted area. It is very important to note that the high statistical fluctuation of the amplification occurs due to dependency on the probability of the secondary ionization and the deepness of the interaction. Such conditions exclude the possibility of getting the quality of amplification necessary to resolve the single photon spectra.

In the breakdown avalanche process, which is the second case, the gain of multiplication is not limited by the depletion thickness. A thickness comparable with the secondary ionization path in the Silicon is sufficient, and the gain became infinite even on the limited depleted thickness of the Silicon. The different charge carriers undo the electric field moving in opposite directions, and the thickness of the amplification region could be just equivalent to the length of the ionization of electrons or holes under defined electric field. This gives the possibility of getting the intrinsic multiplication factor, enough for a suitable signal before the electronics detect a very small photon flux. The statistical fluctuation is much smaller because it is defined not by the statistical fluctuation of the avalanche process but by the characteristic of the quenching process.

The target of this condition is to provide a very high electric field in semiconductor-detecting structures to bring the ionization length of electrons and holes less of the depleted thickness of *pn*-junctions.

### 2.2. Silicon photomultiplier principle of operation

The detection principle of the Silicon Photomultiplier is based on the detection of the space- and time-distributed photons (photon flux) by the space-distributed array of the semiconductor microsensors – microcells with high intrinsic multiplication gain, allowing the detection of a single photon with high efficiency.

The output of the Silicon Photomultiplier is organized by the analog sum of the signals from the microcells. The analog sum of the signals from the microcells, which detected the photons, provides the common analog output signal proportional to the incoming photon flux.

The operational principle of the Silicon Photomultiplier is based on the use of the quenched breakdown avalanche processes in the silicon microstructure elements – micro-cells as *pn*-junctions with implemented special quenching elements. The common electrode structure connected to the microcells, in parallel, is implemented to analogs summing the signals from the microcells.

Sensor microcells are a special type of *pn*-junctions, which operate in breakdown avalanche mode, providing the intrinsic multiplication of the charges created by photons in the sensitive area of microcells by secondary ionization. Above the breakdown voltage, the *pn*-junction can be in a stable state for an infinite time, unless a single carrier is created in the depletion region as a result of photon interaction or of thermal excitation and initiates the avalanche multiplication, producing a self-sustaining flow of the charges.

The amplification gain of the breakdown avalanche process is theoretically infinite. A quenching mechanism is implemented in the sensitive structure to quench the breakdown avalanche process. One solution, in the case of the Silicon Photomultiplier, is the serial resistor for each sensor microcell. After the initiation of the avalanche breakdown process by the photoelectron of thermal electrons, the current rises through the structure and causes the voltage drop on the quenching resistor, and a consequent drop of the voltage is also applied to the *pn*-junction of the sensitive microcell. The quenching starts when the voltage drop on the quenching resistor lowers the voltage applied to the *pn*-junction to a value lower than the breakdown voltage, stopping the avalanche process. After the structure is quenched, a recovery time is then necessary to allow any free or stored charge to be swept from the active region of the device, followed by a recharging wherein the excess bias across the sensitive microcell is restored.

An important aspect of the described process is a significant reduction of the statistical variation of the value of the signal. For Silicon Photomultiplier structures, the amplification factor is defined not by the statistics of the avalanche processes, as in the conventional avalanche photodetectors (APD), but only by *pn*-junction characteristics and quenching circuits. The output of the microcells is identical in charge pulse for every photon detected, and the overall resolution is defined by the characteristics of the structure and the quenching element. Another very important aspect of the overall structure is the uniformity of microcells characteristics across the sensitive area of the Silicon Photomultiplier. This is provided by the modern semiconductor technology. The requirements for the uniformity define the precision of the charge pulse from the different microcells detecting the photons.

A more detailed discussion of the Silicon Photomultipliers and the study of the performances are available in [2,7].

### 2.3. The binary (digital) nature of the silicon photomultiplier cell signal

The Silicon Photomultiplier microcell in principle is the binary (digital) source of the signal, i.e., the signal produced by the microcell is represented by a two level of electronic signal.

Fig. 3 shows the schematic nature of the conventional Silicon Photomultiplier – the oscilloscope signal of the detection of the low photon flux and the statistical distribution of the signals, as detected by the Silicon Photomultiplier structure, with approximately 1,000 microcells at room temperature. Even with the analog summation of the signals, the well-resolved signals (peaks) of the single photon are clearly visible. It is necessary to mention that the second peak of the statistical distribution (peak of the exactly single photon detection in different microcells) is represented and that the statistical width of the single photon signal during the detection of the single photon by the randomly selected microcell from the 1,000 microcells includes the parameter of the identity of the Silicon Photomultiplier microcells. The separation of the signals with the high resolution is clearly visible and shows the high equality of the microcells over the few thousand microcells, confirming the binary (digital) nature of the signal over all the Silicon Photomultiplier structure.

The schematic nature of the single microcell of the Silicon Photomultiplier and the histogram of the Silicon Photomultiplier microcell signal during the detection of the photon (the single microcell is capable of detecting only one photon) are presented in Fig 4. The measurements are performed at room temperature. The histogram consists of two well-separated peaks. Due to the statistical nature of the photon detection, the first peak represents the pedestal – signal level of the noise when the microcell cannot detect the photon. The second peak of the histogram represents the signal level of the photon detection i.e., electronic signal correspondent to the photon detection. Signals "zero" and "one" have well-separated levels and binary (digital) nature and can be well resolved by the electronics discrimination.

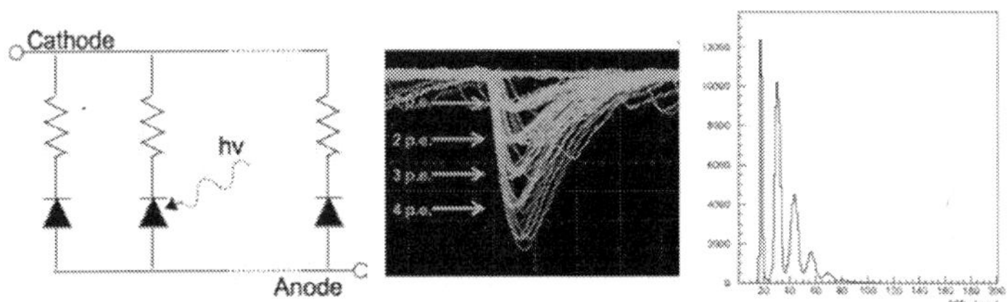


**Figure 3.** The signal characteristics of the conventional Silicon Photomultiplier: (left) schematic view of the array of the microcells with quenching resistors, (center) scope signal of the Silicon Photomultiplier, (right) the statistical distribution of the signals during the detection of the low photon flux with the resolution of the single photon spectra.

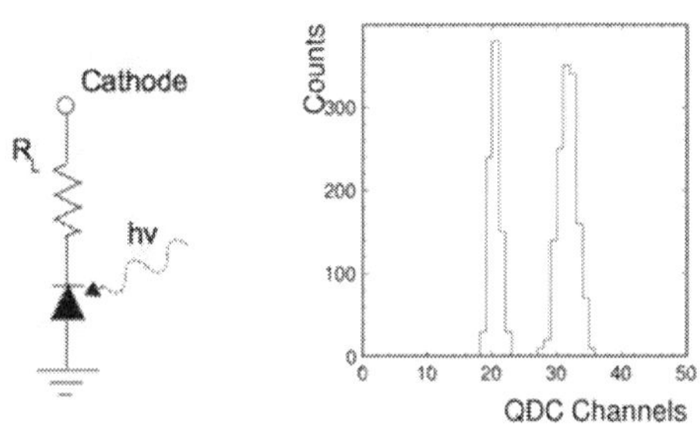


**Figure 4.** The signal characteristics of the single sensitive microcell, such as the microcell of the Silicon Photomultiplier, (left) schematic view of the single microcell and detection of the photon in the microcell, (right) statistical distribution of the signals of the detection of the photon by the single microcell.

The shown characteristics of the single Silicon Photomultiplier microcell and also the characteristics of the array of the Silicon Photomultipliers microcells – conventional Silicon Photomultipliers – demonstrate the real performance of the processes of the photon detection. The single photon represents a source of the binary (digital) information from a single cell and cells array (the uniformity of the characteristics of the microcells, in the arrays of the microcells up to a few thousands cells, satisfies the possibility to resolve the signal of the single photon). The information consists of one bit, i.e.,. the photon detected is "one" and no-photon detected is "zero" for every individual microcell. The amplification gain of the signal, "one photon", is high enough (about $10^6$) to perform the discrimination of the microcell signal to the digital signal, without any additional analogous front-end electronics for the modern standard microelectronic components.

## 3. The digital counting silicon photomultiplier

The binary (digital) nature of the information provided by the Silicon Photomultiplier paves the way for the more advanced structure of the Silicon Photomultipliers, more suitable to the modern way of development in the detection systems. The information provided by the Silicon Photomultiplier cells is quite simple. It consists of a binary signal from the cell, and it is ready for the direct acceptance by the digital processing electronics.

One of the important aspects of this development is the exclusion of the analog front-end electronics, which is usually quite complicated and power consuming. One of the important application of the Silicon Photomultipliers is the medical imaging systems, which the SiPMs will bring to the new fundamental level.

It is not a surprise that the first important step to the digital Silicon Photomultiplier was done in the area of the Silicon Photomultipliers for the medical imaging systems, where the required information is more simplified, as counting of the total number of the photons and time of the detection.

### 3.1. The operational structure and principle of the digital counting silicon photomultiplier

The first approach to the development of the digital Silicon Photomultiplier (dSiPM) was made by the Philips Photon Counting Group for medical applications – Digital Counting Silicon Photomultiplier[8]. The necessary information for the medical imaging system applications, specifically the Positron Emission Tomography, is specific and is defined as the total number of the photons (as information of energy for the identification of the detected gamma from the annihilation process) and the precise time of the arriving of the photons (as the information for the identification of creating the two gammas at the same time from the same annihilation process). These two main parameters are required for the reconstruction and analysis.

The development of the Digital Counting Silicon Photomultiplier was the basis for those requirements. They provide the counting mode as the digital information in the form of the total number of photons detected on the area of SiPM and the arrival time related to the time of the signal.

The dSiPM solves this by the early digitization of the avalanche breakdown microcell output (cell electronics) and integrated electronics (digital summation, trigger network, time-to-digital converter (TDC), pixel controller electronics on the chip).

In contrast to conventional SiPMs, the Digital Counting SiPM is, therefore, a digital (digital-in/digital-out) device that does not use the analog signal processing. As a result, it produces faster and more accurate photon counts with extremely well-defined timing in reference to the first photon detection, both of which are important factors in applications such as medical imaging scanners and high-energy nuclear particle detectors.

Moreover, the digital SiPMs can be fabricated using a standard high-volume CMOS process technology available at the mass production Silicon technology facilities.

Another important aspect of the digital approach is the possibility to improve the noise characteristics of the Silicon Photomultiplier structures. The dark count rate is a critical point of the Silicon Photomultiplier physics and noise characterization, especially under operating at room temperature, which could disturb the low photon flux measurements. The important feature of the dark count rate sources of SiPM is that they are usually localized in the few limited regions over the full Silicon Photomultiplier area. The digital option of the realization of the Silicon Photomultiplier detection structure and the possibility of the direct access to the digital information in the microcell electronics from the processing electronics can dramatically improve this feature. The improvement could be achieved by masking of the majority of the noise microcells by the function of the processing electronics.

The general structure of the digital counting SiPM is shown in Fig. 5, which shows the microcells with active quenching and recharging, cell electronics, readout chain consisting of photon counter, and trigger network including Time Digital Converted (TDC). The output information represents the total number of photons, triggered microcells, as equivalent of energy, and time of the photon detection.

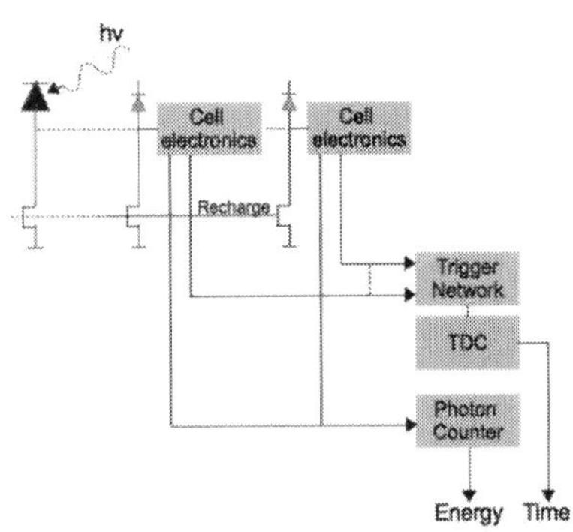


**Figure 5.** General structure of the Digital Counting Silicon Photomultiplier

### 3.2. The cell electronics

In the digital Counting SiPM, each individual microcell is equipped with its own quenching element (active), recharged element for the recovery of the sensitive state after breakdown, and one-bit on-chip analog-to-digital converter (ADC) in the form of a CMOS inverter. The quenching circuitry is added to each microcell to actively (rather than passively) quench and recharge the microcell after triggering. This active quenching/recharging in the device improves the detector's recovery time and respectively the dead time of the detection system dramatically – order of magnitude, as well as reducing its power consumption due to the exclusion of high-resistive elements from the main current line. The implementation of the active quenching circuits solved another technological specific problem as the high-resistivity elements – quenching resistor in the CMOS technology.

Each microcell that experiences an avalanche breakdown process produces its own digital output that is captured, along with the digital outputs from all other triggered microcells, by an on-chip counter. Finally, the output information represented the number (digital form) of the photons detected in a particular time. Hence, the digitally counting SiPM converts the digital events (photon detections) directly into a digital photon count.

Another very important feature of the microcell electronics is the function of the masking. The well-known problem of the conventional Silicon Photomultiplier is the dark count rate, particularly, at room temperature. The dark count rate is the spontaneous generation of the avalanche breakdown processes due to thermally created electrons in the depleted area. The dark count rate signals are undistinguished from the signal of the detection of photons in the microcell (digital nature of the signal). Nevertheless, it is well known that the dark count rate is generated not uniformly around the full sensitive area of the Silicon Photomultiplier and is localized in particular microcells, as the dark count rate map of the digital SiPM shown in Fig. 7. To overcome the dark count rate problem associated with conventional SiPMs, each avalanche breakdown microcell in the Digital Counting SiPMs is also equipped with an addressable static memory cell that can be used to disable or enable the microcell signal transmitting to the output net (masking). The avalanche breakdown microcells that show high dark count rate levels prevented from contributing false counts to the digital Counting SiPMs output. This facility allows the Digital Counting SiPM to achieve better signal-to-noise ratios than conventional devices. As the defective microcells in the array can be disabled, the production yield is also improved.

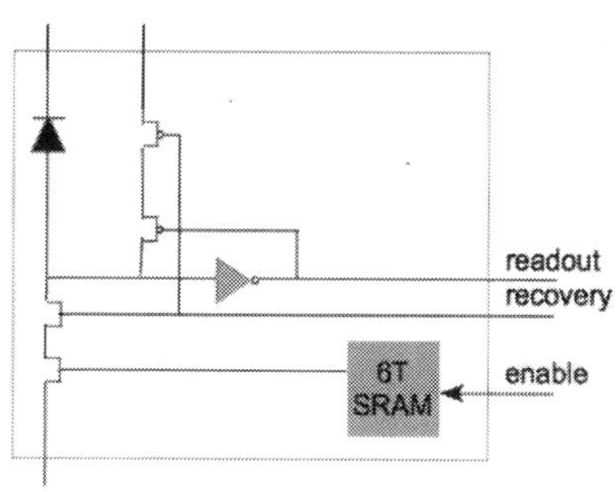


Figure 6. General structure of the avalanche breakdown microcell electronics of the Digital Counting Silicon Photomultiplier

### 3.3. The processing circuits

In contrast to conventional analog SiPMs, all microcells in the Digital Counting SiPM are connected via a low-skew balanced trigger network to an on-chip time-to-digital converter.

The timing resolution of this converter is 20 ps, thus preserving the excellent intrinsic timing performance of the avalanche breakdown microcells of the Silicon Photomultiplier. In implementing this new Digital Counting SiPM technology, the challenge was to integrate the relatively high-voltage avalanche breakdown microcell, which must be reverse-biased to around 30 V, alongside the low-voltage CMOS logic circuits on the same silicon chip, while maintaining the dark count and photon-sensitivity performance.

As an example, one of the realizations has a fully integrated 64 pixel sensor, with a sensing surface greater than 10 $cm^2$ and power consumption less than 15 $mW/cm^2$, using 180 nm CMOS technology on 8-inch Silicon wafers. The device's photon-detection efficiency is 30% at 450 nm; it has an optical crosstalk of 8% and a dark-count rate of 100 $kHz/mm^2$ at 20°C and 900 Hz/mm2 at -40°C.

The implemented processing electronics could be much more intelligent and provide many additional functions.

The Digital Counting Silicon Photomultiplier is an extremely important step for the development of the advanced Silicon Photomultiplier Structures, which used the digital nature of the sensor microcells and possibly processed the sensor microcell signals directly in digital form. This gives the most important evidence that could exclude the most complicated and power consumed analogs front-end electronics and processing of the analogous signals to digital forms.

The missing part of the signal processing fixes the coordinate information of each photon in the photon flux.

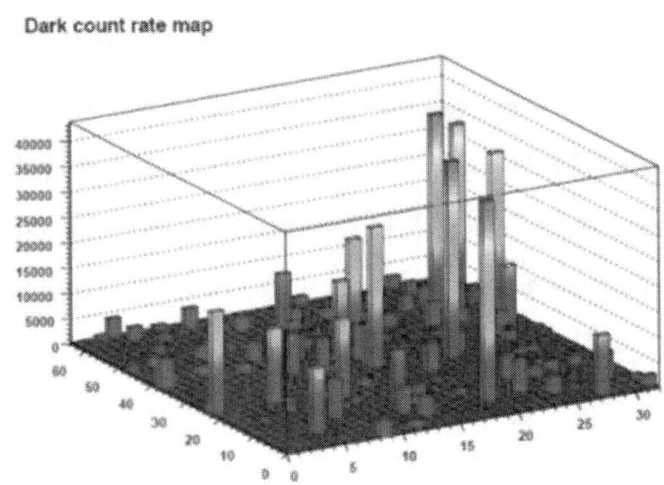


**Figure 7.** The dark count rate map of the non-masked Digital Counting Silicon Photomultiplier

## 4. The digital silicon photomultiplier (Imager)

The other area of the application of the Digital Silicon Photomultiplier is the development of the detecting structures that does not restrict the information to the number of the photons detected in the particular time interval, which is the most important information for the medical imaging system applications. The full information that the Digital Silicon Photomultiplier could provide includes the space distribution of the detected photons in the detection area, i.e., the function of the Digital Imaging of the photon flux[9]. The Digital Silicon Photomultiplier Imager could be referred to the CMOS Image Sensors and covers especially the very important areas for the digitalization of the images with extremely low photon flux such as space, military, homeland security, medical imaging systems, and other scientific applications[10]. The extremely high sensitivity of the Silicon Photomultiplier up to a single photon is very attractive for the many special applications wherein the photon flux is very low.In Fig. 8, a progress is shown in the astrophysics space application[11], where the photon flux from the far astronomical objects can be extremely low and the resolution plays a very important role.

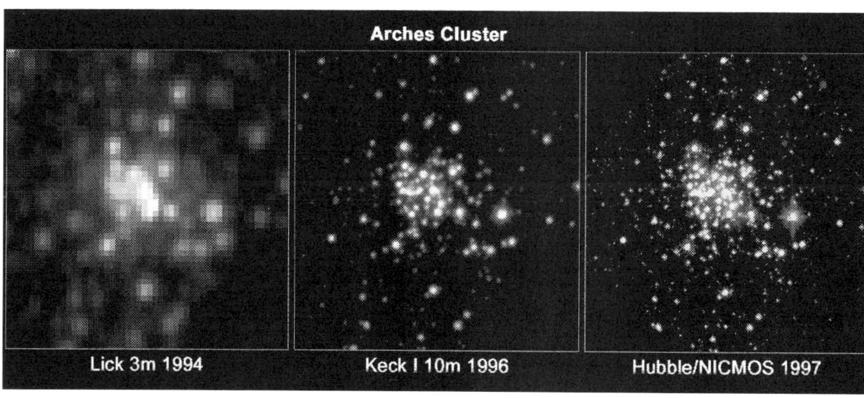


**Figure 8.** Progress in the astrophysics digital imaging performance

The picture shows the progress in the images taken by a "Hubble" telescope after the upgrade of the detection system. It shows the importance of space resolution detection and also the high sensitivity to the light flux, which in the astrophysical study could be extremely low on long distances.

The necessity of the detection of extremely low photon flux images is the main goal for the development of the advanced structures of the Digital Silicon Photomultiplier Imagers of the photon detection device. This will provide full information about the low photon flux such as distribution of the photons in space and time, i.e., the number of photons (the detection of the photons with different energies can be included – i.e., provide the color images), their position in space, and arrival time of the photons at detection plane.

The Digital Silicon Photomultiplier detection system could also provide the color information by the combination of the microcells with different spectral sensitivity.

The fully digital devices (Digital Silicon Photomultiplier Imagers) for the detection of the low-light photon flux provides full information that is carried by the low photon flux - the number of the photons, space distribution of the photons, and time information of the photons arriving, (color information) and will provide this information directly in digital form on the level of each single photon in the low photon flux.

The goal is to develop the fully Digital Silicon Photomultiplier Imager.

### 4.1. The general structure of the digital silicon photomultiplier imager

As mentioned above, the great advantage of the technology of Silicon Photomultipliers, particularly the sensor technology, is the full compatibility with the CMOS technology, which was mainly developed for the microelectronics, i.e., technology gives the possibility of the implementation of the sensors with the same structure and same technology as the electronic elements.

The second argument is the intrinsically binary (digital) nature of the SiPMs microcell signals. As mentioned before, the SiPM cell explicitly detects the single photon. The signal has a two-level feature, which is not significantly disturbed by the electronic noise and does not require the analog processing. This gives the possibility to create the elegant fully digital detection structure with individual microcell addressing. The readout signal can be associated with the coordination of microcells with the detected photon, i.e., it is possible to create the fully functional pixel structure or pixel imager structure.

The first attempt to describe the fully digital Silicon Photomultiplier was taken from the US patent "Quantum Detector Array"[12]. An advanced Digital Silicon Photomultiplier includes a semiconductor substrate, a binary breakdown avalanche sensor element situated on the semiconductor substrate, the binary breakdown avalanche sensor element having a radiation-sensitive area comprising a drift region and amplification region for detecting radiation from a radiation-emission source, particularly electromagnetic photons, and a logic electronics electrically interconnected to the binary avalanche breakdown sensor element for quenching and resetting. Preferably, the logic schematic elements are positioned, at least, partially outside of the radiation-sensitive area to leave at least a portion of the radiation-sensitive area available to detect radiation without an obstruction from the logic schematic elements (geometrical factor of the detection efficiency). This statement is important for the traditional CMOS technology. The modern 3D integration technology gives a possibility to create the advanced Silicon Photomultiplier structures with enhanced geometrical factor, which will be discussed below.

The required characteristics of the sensors signal are (1) the binary avalanche breakdown sensor should provide a signal with the characteristics and (2) provide the efficient function of the integrated logic electronics, i.e., the binary form of the signals correspondent to the modern microelectronics development and the possibility of the integration of the sensor/electronics technology, which is available in modern technology processing CMOS technology.

Referring to the description, a binary avalanche breakdown sensor array, according to a preferred embodiment of the invention, is described. As shown, the binary avalanche breakdown sensors array features a sensitive area, which in a preferred embodiment comprises an epitaxial layer and a common substrate, both preferably made of silicon, although other semiconductor materials may be selected such as gallium arsenide, silicon-germanium, etc.

A binary avalanche breakdown sensor array includes a plurality of sensor elements, which may have submicron dimensions. The sensor elements comprise an $n^+$-type region or layer and a p-type region or layer, which is immediately adjacent to the $n^+$-type region to establish a pn junction as an amplification region and a p-type region as a drift region (The terms "layer" and "region" are used interchangeably herein.) Creation of an $n^+$-type region and a p-type region may be accomplished, for example, using standard doping techniques. The preferable processing of the production is integrated sensor/logic electronics elements technology based on standard CMOS technology.

The sensor elements are separated from one another by trenches formed in the epitaxial layer. Logic electronics elements on the same substrate are preferably positioned in the area of trenches between the sensor elements, outside of the radiation-sensitive area, to leave an entirety of the area available to detect the radiation without obstruction from the logic element. This method provided the suppression of the optical crosstalk in the neighborhood binary avalanche breakdown sensors from the sensor that detected the photon and prevented the decreasing detection efficiency of the low photons flux, especially on the single photon level.

A reverse structure and bias operate the binary quantum sensor element in a breakdown mode, which provides the possibility to have the sensitivity for the different areas of visible light.

Sensor elements, arrays, pixels, meta-arrays, and readout components and systems useful with this invention are described in the U.S. patent application Serial No. 11/783,613 entitled "Quantum Photodetectors, Imaging Apparatus and Systems, and Related Methods"[13].

As this invention has been described as having a preferred design, it is understood that it is capable of further modifications, uses, and/or adaptations of the invention, generally following the principle of the invention and including such departures from the present disclosure within known or customary practice in the arts to which the invention pertains and as may be applied to the central features herein before set.

The present approach to the development of the advanced Digital Silicon Photomultipliers is in pixelated structures that are more suitable to the photodetector structure including the sensor (SiPM microcell), front-end electronics, and control electronics. Also, the structure of the advanced Digital Photomultiplier will include the readout processing electronics (multiplexer) and digital output.

One of the very important areas of application covered by this approach is the digital imager for the extremely low photon flux detection.

General processing of the advanced digital SiPMs will provide the following general functions after the sensor signal: hit decision, noise suppression, hits storage, addressing of the individual pixels, and readout chain with flexible architecture for the processing of the digital signals on the same substrate with sensors.

The choice of the certain readout architecture for the pixel structures is strongly influenced by the purposes of the applications, photons rate, trigger latency, and trigger selectivity which in turn define the readout rate. If the main particular readout functions are performed locally in the pixels, then, the data volume to be read out should be defined by the application and usually have to be strongly reduced.

Schematically, the general structure of the advanced digital SiPM is shown in Fig. 9.

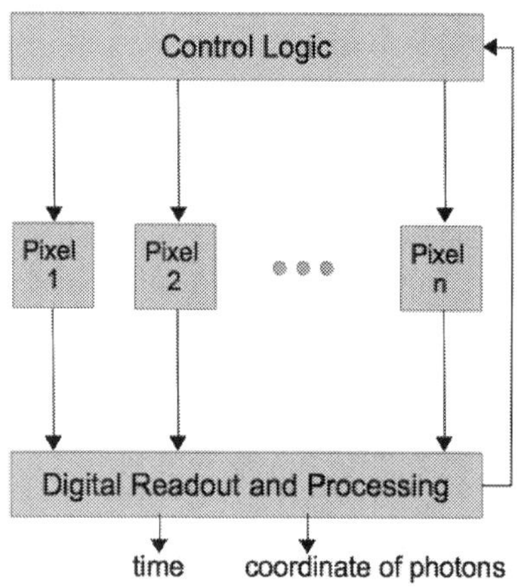


**Figure 9.** General structure of the Digital Silicon Photomultiplier Imager

The main elements are avalanche breakdown micro-cells, quenching elements (generally passive or could be active), logical element (one-bit digital memory) with ideally individual addressing of the each micro cell, and readout and processing net, including the flexible readout and buffers for store the information.

Architecture of microcells readout and processing could be organized in different ways correspondent to the applications.

Although photon counting is by definition a digital task, conventional SiPMs combine the electrical pulses generated by multiple photon detections into a single analog output signal. As previously mentioned, this signal has to be processed by expensive analog power-consuming electronics to recover the photon count (digital form).

By integrating low-power CMOS electronics into the SiPM chip, a digital SiPM developed in which each photon detected is converted directly into an ultrahigh-speed digital pulse that can be directly used by on-chip intelligent processing circuits. In contrast to conventional SiPMs, the digital counting SiPM is, therefore, an all-digital (digital-in/digital-out) device. As a result, it produces faster and more accurate photon counts with extremely well-defined timing of the first photon detection and digital information of addressing the triggered micro-cells individually.

This digital design provides several advantages, such as the pixel and the pixel controller being highly configurable, for the application of SiPM sensors. The individual breakdown avalanche micro cells can be switched on or off, depending on their dark count performance, validation, and integration times. Readout schemes can as well be set according to the application needs. In addition, the digital nature and independence from analog effects, such as gain or amplification, reduces the temperature sensitivity of the device. As only digital signals (photon count and time) are provided, subsequent processing electronics are greatly simplified.

### 4.2. The pixel electronics

Principally, the sensor of the Digital Silicon Photomultiplier is extremely simple and robust. The information is just binary (one-bit/cell), and a very high intrinsic gain allowed to use the output signal directly to the CMOS logic circuits (not necessarily the analog electronics), which simplified the cell electronic chain to a minimum. The advanced Digital Silicon Photomultiplier structure includes the breakdown avalanche microcell with a quenching element (passive or active) and a one-bit digital memory on-chip for storing the hits and addressing and reading out of the net.

An important part of the cell electronics is the circuit for the dark rate suppression.

The general cell electronics is shown in Fig. 10. Each microcell that experiences avalanche breakdown, therefore, produces its own digital output that is captured, along with the digital outputs from all other triggered microcells, and can be readout by a microcell on-chip readout net. Hence, the advanced digital SiPM converts digital events (photon detections) directly into a digital photon count and coordinates the position of each photon count and the time of arrival.

With the same very effective function of the dark rate suppression, each micro-cell in the advanced digital counting SiPM is equipped with an addressable access to the microcell memory that can be used to disable or enable the microcell. Microcells that show high dark count levels can thus be prevented from contributing false hits to the advanced digital SiPMs output.

The advanced digital SiPM equips each individual microcell with its own one-bit on-chip digital memory. Each microcell that experiences avalanche breakdown therefore produces its own digital output that is captured, along with the digital outputs from all other triggered microcells, and can be readout by an on-chip multiplexer. Hence, the advanced digital SiPM converts digital events (photon detections) directly into a digital photon count and coordinate the position of each photon count.

To overcome the dark count problem associated with conventional SiPMs, each microcell in the digital counting SiPM is also equipped with an addressable static memory cell that can be used to disable or enable the microcell. Microcells that show high dark count levels can thus be prevented from contributing false counts to the SiPMs output. This facility allows the digital SiPM to achieve better signal-to-noise ratios than conventional devices. As defective microcells in the array can be disabled, they also help to improve production yield. Additional circuitry is added to each microcell to actively (rather than passively) quench and recharge the microcell after triggering. This active quenching/recharging in the device improves the detector's recovery time, as well as, reduces its power consumption.

### 4.3. The processing electronics

The main feature of the advanced digital Silicon Photomultipliers and Digital Silicon Photomultiplier Imager is the realization of the usage of the digital nature of microcell signal and the two-dimensional space distributed structure of the microcells for the coordinate detection of the signals.

The readout chain of the signals from the two-dimensional one-bit memory in this case, could be realized by a number of flexible readout architecture: counting mode, arrival time mode, and individual cell readout depending on the applications.

From the readout point of view, the pixel structure (cells with cell electronics) represents the two-dimensional one-bit memory elements. The amount of the information is not extremely high.

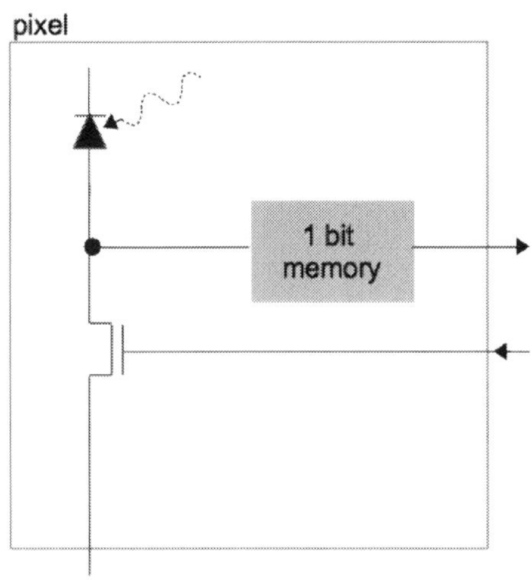


**Figure 10.** General structure of the Digital Silicon Photomultiplier Imager pixel

The organization of the readout could be realized in different ways depending on the performance and applications, and is not covered by this publication. It is referred to many detailed reviews, practical principles, and schematics of the pixel sensors readout.

An important comment follows about the timed performance of the Digital Silicon Photomultipliers Imager, which could be very high and used for the High Speed Imaging applications. As an example, the area of the Digital Silicon Photomultiplier Imager with 10x10 microns pixels with the total digital information is 10000 bits. The readout of this amount of information could be performed extremely fast, even with many methods of the access readout of all pixels.

## 5. The 3D structures of the digital silicon photomultiplier imager

The new developments in the semiconductor technology open new perspectives in the development of the advanced photon-detection structures, with improved detection efficiency and powerful processing electronics up to final information analysis. An extremely important factor is the progress in the three-dimensional (3D) integration technology proposed by a few standard CMOS technology facilities [14].

The main stream is the change in position of the functional elements from the planar view 2D to 3D structure as illustrated in Fig. 11.

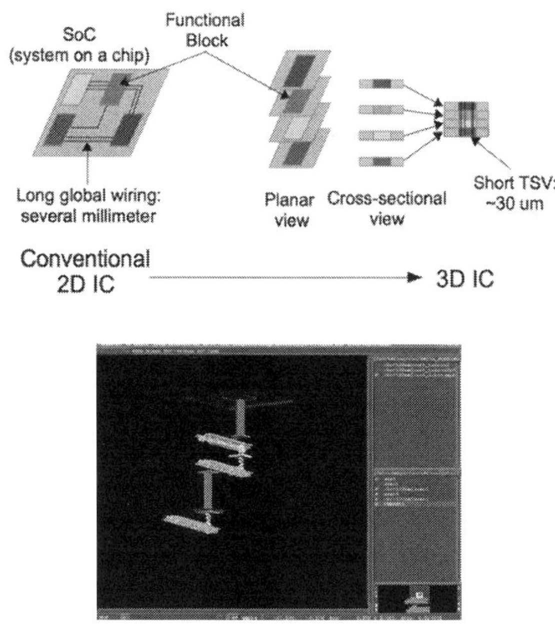


**Figure 11.** General structure of the 3D Integration Technology and details of the 3D integration technology design

One of the modern directions was proposed by the Tezzaron research group. The root of Tezzaron's® technology proposition is delivering the highest vertical interconnect density in the industry. Tezzaron can stack more *wires* vertically through a stack of die in lesser space than any other industry. This allows for contact with more and smaller sensor elements and to put more memory and processing power behind them. The sensors can start processing data right away before they send it off for further processing. Those sensor arrays may *see* anything – from visible light to cosmic rays. They may even *see* the molecules that represent a specific smell, virus, or DNA sequence. Hence, they can *see* not just images, instead, they *see* just about any tiny thing needed to detect.

### 5.1. The general structure

Such modern 3D technology is very important and fully applicable to the sensor technology of Silicon Photomultipliers.

The general structure is shown in Fig. 12.

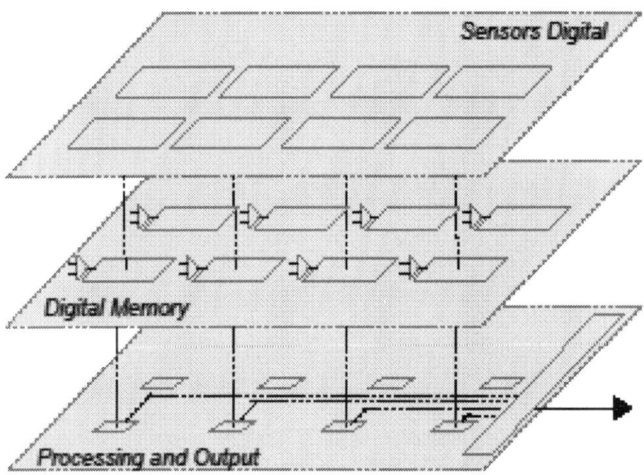


**Figure 12.** General structure of the 3D Structure of the Digital Silicon Photomultiplier Imager

Sensor arrays fully occupy the top layer of the 3D Digital Silicon Photomultiplier Imager structure. This significantly improves the detection efficiency and in the first view, the geometrical factor, which is one of the general problems of the planar pixel sensor structures.

The second layer is dedicated to the distributed memory. One-bit memory corresponds to the pixel and is directly connected with vertical interconnections to the sensor layer. The organization of the memory will include the readout chain and the control circuit, which controls the digital representation of the pixel signal and, for example, the dark rate suppression.

The third layer is dedicated to the processing electronics and communication input /output functions, which could be highly intelligent and may even include the processors for a detailed analysis of the information.

## 6. Conclusion

The fully Digital Silicon Photomultiplier Structures are Digital Imagers for the detection of the low-light photon flux, providing full information, which is carried by the low photon flux – the number of the photons, space distribution of the photons, and the time information of the photons arriving. This detector will provide this information in a digital form on the level of each single photon in the low photon flux. The development of the fully Digital Silicon Photomultiplier Imager is very perspective of the many important applications.

The novel semiconductor technology, such as 3D integration CMOS technology, gives the fundamental improvement of the overall characteristics of the Digital Silicon Photomultiplier Imager with powerful digital information processing and analysis.

## Acknowledgements

The authors acknowledge the support of the project 61210003 by the National Natural Science Foundation of China, the International S&T Cooperation Program of China and the National Key Scientific Instrument and Equipment Development Project of China.

## Author details

N. D'Ascenzo [1,2], V. Saveliev[1,2], Q. Xie[1*] and L. Wang[1]

*Address all correspondence to: qgxie@hust.edu.cn

1 Huazhong University of Science and Technology, Wuhan, China

2 Institute of Applied Mathematics, Russian Academy of Sciences, Russia

## References

[1] V.Saveliev, V.Golovin: Silicon Avalanche Photodiodes on base Metal Resistor Semiconductor (MRS) Structures, Nuclear Instruments and Methods A 2000: 442: 223-229. DOI:10.1016/S0168-9002(99)01225-5

[2] V.Saveliev: Silicon Photomultipliers – New Era of Photon Detection. In Ed. Ki Young Kim, Advances Optical and Photonic Devices: Intech; 2010. p. 249-272. ch 14

[3] N.D'Ascenzo, V.Saveliev: The New Photodetectors for High Energy Physics and Nuclear Medicine. In Ed. Jin-Wei Shi, Photodetectors: Intech; 2012. p. 261-284. ch 13

[4] International Technology Roadmap for Semiconductors, 2013 edition, Process Integration, Devices, and Structures Summary, ITRS, 2013

[5] Y.Haemisch, T.Frach, C. Degenhardt and A. Thom: Fully Digital Arrays of Silicon Photomultipliers (dSiPM) - a Scalable Alternative to Vacuum Photomultiplier Tubes (PMT), Physics Procedia 2012; 37: 1546-1560. DOI:10.1016/j.phpro.2012.03.749

[6] N.D'Ascenzo, P.S. Marrocchesi, C.S.Moon, F.Morsani, L.Ratti, V.Saveliev, A.Savoy-Navarro, Q.Xie: Silicon Avalanche Pixel Sensor for High Precision Tracking, JINST 2014; 9: C03027. DOI:10.1088/1748-0221/9/03/C03027

[7] A. G. Stewart, V. Saveliev, S. J. Bellis, D. J. Herbert, P. J. Hughes, and J. C. Jackson: Performance of the 1 $mm^2$ Silicon Photomultiplier, IEEE Journal of Quantum Electronics 2008; 44,157. DOI:10.119/JQE.2007.910940

[8] T.Frach, G.Prescher, C.Degenhardt, R de Gruyter, A.Schmitz: The Digital Silicon Photomultiplier – Principle of Operation and Intrinsic Detector Performance, 2009 IEEE Nuclear Science Symposium Conference Record 2009; 25-9, 1959-1965. DOI:10.1109/NSSMIC.2009.5402143

[9] J.Sachs: The Basics of Digital Images, White Paper [Internet] 2003. Available at http://www.dlc.com/Temp/downloads/Whitepapers/Basics.pdf

[10] A.J.P. Theuwissen: CMOS Image Sensors: State of the art, Solid State Electronics 2008; 52, 1401-1406

[11] Hubble [Internet]. 1997. Available at http://hubblesite.org/gallery/

[12] V.Saveliev: Quantum Detector Arrays 2010; US Patent US 7,825,384

[13] V.Saveliev: Avalanche Pixel Sensors and Related methods 2012; Us Patent 8,269,181

[14] 3D Technology [Internet]. 2012. Available at http://www.athenis3d.eu/technology.php